AF890967

Editors

E. H. Hirschel/München
K. Fujii/Kanagawa
W. Haase/München
B. van Leer/Ann Arbor
M. A. Leschziner/London
M. Pandolfi/Torino
J. Periaux/Paris
A. Rizzi/Stockholm
B. Roux/Marseille

Springer
Berlin
Heidelberg
New York
Barcelona
Hong Kong
London
Milan
Paris
Tokyo

Engineering ONLINE LIBRARY
http://www.springer.de/engine/

New Results in Numerical and Experimental Fluid Mechanics III

Contributions to the 12th STAB/DGLR Symposium Stuttgart, Germany 2000

Siegfried Wagner, Ulrich Rist, Hans-Joachim Heinemann and Reinhard Hilbig (Editors)

Springer

Prof. Dr.-Ing. Siegfried Wagner
Dr.-Ing. Ulrich Rist
Universität Stuttgart
Institut für Aerodynamik und Gasdynamik
Pfaffenwaldring 21
D - 70569 Stuttgart

Dr.-Ing. Hans-Joachim Heinemann
DLR
Bunsenstraße 10
D - 37073 Göttingen
e-mail: hajo.heinemann@dlr.de

Dipl.-Ing. Reinhard Hilbig
DaimlerChrysler Aerospace Airbus
Head; Technology Programmes "Flight Physics"
Hünefeldstraße 1 - 5
D - 28199 Bremen

Library of Congress Cataloging-in-Publication Data applied for

Die Deutsche Bibliothek - Cip-Einheitsaufnahme
New results in numerical and experimental fluid mechanics. - Braunschweig ; Wiesbaden ; Vieweg ; [Berlin ; Heidelberg ; New York ; Barcelona ; Budapest ; Hongkong ; London ; Mailand ; Paris ; Singapur ; Tokio] : Springer
(Contributions to the ... AG STAB/DGLR symposium ... ; ...)
3. Stuttgart, Germany 2000 / Siegfried Wagner ... ed.. - 2002
(Contributions to the ... AG STAB/DGLR symposium ... ; 12)
(Notes on numerical fluid mechanics ; Vol. 77)
(Engineering online library)
ISBN 978-3-540-42696-7

ISSN 1612-2909
ISBN 978-3-540-42696-7 Springer-Verlag Berlin Heidelberg New York

This work is subject to copyright. All rights are reserved, whether the whole or part of the material is concerned, specifically the rights of translation, reprinting, re-use of illustrations, recitation, broadcasting, reproduction on microfilms or in any other way, and storage in data banks. Duplication of this publication or parts thereof is permitted under the provisions of the German Copyright Law of September 9, 1965, in its current version, and permission for use must always be obtained from Springer-Verlag. Violations are liable for prosecution under the German Copyright Law.

Springer-Verlag Berlin Heidelberg New York
a member of BertelsmannSpringer Science+Business Media GmbH

http://www.springer.de

© Springer-Verlag Berlin Heidelberg 2002

The use of general descriptive names, trademarks, etc. in this publication does not imply, even in the absence of a specific statement, that such names are exempt from the relevant protective laws and regulations and therefore free for general use.

Typesetting: camera-ready by authors
Cover design: de'blik, Berlin
Printed on acid-free paper SPIN: 10837386 62/3020/M – 5 4 3 2 1 0

NNFM Editor Addresses

Prof. Dr. Ernst Heinrich Hirschel
(General editor)
Herzog-Heinrich-Weg 6
D-85604 Zorneding
Germany
E-mail: e.h.hirschel@t-online.de

Prof. Dr. Kozo Fujii
Space Transportation Research Division
The Institute of Space and Astronautical Science
3-1-1, Yoshinodai, Sagamihara, Kanagawa,
229-8510
Japan
E-mail: fujii@flab.eng.isas.ac.jp

Dr. Werner Haase
Höhenkirchener Str. 19d
D-85662 Hohenbrunn
Germany
E-mail: wug.haase@t-online.de

Prof. Dr. Bram van Leer
Department of Aerospace Engineering
The University of Michigan
Ann Arbor, MI 48109-2140
USA
E-mail: bram@engin.umich.edu

Prof. Dr. Michael A. Leschziner
Department of Engineering
Queen Mary & Westfield College (QMW)
University of London
Mile End Road
London E1 4NS
Great Britain
E-mail: mike.leschziner@qmw.ac.uk

Prof. Dr. Maurizio Pandolfi
Politecnico di Torino
Dipartimento di Ingegneria Aeronautica e
Spaziale
Corso Duca degli Abruzzi, 24
I - 10129 Torino
Italy
E-mail: pandolfi@polito.it

Prof. Dr. Jaques Periaux
Dassault Aviation
78, Quai Marcel Dassault
F-92552 St. Cloud Cedex
France
E-mail: periaux@rascasse.inria.fr

Prof. Dr. Arthur Rizzi
Department of Aeronautics
KTH Royal Institute of Technology
Teknikringen 8
S-10044 Stockholm
Sweden
E-mail: rizzi@aero.kth.se

Dr. Bernard Roux
IRPHE-IMT
Technopole de Chateau-Gombert
F-13451 Marseille Cedex 20
France
E-mail: broux@irphe.univ-mrs.fr

FOREWORD

This volume contains the papers presented at the 12th DGLR/STAB-Symposium held at the University Stuttgart, November, 15 to 17, 2000. STAB is the German Aerospace Aerodynamics Association, founded towards the end of the 70's, whereas DGLR is the German Society for Aeronautics and Astronautics (Deutsche Gesellschaft für Luft- und Raumfahrt - Lilienthal Oberth e.V.).

The mission of STAB is to foster development and acceptance of the discipline "Aerodynamics" in Germany. One of its general guidelines is to concentrate resources and know-how in the involved institutions and to avoid duplication in research work as much as possible. Nowadays, this is more necessary than ever. The experience made in the past makes it easier now, to obtain new knowledge for solving today's and tomorrow's problems.

STAB unites German scientists and engineers from universities, research-establishments and industry doing research and project work in numerical and experimental fluid mechanics and aerodynamics for aerospace and other applications. This has always been the basis of numerous common research activities sponsored by different funding agencies.

Since 1986 the symposium has taken place at different locations in Germany every two years. In between STAB workshops regularly take place at the DLR in Göttingen. The changing meeting places were established as focal points in Germany's Aerospace Fluid Mechanics Community for a continuous exchange of scientific results and their discussion. Moreover, they are a forum where new research activities can be presented, often resulting in new commonly organised research- and technology projects.

It is the third time now that the contributions to the Symposium are published after being subjected to a peer review. Many of the contributions present results from the "Luftfahrtforschungsprogramm der Bundesregierung (German Aeronautical Research Programme) 1995-1998". Some of the papers report on work sponsored by the Deutsche Forschungsgemeinschaft (German Research Council), DFG. Therefore, this volume gives an almost complete review of the ongoing work in Germany. The present Symposium has focused on "Laminar/Turbulent Boundary-Layer Transition" and "Hypersonic Flows", two programmes sponsored by the DFG in a Priority Program (Verbundschwerpunkt-Programm) and in three Collaborative Research Centres (Sonderforschungsbereiche).

The order of the papers in this book corresponds nearly to that of the sessions of the Symposium.

The Review-Board, partly identical with the Program-Committee, consisted of J. Ballmann (Aachen), K.-A. Bütefisch (Göttingen), U. Ch. Dallmann (Göttingen), H.-J. Dietrichs (München), G. Eitelberg (Emmeloord), R. Friedrich (München), R. Grundmann (Dresden), P. Hakenesch (München), St. Hein (Göttingen), D. K. Hennecke (Darmstadt), P. Hennig (Unterschleißheim), H. Hönlinger (Göttingen), D. Jacob (Aachen), E. Janke (Göttingen), M. Kloker (Stuttgart), H. Körner (Braunschweig), W. Kordulla (Noordwijk), D. Kröner (Freiburg), G. E. A. Meier (Göttingen), W. Nitsche (Berlin), H. Olivier (Aachen), R. Radespiel (Braunschweig), M. Raffel (Göttingen), G. Redeker (Braunschweig), U. Rist (Stuttgart), W. Rodi (Karlsruhe), G. Schrauf (Bremen), W. Schröder (Aachen), J. Schumacher (Hamburg), J. Szodruch (Hamburg), P. Thiede (Bremen), C. Tropea (Darmstadt), C. Wagner (Göttingen), S. Wagner (Stuttgart) - Chairman - and C. Weiland (München). Nevertheless, the authors sign responsible for the contents of their contributions.

The editors are grateful to Prof. Dr. E. H. Hirschel as the General Editor of the "Notes on Numerical Fluid Mechanics" and to the Springer-Verlag for the opportunity to publish the results of the Symposium.

S. Wagner, Stuttgart
U. Rist, Stuttgart
H.-J. Heinemann, Göttingen
R. Hilbig, Bremen

July 2001

CONTENTS

Hypersonic Flows

Fluid-Structure-Coupling

Fundamentals of Fluid Flow

Transition and Fluidmechanical Modelling

Mathematical Fundamentals / Numerical Simulation

Bluff Bodies

Turbomachineri

Airframe Noise

Laminar Flow Control

Measuring Techniques

High Aspect-Ratio Wings

Design Work for the A3XX High-Lift-Wing

D. RECKZEH[1]

EADS Airbus GmbH
Hünefeldstr. 1-5
28183 Bremen
Germany

Summary

The A3XX configuration being dimensioned larger than any flying aircraft must be operated in the given airport infrastructure. The high-lift system has to provide this ability while being designed to minimum complexity and weight. The aerodynamic design work conducted by EADS Airbus Germany for the design of the A3XX high-lift wing is carried out based on CFD and windtunnel experiments as major tools for configuration development. The available and developed CFD-design environment and the windtunnel infrastructure led to a configuration providing an aerodynamic performance fulfilling these challenging requirements. In parallel least system complexity of the leading edge and trailing edge system was maintained finally leading to a high-lift system which represents an optimised solution under multidisciplinary constraints.

Tools for the aerodynamic design of Airbus high-lift wings

The aerodynamic design of the high-lift wing is dominated by the requirement of an optimised configuration design. It is strongly influenced by the effects on other disciplines, e.g. systems and structures [1], [2]. EADS Airbus Germany is responsible for the aerodynamic design of the Airbus high-lift systems. Based on the cruise wing geometry the high-lift wing has to be designed for meeting the performance requirements. The tools for this complex task with its variety of effects are theoretical methods and windtunnel experiments. In the development of previous Airbus high-lift wings the major work was based on windtunnel experiments. Theoretical methods suited for the use under the tight time constraints of the design process were only available for section design up to now. The assessment and optimisation of the 3D-wing was performed on experimental basis but for A3XX also 3D-CFD methods were established in this task. In the aerodynamic design process of high-lift wings theoretical methods providing a high quality of results with fast computing time and high user friendliness are necessary. Therefore, primarily used methods are CFD codes with low effort concerning time and complexity.

---

[1] High-Lift Wing Design Engineer, Aerodynamic Design Dept. EFP

At the aerodynamic design department at EADS Airbus Germany a CFD infrastructure was established in the last years for being capable of capturing the task of A3XX high-lift wing design [3], [4]. The principle for CFD-based design is the use and combination of methods with complexity end expense appropriate to the momentary design task in a 'chain of methods' (Fig 1). Several methods were established in the last years including a fast quasi-3D-method, an advanced 3D panel-method, a 2D Navier-Stokes method and automated optimisation tools. The use of 3D-Euler and 3D-Navier Stokes methods has only supporting character at the moment. One task of the work on CFD-methods is a further integration of these advanced methods in the design infrastructure.

Besides the available advanced CFD-infrastructure windtunnel experiments have still the central importance for the assessment of the theoretically designed high-lift wings and the final configuration selection. In Airbus high-lift wing design several windtunnels are used. The EADS Airbus Germany low speed windtunnel in Bremen/Germany is a flexible and available tool for first configuration evaluation and selection. The high-lift configuration can be tested with powered turbine simulation as a reflection plane model. In the German-Dutch Windtunnel DNW in Emmeloord/NL tests follow on a large complete model (Fig 2). Powered turbine simulation is also possible as well as tests with empennage and tests under sideslip conditions and in ground effect. For tests at high Reynolds numbers pressurised (e.g. F1 in Fauga/France) or cryogenic windtunnels (e.g. KKK in Cologne/Germany) or a combination of both (ETW in Cologne/Germany) are used.

Requirements of the A3XX

The A3XX represents a completely new aircraft design. In contrast to the A340-500/-600 it is not a derivative of an already flying design and therefore the use of existing manufacturing parts is no constraint to the aerodynamic wing design. The A3XX with its planned derivatives rather represents an own aircraft family which serves the market for aircraft larger than 450 seats. The A3XX will have a take-off weight of up to 590 tons and in the basic version 555 passengers will find place. The cruise speed is defined with a design-Mach-number of Ma=0.85. The wing span was limited to 80m by the requirement to operate in the existing airport infrastructure.

The design of the cruise wing had to be adapted to these constraints. The wing area resulting from the design weight is about 850m^2. However, because of the span limitation the possible aspect ratio of 7.5 is lower than on previous Airbus wing designs. The optimisation process of the cruise wing finally led to a load distribution which implies a higher inboard loading than the optimal elliptic distribution. This increases the induced drag but the advantage in structural weight by the lower bending moment far overcompensates this aerodynamic penalty. Compared to the existing Airbus wings the high design-Mach-number requires a higher sweep angle and, especially in the outboard region of the wing, a reduced relative thickness of the profiles. Thereby the design constraints for the integration of the high-lift system into the given cruise wing shape became clearly more critical.

The A3XX high-lift wing

The design of the A3XX high-lift wing was performed under the previously discussed requirements and resulted in a simple layout with sufficient performance for all members of the A3XX-family [5]. Based on the experience gained especially on the A340 wing a slat as leading edge device and a single slotted Fowler flap as trailing edge device were chosen in the pre-design as the baseline solution. For the first time the CFD high-lift design work was performed as a real wing design instead of section design. Especially the availability of the quasi-3D-method (coupling of 2D-method with lifting surface method, Fig 3) allowed the evaluation of a large number of high-lift wing concepts and variations without any experimental work. CFD helped to start testing with an already pre-optimised design which helped to reduce the expense in testing respectively allowed a deeper and more targeted optimisation in windtunnel. Parallel to the windtunnel-based work CFD-studies of alternative designs and design variations were performed in order to ensure 'being on the right track'.

When a wing design is selected the evaluation of the complete aircraft in high-lift configuration follows. A commercial 3D-panel method (VSAERO) with 'in house'-developed extensions is used (Fig 4). A fast input geometry generation for high-lift wings enables the quick assessment of the aircraft, e.g. for the prediction of the complete aircraft aerodynamic performance, aerodynamic loads for structural design, stalling behaviour, ground effects and the flow field around the configuration. In support to the windtunnel tests and for scaling of the experimental results to the flying aircraft Reynolds number effects and windtunnel-wall interferences can be studied.

While the 3D-methods were established as standard tools for the design process advanced 2D-Navier-Stokes solvers on structured and on hybrid meshes have also been introduced. Especially for the understanding of flow physics and for the prediction of complex flow conditions they complete the capabilities of the fast viscous-inviscid coupled methods used in shape definition. At the moment the coupling of CFD-methods with automated optimisation routines is introduced in high-lift wing design. Being already able to evaluate the possibilities of these methods it is clear that they will gain a great importance already in near future.

The design work for A3XX was characterised and limited by the cruise wing shape. Because of the high design-Mach-number the outer wing profiles have a small relative thickness resulting in a sharp leading edge radius (Fig 5). Furthermore, a significant rear-loading was realised resulting in a small trailing edge thickness which limited the maximum thickness more severely than in previous wing designs (A320, A340). The possible flap contour includes therefore a sharper leading edge radius which results in a pressure distribution with a dramatically higher suction peak on the flap nose. A higher boundary layer loading and a worse separation tendency follow and lead to an earlier decrease of flap efficiency with deflection angle as for a comparable 'A340-type'-flap due to earlier separation. This effect could be reduced by a more effective flap system, e.g. a double-slotted flap. A strong disadvantage however results for the flap system weight which is increased dramatically by the small thickness. Also an acceptable low deformation for maintaining the required setting can not be reached. As shown in Fig 6 the flap thickness can be increased with a thinner spoiler. The suction peak and the boundary layer load can be lowered leading to improved separation behaviour. On the other hand the reduction of the spoiler thickness increases the weight and the deformation tendency of the spoiler. This partially compensates the advantage from the increased flap thickness.

The early designs of the A3XX cruise wing shape had to be modified to enable the integration of a realisable high-lift system. The trailing edge region of the final cruise wing profile shapes was significantly thickened which required a reduction of the rear loading. The outcome for the flap system was the possibility for increased flap thickness with the required positive influence on flow quality, deformation and weight. With regard to a multidisciplinary optimum also the spoiler was thickened in order to improve spoiler weight and deformation although this slightly worsens the flap pressure distribution.

On the leading edge a similar problem linked to the sharp leading edge radius resulting from thin profiles needed to be solved. The suction peak on the A3XX-slat appeared considerably higher than on an A340-type slat. This results in a higher boundary layer loading and separation tendency, as explained before. As a severe separation on the slat normally leads to stalling of the complete wing in this area maximum lift was limited to an unsatisfying level. For maximum lift improvement a further deflection of the slat could have been applied. The larger deflection would have required longer tracks with the resulting higher weight and the difficulty of stowing them in retracted position in the thin cruise wing profile. Therefore, the maximum slat angle was limited. Also the clean wing stalling behaviour was predicted as only marginally acceptable due to the separation onset in the outer wing region caused by the high boundary layer loading from the sharp leading edge in this region.

The wing leading edge had to be redesigned considering high-lift performance. An increase of the outer wing profiles leading-edge radius reduced the slat suction peak and resulted in a slat angle feasible on the aircraft. The demand of a large leading edge radius however stands in direct contrast to requirements from the high-speed flight. Especially in the inboard regions of the pylons a small leading edge radius is necessary to avoid unsteady shock induced separations on the pylon and the wing lower surface which can excite oscillations of the wing structure. A compromise providing acceptable flow conditions for both flight regimes had to be found.

While optimisations of the design were performed based mainly on CFD, the windtunnel represents the central tool for performance determination of the realistic 3D aircraft design. The selected design is tested and varied. Based on these results the final layout of the high-lift wing is decided. The windtunnel offers the possibility for the check of a large amount of variations on the high-lift wing. Although model construction and testing is very expensive the windtunnel represents the tool best suited for the production of the data base for the aircraft and also for the detailed optimisation of the configuration with reliable results.

The windtunnel experiments led to the final selection of the high-lift elements of the A3XX wing. Parallel to the design of the baseline solution other elements were designed and checked with CFD and windtunnel test. Especially for the reduction of the take-off drag Krueger-flaps and a droop-nose device for the inboard wing appeared to be promising alternatives. But even with an aerodynamic advantage the impact on the complete design has to be evaluated. Higher system weight and no experience in design and application of a totally new system balanced out the advantage and led to staying with the slat design Airbus is best experienced in. Of course, the slat was also optimised for least weight with sufficient performance. In windtunnel testing it was found that the wing still provided the necessary maximum lift performance with a cutback of the slat in the region near to the fuselage. This resulted in a reduction of the inboard slat span and the following weight benefit.

On the trailing edge system extensive concept variations were also performed with CFD and windtunnel testing. The problem of maintaining sufficient aileron effectiveness in high-speed flight initiated studies of an additional inboard 'all-speed'-aileron (ASA) comparable to Boeing wing layouts (Fig 6). The flap span was extended to maintain the flap system performance and the outer aileron span was reduced because this device was not needed in high-speed flight any more. As this configuration did not provide the necessary lift performance a double-slotted flap on the inboard wing was examined. In order to avoid the double slotted flap with its negative impact on weight and cost a 'taberon'-alternative was also studied. Here an additional plain flap on the flap system in the engine station would work as a high-speed roll-control device without compromising the performance of the high-lift wing by avoiding the large gap in the deployed trailing edge.

The high-lift system had to be adapted to the changing requirements of the actual aircraft performance status. The initial design showed better aerodynamic performance than needed which was caused by the change of requirements but also from the good performance of the design itself. This resulted in several steps of downsizing of the trailing edge system. The flap chord relative to the cruise wing profile chord could be reduced from 28% in first design to 24% in final design. At the same time the shroud position (i.e. spoiler length) could be reduced. Also the flap span could be reduced from 73% to 65% of wing span. With this flap span reduction and the possible extension of the aileron in the inboard direction sufficient roll control efficiency in high-speed flight can be maintained. The trailing edge solutions with the additional inboard aileron are therefore avoided. All these 'downsizing'-adaptations resulted finally in a big benefit to the wing-weight and the aircraft performance.

Conclusions

The aerodynamic design of the A3XX high-lift wing represents a complex task which can not be considered independently of the complete aircraft design process which is influenced from multidisciplinary constraints. The design resulted in a high-lift wing providing the necessary performance with the lowest possible system complexity (Fig 7). While on the B747 as a comparable aircraft a very complex system with triple-slotted flaps and flexible Krueger-flaps was chosen, the A3XX high-lift wing provides its performance with slats and single slotted flaps only. In the evolution of the A3XX design the high-lift system was adapted several times to the changed requirements in order to be 'well-dimensioned' for each member of the A3XX-family. Intensive use of CFD for the design process led to an already pre-optimised design when first entering the windtunnel. The combination of CFD and windtunnel testing led to a deeply optimised design. The use of advanced CFD methods will be intensively increased in the future.

References

[1] Flaig A., Hilbig R., High-Lift Design for Large Civil Aircraft, AGARD-CP-515, 1992

[2] Rudolph P., High-Lift Systems on Commercial Subsonic Airliners, NASA-CR 4746, Sept. 1996

[3] Reckzeh D., Rechenverfahren für den Entwurfsprozess von Hochauftriebskonfigurationen, DGLR-Jahrestagung 1998, DGLR-JT98-240

[4] Hansen H., Überblick über das Technologieprogramm Hochauftriebskonzepte (HAK), DGLR-Jahrestagung 1998, DGLR-JT98-042

[5] Dreßler U., Auch einfach ist schwierig - Flugphysikalische Auslegung der Hochauftriebskonfiguration A3XX, Vortragsreihe Start+Landekonfiguration, DaimlerChrysler Aerospace Airbus, Bremen, 1999

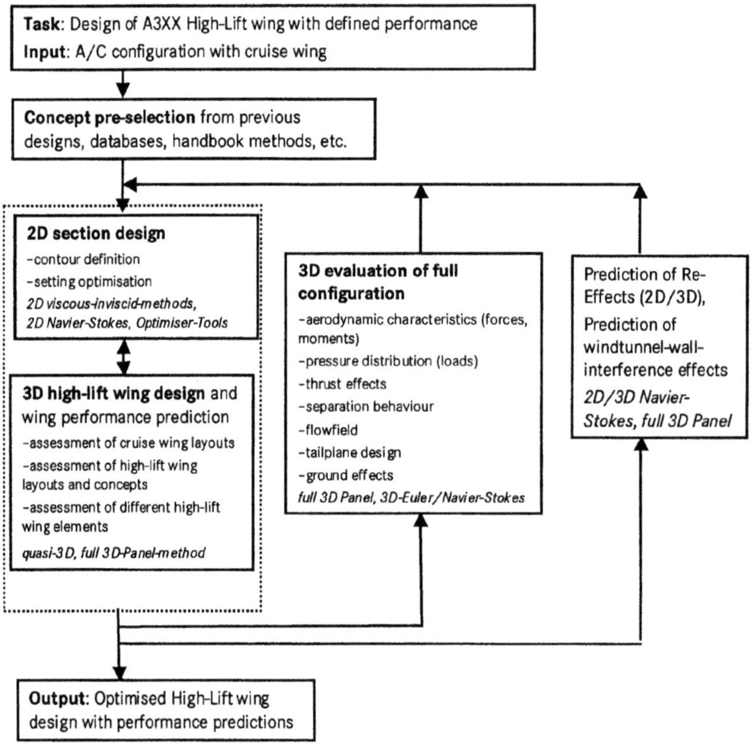

Figure 1: Sequence of the CFD-based high-lift design process

Figure 2: A3XX-complete model in DNW windtunnel

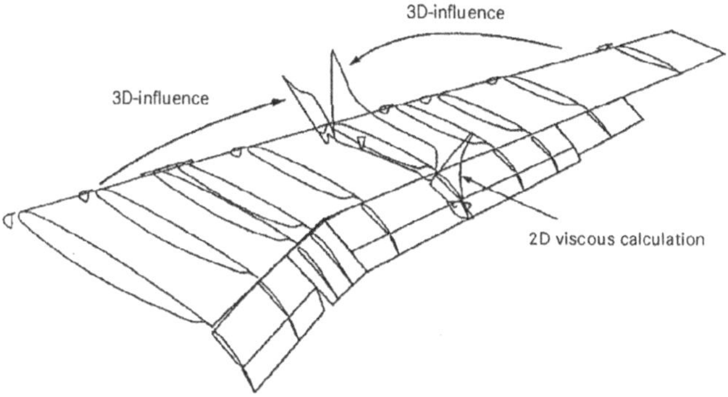

Figure 3: Principle of the quasi-3D-method

Figure 4: Flow analysis for the complete A3XX-configuration with 3D-panel method

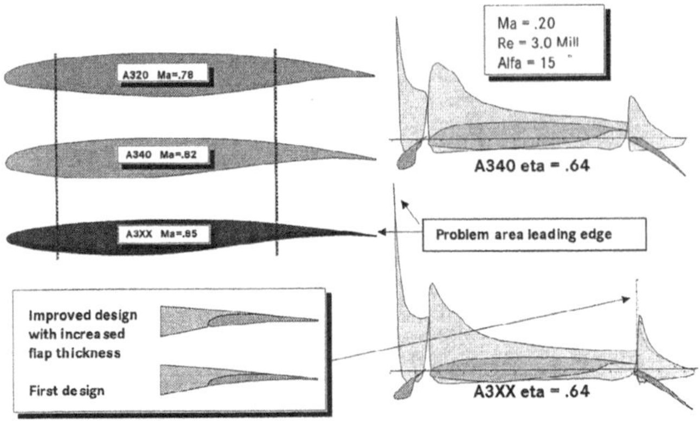

Figure 5: Problem areas on the A3XX high-lift wing profiles

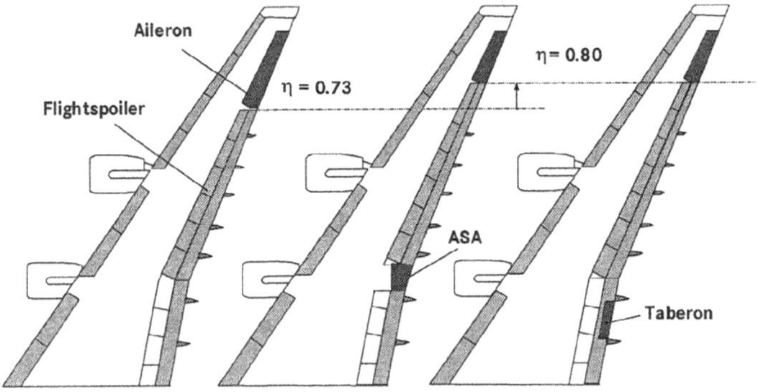

Figure 6: Variation of the A3XX-trailing edge and roll control concept

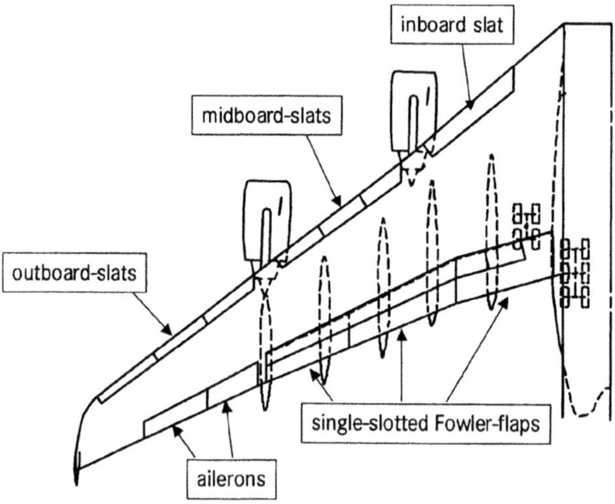

Figure 7: The final A3XX high-lift wing layout

Recent Improvements in the Measurement of Aerodynamic Damping Derivatives

A.-R. HÜBNER[1], T. LÖSER[2]
[1]Institut für Entwurfsaerodynamik, DLR Braunschweig, Lilienthalplatz 7, D-38108 Braunschweig, Germany
[2]DNW-NWB, Lilienthalplatz 7, D-38108 Braunschweig, Germany

Summary

The Oscillatory Motion Support (OMS), used in the German-Dutch Wind Tunnels – Low Speed Wind Tunnel Braunschweig (DNW-NWB) for the measurements of damping derivatives has undergone substantial changes and improvements. The changes in the mechanical system include a new hydraulic 6 degree-of-freedom platform for static model positioning and yaw, and a hydraulic actuator for pitch and roll. An active strut has been employed to suppress the wind tunnel model's natural oscillations. A new data acquisition system along with new data evaluation software is used. For increased accuracy, the model position is measured by a high speed video system. The new components have demonstrated their suitability in a recent wind tunnel entry with an Airbus Megaliner model.

1. Starting situation

Since 1998 wind tunnel experiments for ascertaining dynamic derivatives on civil transport aircraft have been performed in the DNW-NWB. The first measurements have been performed in collaboration with the DLR Institute for Design Aerodynamics within the MEGAFLUG project. MEGAFLUG was established in order to improve the assessment of the aerodynamic behaviour and the influence of the aircraft's components of the new Airbus Megaliner. The most recent wind tunnel model of the Megaliner is shown in fig 1.

The general setup is as follows: the wind tunnel model is mounted on a belly sting in the open test section and performs forced sinusoidal motions in one of the three possible modes pitch, yaw and roll. These oscillations are induced by an electric motor located below the test section floor. In the case of the yawing motion the complete belly sting oscillates, in case of the pitching and the rolling motion the motor drives a pushrod located behind the belly sting; the translatory displacement is transformed into rotatory motion by means of special balance shoes. A Emmen 196-6 6-component internal balance is used for force and moment measurements, while a spring loaded strain gauge beam, located close to the motor, is used for the measurement of the model position. Data acquisition and evaluation is performed by two SUN workstations, one of them controlling the TEDAS data acquisition system, as described in more detail in [5] and. [6].

The data files from wind-on and wind-off measurements are then processed in a program based on sources described in [1].

2. Current status of the measuring system

The basic concept of determining the damping derivatives has remained unchanged; balance and model position signals are repeatedly recorded for a large number of model oscillations in wind-on as well as wind-off conditions. Data from wind-off measurements are subtracted from wind-on results in order to get rid of mass and inertial forces. By Fourier transformation, only the part of the signals having the model oscillation frequency is taken into account. In the following sections, the changes with respect to the starting situation are described.

2.1 Slotted test section

DNW-NWB applies wind tunnel wall corrections to static measurements in the open and closed test sections according to AGARDograph 109 [2]. Wall corrections for dynamic tests are not available at DNW-NWB. In order to minimise wall effects in dynamic measurements, static measurements have been performed in the open as well as in the closed test section. The corresponding wall correction algorithms have been applied. The corrected curves don't match exactly and the adjustable slots have been opened so much, that the uncorrected static results from the slotted test section lie between the corrected static results of the open and the closed test section. The results of the different test sections for the lift coefficient C_L are shown in fig. 3. All subsequent static and dynamic tests are performed with the slot opening determined that way.

2.2 Mechanical system

Nearly the complete mechanical system described in section 1 has been replaced by new components, which have been either manufactured for this purpose or have been rented from third parties to demonstrate the suitability of the new concept.
The main component to be replaced was the driving mechanism for the different modes of oscillation. Undersized, non-adjustable bearings in the balance shoe for pitching mode had to be replaced quite often and play in other areas of the mechanical system falsified the balance signals. The new balance shoe for pitch mode, which incorporates adjustable high precision bearings and an longer lever arm for smaller bearing loads has proved to be absolutely slop free during the four week test period.
The new model support consists of a 6 degree-of-freedom hydraulic platform, consisting of a wind tunnel fixed lower frame and, connected via 6 hydraulic actuators, an upper frame, onto which the belly sting is attached. This platform, known as Stewart platform or Hexapod, has been rented and was used for the yawing oscillation. It is shown in fig. 4. Heave and lateral oscillation have not yet been performed because of insufficient hydraulic power of the integrated pump. Pitch and roll oscillations are driven from a 7th hydraulic actuator, stemming from a rotor test setup, which is attached on the Hexapod's upper frame and drives the pushrod, as depicted in fig. 2.

2.3 Active strut

In the frequency spectrum of the balance signals belonging to forced sinusoidal model oscillations, especially signals belonging to wind-off measurements, distinctive peaks can be found at several frequencies: the fundamental (= oscillation) frequency, the next 2 -3 higher harmonic frequencies, resulting from the geometry of the driving mechanism and the model's natural frequency of the corresponding bending mode. The typical oscillation frequency lies between 1.5 and 3.0 Hz, the natural frequencies of the models used for dynamic tests lie in the range of 17 to 25 Hz.
One way to reduce the amplitudes of the natural frequencies is the use of a -so called- active strut, mounted in the driving mechanism as a part of the pushrod (see fig. 2). The active strut is a cylindrical element capable of changing its length by about 50 μm because of internal piezoelectric elements. Further details can be found in [9]. This strut was lent to DNW-NWB for the time of the test by the manufacturer, Schütze GmbH. Control of the strut was developed by DNW-NWB staff and uses one balance component as an input signal. The frequency of balance input and signal output is 500 Hz. The control setup is shown in fig 5. The effect of the active strut on model position and the balance signals is shown in fig. 6, the wind-off X component of the balance is displayed because of its high sensitivity with respect to natural oscillations. The amplitude of the natural frequency has been reduced in the order of one mag-

nitude. The position signal, taken from the video setup described in section 2.4, has been smoothed considerably. One side effect, which still has to be worked out, is the introduction of other frequencies into the model position. The newly introduced frequencies are of much smaller magnitude and don't affect the model positioning in an adverse way.

2.4 Position measurement system

Pre-tests with an optical position measurement system have indicated, that the spring-loaded strain gauge system leads to errors because elasticity as well as slop in the driving mechanism are not registered. Furthermore, due to the open strain gauges, the system's signal temperature drift was less than ideal.

The optical system currently used consists of a Pulnix TM-6710 digital video camera in conjunction with a Coreco Viper frame grabber, capable of up to 350 frames per second at 648x100 pixel resolution. The camera uses standard 28mm or 35 mm photographic lenses to look at two markers on the wind tunnel model, which indicate the model's position. The camera is looking through the slots into the test section. Illumination of the region of interest is accomplished by a mercury vapour lamp outside the test section, attached to a PMMA fibre and a focussing optic, the latter being inside the test section. Camera installation as well as typical marker positions for pitching motion can be seen in fig. 7. The pictures taken are stored in the computer's Random Access Memory (RAM). For a measuring time of 30 seconds at 330 Hz 1 GB of RAM is required. Camera and frame grabber control, as well as detection of the markers and calculating the co-ordinates of the marker's centres of gravity is performed by a special version of the program picCOLOR by F.I.B.U.S (who assembled the complete video system); more information on this program is available in 7 and 8.

Differences between the values of the oscillation amplitudes as recorded by the video system and an inductive displacement sensor located below the test section floor at the hydraulic actuator are given in table 2. The following conclusions can be drawn from these values:

- The amplitude of the hydraulic actuator is independent of the imposed (aerodynamic) load.
- Wind-off amplitudes of the model are higher than indicated by the inductive sensor. The elasticity of the mechanical system enables the model to overshoot because of the inertial forces acting during the oscillation. The greater mechanical complexity of the balance shoe used for roll leads to a higher overshoot compared to pitch.
- Wind-on amplitudes of the model measured with the video system are smaller than the corresponding wind-off amplitudes. This is the result of aerodynamic forces, which act in opposite direction to the inertial forces („damping derivatives“ Cl_p, Cm_q, Cn_r).

In case of the rolling motion the actual amplitude of the model is approximately 0.25° or 5.5% larger than indicated by the sensors used before the video system, so that derivatives based on the actual model amplitude are about 5% smaller than the values based on the inductive sensor (respectively the strain gauge sensor used in previous tests). This is shown in fig. 8.

2.5 Data acquisition and evaluation system

The transputer and workstation based TEDAS data acquisition system has been replaced with a PC based Hottinger Baldwin MGCplus system. The main reason for replacement was not the higher resolution of the MGCplus, but the difficulty of obtaining hardware and software support for the TEDAS system and the striving at DNW-NWB towards a single operating system for all computer based applications. The operational range of both systems exceeds the current need of sampling 9 data channels at 600 Hz for 30 seconds.

The derivatives, which can be calculated from balance and position data by means of a PC program are given in table 1. A new program has been written in Visual Basic, it is based on the flight mechanic equations given in [1], which have also been used in the old FORTRAN 4 program. The repeatability of results of the whole system can be assessed from the standard deviations calculated from 5 repetitive measurements for the roll damping Cl_p depicted in fig. 8. The standard deviation is indicated by the error bars. Fig. 8 also shows the results of VSAERO (potential flow) calculations on the same geometry. The agreement between calculation and experiment is very good at $\alpha = 0°$. The increasing difference at higher angles of attack can be explained with different axes of oscillation (experiment: wind-fixed, calculation: body-fixed).

3. Further Development

The next wind tunnel entry with the OMS will take place in January 2001. In that tunnel entry the Hexapod platform previously used will be replaced by a larger model. Increased size of upper and lower frames as well as increased diameter of the hydraulic cylinders of the new platform (still on rental basis) lead to a stiffer system. The platform control will be improved and a more powerful hydraulic pump will enable the platform to perform heave and lateral oscillation with sufficient amplitudes. A new platform control software will make oscillations about body-fixed axes possible.

In 2001 DNW-NWB will acquire an own platform of that type, which will be built based on the experiences gained in the wind tunnel entries with the rented platforms. The platform control software will enable the platform to perform not only sinusoidal but arbitrary motions, platform control will be integrated into the wind tunnel control.

The actuator used for pitch and roll will be equipped with an adapted servo valve and a high resolution magnetostrictive position sensor for improved accuracy and reduced tendency to excite the model's natural frequencies. An important aspect to be improved on both hydraulic systems is the safety of balance and wind tunnel model in case of loss of electrical or hydraulic power or in case of control failure of any kind.

The relay trigger mechanism for video position measurement system and balance data acquisition will be replaced by an electronic device, improving the synchronisation of the two systems. Long term plans for the video system include real time position measurement and possibly three dimensional position measurement by use of a second camera.

Since the first tests with the active strut showed promising results, DNW-NWB plan to use an own active strut in further tests. Control of the active strut will be further refined, a DSP board will be used to speed up the required data processing.

It is intended that the test setup described will belong to the standard wind tunnel testing tools in DNW-NWB.

References

[1] Dornier GmbH (Ed.). Oszillierende Derivativawaage für die 3m-Unterschall-Windkanäle der BRD / Entwicklungsphase IV Benutzerhandbuch für die Meßdatenerfassungsanlage der MOD. Forschungsbericht 76/13 B, 1976.

[2] H.C. Garner (Ed.). „Subsonic Wind Tunnel Wall Corrections". AGARDograph 109, 1966

[3] A.-R. Hübner, S. Peters. Ermittlung aerodynamischer Dämpfungsderivativa zweier Airbus-Typen. AG STAB Jahresbericht 1999, pp 37 - 38

[4] A.-R. Hübner. Vergleich experimentell ermittelter aerodynamischer Dämpfungsderivativa zweier Airbus-Typen. Diplomarbeit Nr. 266, Institut für Strömungsmechanik, TU Braunschweig, 1999.

[5] T. Löser. „Dynamic Force and Pressure Measurements on an Oscillating Delta Wing at Low Speeds". DLR IB 129-96/6, 1996.

[6] T. Löser. „Kraftmessungen an einem A 340-300 Modell zur Bestimmung dynamischer Derivativa". DLR IB 129-98/24, 1998.

[7] R.H.G. Müller, K. Pengel. „Blade Deflection Measurement at the Low Noise ERATO Rotor". 26th European Rotorcraft Forum, The Hague, Netherlands, 26th - 29th September 2000, paper 104.

[7] R.H.G. Müller, K. Pengel. „Blade Deflection Measurement at the Low Noise ERATO Rotor". 26th European Rotorcraft Forum, The Hague, Netherlands, 26th - 29th September 2000, paper 104.
[8] R.H.G. Müller et. al. „Investigation of Large Scale Low Speed Air Condition Flow using PIV". 9th Int. Symp. on Flow Visualization, Heriot-Watt-University, Edinburgh, 2000, pp 218-1 – 218-12.
[9] R. Schütze. "Leichte, hochbelastbare CFK-Stäbe für Leichtbaustrukturen der Luft- und Raumfahrt". In: Verbundwerkstoffe und Werkstoffverbunde, WILEY-VCH Verlag, Weinheim 1999

Table 1: Derivatives obtained from dynamic measurements (Heave and lateral oscillation scheduled for 2001)

Pitch	C_{D_α}	$C_{D_{\dot\alpha}}+C_{D_q}$	C_{L_α}	$C_{L_{\dot\alpha}}+C_{L_q}$	C_{m_α}	$C_{m_{\dot\alpha}}+C_{m_q}$
Yaw	C_{l_β}	$C_{l_r}-C_{l_{\dot\beta}}$	C_{n_β}	$C_{n_r}-C_{n_{\dot\beta}}$	C_{Y_β}	$C_{Y_r}-C_{Y_{\dot\beta}}$
Roll		C_{l_p}		C_{n_p}		C_{Y_p}
Heave		$C_{D_{\dot\alpha}}$		$C_{L_{\dot\alpha}}$		$C_{m_{\dot\alpha}}$
Lateral osc.		$C_{l_{\dot\beta}}$		$C_{n_{\dot\beta}}$		$C_{Y_{\dot\beta}}$

Table 2: Oscillation amplitudes as measured by different sensors

Oscillation mode	Sensor	Amplitude wind-off	Amplitude wind-on
Pitch	inductive	4.02°	4.08°
	Video	4.20°	4.07°
Roll	inductive	4.37°	4.37°
	Video	4.66°	4.57°

Figure 1 Megaliner wind tunnel model in the DNW-NWB

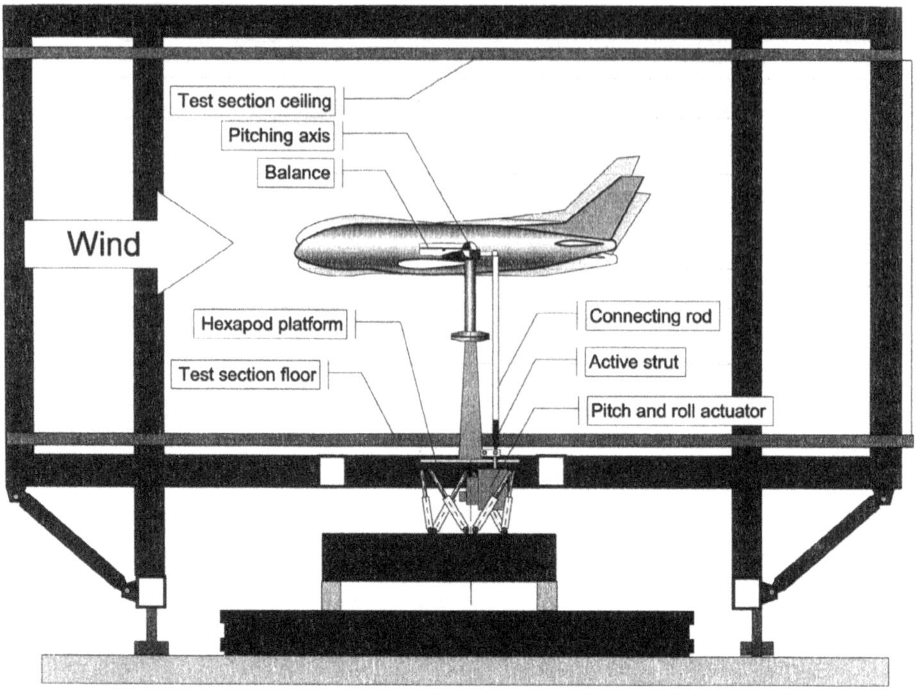

Figure 2 Schematic view of the test setup using hydraulic actuators

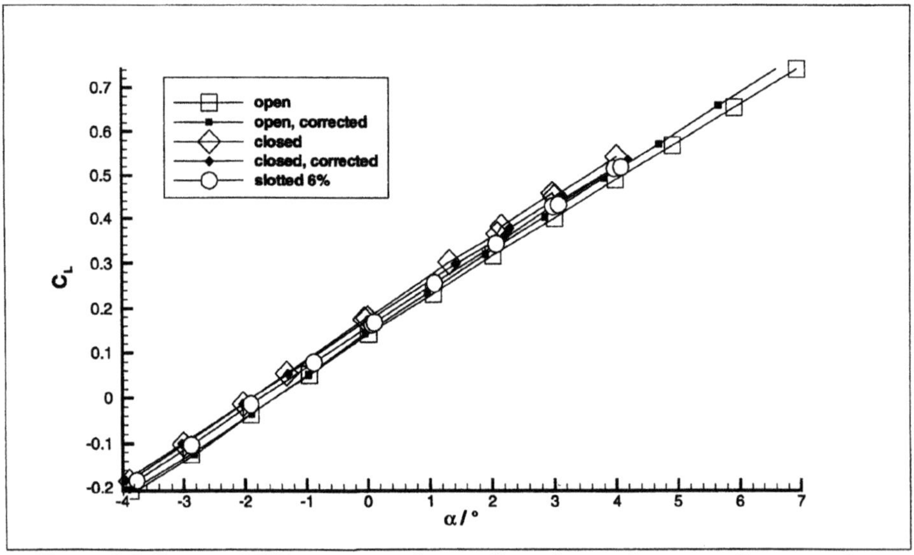

Figure 3 Influence of different test section configurations and corresponding correction methods on the lift coefficient C_L of the Airbus A340-300 model.

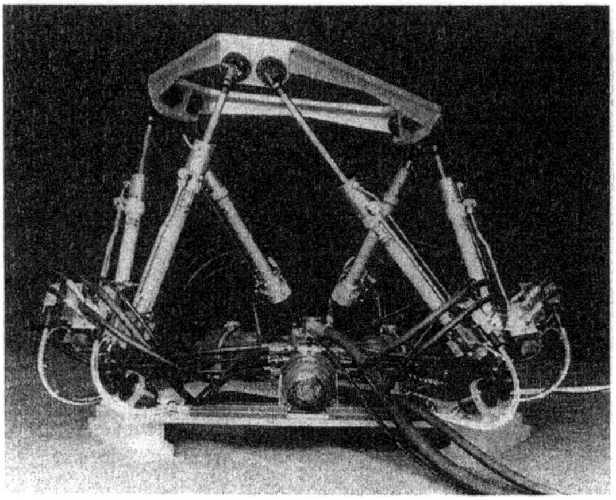

Figure 4 Hexadrive hydraulic platform.

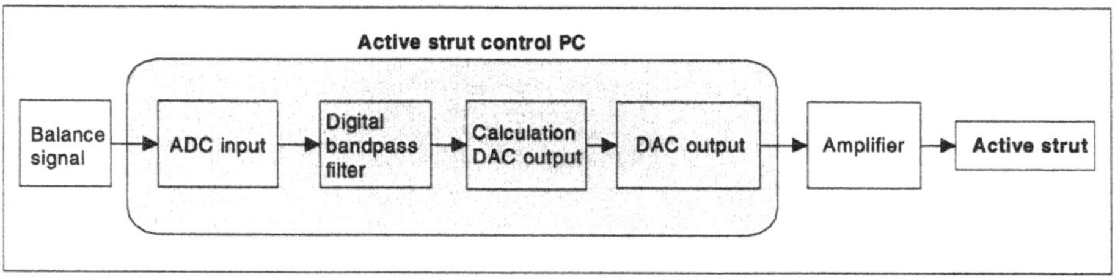

Figure 5 Principle of active strut control.

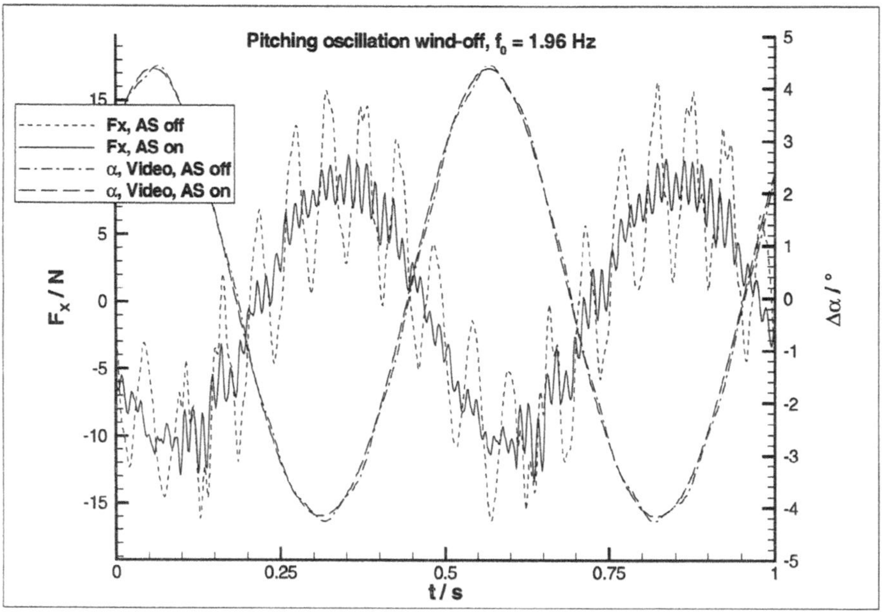

Figure 6 Effect of the piezoelectric active strut (AS) on model position (α) and X component of the balance signals.

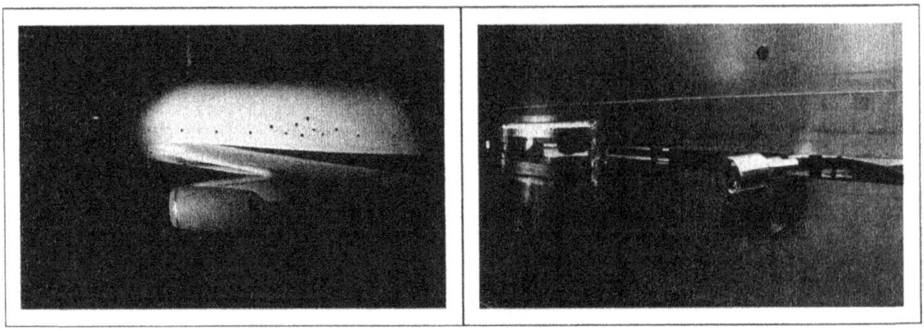

Figure 7 Installation of video camera and light source (right) and position markers on the model's fuselage.

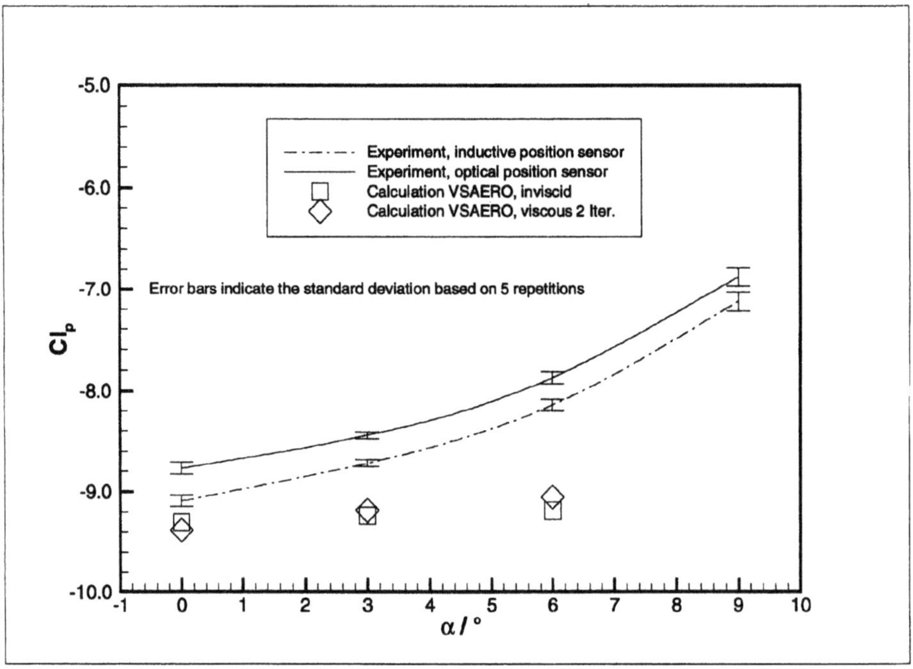

Figure 8 Roll damping derivative Cl_p as a function of angle of attack; effect of different position sensors; comparison of experimental results with CFD calculations.

Calculation of Horizontal Stabilizer Efficiency with a Navier-Stokes Code

C. Nebel, D. Hummel
Institut für Strömungsmechanik, TU Braunschweig
Bienroder Weg 3, 38106 Braunschweig
Germany

Summary

In recent years Navier-Stokes methods have rapidly improved. Pressure distributions and aerodynamic coefficients for transport aircraft in crusing flight conditions can be calculated in a sufficient way. Encouraged by these achievements aircraft designers aim to use these methods to determine aerodynamic derivatives in order to make predictions for the handling qualities of a configuration at the stage of design. In this paper a method for the determination of the horizontal stabilizer efficiency is described. The steps of mesh generation and validation of the calculations are shown. It is analyzed whether the quality of numerical results is sufficient to determine horizontal stabilizer efficiency.

1. Introduction

The "Institut für Strömungsmechanik" was involved in the MEGAFLOW project with the grid generation for a configuration consisting of wing, body, horizontal stabilizer and vertical fin [1], [2]. Navier-Stokes calculations for a transport aircraft configuration in cruising flight condition were carried out. The results were validated mainly based on experimental pressure distributions of the horizontal stabilizer. The structure of the flowfield was analyzed based on the numerical results.

In a continued co-operation with EADS Airbus Bremen new experimental data of a Megaliner configuration became available. These were needed to have a closer look at the quality of the calculations. Besides the determination of lift coefficients more sensitive quantities such as horizontal stabilizer efficiency were analyzed.

2. Horizontal Stabilizer Efficiency

According to [3] horizontal stabilizer efficiency can be described as

$$\left(1 - \frac{\partial \alpha_W}{\partial \alpha}\right) = \frac{\frac{\partial c_{LT}}{\partial \alpha} - \frac{\partial c_{LWB}}{\partial \alpha}}{\frac{q_H}{q_\infty}\frac{\partial c_{HV}}{\partial i_H}} = \frac{\frac{\partial c_{HV}}{\partial \alpha}}{\frac{\partial c_{HV}}{\partial i_H}} \tag{1}$$

with the angle of attack α, the downwash angle α_W, the setting angle i_H of the horizontal stabilizer, the stagnation pressures of the free stream q_∞ and at the horizontal stabilizer q_H, the lift coefficient c_{LT} of the complete configuration consisting of wing, body, horizontal stabilizer and vertical fin, the lift coefficient c_{LWB} of the wing-body configuration and the lift coefficient c_{LHV} of the horizontal stabilizer and the vertical fin. The influence of the vertical fin is rather small. In contrast to [3] the component of the vertical fin is added to a combined component of horizontal stabilizer and vertical fin because experiments were carried out only for the complete configuration and the wing-body configuration. For the determination of the derivatives a minimum of six calculations is needed. $\partial c_{LT}/\partial\alpha$ is determined by two calculations for the complete configuration at the same i_H and different α, $\partial c_{LWB}/\partial\alpha$ by two calculations for the wing-body configuration and $\partial c_{HV}/\partial i_H$ by two calculations for the complete configuration at the same α and different i_H.

3. Grid Generation

For the determination of derivatives fine grids are needed allowing the same quality of results at different angles of attack and setting angles. For this purpose one grid for the wing-body configuration with 3.8 million gridpoints and two grids for the complete configuration at two different setting angles with 4.9 million gridpoints were generated with the grid generation programme MegaCads which is part of the MEGAFLOW system [4]. The grid for the complete configuration is shown in **Fig. 1**.

4. Validation

The calculations were carried out with version 116 of the Navier-Stokes code FLOWer, which is also part of the MEGAFLOW system [5]. As a part of the validation procedure the Baldwin-Lomax turbulence model [6], the Wilcox $k-\omega$ turbulence model [7] and the $k-\omega\,(LEA)$ turbulence model [8] were compared. With the latter complex pressure distributions consisting of a pressure minimum in the front part of a profile and a second one in the rear were described best. Therefore it was used for the investigations of horizontal stabilizer efficiency.
Another part of the validation was the analysis of the effect of numerical parameters such as the numerical viscosity. As a result the pressure distribution in the section $\eta = 0.32$ is shown for different angles of attack in cruising flight condition and below ($\alpha = 0.0°; 1.4°; 2.0°$) and two sets of parameters for the numerical viscosity ($Vis1 : k^{(2)} = 1/2, k^{(4)} = 1/32; Vis2 : k^{(2)} = 1/4, k^{(4)} = 1/64$) in **Fig. 2**. At $\alpha = 0.0°$ and $\alpha = 1.4°$ the shock on the upper surface is located slightly more rearward in the pressure distributions calculated with the lower viscosity $Vis2$, but the experimental data are not expressive enough to determine which one is better. At $\alpha = 2.0°$ an expansion in the rear part of the pressure distribution is calculated with $Vis2$ which is not shown by the experiment.

In **Fig. 3** the lift coefficient c_{LWB} of the wing-body configuration at $Ma_\infty = 0.85, Re_\infty = 3 \cdot 10^6$ is shown for the two settings of the numerical viscosity. The separation of the contributions of the components shows, that the different settings of the numerical viscosity influence the wing and have almost no effect on the body. The differences in the calculation of c_{LWB} are caused by the little discrepancies in the pressure distributions shown in Fig. 2. Because there is a better conformity between c_{LWB} from the experiment and the calculations with the lower numerical viscosity, $Vis2$, this setting was used for the following cases. With $Vis2$ the calculated lift coefficient is a little smaller than the one from the experiment at small angles of attack $\alpha < 1.0°$ and bigger at larger ones. The nonlinearity of the curve starting at $\alpha = 1.4°$ which is caused by the growth of supersonic areas can be seen in calculation and experiment.
As a result of the calculations for the complete configuration at $i_H = -1.5°$ the pressure distribution in two sections of the horizontal stabilizer $\eta = 0.12$ and $\eta = 0.75$ is shown in **Fig. 4**. At $\eta = 0.12$ there are some differences between experiment and calculation, which are probably caused by the representation of the boundary layer on the body in the calculations. At $\eta = 0.75$ there is a good conformity between experiment and calculation.
The lift coefficient for the complete configuration at $i_H = -1.5°$ and $i_H = 0.0°$ can be seen in **Fig. 5a** and **5b**. Both figures show a good conformity between calculation and experiment. The lift is divided into the contributions of the components again. The contributions of the wing and the body have almost the same amount as for the wing-body configuration. The contribution of the vertical fin is almost zero. At $i_H = -1.5°$ the stabilizer contributes negative lift at all investigated angles of attack. At $i_H = 0.0°$ the lift coefficient of the stabilizer is negative for the angles of attack $\alpha < 0.7°$ and positive for larger values of α.

5. Results

The lift coefficient c_{LHV} of the horizontal stabilizer and the vertical fin is determined by the difference of two calculations vs. experimental data for a configuration with and without stabilizer and fin. The amount is small compared to the lift coefficient of the complete configuration. c_{LHV} depending on the angle of attack for the two setting angles $i_H = -1.5°$ and $i_H = 0.0°$ is shown in **Fig. 6a**. The slope of the curves determines the numerator of Eq.(1). The amount of c_{LHV} at small angles of attack can be calculated in good agreement with the experiment, but at higher angles of attack some discrepancies occur. One of the possible reasons is the grid. The presence of the horizontal stabilizer and of the vertical fin in the grid for the complete configuration leads to a different resolution and structure of the grid in the rear fuselage region compared to the grid for the wing-body configuration. The different grids can cause different discretization errors for c_{LT} calculated on the grid for the complete configuration and for c_{LWB} calculated on the grid for the wing-body configuration. The other reason for the discrepancies may be the effect of the numerical dissipation on the calculated downwash.

The lift coefficient c_{LHV} depending on the setting angle i_H for different angles of attack α is shown in **Fig. 6b**. The slope of the curve is the denominator of Eq.(1). The figure shows that a change of i_H is described with a good accuracy.
Calculating the horizontal stabilizer efficiency from the derivatives described above leads to $(1 - \partial\alpha_W/\partial\alpha) = 0.34$ for the experiment and $(1 - \partial\alpha_W/\partial\alpha) = 0.45$ for the calculation. Concerning the experimental side comprehensive corrections of the data have been applied, which took into account the windtunnel wall and sting effects as well as the deformation of the model under aerodynamic loading. This standard of the experiments suggests that at present most of the discrepancies in the determination of $\partial c_{LHV}/\partial\alpha$ seem to result from the calculations. In order to improve the quality of the numerical investigations much more details of the flowfield have to be studied both numerically and experimentally: A comparison of measured and calculated velocity distributions in the boundary layer and in the wake of the wing should show, that physical phenomena are well predicted by the calculations. Unfortunately experiments of this kind are very rare. As a basic step the comparison of oilflow-patterns from the experiment and shear stress directions from the calculation are shown for the rear fuselage section in **Fig. 7**. The stagnation point in front of the horizontal stabilizer and the separation line caused by the horseshoe vortex originating from the intersection of the rear fuselage and the horizontal stabilizer have the same position in calculation and experiment.

6. Conclusions

The numerical results described in this paper confirm that a high quality of results is achieved concerning the calculation of pressure distributions and aerodynamic coefficients. The efficiency of the horizontal stabilizer is influenced by two derivatives. The change of the lift coefficient c_{LHV} with the the setting angle i_H can be calculated in a good quality, but further action has to be taken to determine the change of c_{LHV} with the angle of attack α. One potential for improvement is the minimization of the influence of different discretization errors made by performing calculations on the different grids for the complete and the wing-body configuration. Another one is the reduction of numerical dissipation in the wake of the wing in order to insure that the effects on the stabilizer are considered in the required accuracy.

Acknowledgements

The authors would like to thank the BMBF for funding throughout the project MEGAFLOW, the MEGAFLOW partners for the development of the code, EADS Airbus Bremen for the contribution of the experimental data and interesting discussions as well as HLRS Stuttgart for providing the possibility of performing these time consuming calculations on their SX5.

References

[1] Kroll, N.; Rossow, C.-C.; Becker, K.; Thiele, F.: *MEGAFLOW - A numerical flow simulation system.* ICAS-Paper 2.7.4, 1998.

[2] Nebel, C.: *Navier-Stokes Netze für komplexe aerodynamische Konfigurationen.* Institut für Strömungsmechanik der TU Braunschweig, Bericht 99/3, 1999.

[3] Schlichting, H.; Truckenbrodt, E.: *Aerodynamik des Flugzeugs.* Vol. 2, 2nd Ed., Springer-Verlag, 1969.

[4] Brodersen, O.; Ronzheimer, A.; Ziegler, R.; Kunert, T.; Hepperle, M.; Wild, J.; Heinrich, R.: *Aerodynamic applications using MegaCads.* Proceedings of the 6th International Conference on Numerical Grid Generation in Computational Field Simulations, London 1998, pp. 793-802.

[5] Kroll, N.; Radespiel, R.; Rossow, C.-C.: *Accurate and efficient flow solver for 3D applications on structured meshes.* AGARD R-807, 1995, pp. 4.1-4.59.

[6] Baldwin, B.; Lomax, H.: *Thin layer approximation and algebraic model for separated turbulent flows.* AIAA-Paper 78-0257, 1978.

[7] Wilcox, D.C.: *Turbulence modelling for CFD.* DCW Industries Inc., 1998.

[8] Rung, T.; Lübcke, H.; Franke, M.; Xue, L.; Thiele, F.; Fu, S.: *Assessment of explicit algebraic stress models in transonic flows.* In: W. Rodi, D. Laurence (Eds.): Engineering Turbulence Modelling and Experiments 4, Elsevier, Amsterdam, 1999, pp. 659-668.

Figures

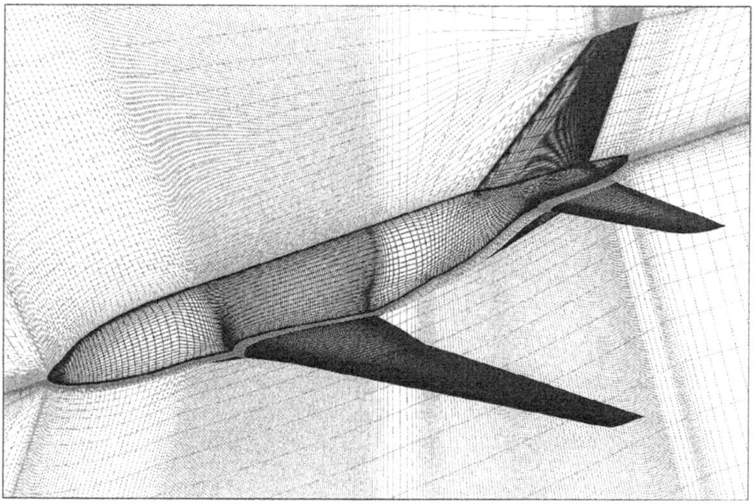

Figure 1 Grid for a Megaliner configuration; HH-topology, 4.9 million gridpoints

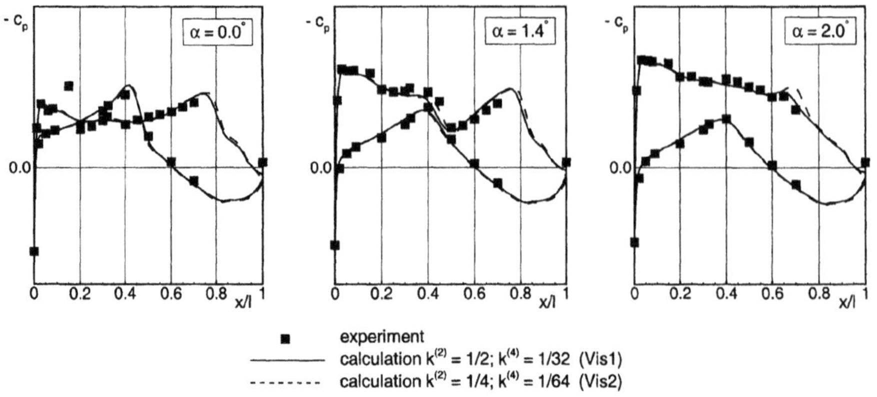

Figure 2 Pressure distribution in section $\eta = 0.32$ of the wing; wing-body configuration, $Ma_\infty = 0.85$, $Re_\infty = 3 \cdot 10^6$

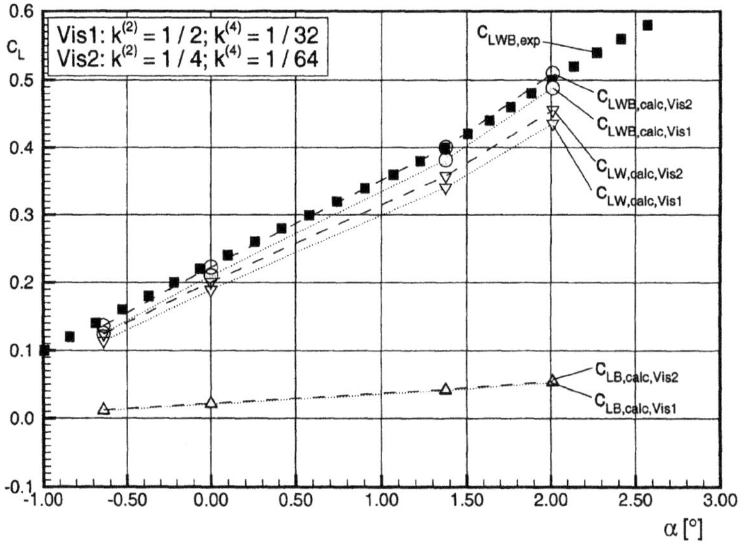

Figure 3 Calculated lift coefficient $c_{LWB,calc}$, contributions of wing $c_{LW,calc}$ and body $c_{LB,calc}$, measured lift coefficient $c_{LWB,exp}$; influence of numerical viscosity, wing-body configuration, $Ma_\infty = 0.85$, $Re_\infty = 3 \cdot 10^6$

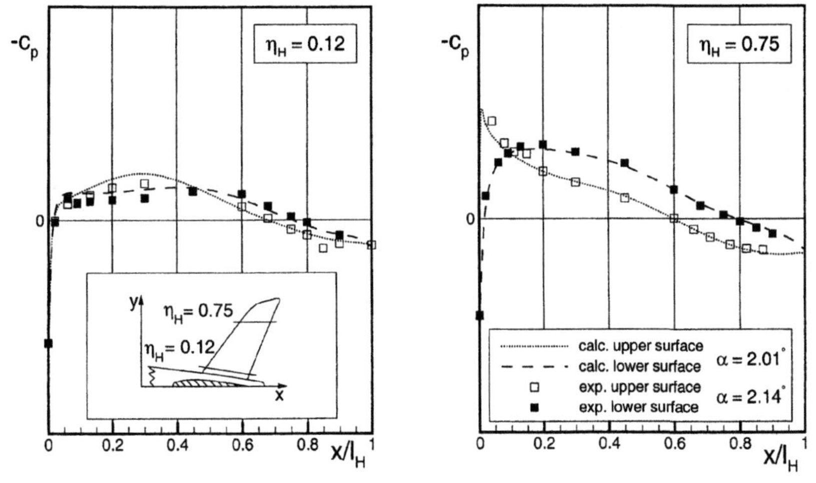

Figure 4 Pressure distribution in sections $\eta_H = 0.12$ and $\eta_H = 0.75$ on the horizontal stabilizer; complete configuration, $i_H = -1.5°$, $Ma_\infty = 0.85$, $Re_\infty = 3 \cdot 10^6$

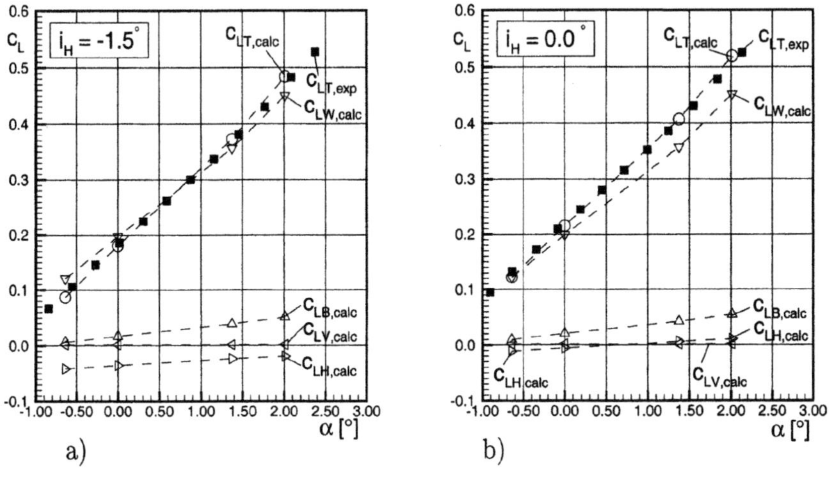

Figure 5 Calculated lift coefficient $c_{LT,calc}$, contributions of wing $c_{LW,calc}$, body $c_{LB,calc}$, horizontal stabilizer $c_{LH,calc}$ and vertical fin $c_{LV,calc}$ compared with measured lift coefficient $c_{LT,exp}$; $Ma_\infty = 0.85$, $Re_\infty = 3 \cdot 10^6$ a) $i_H = -1.5°$; b) $i_H = 0.0°$

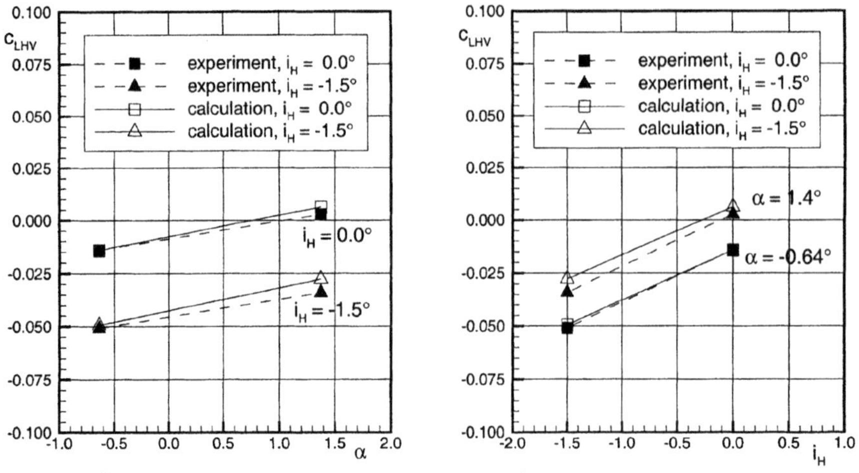

a) b)

Figure 6 Lift coefficient c_{LHV} for horizontal stabilizer and vertical fin;
$Ma_\infty = 0.85$, $Re_\infty = 3 \cdot 10^6$
a) dependent on α for different i_H
b) dependent on i_H for different α

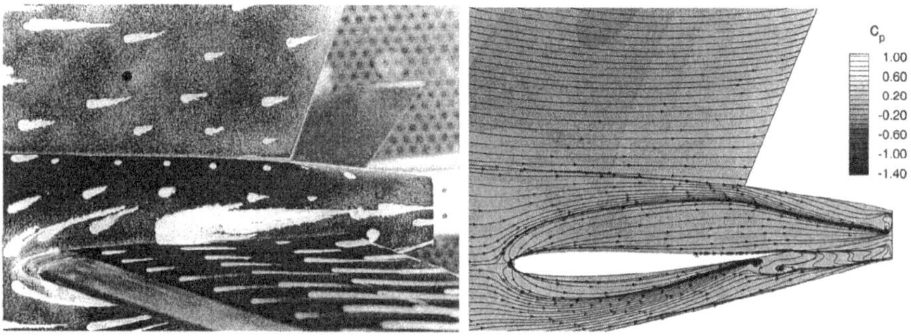

a) b)

Figure 7 Rear body of complete configuration;
$Ma_\infty = 0.85$, $Re_\infty = 3 \cdot 10^6$, $\alpha = 2.0°$, $i_H = -1.5°$
a) oil-flow-pattern (experiment)
b) surface pressure distribution and shear stress directions (calculation)

3D RANS Simulations for High-Lift Analysis of Transport Aircraft Configurations

S. MELBER
German Aerospace Center (DLR), Institute of Design Aerodynamics
Lilienthalplatz 7, D-38108 Braunschweig, Germany

Summary

This paper presents the numerical simulation and the analysis of viscous high-lift flows around a complex wing/body configuration (DLR ALVAST) with deployed high-lift devices. The solution of the Reynolds averaged Navier-Stokes equations (RANS) is carried out using the unstructured DLR TAU-code. The results of the computation are compared to measurements. The investigations aim at a better understanding of the aerodynamics at the wing root and the lift breakdown.

1 Introduction

The optimization of a transport aircraft at high incidence with respect to low speed take-off/landing capabilities and handling qualities near wing stall is a complex aerodynamic problem. The flow around the wing with deployed high-lift devices in this condition is characterized by the existence of areas with separated flow and a strong wake/boundary layer interaction. There are in addition some critical areas like the engine/nacelle integration or the wing/fuselage junction, which trigger premature flow separation. A properly triggered wing stall is usually advantageous for handling quality reasons, but on the other hand it limits the maximum lift. The understanding of the flow physics concerning these cases is very important for the design of a complex high-lift configuration.

The design of the AIRBUS A321 [1] is an example for such optimization. In this case a focus is laid on the wing root flow, especially on the inner slat-end/fuselage juncture. An efficient way to stabilize and strengthen the inner slat end vortex, which delays the wing root stall, is to fasten small devices at the inner slat end [2].

The two main possibilities to study such aerodynamic effects are on one side the wind-tunnel testing and on the other side the application of computational fluid dynamics (CFD). To improve the accuracy, reliability and performance of Reynolds-averaged Navier-Stokes flow solvers especially for high-lift analysis of realistic three-dimensional aircraft geometries several studies have been carried out [3] - [8]. Compared to wind-tunnel experiments the use of CFD holds promise to substantially accelerate the aerodynamic design, save costs and to allow simultaneously a detailed insight into the complete flow field around the considered configuration.

This paper continues the investigations of the DLR ALVAST wing/body transport aircraft configuration in high-lift condition [3], [4]. The focus is mainly on an improved understanding of the complex flow field, which is dominated by effects caused by the three-dimensionality. Moreover, the current reproduction of the stall effects and the lift breakdown as were found in the wind-tunnel tests is the other objective. To reach this, the stall characteristics of the ALVAST configuration will be investigated with the help of lift- and drag polars computed with the hybrid unstructured flow solver DLR-TAU. Furthermore, the influence of the wing root flow field on the lift breakdown will be discussed.

2 Aircraft Configuration and Computational Grids

The DLR ALVAST wing/body wind-tunnel model [9] with deployed slat and flaps in take-off configuration is selected as the baseline high-lift configuration. The geometrical relations of the ALVAST model are similar to the AIRBUS A320 transport aircraft configuration (fig. 1). For the past number of years suitable validation data for the CFD investigations have been produced during various national and European wind-tunnel campaigns with the full- and half-model [10], [11].

Two different CAD-descriptions form the basis for the present investigations. The first one is simplified compared to the wind-tunnel model to allow for block-structured grid generation in preceding studies. In this geometry the slot between the slat and fuselage in the area of the wing-junction was not modeled with all details like slat-horn and slat-stump at the leading edge. The wing-root fairing on the upper-side of the wing/fuselage junction was likewise not modeled. On this geometry two different grid types were built. The first was a block-structured curvilinear grid around the complete configuration, which was generated with the interactive grid generation package MEGACADS and presented in [8] and modified by adding slat/flap gaps on the wing/fuselage junction later on. This grid consists of about $9.2 \cdot 10^6$ grid points in 50 blocks.

The second one is a hybrid unstructured grid (grid I, fig. 2) which was generated using the commercial system CENTAUR [12]. The grid consists of two parts: a quasi-structured prismatic cell layer with a specified number of cells around the surface of the geometry in order to achieve a good resolution of the viscous effects inside of the boundary layer and tetrahedral cells to fill the outer domain of the flow-field. The initial grid consists of about $4 \cdot 10^6$, and after an adaptation during the flow solution process the number of points increases to about $5.8 \cdot 10^6$ points.

Both grids were used for a comparison of the hybrid unstructured flow solver TAU and the block-structured flow-solver FLOWer. Details of this investigation can be found in [3] and [4]. In this paper the results of the computation of the lift-curve with TAU on grid I will be shown.

The second CAD-description captures more details of the wind-tunnel model of the ALVAST (fig. 3, 4). Further on a check of the first CAD-model has shown inaccurate span-wise flap-slot geometries and partly great differences of the slat position compared to the wind-tunnel model. This differences were also corrected in the improved CAD-description. The hybrid unstructured grid II based on the second CAD-description was also built with CENTAUR. To resolve all features of the geometry, the number of points are improved to $10 \cdot 10^6$ in the initial grid. A complete field-adaptation was not done at this time, only the contribution of the points in the prismatic layer was adapted to reach everywhere $y^+ \approx 1$.

3 Flow Solution Method

The Reynolds-averaged Navier-Stokes equations are solved by the hybrid unstructured flow solver DLR TAU, which is based on a three-dimensional finite volume scheme. The governing equations are solved on a dual background grid, which, together with the edge-based data structure, allows to run the code on any type of cells. The edge-based data (dual grid) is directly determined from the initial grid by a preprocessing module. The solver is part of the MEGAFLOW-project [13] and is presented in [14] in more detail.

The flow variables are stored in the centers of the dual grid, i.e. the vertices of the primary grid. The temporal gradients are discretized using a multi-step Runge-Kutta scheme. The inviscid fluxes are calculated either by a Roe- or AUSM-type 2nd-order upwind scheme, or by employing a central method with scalar dissipation, which is used for the present study. The gradients of the flow variables are determined by employing a Gauss-Green formula. The viscous fluxes are discretized using central differences.

The turbulence models available in the TAU-code are the Spalart-Allmaras model, the Wilcox $k\omega$-model and the SST-model according to Menter. Transition can be fixed by assigning a flag (turbulent or laminar) to each surface-node. The ALVAST configuration is computed by using the Spalart-Allmaras [15] turbulence model in a fully turbulent flow field because in the wind-tunnel tests no transition fixing was made. In order to accelerate the convergence to steady state, local-time stepping and a multigrid technique based on agglomeration of the dual-grid volumes are employed. Optimization for different architectures is achieved by vector- or cache-type coloring of the edges, on which most of the work is done. For parallel computations a domain-decomposition is used providing a subset of dual grids.

An adaptation module for unstructured grids is also available in the TAU-code. It detects regions with insufficient grid resolution by gradient sensors of flow variables and performs local grid refinement by bisection of cells. The initial solution is then interpolated to the adapted grid. In addition, the adaptation also allows the redistribution of the prismatic layers to capture the viscous boundary-layer in Navier-Stokes computations adequately.

4 Numerical Results

The following flow computations were performed for a free stream Mach-number of $M_\infty = 0.22$ and a Reynolds-number of $Re_\infty = 2 \cdot 10^6$ related to the mean aerodynamic wing-chord. Corresponding experimental data from measurements on the full-model from the large low speed facility of the German-Dutch wind-tunnel DNW are used for comparisons to numerical results.

4.1 Simplified Geometry

To investigate the high-lift behavior of the ALVAST configuration the lift- and drag polar was computed with the unstructured flow solver TAU on grid I. After an initial convergence the grid was adapted on the pressure loss Δp_{tot} and a redistribution of the points in the boundary layer to reach $y^+ \approx 1$ was done at each angle of attack. Each case was converged to a point with a constant lift coefficient except the cases with strong separated flow.

The resulting lift coefficients C_L in the linear range of the $C_L(\alpha)$-curve are consistently over-predicted by about 5 % (fig. 5, experiment and grid I). Beginning with an angle of attack of $\alpha \geq 22^o$ the computed lift breaks down. However, the angle of attack for maximum lift agrees well with the measured data. A similar behavior can be observed for the drag coefficient, which is also over-predicted by the numerical method in the complete range of angle variations (fig. 6).

For a descriptive discussion of the computed lift breakdown in the figures 7-9 the ALVAST configuration is shown with stream lines in the flow-field. The vortices on the flap- and wing side edges and in the area of flow separation are shown at angles of attack of 4°, 12° and 22°.

As supposed by the remarkably linear range of the lift polar, the flow does not separate up to an angle of attack of $\alpha = 21^o$. From this point there is a separation in the area of the wing root, which extends stream-wise over the whole wing chord. Adjacent to it there is a weak separation on the outer end of the outboard flap beginning at an angle of attack of $\alpha = 22^o$, which extends to approximately 30% of the local wing chord and which moves with an increasing angle of attack to the inboard-side of the wing.

4.2 Improved Geometry

The differences between the wind-tunnel model and the CAD-description of the simplified geometry (grid I) can have a considerable influence on differences between the experiment and the simulation. To investigate this and to have a geometry as close as possible to the experiment a computation on grid II near the angle of attack of maximum lift ($\alpha = 21^o$) was performed. During the flow solution process the contribution of the points in the prismatic layer was adapted to reach $y^+ \approx 1$. The solution was fully converged to obtain a constant lift coefficient and will be discussed subsequently.

Aerodynamic Coefficients and Flow Separations

In the lift polar (fig. 5, grid II) the lift coefficient of the computation is much closer ($\Delta C_L = 1.7\%$) to experiments as the lift coefficient of grid I ($\Delta C_L = 4.0\%$) at an angle of attack of $\alpha = 21^o$. This must be an effect of the improved geometry because the flow solver or its configuration was not changed between the two cases.

In figure 10 the pressure distributions on the elements of the high-lift configuration at 46% half span are shown. On the slat there is a good agreement in the pressure distribution of the computation and the experiments except for the suction peak in the computation which is higher than the suction peak in the experiments. A similar behavior is found for the main wing. The computation almost lies on the experiment, only the double suction peaks are higher than experimental ones. This double peak is determined by variation in the curvature on the surface in a small area where the retracted slat ends. It is not resolved in the experiments because there are not enough pressure tappings in this area of the wing. The pressure distributions on the flaps show likewise a good correlation between experiment and simulation.

To detect flow separation on the ALVAST configuration calculated with grid II in figure 11 stream lines in the flow-field are shown. There are two unconnected areas of separated flow: First near the trailing edge of the main wing in the area of the wing root, which extends approximately over 30 % of the wing chord. A second separation zone can be found at the trailing edge of the main wing in the area of the whole outboard flap and the outer unflapped wing. The inner flap cut out has also a weak separation. A remarkable point in fig. 11 is the strong span-wise flow in the separated areas from the inner flap cut out to the outboard end of the outboard flap. This span wise flow cannot be found in the rest of the flow field.

Wing Root Flow and Lift Breakdown

In the experiments with the ALVAST half-model in the low-speed wind-tunnel in Braunschweig [10] there was an early lift breakdown at low angles of attack with the original configuration (similar to the second CAD-description without slat-horn and wing-root fairing). Flow visualization tests with tufts indicate that flow separation starts at the wing fuselage junction and then spreads suddenly over the whole inboard wing. On the other hand the outboard wing area retains attached flow up to high angles of attack.

To improve the stall angle of the ALVAST two modifications on the geometry have been carried out: First at the wing/fuselage junction a wing-root fairing on the upper wing was added. And second, a slat end plate was added on the inboard slat end. With this modifications the lift breakdown in the experiments starts at higher angles of attack with flow separation on the outboard wing near the trailing edge. Tests were also made either with the wing-root fairing or the slat end plate and it was found out that the wing-root fairing is clearly the decisive factor for the prevention of premature flow separation. The combination of fairing and slat end plate shows the best improvement in the high-lift behavior of the ALVAST configuration.

To understand the lift breakdown a closer view of the flow features in the area of the wing fuselage junction is necessary with the help of the computations. In figure 12 the pressure distribution on the surface and stream lines are shown at the trailing edge of the main wing close to the wing/fuselage junction. The planes show areas of velocity loss against the surrounding flow. This planes emphasize a small area of separated flow at the trailing edge of the main wing along the wing/fuselage junction. The fillet along this sharp corner on the upper wing should reduce this flow separation and with it the early lift breakdown. In the calculations on grid I without a wing-root fairing the area of separated flow is substantialy bigger and the lift breakdown is therefore triggered by this flow separation (fig. 9).

The second device to improve the stall incidence triggered by wing/fuselage junction is the slat end plate as shown in figure 13 together with the pressure distribution on the surface and stream lines. On the upper side of the slat end plate a strong vortex forms and develops into a straight line on the upper side of the wing close to the fuselage. In the pressure distribution of the fuselage the beginning of the vortex path is clearly visible. As the rotation disappears the stream lines turn into the surrounding stream direction. The velocity loss of the vortex can likewise be found in figure 12 along the fuselage. The straight path of the vortex near the slat end plate is caused by the accelerated flow in the shortened slat gap between the slat and the fuselage.

This slat end vortex provokes two opposed effects in the flow field: First, it shields the wing root region against the down-wash of the surrounding flow field and so enlarged the trend of flow separation in this area. And second, this negative effect is compensated by the induced rotating flow, which adds a flow component towards the root corner and thus induces higher-energy air into the root area. In the calculation on grid II the slat end vortex seems to burst after a short distance (fig. 13). At the moment it is not clear, if this is an effect of flow physics or of grid resolution, which is maybe too coarse to resolve the vortex in this area. An adaptation of the grid along the vortex should answer this question. Another question is to investigate if the vortex is more stable at lower incidences and if an enlarged slat plate or horn could strengthen and stabilize the slat vortex [2] and thereby improve the high-lift behavior.

5 Concluding Remarks

The numerical flow simulation of transport aircraft in high-lift configuration with the unstructured TAU-Code shows a good prediction of the aerodynamic parameters in comparison to wind-tunnel experiments up to the maximum lift. On a simplified CAD geometry the lift coefficient is about 5 % higher than the experimental value. An improved CAD geometry closer to the wind-tunnel model reduces the deviation between computation and experiment to an 1.7 % over-predicted lift coefficient at an angle of attack of $\alpha = 21^o$ and a good agreement in the pressure distribution of the computation and the experiments. From experiments [2], [10] it is well known that the flow at the wing/fuselage junction has a considerable influence on the stall behavior. In the computations this area is characterized by two effects: First, an area of separated flow at the wing trailing edge near the wing/body junction, which can trigger the maximum incidence in this area. And second, a vortex from the slat end plate, which plays an important role in stabilizing the critical corner flow in the experiments. Further investigations using the unstructured approach will follow. Overall, the numerical flow simulation of RANS equations can be considered as a powerful tool to investigate complex high-lift configurations of transport aircraft.

References

[1] Flaig, A.; Hilbig, R.:"High-Lift Design for Large Civil Aircraft" AGARD-CP-515, Sep. 1993.

[2] Schwetzler, D.: "Improvement of Maximum Lift in Flight by Optimization of Vortex Flow." High-Lift and Separation Control Conference Proceedings, Univ. of Bath, United Kingdom, Mar. 29-31, 1995, pp. 4.1-4.10.

[3] Rudnik, R.; Melber, S.; Ronzheimer, A.; Brodersen, O.: "Aspects of 3D RANS Simulations for Transport Aircraft High-Lift Configurations" AIAA paper 2000-4326, 2000.

[4] Melber, S.; Rudnik, R.; Ronzheimer, A.: "Structured and Unstructured Numerical Simulation in High-Lift Aerodynamics." Workshop on EU-Reserach on Aerodynamic Engine / Aircraft Integration for Transport Aircraft, 26-27 September 2000, DLR Braunschweig, 2000, pp. 13-1 - 13-10.

[5] Rogers, S.E.; Roth, K.; Baker, M.D.; Nash, S.M.: "Navier-Stokes Analysis of a High Wing Transport High-Lift Configuration with externally blown Flaps." AIAA paper 2000-4219, 2000.

[6] Rogers, S.E.; Roth, K.; Cao, V.Hoa; Slotnick, J.P.; Whilock, M.; Nash, S.M.; Baker, M.D.: "Computation of viscous Flow for a Boeing 777 Aircraft in Landing Conf." AIAA paper 2000-4221, 2000.

[7] Mavriplis, D.J.: "Parallel Performance Investigations of an Unstructured Mesh Navier-Stokes Solver" ICASE Report No. 2000-13, March 2000.

[8] Rudnik, R.; Ronzheimer, A.; Raddatz, J.: "Numerical Flow Simulation for a Wing/Fuselage Transport Configuration with Deployed High-Lift system" in Notes on Numerical Fluid Mechanics, Vol. 72, pp. 363-370, Vieweg-Verlag, Braunschweig/Wiesbaden, 1999, Proceedings of the 11th AG STAB/DGLR Symposium Berlin, 1998.

[9] Kiock, R.: "The ALVAST Model of DLR." DLR IB 129-96/22, 1996.

[10] Puffert-Meissner, W.: "ALVAST Half-Model Investigations in the Low-Speed wind-tunnel Braunschweig." DLR IB 129-95/11, 1995.

[11] Puffert-Meissner, W.: "ALVAST Half-Model wind-tunnel Investigations and Comparison with Full-Span Model Results." DLR IB 129-96/20, 1996.

[12] Kallinderis, Y.: "Hybrid Grids and Their Applications." Handbook of Grid Generation, CRC Press, Boca Raton / London / New York / Washington, D.C., pp. 25-1 - 25-18, 1999.

[13] Kroll, N.; Rossow, C.-C.; Becker, K.; Thiele, F.: "MEGAFLOW - A Numerical Flow Simulation System." 21st ICAS congress, 1998, Melbourne, 13.09-18.09.1998, ICAS-98-2.7.4, 1998.

[14] Galle, M.: "Solution of the Euler and Navier-Stokes Equations on Hybrid Grids" AGARD F.D.P. Symposium on Progress and Challenges in CFD, Methods and Algorithms, Seville, 2-5 Oct., 1995.

[15] Spalart, P. R., Allmaras, S.R.: "A One-Equation Turbulence Model for aerodynamic Flows." AIAA-paper 92-0439, 1992.

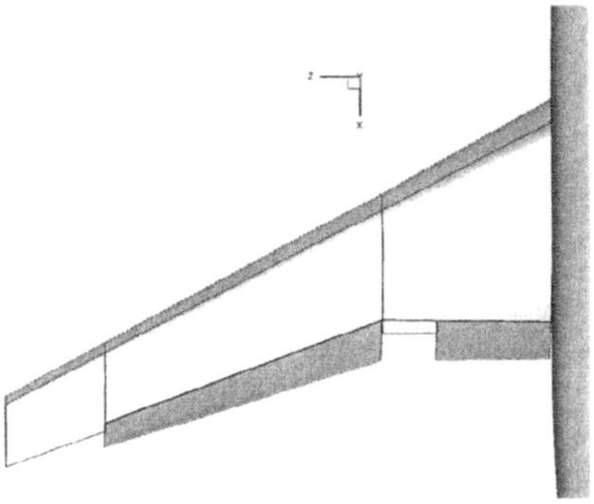

Figure 1 Planform of DLR ALVAST model in high-lift configuration ($\Psi_{LE} = 27.1^o$, $\Lambda = 9.5$, $\lambda = 0.3$, $\delta_s = 20.0^o$, $\delta_f = 19.5^o$)

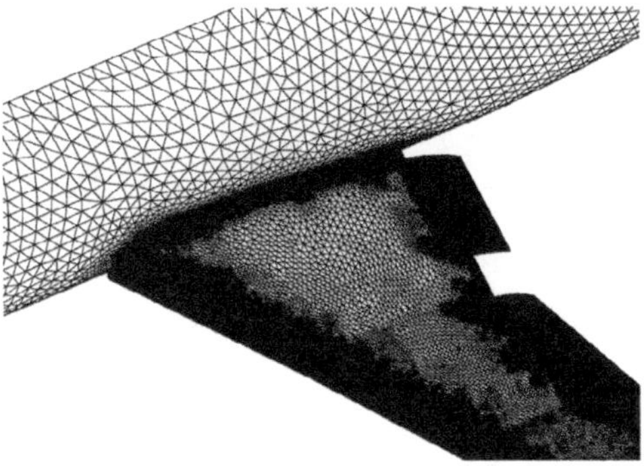

Figure 2 Hybrid unstructured surface grid I

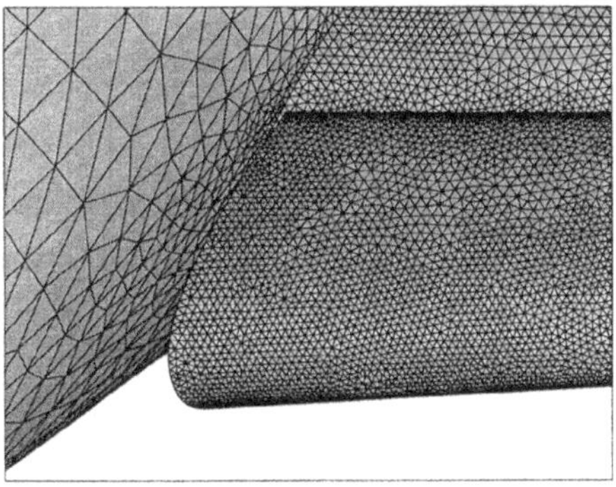

Figure 3 Unstructured surface grid I in the area of wing/fuselage junction

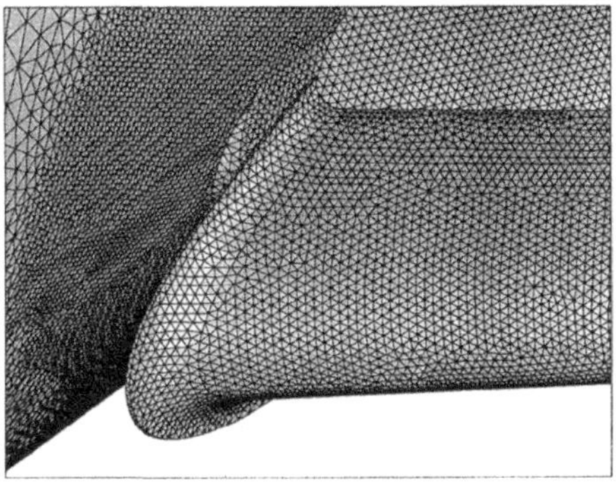

Figure 4 Unstructured surface grid II in the area of wing/fuselage junction

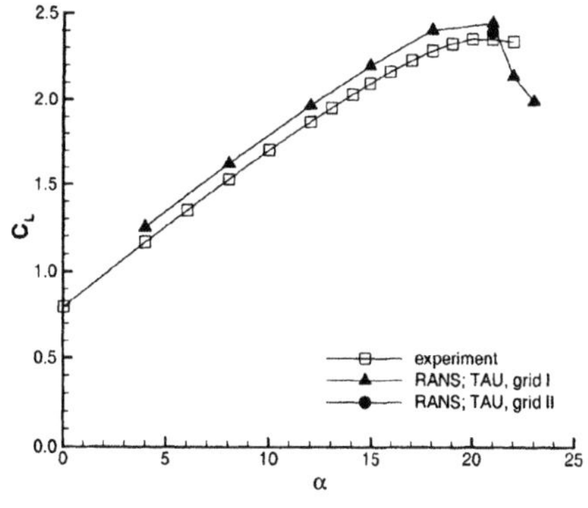

Figure 5 Lift polar

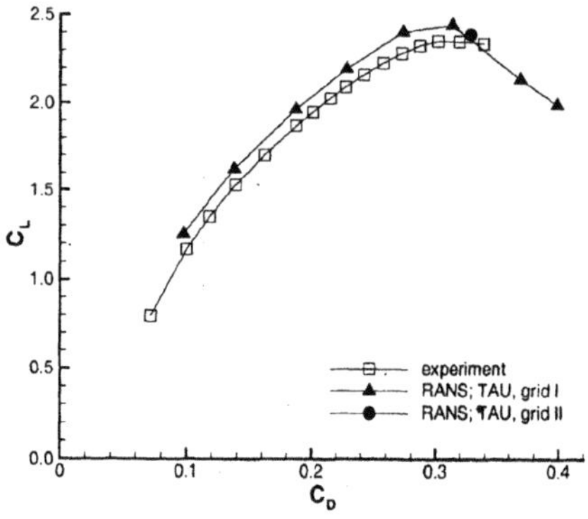

Figure 6 Drag polar

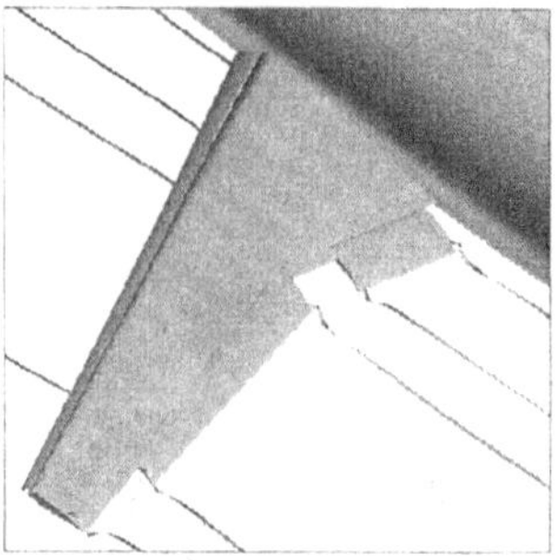

Figure 7 Streamlines off the surface, $\alpha = 4.00^o$, grid I

Figure 8 Streamlines off the surface, $\alpha = 12.03^o$, grid I

Figure 9 Streamlines off the surface, $\alpha = 22.00^o$, grid I

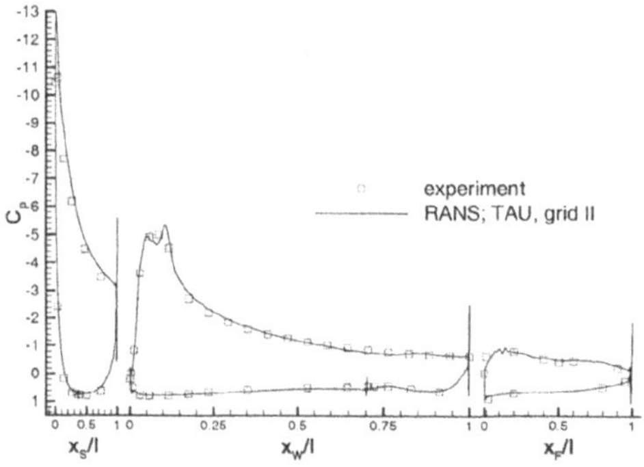

Figure 10 C_p distribution at 46% half span, $M_\infty = 0.22$, $Re_\infty = 2 \cdot 10^6$

Figure 11 C_p on the surface, flow separation shown by stream-lines, grid II

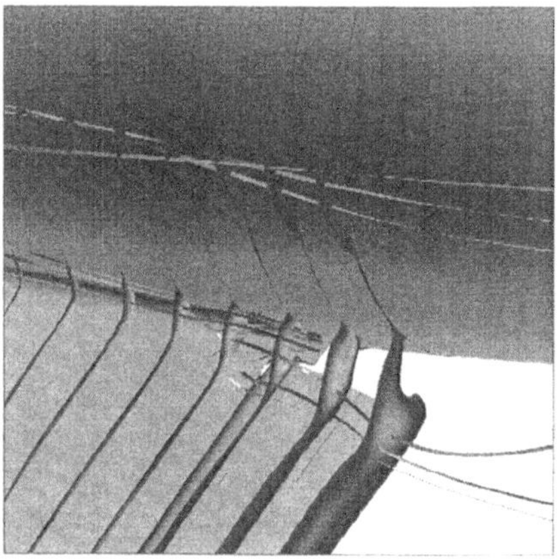

Figure 12 Wing/fuselage junction with wing-root fairing and stream-lines, areas of velocity loss in cut-planes, grid II

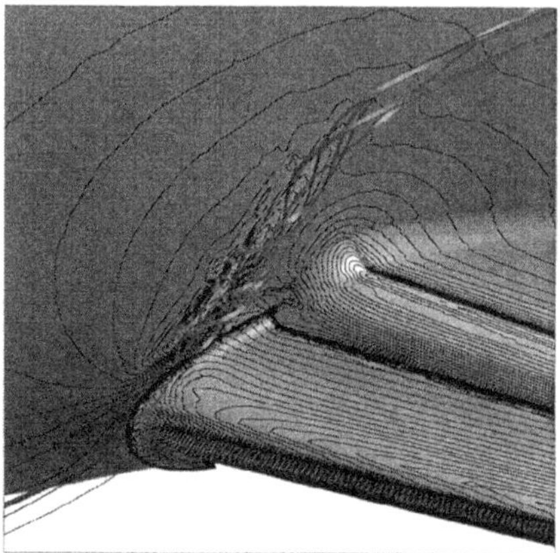

Figure 13 C_p on the surface, slat end plate and stream-lines, grid II

Numerical Investigation of the Reynolds-Number-Effect on the Setting Sensitivity of a 2D High-Lift Multi-Element Airfoil

J. Wild
DLR, Institute of Design Aerodynamics,
Lilienthalplatz 7, 38108 Braunschweig, Germany

Summary

The investigation focuses on the numerical sensitivity study of different slat settings on the maximum lift coefficient of a 2D high-lift multi-element airfoil. In addition to the influence of gap and overlap on the maximum lift coefficient, the Reynolds number effect on the optimal setting is investigated and compared with experimental results.

1 Introduction

High-Lift design of transport aircraft today is still mostly performed in the wind-tunnel. But this procedure has some shortcomings. First, due to model construction and testing time, the number of different settings of the high-lift components that can be tested within a wind-tunnel campaign is usually limited. The multidimensionality of the setting selection especially for multi-element wings leads to an early preselection of some setting values. In most cases for example the deflection angles of the high-lift devices are fixed before testing starts. Another disadvantage is that flight conditions for a large transport aircraft, mainly the Reynolds-number can not be achieved, except in cryogenic wind-tunnels, like the DNW-KKK or the ETW. But there, because of the cryogenic conditions, testing of a wide range of high-lift settings is quite cost-intensive.

On the other hand the developments and improvements achieved in Reynolds-averaged Navier-Stokes methods (RANS) can lead a way out of the restrictions mentioned above. Efficient and accurate flow simulation today is able to accompany experiments in the sense of leading to the proper experimental setup. Additionally, with the deep insight into the flow field provided by CFD, special interesting regions in the flow domain can be detected for the detailed analysis in the wind-tunnel. With todays flow solvesr there are tools available that can complement experiments at an early stage of aircraft design.

The fact, that these modern CFD methods have not yet found their way into industrial high-lift design, results mainly from the not yet provided validation of flow solvers for this type of flow. So the main effort to increase the acceptance of CFD methods for industrial design is to show, that these methods are able to not only simulate the flow around a given configuration, but that they are able to properly predict sensitivities for changing configurations, and further to predict the trends of the sensitivities with the increase of Reynolds-number up to flight conditions. To close this gap in CFD validation is the main objective of the present investigation.

2 Geometry description and flow conditions

The configuration used in this investigation is the H8Y 3-element airfoil of EADS Airbus, which was measured in the DNW-KKK cryogenic wind-tunnel within the German BMBF-founded technological program HAK 2 by Hansen and Szabo [1] at different Reynolds-Numbers. This high-lift configuration consists of a leading edge slat and a single slotted fowler flap (**Fig. 1**). The base airfoil was designed for a generic very large transport aircraft of a size similar to A3XX, which is expressed by a very sharp leading edge, causing a maximum lift limit by a leading edge stall.

For the comparison with the experimental data, four different slat settings, which were measured at two Reynolds-numbers, are calculated (**Fig. 2**). Relative to the reference position 4 these positions include two variations of different gaps and one of increased overlap. With this setup the sensitivity of the H8Y configuration can be analyzed according to the gap-, the overlap- and the Reynolds-number-effect.

The calculations are performed for the complete lift versus angle-of-attack curve at $M_\infty = 0.178$ for the two Reynolds-numbers $Re_\infty = 2.88 Mio.$ and $Re_\infty = 9.0 Mio.$. In the presented framework only the maximum lift coefficient values from the calculated data are used. Since the position of transition was neither measured nor fixed in the experiments, the computations are performed assuming fully turbulent flow.

3 Flow Simulation

The calculations are performed using the DLR flow solver FLOWer [2], which solves the compressible Reynolds-averaged Navier-Stokes-equations on multi-block structured grids by a finite-volume approach using an explicit Runge-Kutta time-stepping scheme. Convergence is accelerated by well-known methods like local-time-stepping, residual smoothing and multi-grid. For turbulence modeling the Spalart-Allmaras one-equation-model [3] with the modifications of Edwards and Chandra [4] is applied.

The used multi-block grids (**Fig. 3**) in C-topology around each solid wall boundary consists of approximately 80000 cells. Appropriate resolution of the boundary layers is achieved by a first cell thickness of 10^{-5} chord lengths, which results in a y^+-value between 0.75 and 2 depending on the Reynolds-number. Due to the parametric philosophy of the DLR grid-generator MegaCads [5] it was possible to generate grids of equal resolution for the different slat settings in a short time, so that it can be assumed, that differences in the solution are not triggered by differences of the grids.

4 Setting Sensitivities

In general the computations show a slight over-prediction of the maximum lift coefficient compared with the measurements of $\Delta C_{L,max} \approx 0.1$ for the low Reynolds-number and $\Delta C_{L,max} \approx 0.05$ for the high Reynolds-number. However, since this behavior is the same for all configurations, the sensitivities should not be influenced.

Fig. 4 shows the tendencies of a decreasing slat gap on the maximum lift coefficient. For the low Reynolds-number the optimum slat gap at the middle position is predicted correctly. Also for the high Reynolds-number the decrease of maximum lift coefficient with decreasing gap is captured by the calculations. For the gap sensitivity the computations show a good prediction capability of the CFD method at both Reynolds-numbers. The overlap sensitivities, as shown in **Fig. 5**, show the same good agreement of experimental data and the numerical simulation. Besides the overall tendency of decreasing maximum lift with increasing overlap, also the decrease of the gradient with the increasing Reynolds-number is captured by the calculations.

Prediction of the sensitivities of the global coefficients is only one aspect of validation. Especially regarding multi-element-airfoils it is of interest, whether the reason for changes in the global coefficients due to geometry changes agree in detail with the measured flow physics. Therefore, a detailed comparison of the pressure distributions will give an insight into the prediction of local changes in the flow field. A look at the suction peak of the main airfoil element, which is nearest to where the setting change takes place, shows a good agreement between experiment and calculation, either for the low Reynolds-number (**Fig. 6**) as well as for the high Reynolds-number (**Fig. 7**). The gap-effect on the pressure distribution (Pos. 1-4-7) can be detected as a nearly linear reduction of the suction peak with decreasing gap, that match also the differences in the experimental data. This reduction of the suction is according to Smith's slat-effect [6], that the presence of a lifting element in front of an airfoil reduces the over-speed at the leading edge by its circulation. Consistently the reduction of the over-speed is increased with a closer positioning. Since an overlap variation (Pos. 4-5) does not change this parameter, the suction stays nearly constant for this case. To complete the setting effects on the pressure distribution a view on the slat trailing edge shows (**Fig. 8** and **9**), that the gap variation mainly effects the lower side pressure in front of the trailing edge, while the upper side pressure nearly stays constant. Overlap variation in contrast influences the trailing edge pressure either on lower and upper side. It can be examined, that increasing overlap moves the point of maximum flow velocity in the gap upstream, so that the gap flow is slightly decelerated towards the trailing edge.

5 Reynolds-Number effect

The dependency of the setting sensitivity to the Reynolds-number can be derived from the comparison of the maximum lift coefficients as a function of the Reynolds-number (**Fig. 10**). The comparison with the experimental data shows, that most of the tendencies between the different configurations are correctly calculated. It can be seen that for all configurations the maximum lift coefficient increases with increasing Reynolds-number. So there is no configuration tested, that wouldn't work under flight conditions. Only the computed drop in maximum lift value for position 1 at the low Reynolds-number gives a sensitivity too low compared to the experiment.The reversal of the gap sensitivity between position 1 and position 4 for the increase of Reynolds-number is predicted correctly. The reason for this reversal can be located on the slat when regarding the difference in pressure between position 1 and position 4, where $\Delta c_p = c_{p,pos.1} - c_{p,pos.4}$ (**Fig. 11**). While for the low Reynolds-number the suction on the complete upper side of the slat is decreasing from position 4 to position 1, the suction is increasing at the high Reynolds-number for the same change in position. Even though the effect is not so pronounced in the calculations than in the experimental data the effect is well captured by the numerical simulation.

6 Conclusion

This investigation shows, that the application of RANS-solvers for high-lift flows is feasible also for sensitivity analysis and trends are predicted correctly. It can be supposed, that the main flow physics responsible for high-lift setting sensitivities are appropriately modeled by the numerical simulation. The big advantage is that also the dependency on the Reynolds-number is predicted correctly, so that the numerical simulation is able to give answers for the change of sensitivities for higher Reynolds-numbers.

7 Acknowledgement

The author would like to thank Mr. Hansen form EADS Airbus Bremen for the opportunity of using the described configuration and providing the experimental data for the presented investigations.

References

[1] H. HANSEN, I. SZABO: *"Investigation of Stall Characteristics of an A3XX Relevant Airfoil up to High Reynolds Numbers in the Technology Program HAK 2"*. In: W. NITSCHE ET. AL.: *Notes on Numerical Fluid Dynamics*, Vol. 72, Vieweg (1999), pp. 218-225

[2] DLR (ED.): *"FLOWer Installation and User Handbook - Rel. 116"*. DLR - Institute for Design Aerodynamics, Braunschweig (2000)

[3] P.R. SPALART, S.R. ALLMARAS: *"A One-Equation Turbulence Model for Aerodynamic Flows"*. AIAA Paper 92-0439 (1992)

[4] J.R. EDWARDS, S. CHANDRA: *"Comparison of Eddy Viscosity-Transport Turbulence Models for Three-Dimensional, Shock-Separated Flow"*. *J.Aircraft,* Vol. 34, No. 4 (1996), pp. 756-763

[5] O. BRODERSEN, M. HEPPERLE, A. RONZHEIMER, C.-C. ROSSOW, B. SCHÖNING: *"The Parametric Grid Generation System MegaCads"*. In: B.K. SONI, J.F. THOMPSON, J. HÄUSER, P. EISEMAN: *5th International Conference on Numerical Grid Generation in Computational Field Simulation,* NSF (1996), pp. 353-362

[6] A.M.O. SMITH: *"High-Lift Aerodynamics"*. AIAA Paper 74-939 (1974)

Figure 1 H8Y 3-element airfoil

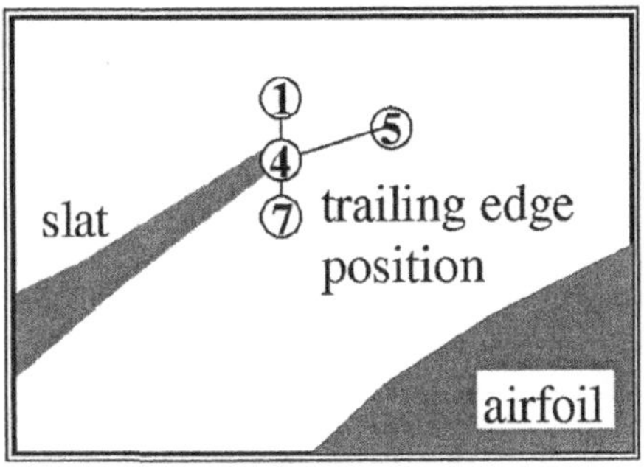

Figure 2 Definition of setting variation positions referenced by slat trailing edge position for the EADS H8Y 3-element airfoil as tested in DNW-KKK

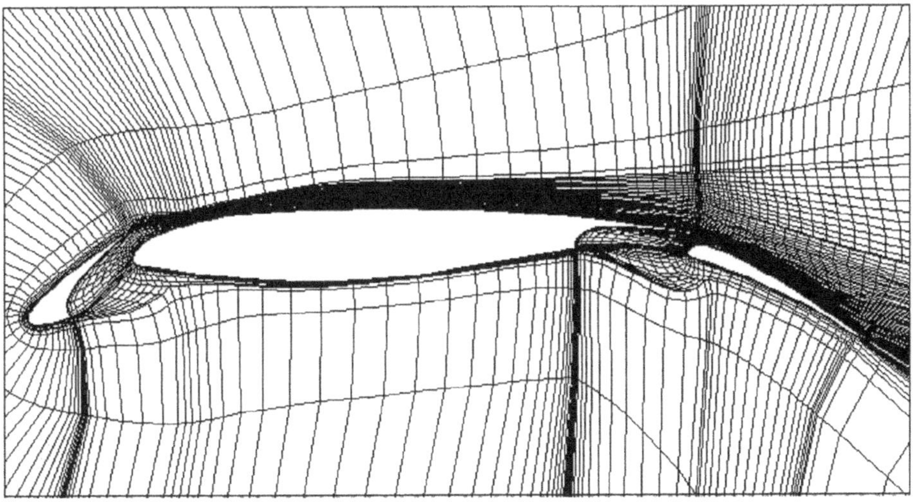

Figure 3 9 block grid around H8Y 3-element airfoil in C-topology around each solid wall (only every 4th line drawn)

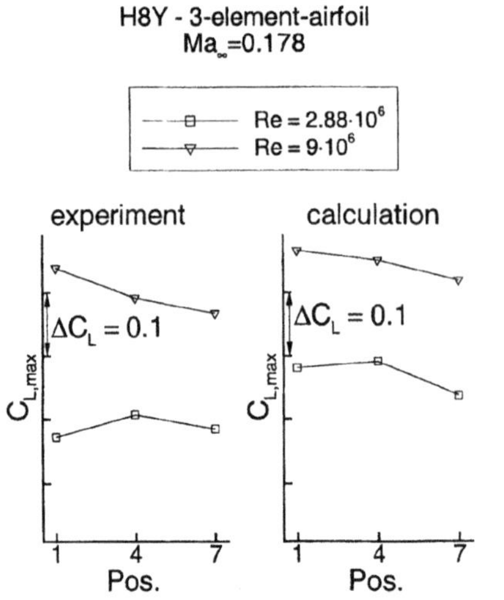

Figure 4 Dependency of the maximum lift coefficient from the slat gap of the H8Y configuration for different Reynolds-numbers

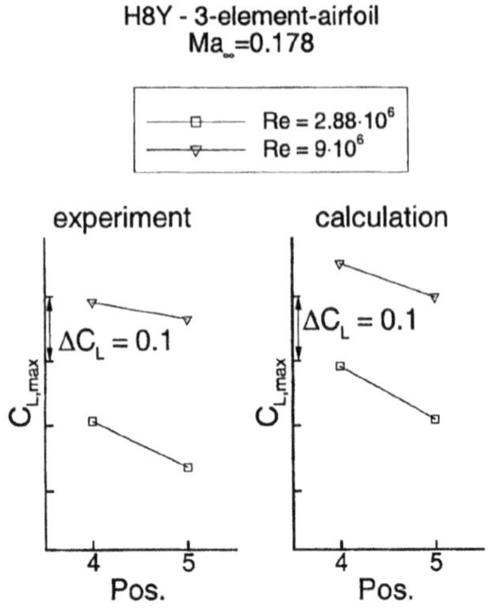

Figure 5 Dependency of the maximum lift coefficient from the slat overlap of the H8Y configuration for different Reynolds-numbers

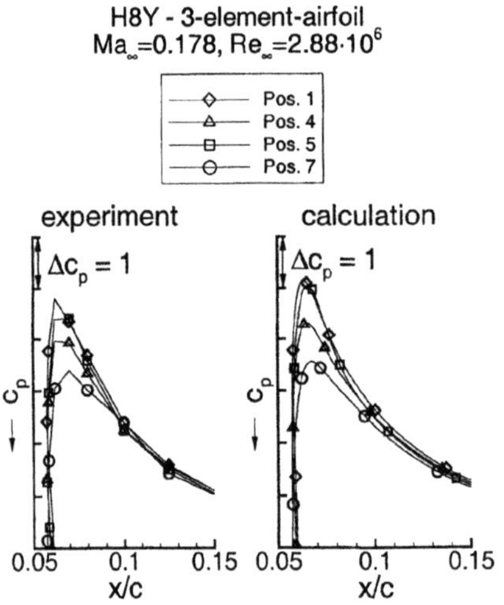

Figure 6 Differences of the suction peak at the main airfoil element in dependency of the slat position of the H8Y configuration for $Re = 2.88 Mio.$

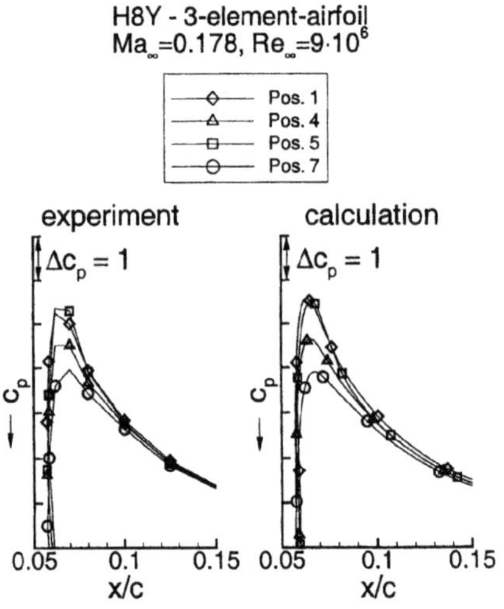

Figure 7 Differences of the suction peak at the main airfoil element in dependency of the slat position of the H8Y configuration for $Re = 9 Mio.$

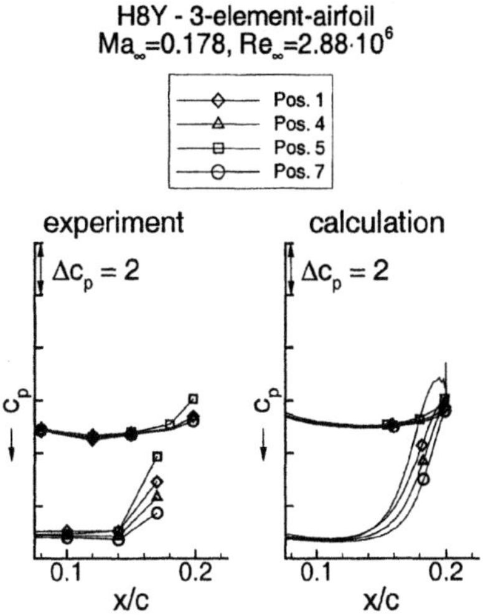

Figure 8 Differences of the trailing edge pressure of the slat in dependency of the slat position of the H8Y configuration for $Re = 2.88 Mio.$

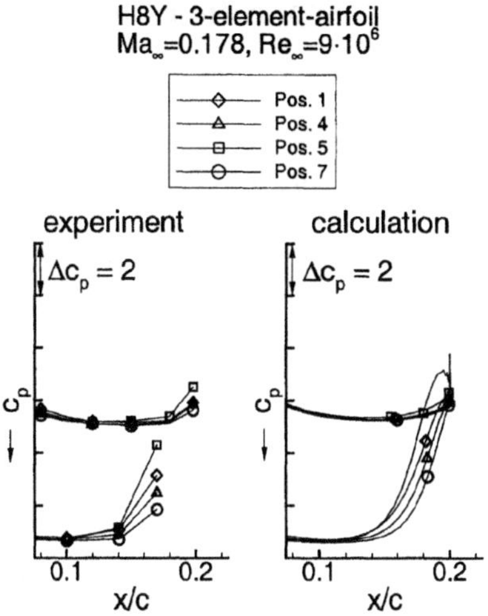

Figure 9 Differences of the trailing edge pressure of the slat in dependency of the slat position of the H8Y configuration for $Re = 9 Mio.$

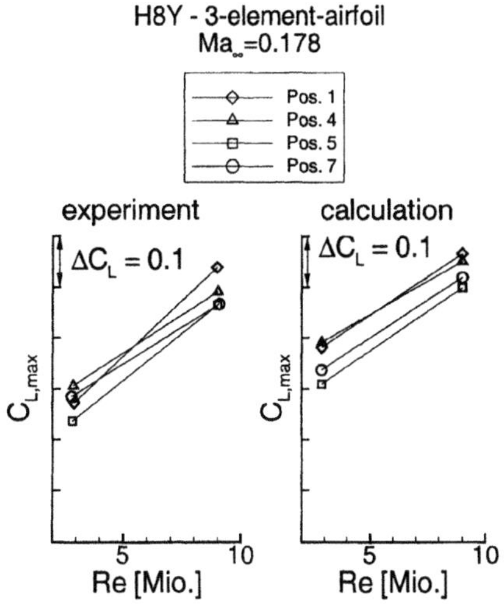

Figure 10 Dependency of the maximum lift coefficient on the Reynolds-number for different slat settings of the H8Y configuration

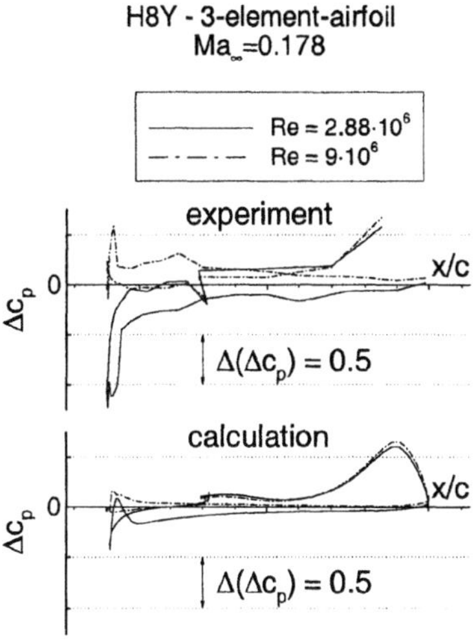

Figure 11 Difference of pressure coefficient between position 1 and position 4 of the slat of the H8Y configuration: $\Delta c_p = c_{p,pos.1} - c_{p,pos.4}$

Separation Control on a High-Lift Configuration by Periodic Exitation

F. Tinapp, W. Nitsche
Institut für Luft- und Raumfahrt, Technische Universität Berlin
Sekr. F2, Marchstr. 14, 10587 Berlin, Germany

Summary

This paper describes some experimental results, aiming at separation control on a simple high-lift configuration. A separated flow can be forced to reattach by introducing periodic excitation. These are of fixed frequency and amplitude, using periodic blowing and suction through a narrow slot aligned in spanwise direction on the flap.

The paper concentrates on results achieved with a fixed geometric configuration of the test model. At this configuration the flow on the flap was fully separated in the case without excitation. Full reattachment of the separated shear layer could be obtained due to excitation in a frequency range of 50 Hz - 400 Hz.

1 Introduction

High-lift systems are of complex mechanics, generally consisting of a combination of leading-edge slats and multiple trailing-edge flaps. At high angles of attack the flow over high-lift wings may separate, resulting in a severe lift reduction and a strong increase of drag. If the onset of separation could be delayed towards a higher angle of attack, it will either be possible to achieve a higher lift or to reduce the mechanical complexity of the high-lift system.

Recent investigations showed clearly that periodic excitation of the separated shear layer results in a partial reattachment and thus in an increase of lift and decrease of drag [1] [2] [3] [4] [5]. The investigation presented in this paper discusses this problem and deals with experiments aimed at separation control. This is done via excitation of the separating boundary layer on the flap of a simple high-lift configuration, by means of a suction/blowing slot.

2 Experimental Setup

2.1 Test model

The test model used in our experimental investigation (see figure 1) was a 2-element high-lift configuration [6], consisting of a main airfoil with NACA 4412 profile (chordlength: 180 mm) and a trailing edge flap with a NACA 4415 contour (72 mm chordlength). The flap was placed underneath the trailing edge of the main airfoil, forming a gap of a fixed height and overlap. The angle of attack of the whole configuration (α) and the flap angle (η) were varied independently. The model was installed in a wind tunnel with a height of 1.4 m in order to avoid the influence of the tunnel walls on the flow around the test model. Due to the span of 400 mm, the model had to be mounted between two sidewalls, while the wind tunnel presents a width of 2 m.

The free stream velocity u_∞ was 14 m/s, yielding a Reynolds number based on the chordlength of the main-airfoil of 160.000. To ensure turbulent separation of the flow, turbulator strips were placed close to the leading edges of the main airfoil as well as on the flap. The test model was mounted on a force balance to be able to determine the aerodynamic forces and it was equipped with static pressure taps. The flow field around the test model was measured using hot wire probes that were mounted on a 3D-traverse system.

2.2 Excitation system

The flap was equipped with a 0.3 mm wide slot, aligned in spanwise direction. The slot is located at 3.5% of flap-chord, where the separation occurred (as can be seen in the results). This slot was connected to a pressure chamber inside the flap that was fed by periodic pressure signals, generated by two loudspeakers mounted outside of the flap. This resulted in an oscillating jet that emerged from the narrow slot in the flap (see figure 2). The two excitation parameters, frequency and intensity can be written non-dimensional as Strouhal number and an impulse coefficient:

$$St = \frac{F \cdot l_{char}}{u_\infty} \tag{1}$$

$$c_\mu = 2 \cdot (\frac{H}{c}) \cdot (\frac{v'}{u_\infty})^2 \tag{2}$$

With the flap chord as characteristic length l_{char}, H = slot width, c = chord of main airfoil and v' = velocity fluctuation at the slot exit.

As the impulse of excitation jet was not constant over the whole frequency range, the excitation apparatus had to be calibrated as a function of jet impulse against the pressure fluctuation inside the pressure chamber for different frequencies (see figure 3).

3 Results

3.1 Basic flow

The airfoil was mounted on the force balance to determine the lift coefficient c_L, to get an insight into the flow behaviour in the case of natural flow (without excitation). As an example, in figure 4 the lift coefficient c_L is plotted against the flap angle η for a fixed angle $\alpha = 7°$ with three corresponding static pressure distributions on the suction side. The third one (case c, $\eta = 36°$) indicates separated flow over the flap airfoil at a low c_L, as shown in the polar diagram. Results of this testcase will be presented in this paper.

Figure 5 shows the velocity distribution above the flap. The jet that emerges from the gap between the main airfoil and the flap separates near to the flap's leading edge and forms a recirculation area above the flap. On the right side, three power spectra of the velocity fluctuations at three different positions in the flow field are plotted. While there are only negligible velocity fluctuations in the jet flow, the shear layer between the jet and the separated flow area contains low frequency fluctuations of high amplitude, indicating the presence of big vortex structures. As these structures had to be excited, in order to achieve reattachment of the flow, periodic excitation was applied on the flow.

3.2 Excited flow

Exciting the separated shear layer with a frequency of $F = 100$ Hz ($St = 0.51$) and a sufficiently high amplitude, results in a completely changed flow field, as can be seen in figure 6. Graph a shows the still separated flow above the flap at an excitation amplitude of $c_\mu = 12 \cdot 10^{-5}$ while in graph b the amplitude is $c_\mu = 45 \cdot 10^{-5}$. In the case of the higher excitation amplitude, the flow reattaches completely over the flap, thus restoring the lift c_l of the test configuration.

To determine the influence of the excitation, a periodic jet was applied to the separated flow which could be varied in frequency and amplitude. In figure 7 the achieved lift c_l is plotted against the excitation intensity c_μ for different excitation frequencies. When increasing the intensity, the lift remains low (separated flow) until reaching a certain level of excitation intensity. Here the lift increases quickly and reaches a value much higher than before (reattached flow). When decreasing the amplitude after reattachment, the flow remains attached until reaching another, much lower limit were it re-separates again. This hysteresis effect can be found at each excitation frequency. Reattachment could be obtained at each tested frequency, but with higher achieved lift at low frequencies. Therefore, an excitation frequency of $F = 100$ Hz together with an intensity of $c_\mu = 50 \cdot 10^{-5}$ was chosen for further experiments. The next graphic (figure 8) depicts the same region as shown already in figure 5. At three different sections above the flap, the mean velocity and the velocity fluctuations of the flow are plotted in the graphics below. The solid line represents the case of basic flow (without excitation, $c_\mu = 0$) while the dashed lines stand for different excitation intensities at a fixed frequency of $F = 100$ Hz. A small increase of the amplitude, results only in a slight change of the mean velocity (section a) but there is an enhanced velocity fluctuation in the shear layer between the separated jet and the recirculation zone. When reaching the excitation level that leads to reattachment, the flow field is changed dramatically. The former separated jet reattaches completely and remains attached up to the trailing edge of the flap (section c).

The last figure presents the results of lift enhancement due to periodic excitation for different flap angles η. The curve marked with open squares represents the case without excitation, while the curve marked by circles shows the achievable lift when applying excitation of $F = 100$ Hz and an amplitude that was adapted to the respective requirements. It shows that the lift can be enhanced by periodic excitation especially in the range of high angles of attack, where the basic flow tends to separate. The lift enhancement was up to 20% for the chosen configuration and excitation parameters. At lower angles of attack, where the basic flow is still attached to the surface, the excitation of the flow above the flap has no effect.

4 Conclusions

The paper describes experiments on separation control on a simple 2-element high-lift configuration. The separated flow above the flap could successfully be forced to reattach by introducing periodic excitation through a narrow slot near the separation point of the flow. By doing this, the vortex structures in the shear layer between the separated flow and the recirculation area, were enhanced. For the chosen test configuration, the lift could be restored and enhanced up to 20%

compared to the case without excitation. The reattachment of the flow occurs, exciting the flow at low frequencies ($F = 100$ Hz, $St = 0.5$) and a high amplitude ($c_\mu = 50 \cdot 10^{-5}$).

5 Acknowledgements

The work was financially supported by the German Science Foundation (DFG) in the framework of the SFB 557 "Beeinflussung komplexer, turbulenter Scherschichten".

References

[1] Dovgal, A., Control of Leading-Edge Separation of an Airfoil by Localized Excitation. DLR-FB-93-16 (1993), DLR-Forschungsbericht.

[2] F.B. Hsiao, C. L. and Shyu, L., Control of Wall-Separated Flow by Internal Acoustic Excitation. *AIAA Journal*, 28, No.8, 1440 (1990).

[3] A. Seifert, A. D. and Wygnanski, I., On the Delay of Airfoil Stall by Periodic Excitation. *Journal of Aircraft*, 33, No.4, 691 (1996).

[4] Tinapp, F. and Nitsche, W., LDV-measurements on a high-lift configuration with separation control. vol. 1, p. 19.1 (1998), 9th International Symposium on Applications of Laser techniques to Fluid Mechanics.

[5] Tinapp, F. and Nitsche, W., On active control of high-lift flow. vol. 4, pp. 619–626. Elsevier Science Ltd. (1999), Engineering Turbulence Modelling and Experiments.

[6] Adair, D. and Horne, W., Turbulent Separating Flow Over and Downstream of a Two-Element Airfoil. *Experiments in Fluids*, 7, 531 (1998).

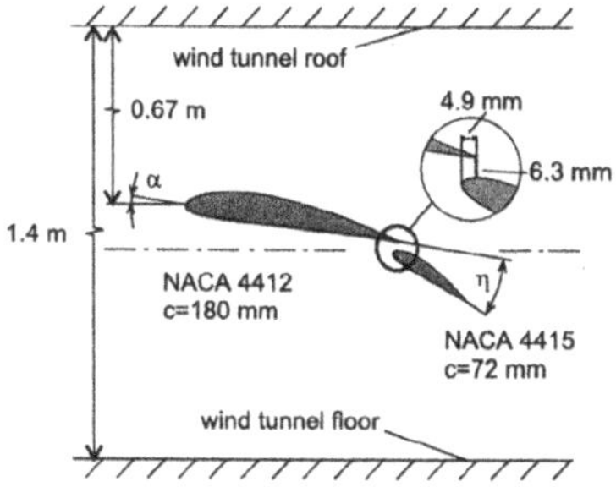

Figure 1 Experimental setup, crossection of the test bed, span = 400 mm

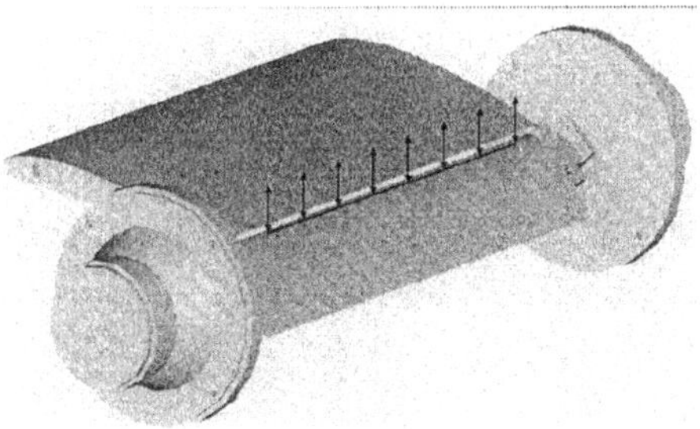

Figure 2 Test model with excitation apparatus, excitation slot (0.3 mm width) near the flap trailing edge (x/c=3.5%), arranged in spanwise direction

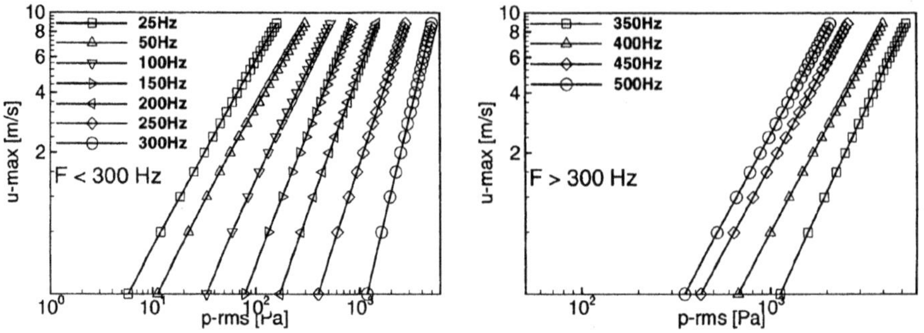

Figure 3 Calibration of the excitation apparatus, maximum excitation jet velocity vs. pressure fluctuations in the pressure chamber for different frequencies

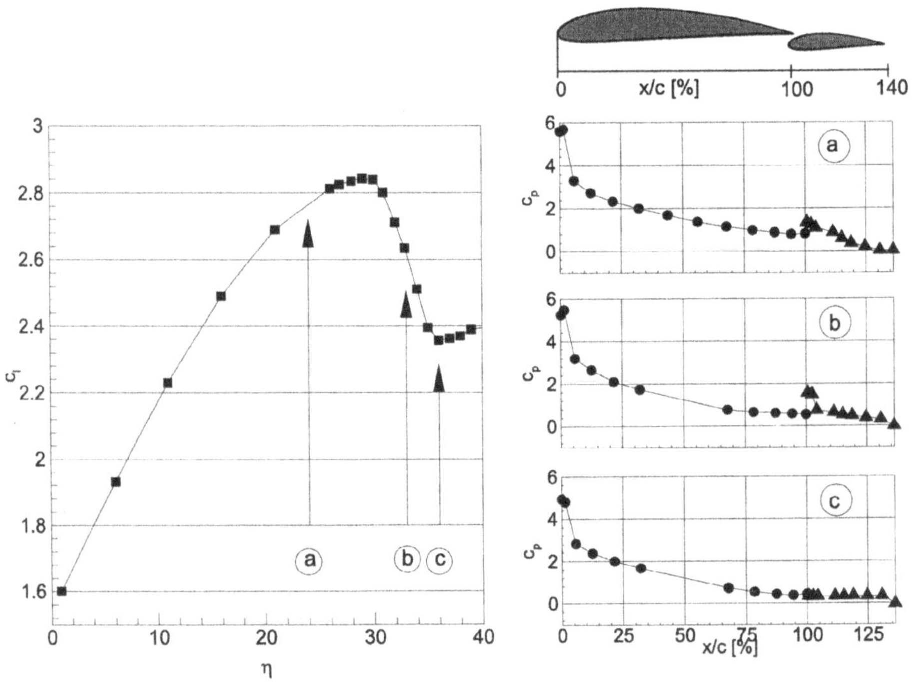

Figure 4 Exemplary polar diagram (c_L vs. η at $\alpha = 7°$) with three static pressure distributions (suction side only): a) attached flow, b) partly separated flow, c) completely separated flow

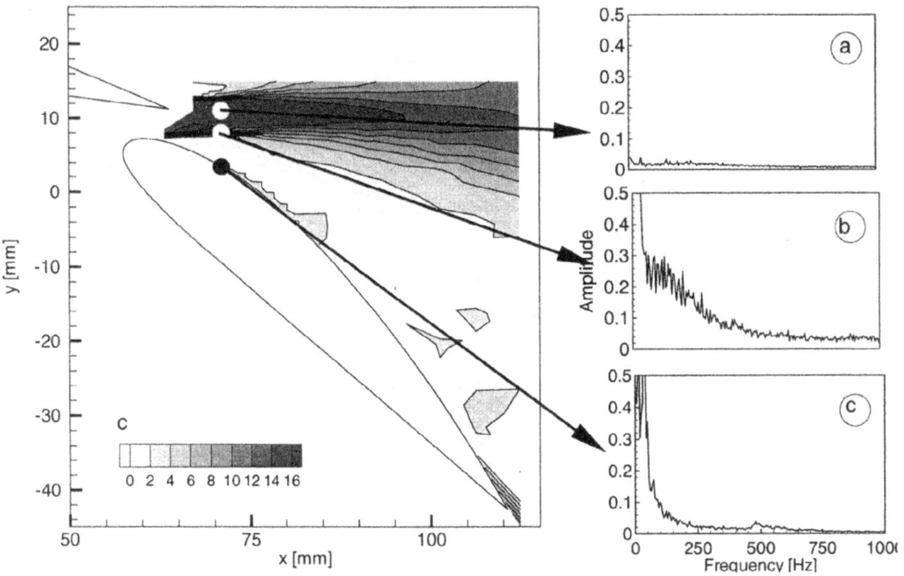

Figure 5 Mean velocity ($c = \sqrt{u^2 + v^2}$) above the flap ($\alpha = 7°$, $\eta = 36°$) and power spectra of velocity fluctuations: a) separated jet flow, b) shear layer between jet and recirculation zone, c) recirculation area

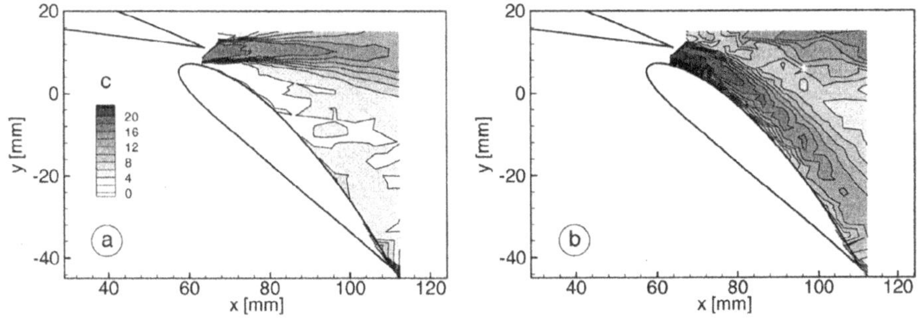

Figure 6 Flow field above the flap at $\alpha = 7°$ and $\eta = 36°$: a) with weak excitation ($F = 100$ Hz, $c_\mu = 12 \cdot 10^{-5}$) b): with strong excitation ($F = 100$ Hz, $c_\mu = 45 \cdot 10^{-5}$)

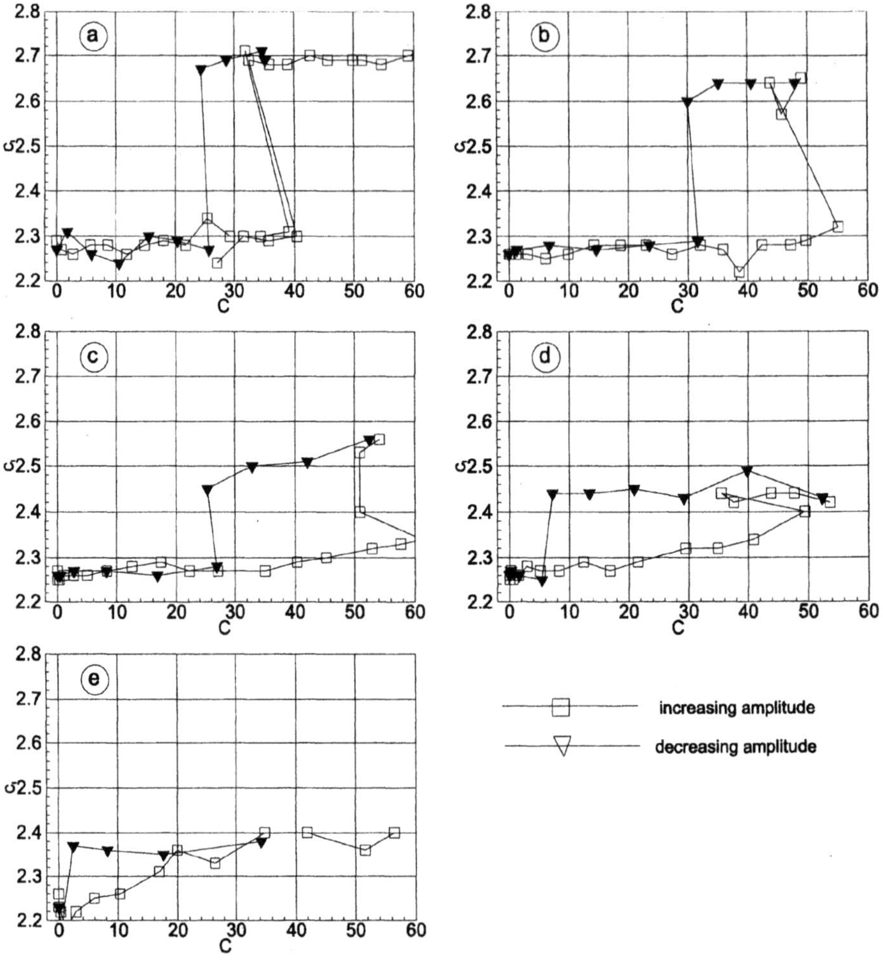

Figure 7 Achievable lift at excitation of varying amplitude (c_μ) and frequency (St) at a fixed geometry ($\alpha = 7°$, $\eta = 36°$): a) $St = 0.26$ ($F = 50$ Hz), b) $St = 0.51$ ($F = 100$ Hz), c) $St = 1.03$ ($F = 200$ Hz), d) $St = 1.54$ ($F = 300$ Hz), e) $St = 2.06$ ($F = 400$ Hz), all c_μ times 10^{-5}.

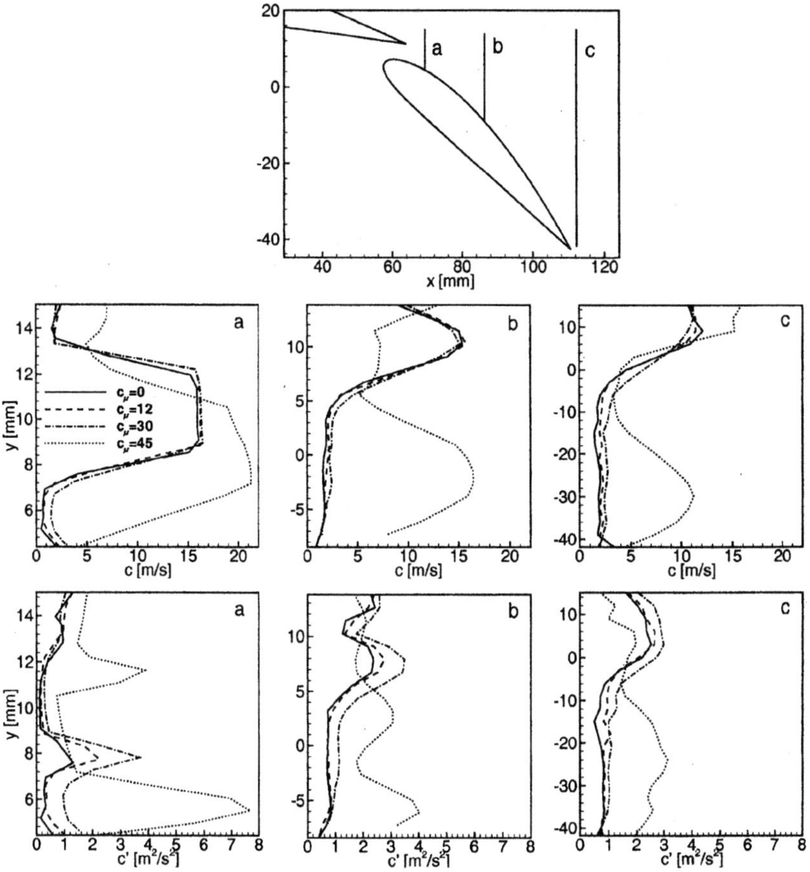

Figure 8 Profile plots of mean velocity and velocity fluctuations at three different sections above the flap for different c_μ

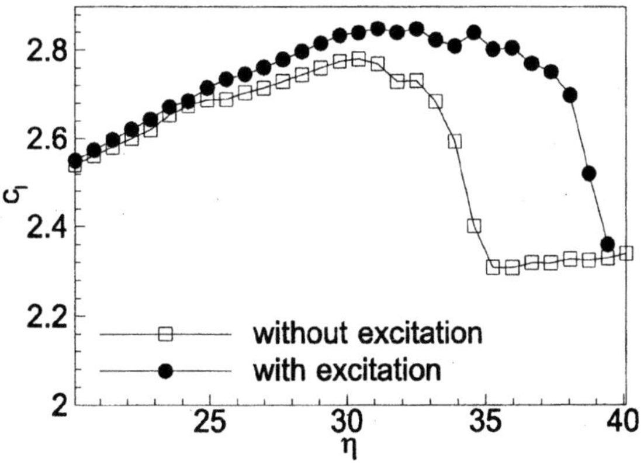

Figure 9 c_L vs. η without and with excitation, $St = 0.51$, c_μ adapted, $\alpha = 7°$

Numerical Investigation of the Effect of the High-Lift Configuration of a Transport Aircraft on its Vortex Wake

E. Stumpf
German Aerospace Center (DLR),
Institute of Design Aerodynamics,
Lilienthalplatz 7, 38108 Braunschweig, Germany
e-mail: eike.stumpf@dlr.de

Summary

A numerical investigation, based on the Euler equations, of the nearfield wake vortex of four high-lift configurations of a narrow body airliner is carried out. The main focus is to study to what extent the configurations meet a criterion for the accelerated destruction of wake vortices proposed by Rennich/Lele. Additionally a spatial direct numerical simulation of the wake up to 35 spans is presented for the airliner model in standard landing configuration.

1 Introduction

World airports face a long-term capacity problem due to the air traffic control separation regulations presently in use. These empirically found separation standards determine the minimum longitudinal distance between two aircraft on the same flight path in order to prevent the following aircraft from encountering potentially hazardous wake turbulence. Despite extensive experimental and numerical investigations (e.g. see review articles [9, 10]) a full understanding of how to design future aircraft with less hazardous wakes is remote. A first promising design guideline is proposed by Rennich and Lele in [8]. They showed that the amplification of inherent instabilities of a vortex wake is highest and hence the lifetime is minimized if the vortex system translates rigidly downwards under its own influence. For a four-vortex system, consisting of two pairs of counterrotating vortices, a point vortex analysis gives the condition for equal sink speed of all vortices:

$$D^3 + 3GD^2 + 3G + D = 0 \tag{1.1}$$

with $D = d_2/d_1$, d_1 resp. d_2 being the spacing of the counterrotating vortices of the two pairs and $G = \Gamma_2/\Gamma_1$ the ratio of their circulations. As the vortices lie constantly in one plane during decent the outer vortex pair imposes high strain on the inner one, thereby in the first stage amplifying the inherent instabilities on the inner vortex pair. The second stage of the instability begins as the growing perturbation on the inner vortices induces a Crow-type instability on the outer vortex pair. This mechanism is hereafter termed Rennich-concept. The simple four-vortex model can be applied to aircraft in high-lift configuration if the complex system of discrete vortices is reduced to four vortex centroids. In general four centroid domains comprising vortices which eventually merge or collapse far downstream may be specified.

Following computations carried out by Das and Longo [1, 2] it can be expected that the Euler equations yield the main flow phenomena of spiraling vortical flows. Moreover Donaldson

[3] and Hoeijmakers [5] showed the processes in the roll-up region of a wake to be essentially inviscid if no separation is present. Thus, in the present investigation the DLR-τ-Code [4], based on the Euler equations, is used for the nearfield calculations. Its predictive capability for nearfield vortex wakes of high-lift configurations in quiescent atmosphere was shown in [12]. In the present paper wake calculations of four different high-lift configurations of a narrow body airliner model up to a half span downstream of the model wing tip are compared. The main focus of the work is to study to what extent the standard flap and slat setting of the Airbus A320-similar model meets the Rennich-criterion (equation 1.1) and what kind of modification is needed to come close to the fulfillment. Additionally as a first step towards the feasibility evaluation of the Rennich-concept a spatial direct numerical simulation of the flow downstream up to 35 spans is carried out with the DLR -Mesoscop-Code for the airliner model in standard landing configuration. A proof of the Rennich-concept in this context remains future work.

2 Description of Geometry

The aircraft model used is the DLR-ALVAST (Figure 1). That is a narrow body airliner model without empennage with a span b = 3.4m, sharp edged wing and flap tips and blunt trailing edges. The model is roughly a 1:10 scaled Airbus A320. Detailed geometrical information can be found in [7]. The baseline case is the standard landing configuration with deflection angles δ_{Flap}=32° for the inner and outer flap and the slat δ_{Slat}=27°. Three more configurations are examined: the take-off setting (δ_{Flap}=20°, δ_{Slat}=19.5° with VHBR-TPS (very high bypass engine simulator) , a differential setting with the inner flap in take-off and outer in landing position and a modified landing configuration, modified such that the two single flaps are replaced by one continuous flap.

3 Numerical Set-up

Nearfield simulation

All basic unstructured grids are generated with the commercial tool ICEM Tetra. The numerical results for the nearfield are obtained with the DLR-τ-Code , namely a finite-volume Euler/Navier-Stokes solver working on unstructured grids. Local time-stepping and multigrid for convergence acceleration is included, the inviscid fluxes are calculated by employing a central method with scalar dissipation. A grid adaptation module enables the primary grid to be adapted to the solution by cell division if an enhanced resolution is required in certain regions of the computational domain. As an indicator for the adaptation total pressure gradients were used. Since Euler calculations overpredict lift considerably the angle of attack is adjusted for each configuration individually in order to achieve the measured lift coefficient of $c_L = 1.74$ at $M_\infty = 0.2$ (landing config.: α_{exp}=5°, $\Delta\alpha_{calc}$=$-1.8°$; differential config.: α_{exp}=7.4°, $\Delta\alpha_{calc}$=$-2.6°$; take-off config.: α_{exp}=12°, $\Delta\alpha_{calc}$=$-4.5°$).

Extended nearfield simulation

DLR-MESOSCOP is used to perform three-dimensional spatial direct simulation with constant viscosity of the extended nearfield. The code solves the instationary, incompressible Navier-Stokes equations. The physical fields are discretized on a Cartesian staggered grid and integrated in space and time by second-order finite differencing. Propagation in time is performed by a prognostic step for advection and diffusion using the second-order Adams-Bashforth scheme followed by a diagnostic step which solves the Poisson equations for the dynamic pressure. The integration scheme is non-dissipative and only weakly diffusive. Figure 2 shows the set-up for the extended nearfield simulation which is done stepwise in subsequent boxes. A plane at $x/span = 0.5$ of the Euler-simulation result serves, interpolated on a regular grid, as inflow boundary for the first DNS-box. Subsequently the values of the outflow plane of a DNS-box are interpolated as inflow boundary condition for the following DNS-box. Preceding test runs showed that a resolution of minimum four points is needed within the vortex cores to ensure physical downstream convection of vorticity.

4 Nearfield Simulation Results

The Euler computations are carried out on a NEC SX-5 using Jameson‘s central dissipation scheme [6] with a 4th difference dissipation coefficient of 1/64. The unstructured meshes are adapted in 10 adaptation steps leading to approximately 6.0 x 10^6 grid points. A convergence of the density residual of five orders of magnitude to steady state was achieved for all configurations.

4.1 Comparison with experimental data

For validation wind tunnel measurements are taken which were carried out in the low speed facility DNW-NWB at DLR-Braunschweig. The ALVAST half model was mounted on a peniche of 68mm height in the closed test section (3.25m x 2.8m). A five hole probe rake was installed for scanning the plane $x/b = 0.5$ downstream of the model wing tip. The comparison with experimental data is exemplary shown for the ALVAST in landing configuration.

Core radius

In the roll-up region the discrete vortices are in general not circular. Hence, for comparison the dispersion radius r_d as an integral measure is appropriate. The dispersion radius is defined by

$$r_d = \left\{ \frac{1}{\Gamma} \iint \left((y - \bar{y})^2 + (z - \bar{z})^2 \right) \omega_x \, dy \, dz \right\}^{1/2} \tag{4.1}$$

with ω_x being the x-component of vorticity and $\bar{y}$ resp. $\bar{z}$ the location of the vortex centroid.

$$\bar{y} = \frac{1}{\Gamma} \iint y \, \omega_x \, dy \, dz \qquad \text{and} \qquad \bar{z} = \frac{1}{\Gamma} \iint z \, \omega_x \, dy \, dz \tag{4.2}$$

Table 1 shows a moderate overprediction of up to 55% by the simulation compared at a halfspan downstream of the model wing tip.

Spanwise Lift Distribution

Despite the angle of attack adjustment in order to obtain the measured lift coefficient the simulated lift of the single elements - slat, wing and flaps - is not correct. The lift contributed by the slat is underestimated whereas the lift of the flaps is simulated too high. However, a fortunate cancellation of effects gives a spanwise lift distribution close to experiment as can be seen in Figure 3. Within the region of the thrust gate the good agreement with the measurement is not given since here the underestimated sectional slat lift is not balanced by a flap.

Wake Topology

Figure 4 shows a comparison of the wake structure between wind tunnel measurement and Euler calculation at $x/b = 0.5$ indicated by streamlines - on the left side the five-hole-probe results, on the right side the computational results. The five-hole-probe measurements do not cover the complete cross section and are not continuous due to mechanical restrictions of the rake positioning system. The computational results match the experimental data remarkably well. Only the vertical position of the vortices differ slightly. This can be explained both by the blockage effect of the model in the closed test section and by the added peniche height to the distance to the symmetry plane (here, the test section floor).

4.2 Nearfield Wake Vortex Evolution

Effect of high-lift configuration

The results of the wake calculation are plotted in surveying planes perpendicular to the flight axis. The position is given in percentage of span referenced to the wing tip trailing edge. Contour variable is total pressure loss (10 levels $\Delta P_t/P_t = 0.0007 - 0.09$). The landing configuration is shown in Figure 5 ($\alpha = 3.2°$). The inner counterrotating vortices (i.e. counterrotating with regard to the wing tip vortex), stemming from the inner sides of the outer and inner flaps, rotate slowly (approx. $10°/span$) around the corotating outboard vortex of the inner flap. Whereas the corotating wing tip vortex and outboard vortex of the outer flap are rotating clockwise around each other, reaching about one quarter of revolution at $x/b = 0.5$. Retracting the inner flap on take-off deflection gives the differential setting ($\alpha = 4.8°$, Figure 6). The vortices of the inner flap are weaker, the outer vortices are similar to the landing configuration. In comparison it can be noticed that the rotation of the outer vortices of the single-flap configuration (Figure 7) is less than 80°. That is due to the small angle of attack of $\alpha = 0.25°$ leading to a weaker outer flap vortex.

Effect of TPS

Figure 8 shows the three surveying planes and additionally the engine jet as iso-total-pressure-loss surface ($\Delta P_t = -0.05$). Engine setting is take-off thrust (total pressure ratios: fan jet flow = 1.28, core jet flow = 1.13). The interaction between the thrust gate vortices and the engine jets leads both to a deformation of the jet by the upward propelling effect of the vortices and to an inward displacement of the vortices.

Fulfillment of Rennich-criterion

The reduction of complex wakes to four centroids requires the specification of centroid domains comprising vortices which eventually merge or collapse far downstream. Exemplary for the AL-VAST in landing configuration the integration limits are shown in Figure 9. In Figure 10 the lower curve is the graph of the cubic equation (equation 1.1). Centroids with circulation- and spacing-ratios lying directly on the curve experience equal sink speed. The area below the curve marks the ratios with which the inner vortex pair moves upward. Whereas the area above comprises the ratios with which the inner vortex pair sinks, however with different speed compared to the outer pair. Following an estimation by Stuff [11] the Rennich-concept might work in the complete hatched area. Above, the inner vortex pair is too weak to induce instabilities on the outer one. Both the differential and take-off configuration are far from the fulfillment of equation (1.1) and have too weak inner vortex domains. That is also true for the landing configuration, however it almost fulfils the Rennich-criterion. Of the examined configurations only the single-flap layout bears the potential of inducing instabilities according to the Rennich concept.

5 Extended Nearfield Simulation Results

The results of the DNS calculations are shown in Figures 11/12. The calculation was only done for the portside of the ALVAST model in landing configuration. For clarity the mirror image is included giving a picture of the full wake. Perspective is from below, iso-surface-variable is vorticity ($|\omega_x| = 200s^{-1}$). Figure 11 displays the downstream evolution up to 7 spans. The merging of the wing tip vortices and the outboard vortices of the outer flaps is completed at 3.5 spans. On the inner vortices a short wave instability ($\lambda = 0.3b$) evolves which shows highest amplitudes whenever the three inner vortices lie within the horizontal plane. After 16 spans the inner vortices collapse without triggering instabilities on the outer vortex pair (see Figure 12). This is in accordance to the prediction of the modified Rennich- diagram (Figure 10).

6 Conclusion

Nearfield simulation
It was shown that Euler calculations on unstructured grids with adjusted lift coefficients are capable to simulate the formation and development of nearfield wake vortices of a narrow body airliner in high-lift configuration in good agreement to measurement resuts. Due to the moderate slat and flap deflection angles no separation on the surfaces was to be expected. This fact, in combination with the models‘ sharp-edged wing and flap tips, representing geometrical singularities that lead to Reynolds number independence of the separation at the edges, made it a good natured case for the application of an Euler code for calculations of a nearfield vortex wake.
Effect of Complexity
The engine jet has a major impact on the wake topology. The wake is both displaced and deformed by the jet momentum. In real configurations, depending on the location of the aircraft center of gravity, it also might happen that the vortices of the horizontal tail sink downwards and interact with the wing/flap/pylon wake vortices. Hence, the engines and the horizontal tail have to be included if the feasability of the Rennich-concept is to be evaluated.
Centroid Approach
In general it is possible to reduce complex high-lift wakes to four vortex centroids. That gives the chance to use the simple four-vortex model as proposed by Rennich/Lele.
Outlook
Most promising for the Rennich-concept are configurations without thrust gate since the corotating inner vortex weakens the inner vortex domain such that it collapses before being able to trigger instabilities on the outer vortex pair.

7 Acknowledgments

Parts of the presented results have been obtained by DLR within the DLR-project ‚Wirbelschleppe‘. The author is grateful to Dr. T. Gerz, T. Hofbauer, Dr. F. Holzäpfel, Dr. A. Bergmann and his staff at DNW-NWB for their support.

References

[1] A. Das, "Analysis of Spiraling Vortical Flows around Slender Delta Wings Moving in an Inviscid Medium", *Z. angew. Math. Mech*, Vol. 71, No. 11, 1991, pp. 465-471

[2] A. Das, J.M.A Longo, "Numerical Analysis of the Vortical Flow around a Delta Wing-Canard Configuration", *J. Aircraft*, Vol. 32, No. 4, 1995, pp. 716-725

[3] D. Donaldson, C. duPont, A.J. Bilanin, "Vortex Wake of Conventional Aircraft", AGARD AG 204, 1975

[4] T. Gerhold, M. Galle, O. Friedrich, J. Evans, "Calculations of Complex Three-Dimensional Configurations Employing the DLR-TAU-Code", AIAA 97-0167, 1997

[5] H.W.M. Hoeijmakers, "Computational Vortex Flow Aerodynamics", AGARD CP 342, 1983

[6] A. Jameson, W. Schmidt, E. Turkel, "Numerical Solutions of the Euler Equations by Finite Volume Methods Using Runge-Kutta Time-Stepping Schemes", AIAA 81-1259, 1981

[7] R. Kiock, "The Alvast Model of DLR", DLR-IB 129-96/22, 1996

[8] S.C. Rennich, S.K. Lele, "Method for Accelerating the Destruction of Aircraft Wake Vortices", *J.Aircraft,* Vol. 36, No. 2, 1999, pp. 398 - 404

[9] V.J. Rossow, "Lift-Generated Vortex Wakes of Subsonic Transport Aircraft", *Prog. Aerospace Science*, No. 35, 1999, pp. 507 - 660,

[10] P.R. Spalart, "Airplane Trailing Vortices", *Annu.Rev.FluidMech.*, 1998, pp. 107-138

[11] R. Stuff, "The correlation between Near- and Farfield with Regard to Wake Vortex Collapse", to be published in Notes on Numerical Mechanics (Proceedings of AG STAB/DGLR Symposium Nov. 2000)

[12] E. Stumpf, R. Rudnik, A. Ronzheimer, "Euler Computation of the Nearfield Wake Vortex of an Aircraft in Take-Off Configuration", *Aerosp. Sci. & Tech.* No. 4, 2000, pp. 535 - 543

Table 1 Comparison of dispersion radius r_d

	Wing Tip	*Outer Tip Outboard Flap*	*Inner Tip Outboard Flap*	*Outer Tip Inboard Flap*
$\Delta r_d / r_{d-Exp.}$	55.27%	10.43%	17.86%	46.61%

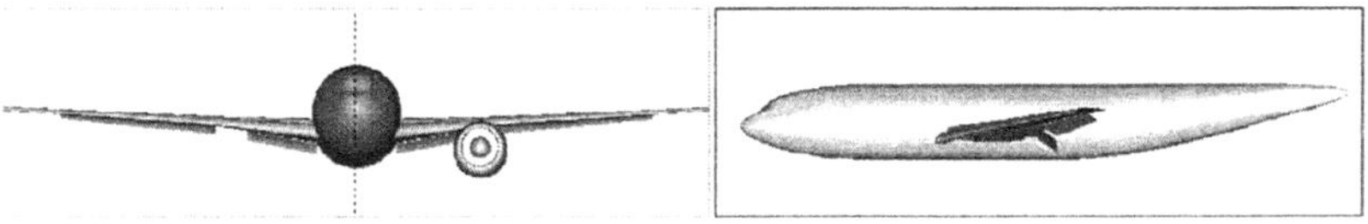

Figure 1 ALVAST Model

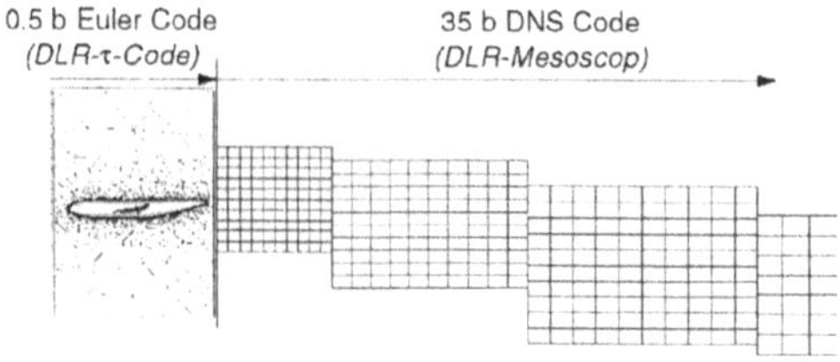

Figure 2 DNS-Simulation Set-Up

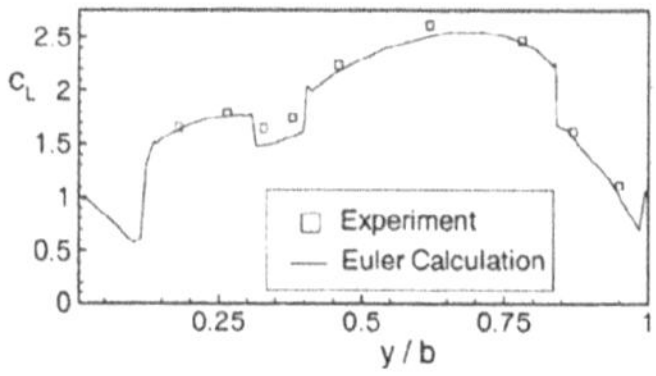

Figure 3 Lift Distribution

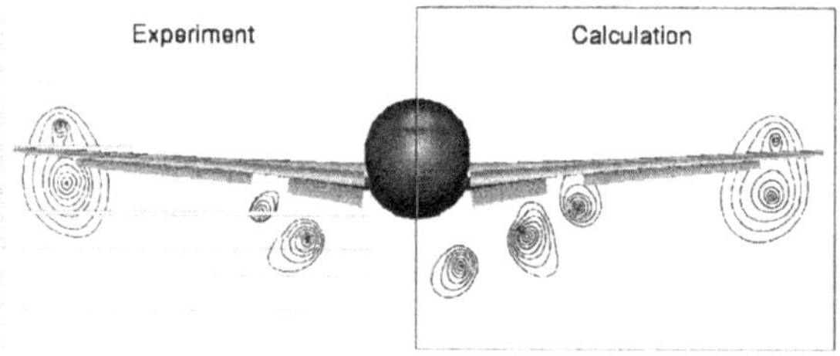

Figure 4 Wake topology comparison at x/b = 0.5 (Streamlines)

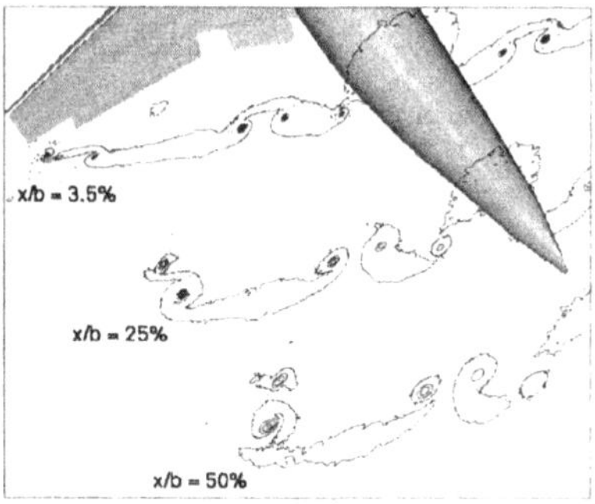

Figure 5 Landing Configuration ($\Delta P_t/P_t = 0.0007 - 0.09$, 10 levels)

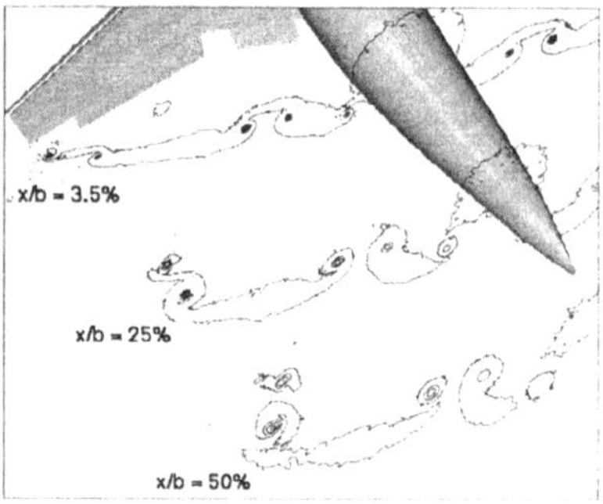

Figure 6 Diff.-Setting Configuration ($\Delta P_t/P_t = 0.0007 - 0.09$, 10 levels)

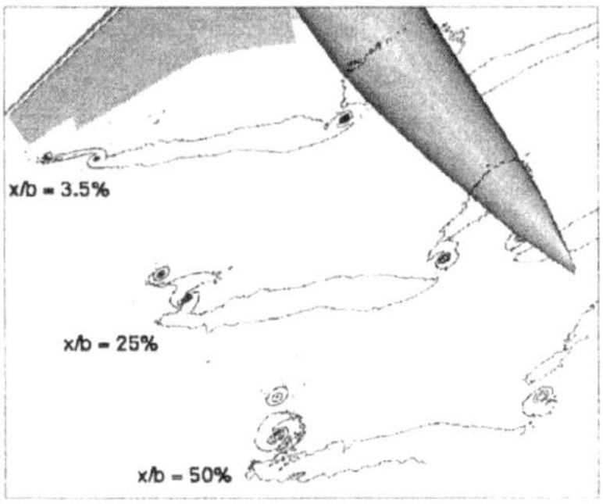

Figure 7 Single-Flap-Configuration ($\Delta P_t/P_t = 0.0007 - 0.09$, 10 levels)

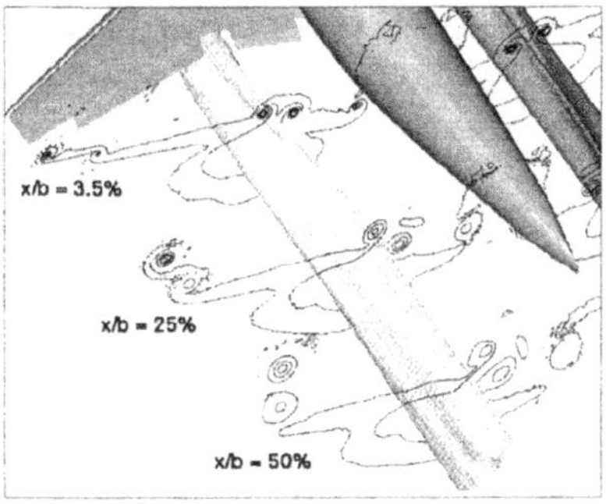

Figure 8 Take-Off TPS Configuration ($\Delta P_t/P_t = 0.0007 - 0.09$, 10 levels, jet iso-surface $\Delta P_t/P_t = -0.05$)

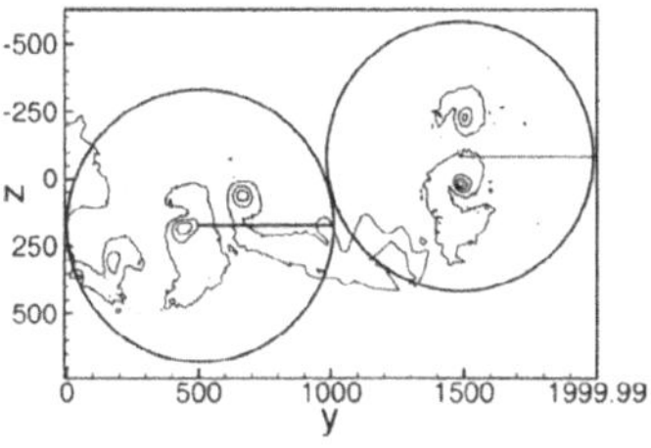

Figure 9 Centroid Domains Landing-Config.

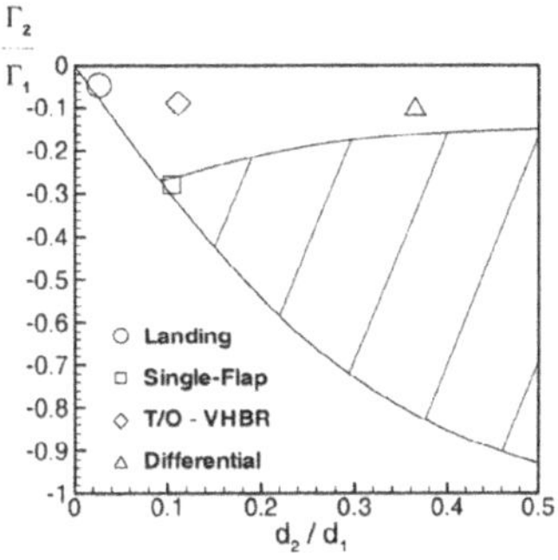

Figure 10 Rennich-Criterion

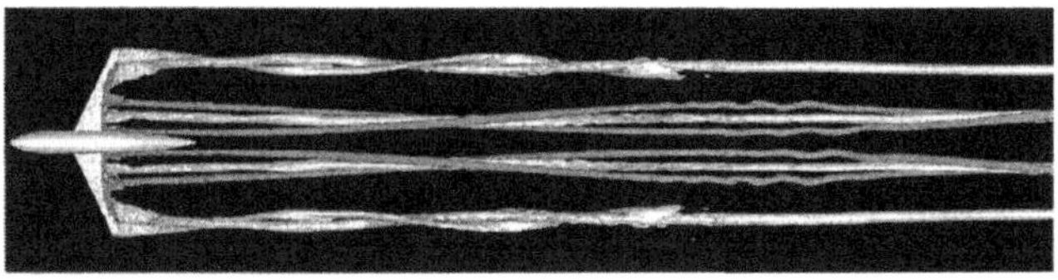

Figure 11 Extended Nearfield (iso-surface $|\omega_x| = 200s^{-1}$, 7 Spans)

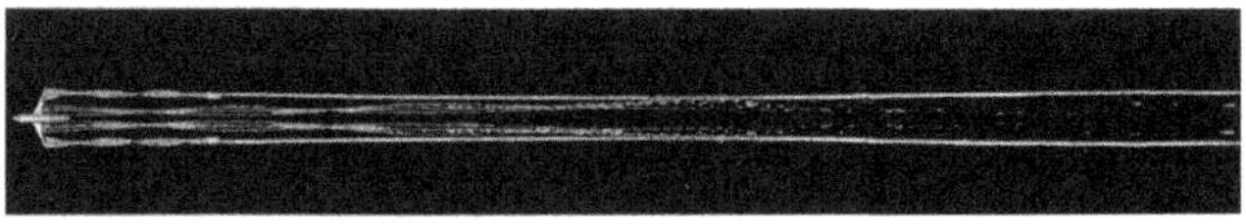

Figure 12 Extended Nearfield (iso-surface $|\omega_x| = 200s^{-1}$, 35 Spans)

The Relationship Between Near- and-Far-Field of Vortex Wakes from Aircraft with High Aspect Ratio Wings

R. Stuff
Deutsches Zentrum für Luft- und Raumfahrt, Institut für Strömungsmechanik, Bunsenstraße 10, D-37073 Göttingen, Germany

Summary

Far field measurements of the vortex wake shed from a transport aircraft indicate that the trailing vortex pair from the main wing undergoes a Rayleigh-Ludwieg instability and premature breakdown, if it encounters another vortex pair turning in the opposite sense. The origin of this latter vortex pair is the trim force of the horizontal tail, and, or, the reduction of lift across the fuselage due to the wing/fuselage junction. The opposite vortex pairs however, due to their self-induction, tend to move upwards whereas the main vortex pair moves downward. The condition, that the former does not happen is introduced as a necessary condition for premature vortex breakdown in the far field. This necessary condition can be checked upon in the near field. The Rayleigh-Ludwieg instability then, is the sufficient condition for premature vortex breakdown.

1 Introduction

In view of limited airspace and increasing air traffic recent vortex wake research has become important again. In particular, the vortices shed from wide bodied transport aircraft are a hazard to following aircraft in approach, climb, and cruise. There are many proposals of how to alleviate the forces and moments induced by trailing vortices. However, a relationship between the near and far vortex field has not been published yet. Analysing far field measurements of vortex wakes and recalling early knowledge of vortex breakdown over delta wings, a relationship between the near and far vortex wake is obtained. The far vortex wake can therefore be described in terms of the aircraft parameters. This is necessary for the assessment and certification of aircraft with respect to the hazard of trailing vortices and for minimizing the vortex strength through aerodynamic design.

2 Analysis of Far-Field Vortex Wake

The far-field vortex wake of a 1/22 scale model of an Airbus A-300 has been investigated in the catapult of ONERA in Lille [1], [2]. In this facility, the model was flying like a glider through the test section passing through a laser light sheet perpendicular to the flight direction, where the vortex wake was measured using 2 Component Particle Image Velocimetry (2C-PIV). The vortex wake was investigated until the descending wake was affected by ground effect, which occured at about 80 wing spans behind the aircraft. Each of the two configurations used had to be longitudinally stable and trimmed for a flight speed of about 23.5 m/s. In the first configuration, designated as the (0,15)-configuration, only the outbard landing flaps were set out at a 15° angle, the second was designated as the (15, 15)-configuration with both, the outboard and inboard landing flaps, set out at a 15° angle. For both configurations the mean chord Reynolds-number was 500000. In order to avoid laminar separation the lift coefficient of the two configurations was chosen to be 1.2. The vortex wake of the two configurations developed differently right from the trailing edge of the model. In the near field of the (0, 15)-

configuration, a vortex pair emanating from the inside edges of the outboard landing flaps was entrained into the descending main vortex pair. The sense of rotation of this flap vortex pair was opposite to that of the main vortex pair. The entrainment of such an opposite vortex pair was not observed in the case of the (15, 15)-configuration, even though for both configurations, opposite vortex pairs must have rolled up from the horizontal tail and the wing/fuselage junction. In the far field the lateral spacing of the main wing vortex pair of the (0, 15)-configuration was greater than for the (15, 15)-configuration. In addition, in the far field, a strong wing vortex pair with an axisymmetric core and no vortex breakdown was observed for the (15, 15)-configuration, whereas for the (0, 15)-configu-ration the vortex wake descended at half the speed of the (15, 15)-configuration and exhibited a non-circular, irregular vortex core in breakdown.

2.1 Vortex breakdown

Between 1960 and 1980 numerous investigations of the breakdown of vortices shed from the leading edge of delta wings of fighter aircraft were undertaken. From these investigations, two different types of vortex breakdown are known, the near-axisymmetric bubble form and the spiral form, see the review by Wedemeyer [3] for example.

The bubble typed breakdown implies, that a vortex bursting occurs due to loss of pressure head and adverse pressure gradient, which is imparted to the vortex by solid surfaces in the environment. As there are no solid surfaces far behind the aircraft, this type of breakdown cannot be expected for a vortex pair trailing far behind a commercial aircraft. The spiral form of vortex breakdown may be affected by an adverse pressure gradient too, as shown by Ludwieg [4] where a leading edge vortex over a delta wing became unstable at the trailing edge. However, Ludwieg [4] also showed that the origin for the spiral form of vortex breakdown is a hydrodynamic instability, which occurs if the vorticity changes its sign. Rayleigh [5] investigated different velocity distributions of a vortex without an axial flow component and found that the vortex becomes unstable, if the swirl velocity in the radial direction decays more rapidly than $1/r$. Then the destabilising centrifugal forces dominate restoring pressure forces. This statement is consistent with the above one of the vorticity changing its sign. The instability criteria of Rayleigh [5] was generalised by Ludwieg [4], [6], [7] to include the radial gradient of the axial velocity component. Therefore, the instability is called the Rayleigh-Ludwieg instability.

In the case of the (0, 15)-configuration of the Airbus A-300, the vorticity is changing sign because a counter-rotating vortex is entrained into the main wing vortex. In this particular case, the vortex breakdown is understood as a Rayleigh-Ludwieg instability, i.e. as the spiral form of vortex breakdown. In addition, the counter-rotating vortex is not only entrained into, but also stretched by the main vortex, which increases its level of vorticity. After the spiral form of vortex breakdown has started, for both cores of the main vortex pair a rapid increase in size is observed. However, for a while the circulation of the main vortex pair still continues to only slowly decay until the cores have become so large spatially that the two legs of the trailing main vortex pair meet and merge. Only after this merging a rapid decrease of the circulation is observed, which finally alleviates the hazards in encountering wake vortex.

Other forms of instabilities occur in the case of corotating, merging vortices. For this case, Crouch [10] found instabilities. However, an instability which is transient may not be of much interest to the aircraft designer. According to the experiments by Chen et al. [11], the total circulation remains almost constant and the vortex cores do not show premature breakdown.

2.2 Self-induction of vortex sheets

Two questions still remain to be answered. Firstly, why did the trailing vortices display a Rayleigh-Ludwieg instability in the case of the (0, 15)-configuration of the A-300, but not for the (15, 15)-configuration of the same aircraft? Secondly, why is the horizontal vortex spacing in the case of the (0, 15)-configuration larger than for the (15, 15)-configuration?

The answer to the first question lies in the rolling up process of the vortex sheet of a wing and horizontal-tail configuration. In a numerical simulation of the rolling up process, the lift is assumed to be elliptical across the span. The lift of the horizontal tail is negative and producing a vortex sheet, the vorticity of which is opposite to that of the wing vortex. Instead of using only two discrete vortex pairs, the two vortex sheets are each represented by 160 small discrete vortices with equal spacing, but different intensity. By using equal spacing of the vortices, unrealistic singularity effects are avoided. The rolling up process now is studied by means of a vortex line method similar to the one described by Rom [8] and as applied by Wedemeyer [9]. The vortex sheets of the horizontal tail and the wing respectively are designated by their circulations Γ_P and Γ_M and their spacings b_P and b_M, where $b_P=\pi/4\bullet$span of the horizontal tail and likewise $b_M=\pi/4\bullet$span of the wing and subscript p stands for perturbation vortex and likewise M for main vortex. Figure 1 shows the instantaneous pictures of the movement of the discrete vortices representing the two vortex sheets for relative spans, $b_P/b_M=0.3,=0.4,=0.5$ and a ratio of $\Gamma_P/\Gamma_M=0.75$. For $b_P/b_M=0.3$, the entire vortex sheet from the tailplane convects upwards as it trails downstream. For $b_P/b_M=0.4$, a fraction of it ascends and the other fraction descends. For $b_P/b_M=0.5$, corresponding to the (0, 15)-configuration, it is completely entrained into the descending wing vortex sheet. Thus, the numerical simulation reveals that in a transitional domain, the counter-rotating vortex sheet does not descend as a whole. Instead, that portion of Γ_P, which is entrained into the descending main vortex pair, gradually increases with the spacing from 0% to 100%. In the case of the Airbus A-300 the spacing ratio is $b_P/b_M=0.363$. According to the above numerical simulation a portion of the vortex sheet of the horizontal tail could be expected in the wake of the descending main vortices. However, the simulation does not include the counter-rotating vortex pair of the wing/fuselage junction. The spacing ratio of the latter is given by the fuselage diameter and much smaller than $b_P/b_M=0.363$. The self-induction of the combined counter-rotating vortex pair supersedes the induction from the main vortex. It moves upward as shown in figure 1 for $b_P/b_M=0.3$. Hence, in the case of the Airbus A-300 (15, 15)-configuration, no trace of the opposite vortex pair is found in the wake of the descending wing vortex pair and no Rayleigh-Ludwieg instability occurs. On the other hand, in the case of the Airbus A-300 (0, 15)-configuration, the spacing ratio of the counter-rotating vortex pair from the inside edges of the outboard landing flaps is with $b_P/b_M=0.5$, see figure 1, so large that the induction of the wing vortex pair on the opposite vortex pair supersedes the self-induction of the latter, which, therefore, is entrained into the main vortex pair and causes a Rayleigh-Ludwieg instability.

The answer to the second question follows from the rules of merging vortices. For the (15, 15)-configuration, only the main wing vortices were found in the descending wake. The vortex spacing is about $b_M\approx\pi/4\bullet$ wingspan. For the (0. 15)-configuration, there is another opposite vortex pair.Then, the common center of rotation, i.e. centroid is on the line connecting the main and opposite vortex, but outside the main vortex, which increases the wake spacing. This happened to be the case in the wake of the (0, 15)-, but not of the (15, 15)-configuration.

3 The Necessary Condition

From the numerical simulation mentioned above, the descending portion of Γ_P can be evaluated by the addition of the circulation from those small discrete vortices of the tailplane vortex sheet which descend with the wing vortex pair. The result is shown in figure 2, where the circulation ratio Γ_P/Γ_M is plotted over the spacing ratio b_P/b_M. A fan of curves is obtained, instead of having one curve as proposed by Donaldson & Bilanin [12] and Rennich & Lele

[13]. Their curve is a yes or no criteria, in that above their curve, the entire opposite vortex pair descends and below their curve, it entirely ascends. The fan in figure 2 represents the transitional domain with the 0% curve at its lower and the 100% curve at its upper boundary. Each curve of the fan indicates a percentage of the circulation, Γ_P, descending with the main vortex pair. The percentage in terms of the circulation Γ_M of the main vortex pair is given by the product of the circulation ratio and the percentage of the respective fan curve. The intersection of the fan curve and the circulation ratio defines a design point which for a particular aircraft should be at a sufficiently high percentage to obtain premature vortex breakdown. In the case of the (0, 15)-configuration of the Airbus A-300, a premature vortex breakdown was observed at a value of 0.15 for Γ_P/Γ_M •percentage. For this particular configuration, the vortices of the main wing vortex pair encountered the vortices of the opposite vortex pair at about 30 wing spans behind the aircraft. At this distance the core of the wing-vortex has already aged which depends on the Reynolds number, see also Ludwieg [4]. In the diagram of figure 2, the design points of all present wide body aircraft are presumably too low and too much to the left to have premature vortex breakdown. This is indicated in figure 2 by the two dashed rectangles, one only for the opposite vortex pair, $\Gamma_P=\Gamma_F$, emanating from the wing/fuselage junction, the other one for the addition of Γ_F with that one emanating from the horizontal tail, $\Gamma_P=\Gamma_H+\Gamma_F$. With the minimum value of the above product for premature vortex breakdown given, the necessary condition is defined. The necessary condition can be calculated by using the potential vortex line method after [8] and [9], if the entire vortex sheet shed from the particular aircraft is known in the near field. It also should be noted that the lift distribution of each aircraft design in one way or another deviates from the above elliptical distribution of the wing and horizontal-tail configuration. Thus, for a more refined approach the particularities of each aircraft design have to be taken into account in the evaluation of the curves shown in figure 2.

4 The Sufficient Condition

In the past, the hydrodynamic Rayleigh-Ludwieg instability has been investigated mainly with respect to the lift of delta wings. Ludwieg [7] found the modes of instability of leading edge vortices becoming unstable over the trailing edge of such a wing. In these cases, the stream-surfaces of the vortex become eccentric. The positions of the minima of total and static pressure no longer coincide, thereby creating a velocity distribution, the minimum of which is close to its maximum. The disturbance increases exponentially downstream and so does the vortex core diameter. For a detailed modal analysis Ludwieg's theory has to be applied to a vortex wake, which is not affected by an adverse pressure gradient in the environment.

In their paper on the stability of a vortex wake, Fabre and Jacquin [14] took an initial configuration of a four-vortex aircraft wake model, for which the transitional domain of figure 2 collapses into one line. Unfortunately, such a wake does not exist in the present case, for which that part of Γ_P which descends is wrapped helically around its respective wing vortex. Which modes of the Rayleigh-Ludwieg instability are to be expected in the wake vortex far behind a transport aircraft still are to be determined. However, from Ludwieg's stability diagram shown in figure 3 unstable areas of the flow field can be found. These are given in figure 3 in terms of the radial gradients of the swirl and axial velocities. These gradients, for example, may be determined as derived quantities from stereoscopic PIV (3C-PIV) data, see also [15] and [16].

5 Acquisition of the Relevant Data

The near field vortex sheet of the entire aircraft serves as an initial solution to the vortex line method after [8] and [9]. It may be taken, for example, from PIV measurements directly behind

the horizontal tail of aircraft models in wind tunnels or water channels. The laser light sheet should be perpendicular to the main wind tunnel velocity and cover at least one half span of the model. To be able to establish the diagram of figure 2 for the particular aircraft, 2 Component Particle Image Velocimetry (2C-PIV) is sufficient. In the near field, the discrete vortices descernible in the far field are not yet rolled up completely. Thus, a vortex line has to be found, which separates the lines of vorticity which roll up into the wing-vortex from those rolling up into the tail-vortex. In analogy to the neutral Mach line, which in supersonic flows separates the Mach lines running into the shock of the leading edge from those running into the shock from the trailing edge, the separating vortex line may be called neutral vortex line. For both families of vortex lines, those on the inboard side and those on the outboard side of the neutral vortex line, the circulation as well as centroid and its location are calculated from the 2C-PIV data. The circulation and spacing ratios determine the design point in figure 2. The vortex line method provides the percentage of Γ_P which descends with the wing vortex pair.
In order to substantiate this method, it should be doublechecked by comparing its predictions to the corresponding far field obtained from PIV measurements in the catapult of ONERA. As the trailing vortices are generated by local spanwise gradients of the lift distribution, the latter may be found from measurements with pressure sensitive paint. This provides a method of checking the spanwise lift which also should be available from numerical methods. For the initial solution of stability calculations, however, the complete velocity vector has to be known. This may be obtained from stereoscopic PIV (3C-PIV) of the entire near field vortex sheet.

6 Measures for a Premature Vortex Breakdown

There are active and passive measures for achieving a premature vortex breakdown. In practice, considering structural specifications and flight mechanics, passive measures do outperform the active ones. Active measures therefore, are ruled out here. Figure 2 shows what kind of passive measures should be taken. The design points of all current wide body aircraft are presumably too low and too much to the left in the diagram. By increasing the span of the horizontal tail, the design points are shifted to the right and upward in figure 2 and therefore into the domain favorable for premature vortex breakdown. Such relevant measures have already been suggested by DLR-Göttingen in a patent [17].
Using the Airbus A-320 as an example, the relative span of the horizontal tail is b_H/b_M=0.365 and its aspect ratio Λ_H=5.0, see also [18]. From the moment coefficient, the circulation ratio of the horizontal tail is found to be Γ_H/Γ_M=0.65. The vortex pair due to the wing/fuselage junction has a relative span of b_F/b_M=0.11, and an estimated relative circulation of Γ_F/Γ_M=0.18. The addition of both vortex pairs yields for the resulting vortex pair, a relative span of b_P/b_M=0.31, and a relative circulation of , Γ_P/Γ_M=0.83. The design point P is in the spot marked in figure 2. The design point is too low and too much to the left for premature vortex breakdown. An increase of the relative span of the horizontal tail from 0.365 to 0.42 corresponds to an increase of its real span from 12.45m to 13.8m. For the same aerodynamic trim force, the relative circulation is reduced to Γ_H/Γ_M=0.56. For the combined vortex pair, we then have Γ_P/Γ_M=0.74, and b_P/b_M=0,35. This measure shifts the design point P of the A-320 to the point P' in the hatched area, where a 47% portion of the opposite vortex pair with an opposite circulation of $(\Gamma_P/\Gamma_M)\bullet 0.47$=0,35 descends with the main vortex. This is sufficient for premature vortex breakdown. Another benefit is obtained from this too, since the aspect ratio of the horizontal tail increases from 5 to 6.6. This is beneficial for performance in take off and climb. As far as the trim force and the reduction of lift across the fuselage is concerned, the calculation depends on estimations. PIV measurements of the wake taken behind the tail will provide more refined data.

7 Further Improvements and Concluding Remarks

The present paper provides to the aircraft designer an outline of how to deal with the wake vortex issue. The outline is obtained using knowledge gathered in the 1960's from investigations of vortices over delta wings and from most recent EU-funded projects, such as, tests using the ONERA catapult, which were part of the WAVENC-project [2]. The objectives are to provide to the traffic authorities, a method of assessment of aircraft concerning their vortex wake and a procedure for minimising the hazard of wake vortices. In this context, it is beneficial to be able to perform parameter variations in the near field of a wind tunnel or water channels and to relate the near to the far field as sketched in the present paper.

A couple of aspects of this outline need further substantiation by supplementary theoretical, numerical, and experimental work. For the far field wake, the original work by Ludwieg [6], [7] should be supplemented by global stability analysis, which considers the entire plane normal to the axes of the trailing vortices as the basic field. The relevant tool is a solution of the partial derivative eigenvalue problem [19], in which the two spatial directions of the normal plane are resolved and the direction into the vortex axis is assumed to be homogeneous. Classical straightforward methods may be applied to the necessary condition. Current numerical methods should be capable of determining independently from experiments the entire aircraft's lift distribution, which is used as an initial solution to the vortex line method after [8] and [9]. Also, the lift distribution may be checked experimentally by using pressure sensitive paint. The near and far field can be investigated in the catapult of ONERA in Lille. The near field may be obtained from 2C-PIV measurements covering at least one half of the width of the wake. Crucial flow phenomena occuring in the far field, such as the onset of the Rayleigh-Ludwieg hydrodynamic instability, should be investigated by using stereoscopic PIV, i.e. 3C-PIV. During vortex breakdown, 2 points in time are significant. The first one is the time at which the instability of the vortex cores starts. The second is the time at which one side of the vortex begins to merge with the other one and the rapid decay of the circulation is initiated. High Reynolds-number effects, such as separated flows may be studied in the cryogenic European Transonic Windtunnel. The wake vortices are very sensitive to separated flows and even very small intrusive probes. Only non-intrusive testing techniques yield consistently reliable data. The trailing vortices are strongly unsteady during vortex breakdown. All these confounding features may be studied experimentally by using the stereoscopic PIV technique.

References

(1) P. Coton. "Caracterisation et Modelisation du Sillage d'un Avion a partir d'Essais en Vol de Maquettes en Laboratoire". In Agard-Cp-584 on The Characterisation and Modification of Wakes from lifting vehicles in Fluids, 1996, pp 28-1 to28-21.

(2) R.Stuff, L. Dieterle, G.R. Schneider, P. Coton, J.F. Lozier, A. Gilliot." Experimental Study of Wake vortex formation and evolution, 2D-PIV measurements with oil droplets on a micrometric scale". WAVENC-TN 4, EU-Contract No. BRPR_CT977-0593, 1999.

(3) E. Wedemeyer. "Vortex breakdown". In AGARD-LS-121 on High Angle of Attack Aerodynamics, 1982, pp 9-1 to 9-17.

(4) H. Ludwieg. "Vortex breakdown". DLR-FB 70-40, 1970.

(5) Lord Rayleigh. "On the dynamics of revolving fluids". Proc. Roy. Soc. London (A) **93**, 1916, pp 148-154, reprinted in Scientific papers 1911-1919, Vol. VI, 1920 pp 447-453.

(6) H. Ludwieg. "Stabilität der Strömung in einem zylindrischen Ringraum". Z. Flugwiss. **8**, 1960, pp 135-140.

(7) H. Ludwieg. "Ergänzung zu der Arbeit: Stabilität der Strömung in einem zylindrischen Ringraum". Z. Flugwiss. **9**, 1961, pp 359-361.
(8) J. Rom. "High Angle of Attack Aerodynamics". Springer, 1992.
(9) E. Wedemeyer. Private communication. 1999
(10) J.D. Crouch. "Instability and transient growth for two trailing vortex pairs". J. Fluid Mech., **350**, 1997, pp 311-330.
(11) A.L. Chen, J.D. Jacob, Ö. Savas. "Dynamics of corotating vortex pairs in the wakes of flapped airfoils". J. Fluid Mech., **382**, 1999, pp 155-193.
(12) C.duP. Donaldson, A.J. Bilanin. "Vortex Wakes of Conventional Aircraft". AGARDograph No. 204, 1975.
(13) S.C. Rennich, S.K. Lele. "Method for accelerating the destruction of aircraft vortices". J. Aircraft, **36**, 1999, pp 398-404.
(14) D. Fabre, L. Jacquin. "Stability of a four-vortex aircraft wake model". Phys. Fluids, **12**, 10, 2000, pp 2438-2443.
(15) M. Raffel, C. Willert, J. Kompenhans. "Particle Image Velocimetry". Springer, 1998.
(16) C. Willert. "Stereoscopic digital particle image velocimetry for application in wind tunnel flows". Meas. Sci. Technol. **8**, 12, 1997, pp 1465-1479.
(17) R. Stuff, H. Vollmers. "Aircraft with means for a premature breakdown of the wing vortex pair". German Patent Office application No: 199 50 403.2-22, British Patent Office appl. No:0025739.4, French Patent Office appl. No:00.13429, US application pending.
(18) "Jane's All The World's Aircraft". 2000-2001 edition, ed. P. Jackson, Jane's Information Group Inc. Alexandria, Va, USA, 2000.
(19) V. Theofilis."Linear instability analysis in two spatial dimensions".Proc.ECCOMAS Conference, Athens, Greece, Sept. 7-11, 1998, editors K. D. Papailiou et al, **1**, 1, pp 547-552.

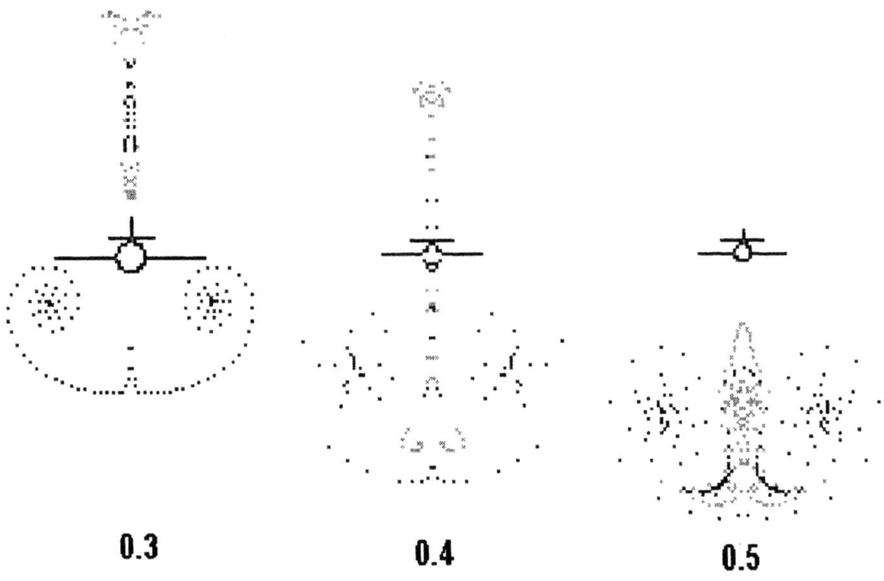

Figure 1 Development of aircraft wake as a function of relative span of horizontal tail. Black points represent vortices from wing, grey points vortices from horizontal tail. The relative spans of the horizontal tail of b_P/b_M=0.3 and b_P/b_M =0.5 correspond respectively to the (0, 15)- and (15, 15)-configuration.

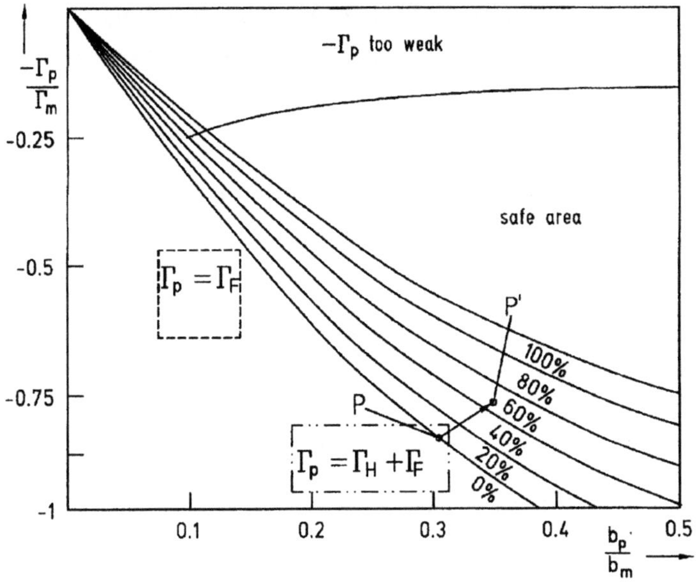

Figure 2 Circulation ratio over spacing ratio. b_P and b_M are the centroids of vorticity respectively inboard and outboard the neutral vortex line, i.e. the middle of the wing.

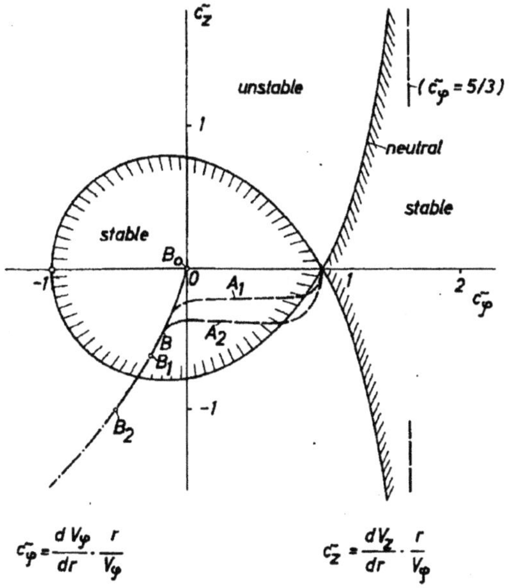

Figure 3 Stability diagram after Ludwieg[4]. The dimensionless radial gradient of axial velocity is plotted over the dimensionless radial gradient of swirl velocoty.

Influence of Different Flap Settings on the Hazard Posed to Following Aircraft

I. SCHELL, E. ÖZGER, D. JACOB
Institut für Luft- und Raumfahrt, RWTH Aachen
Wüllnerstr. 7, 52062 Aachen, Germany

Summary

In the present paper the influence of different flap settings of the wake generating aircraft on the hazard posed to a following aircraft is investigated. The near wake region of a rectangular wing with flaps is computed by means of a 3D panel method while the far wake region is simulated by a 2D Navier Stokes scheme in vorticity form. The hazard posed to a following aircraft is quantified by the value of the maximum induced rolling moment which is calculated by means of 2D strip theory. A preceding experimental investigation showed that the maximum induced rolling moment on the following aircraft related to the wing lift of the generating aircraft is decreased by deflecting the flaps. This result is confirmed by the numerical simulation for moderate flap deflections whereas higher flap angles lead to an increase of the maximum induced rolling moment related to the wing lift. The computation of the far wake region shows that in these cases the merger of the wing tip and flap vortex leads to a reconcentration of circulation and a reduction of the alleviating effect of the flap deflection.

1. Introduction

The vortices in the wake of starting and landing aircraft pose a hazard for following aircraft leading to a limitation of the capacity of airports due to the negatively influenced starting and landing frequency. The most hazardous feature of the wake occurs as an overpowering rolling moment near the center of the vortex. For this reason, research of the characteristics of the vortex wake produced by lift generation has concentrated on the structure of the wakes and, in particular, on the rolling moment that they induce on a following aircraft that encounters them. Although many investigations dealing with this problem have been conducted in the last decades it is still of great importance because of projects like future high capacity transport aircraft (e.g. A3XX) due to the large strength of their shedded vortices.
Experimental and numerical investigations on flapped wings [3-11,13] are difficult to conduct due to the high complexity of the wake structure. In many investigations [3,4,8] a detailed survey of the wake structure behind a flapped wing was carried out but the influence of different flap settings on the maximum induced rolling moment on a following aircraft was not investigated. Orloff et al. [10] and Corsiglia et al. [5,6] quantified the hazard posed to following aircraft in terms of different flap settings by means of windtunnel and flight tests. The results show that the induced rolling moment behind a B747 airplane with fully deployed inboard flap and retracted outboard flap is significantly reduced compared to the standard landing configuration. The authors attributed this to the alteration of the circulation distribution of the generating wing. In experimental investigations of Schell et al. [13] and Özger et al. [11] the near-field wake structure of various wings with flaps was examined for different flap

settings. As a result of these investigations the configurations with retracted flaps showed the highest values of maximum induced rolling moment related to the wing lift and a reduction could be achieved by flap deflection. The reduction was caused by the division of the circulation into several vortices which leads to a decrease of the maximum induced rolling moment related to the lift, at least in the investigated wake field up to $x/b = 1.0$, where no vortex merger of the flap and tip vortex occured. Motivated by the experimental investigation of Schell [13], which was conducted with the rectangular wing with flaps shown in *figure 1*, a numerical investigation is performed to clarify whether there exists a favourable flap setting with a low induced rolling moment and if the alleviation accomplished in the near region (see *figure 2*) is still discernible in the far region where vortex merger takes place.

2. Numerical methods

For the computation of the flow behind the wing the wake is divided into two distinct parts, a near region extending from the trailing edge of the wing where the flow is considered to be three dimensional (one span in length), and a far region where the flow is assumed to be time dependent two dimensional (for details see 2.1 and 2.2). A low-order panel method [1] is used to calculate the three dimensional wake roll-up in the near region behind the different wing configurations. At one span behind the wing a vorticity distribution is evaluated from the calculated wake geometry and circulation distribution. This vorticity distribution serves as an initial condition for the calculation of the far region with a two dimensional unsteady Navier Stokes Solver [7].

2.1 Panel method

Three dimensional irrotational incompressible flows around arbitrary configurations are characterized by a perturbation velocity potential ϕ [1,14,16]. This potential satisfies the Laplace equation

$$\nabla^2(\phi_\infty + \phi) = 0 \quad \text{with} \quad \phi = \underbrace{\frac{1}{4\pi}\int_S \sigma \cdot \left(\frac{1}{r}\right) dS}_{sources} + \underbrace{\frac{1}{4\pi}\int_{S,W} \mu \vec{n} \cdot \vec{\nabla}\left(\frac{1}{r}\right) dS}_{doublets} \qquad (1)$$

in a region surrounding the body with a surface S, whereat ϕ_∞ is the potential of the freestream. In order to obtain an unique solution in lifting case, a potential discontinuity surface, the so-called wake with a surface W, leaving the sharp trailing edge and extending to infinity, has to be introduced. According to Green's third identity, a solution of equation (1) may be expressed at any point p as a potential induced by a combination of source singularities of strength σ and doublet singularities of strength μ, distributed on the surfaces S and W, where r is a distance from the field point p to the surface. A required solution is found by imposing the Neumann condition on S and W. The free wake surface is adjusted iteratively in order to guarantee tangential velocity along the wake surface and zero pressure jump across the wake.

2.2 Two dimensional Navier-Stokes method

The far region is calculated by treating the three dimensional steady flow in the wake as two dimensional and unsteady based on the work of Westwater [15], Rosenhead [12] and Fell [7]. The method assumes a flow with weakly bend vortex-filaments and a dominating vorticity component $\omega_x \approx \omega$. Three dimensional effects caused by vortex stretching and bending and by the axial jets within the vortex cores are neglected. Another simplification is achieved by assuming that the vortices are dominated by viscous effects rather than turbulent phenomena, so that the calculated vortex decay rates are smaller than those found in real wakes. The laminar treatment of the flow is therefore a worst case scenario. The 3D-Navier-Stokes equations are transformed into the 2D-vorticity transport equation (2), where $t = x/U_\infty$. The velocities v and w are obtained by solving the Poisson equation (3) for the stream function Ψ. As an initial condition a vorticity distribution is necessary which is obtained from the panel calculation as discussed in the following chapter.

$$\frac{\partial \omega}{\partial t} + v\frac{\partial \omega}{\partial y} + w\frac{\partial \omega}{\partial z} = \nu \cdot \left(\frac{\partial^2 \omega}{\partial y^2} + \frac{\partial^2 \omega}{\partial z^2}\right) \tag{2}$$

$$\frac{\partial^2 \Psi}{\partial y^2} + \frac{\partial^2 \Psi}{\partial z^2} = -\omega \quad \text{with} \quad v = \frac{\partial \Psi}{\partial z}, \quad w = -\frac{\partial \Psi}{\partial y} \tag{3}$$

3. Numerical results

The configurations of the numerical investigation are presented in *table 1*. The computation of the wake of the configurations was performed at approximately the same lift coefficient $c_L = 1.3$. In order to achieve this, lift curves of the different configurations were evaluated by means of a computation with a frozen wake to establish angles of attack with equal lift. Then, computations with a free wake were performed for a wake length of two spans. The angles of attack and the lift coefficients of the free wake computation are listed in *table 1*. The nondimensional circulation distributions $\Gamma/U_\infty b$ of the investigated configurations with equal inboard and outboard flap deflection are presented in *figure 3a*. For increasing flap angles circulation is redistributed from the wing tip to the flap. An approximately uniform distribution of circulation along the wing span is accomplished for moderate flap angles. *Figure 3b* shows circulation distributions for the configurations with different inboard and outboard flap deflection where the inboard flap angle was chosen to be twice the value of the outboard flap angle. Hereby, four vortices are shed from the wing, a wing tip and three flap vortices. Analog to the configurations with equal inboard and outboard flap deflection the circulation is redistributed for increasing flap angles from the wing tip to the flap vortices.

The wake geometry for the flap setting $\delta_i = \delta_o = 20°$ is shown as an example in *figure 4a*. Three vortices, one coming from the wing tip and two from the flap edges, can be discerned in the regions where the vortex sheet rolls up. As another example the wake geometry for the flap setting $\delta_i = 20°$, $\delta_o = 10°$ is shown in *figure 4b*. Four vortices, one coming from the wing tip and three from the flap edges, can be discerned in the regions where the vortex sheet rolls up.

The initial condition for the 2D-Navier-Stokes computation of the far region is evaluated from the results of the panel computation at $x/b = 1.0$ (*figure 5*). First, the calculated wake geometry and circulation distribution is interpolated for the distance $x/b = 1.0$ (as shown in *figure 5a*).

Then the wake at $x/b = 1.0$ is divided into the wing tip and flap vortices, where the circulation is mainly concentrated, and the vortex sheet connecting them. The circulation of the wing tip and flap vortices is conserved and the positions are evaluated according to the invariance of the first moment of circulation according to Betz [2]. More line vortices are introduced into the vortex sheet in order to obtain a more continous distribution of circulation (*figure 5b*). Finally, the circulation Γ_i of the singular line vortices is distributed by use of Gauss vortices to obtain a continous vorticity distribution $\omega(y,z)$ according to

$$\omega(y,z)=\sum_{i=1}^{n}\frac{\Gamma_i}{2\pi\cdot k^2\cdot R_{c,i}^2}\cdot e^{-\left[\frac{(y-y_i)^2+(z-z_i)^2}{2k^2R_{c,i}^2}\right]},\ k=0.6308348... \quad (4)$$

where y_i and z_i are the positions of the line vortices, k is a constant and $R_{c,i}$ is the core radius of the Gauss vortices. The core radius of the wing tip and flap vortices is evaluated according to the invariance of the second moment of circulation [2]. The core radius $R_{c,i}$ of the vortex sheet hardly influences the wake dynamics and is therefore assumed to be constant and set to $R_{c,i} = 0.1\cdot c$ (*figure 5c*) according to experimental results.

The subsequent 2D-Navier-Stokes computation is performed from a distance behind the wing of $x/b = 1$ to $x/b = 21$ in order to guarantee the merger of the wing tip and the outboard flap vortex. *Figure 6a* shows as an example the wake structure of the far region for the configuration with equally deployed flaps ($\delta_i = \delta_o = 20°$). The wing tip and outboard flap vortex rotate around each other until they merge at a distance of about $x/b = 15$. Because of the mutual induction of the three vortices the wing tip and outboard flap vortex move in positive y- and negative z-direction whereas the inboard flap vortex moves in negative y- and z-direction.

The wake structure in the far region of a configuration with not equally deployed flaps ($\delta_i = 20°$, $\delta_o = 10°$) is shown in *figure 6b*. The wing tip and outboard flap vortex rotate around each other until they merge at a distance of about $x/b = 13$. The midboard flap vortex rotates around the wing tip and outboard flap vortex and does not merge within the computed wake area. The trajectory of the inboard flap vortex is nearly the same compared to the one with equal flap deflection at $\delta_i = \delta_o = 20°$.

The maximum rolling moment induced on a following aircraft with a span ratio of $b_f/b = 0.2$ was evaluated for the 2D-Navier-Stokes wake computation and displayed in *figure 7a* for the configurations with equal inboard and outboard flap deflection. For the configuration with stowed flaps the maximum induced rolling moment does not change significantly and decreases slightly downstream due to the viscosity. The maximum induced rolling moment of the configurations with deployed flaps oscillates due to the rotation of the wing tip and flap vortex around each other in the wake of the wing. With merging of the wing tip and outboard flap vortex this oscillation diminishes. After completion of the merger the curves of maximum induced rolling moment lie parallel to the curve with stowed flaps.

The process of merger depends on the distribution of circulation on the wing tip and outboard flap vortex. For an equal distribution of circulation on the wing tip and outboard flap vortex merger takes place further downstream and the oscillations are more distinct while for an unequal distribution one vortex dominates the flow field so that merger occurs earlier downstream and the oscillations are weaker.

When comparing the maximum induced rolling moments $c_{l,max}$ related to the wing lift c_L of the generating aircraft for the configurations with equal inboard and outboard flap deflection at a distance of $x/b = 1$ it can be seen that for increasing flap angles until $\delta_i = \delta_o = 20°$ the maximum induced rolling moment is decreasing which was also observed in the experimental re-

sults of a previous investigation [13]. Further increase of the flap angle leads to an increase of the maximum induced rolling moment to values that can be even higher than for the configuration with no flap deflection ($\delta_i = \delta_o = 30°$). This can be attributed to the increasing circulation of the flap vortices which become dominant in the wake flow. Downstream at a distance of $x/b = 21$ where the merging process is completed the effect of alleviation by a moderate flap deflection is still discernible but less pronounced than at $x/b = 1$.
The maximum induced rolling moment coefficient $c_{l,max}$ related to the wing lift c_L of the generating aircraft versus the distance to the trailing edge x/b for unequal inboard and outboard flap deflection is shown in *figure 7b*. Again oscillations in the maximum induced rolling moment can be seen downstream which are attributed to the rotation of the wing tip, outboard and midboard flap vortex. The merger of the wing tip and outboard flap vortex reduces the amplitude of the oscillation not completely because of the still rotating midboard vortex. The amount of alleviation increases with higher flap angles at $x/b = 1$ and is comparable to the reduction of the maximum induced rolling moment for the configurations with equal flap deflection. Downstream at a distance of $x/b = 21$ where the merging process is partly completed the amount of alleviation is still discernible but less pronounced than at $x/b = 1$ and nearly of the same value for all configurations with unequal flap settings.

4. Conclusion

A numerical investigation of the wake structure behind a flapped rectangular wing was conducted to examine the influence of different flap settings on the induced rolling moment on a following wing. The numerical simulations were performed by means of a three dimensional panel method for the calculation of the trailing vortices in the near field region and a two dimensional Navier-Stokes method for the calculation of the trailing vortices in the far field region.
The numerical investigation was performed to clarify whether there is an optimal flap setting to achieve maximum alleviation and if the alleviating effect of the flap deflection still exists in the far field region up to twenty spans behind the wing, where the merging of the wing tip and the outboard flap vortex is comleted. The results show that moderate flap angles of the generating aircraft lead to the largest reduction of the maximum induced rolling moment on the following aircraft because of the nearly equal distribution of the circulation on the wing tip and the flap vortices. High flap angles lead to dominant flap vortices and to higher induced rolling moments because of the unequal distribution of circulation on the wing tip and the flap vortices. The results also show that the merging process of the wing tip and outboard flap vortex does not annihilate the alleviating effect of the flap deflection but diminishes it because of the reconcentration of circulation caused by the merging process.

Acknowledgements

The investigation is supported by the German Research Association (DFG).

References

[1] Behr, R., Wagner, S., "Ein nichtlineares Wirbelgitterverfahren zur Berechnung von Interferenzeffekten zwischen freien Wirbelschichten und Tragflächen", DGLR-Bericht 88-05, pp. 303-316, 1988

[2] Betz, A., "Verhalten von Wirbelsystemen", ZAMM, Bd. XII, No. 3, pp. 164-174, 1932

[3] de Bruin, A. C., Hegen, S. H., Rohne, P. B., Spalart, P. R., "Flowfield survey in trailing vortex system behind a civil aircraft model at high lift", AGARD-CP-584, 1996

[4] Chen, A. L., Jacob, J. D., Savas, Ö., "Dynamics of corotating vortex pairs in the wake of flapped airfoils", Journal of Fluid Mechanics, Vol. 382, pp. 155-193, March 1999

[5] Corsiglia, V. R., Dunham, R. E., "Aircraft Wake-Vortex Minimization by Use of Flaps", NASA SP-409, pp. 305-338, 1977

[6] Corsiglia, V. R., Rossow, V. J., Ciffone, D. L., "Experimental study of the effect of span loading on aircraft wakes", Journal of Aircraft, Vol. 13, No. 12, pp. 968-973, December 1976

[7] Fell, S., Staufenbiel, R., "Formation and Structure of Vortex Systems Generated by Unflapped and Flapped Wing Configurations", Z. Flugwiss. Weltraumforschung, Vol. 19, No. 6, pp. 366-379, December 1995

[8] Hueneke, K., "Structure of a transport aircraft-type near field wake", AGARD-CP-584, 1996

[9] Iversen, J. D., Corsiglia, V. R., "Hot-Wire, Laser-Anemometer, and Force Measurements of Interacting Trailing Vortices", Journal of Aircraft, Vol. 16, No. 7, pp. 448-454, July 1979

[10] Orloff, K. L., Ciffone, D. L., "Vortex Measurements behind a Swept Wing Transport Model", Journal of Aircraft, Vol. 11, No. 6, pp. 362-364, June 1974

[11] Özger, E., Schell, I., Jacob, D., "On the Structure and Attenuation of an Aircraft Wake", AIAA Paper 2000-4127, 2000

[12] Rosenhead, L., "The Formation of Vortices from a Surface of Discontinuity", Proc. Roy. Soc. London, A134, pp. 170-192, 1931

[13] Schell, I., Özger, E., Jacob, D., "Influence of Different Flap Settings on the Wake-Vortex Structure of a Rectangular Wing with Flaps and Means of Alleviation with Wing Fins", Aerospace Science and Technology, No. 2, pp. 79-90, April 2000

[14] Sytsma, H. S., Hewitt, B. L., Rubbert, P. E., "A Comparison of Panel Methods for Subsonic Flow Computation", AGARDograph, No. 241, February 1979

[15] Westwater, F. L., "Rolling Up of a Surface of Discontinuity behind an Airfoil of Finite Span", Aeronautical Research Council Reports and Memoranda, No. 1962, pp. 116-131, 1935

[16] Yeh, D. T., Plotkin, A., "Vortex Panel Calculation of Wake Rollup behind a Large Aspect Ratio Wing", AIAA Journal, Vol. 24, No. 9, pp. 1417-1423, September 1986

Table 1. Computed cases

Flap angle δ_i / δ_o	0°/0°	5°/5°	10°/10	15°/15	20°/20	25°/25	30°/30	10°/5°	20°/10	30°/15
AoA α	10°	7.984°	6.089°	4.108°	2.143°	0.212°	-1.627°	7.011°	4.079°	1.234°
Lift c_L	1.284	1.292	1.299	1.310	1.321	1.333	1.347	1.296	1.310	1.329
Number of vortices	1	3	3	3	3	3	3	4	4	4

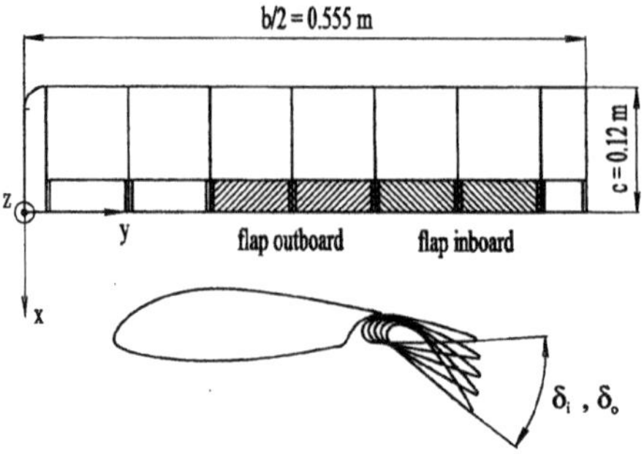

Figure 1. Sketch of flapped wing model

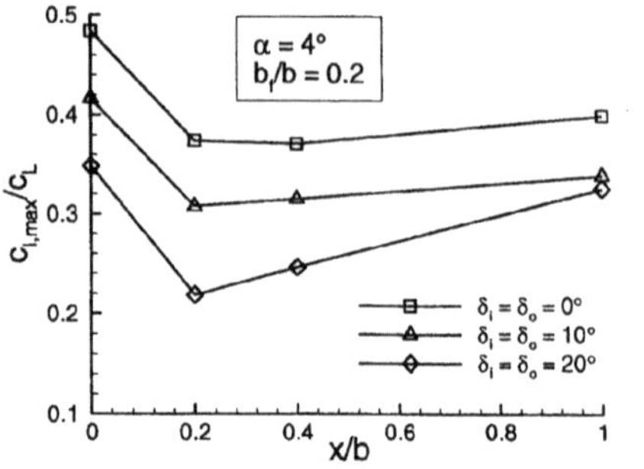

Figure 2. Maximum relative rolling moment $c_{l,max}/c_L$ induced on a following aircraft versus distance to the trailing edge

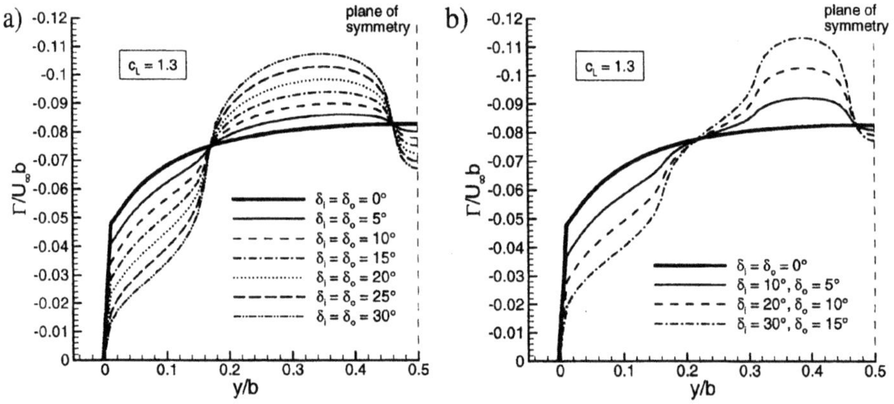

Figure 3. Nondimensional circulation distribution $\Gamma/U_\infty b$ of configurations with a) equal inboard and outboard flap deflection, b) different inboard and outboard flap deflection

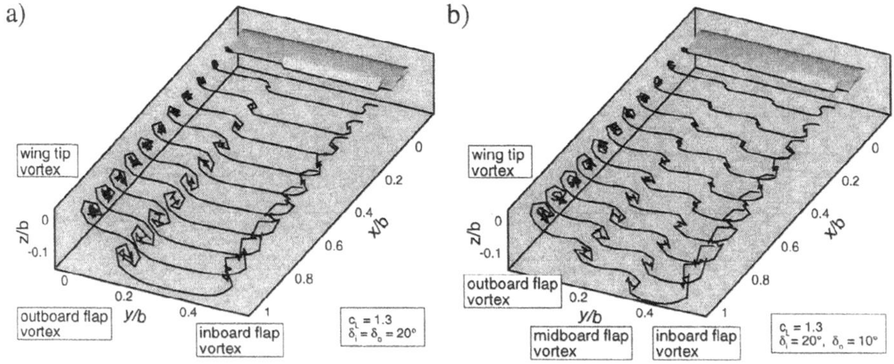

Figure 4. Panel computation of the wake geometry for a) $\delta_i = \delta_o = 20°$ and $\alpha = 2.143°$, b) $\delta_i = 20°$, $\delta_o = 10°$ and $\alpha = 4.079°$

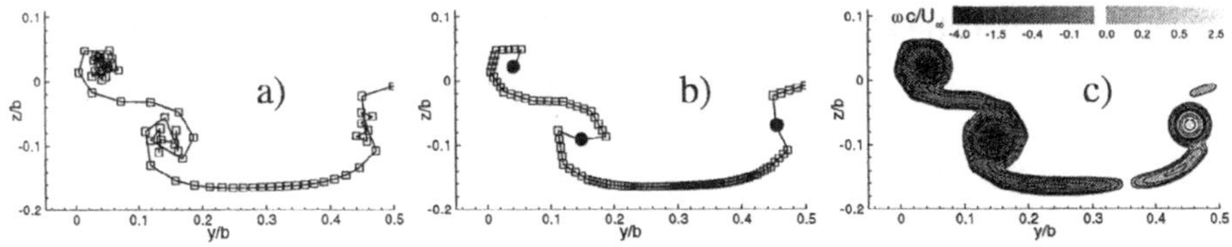

Figure 5. Evaluation of an initial condition for the 2D-Navier-Stokes computation for the configuration $\delta_i = \delta_o = 20°$, $c_L = 1.3$

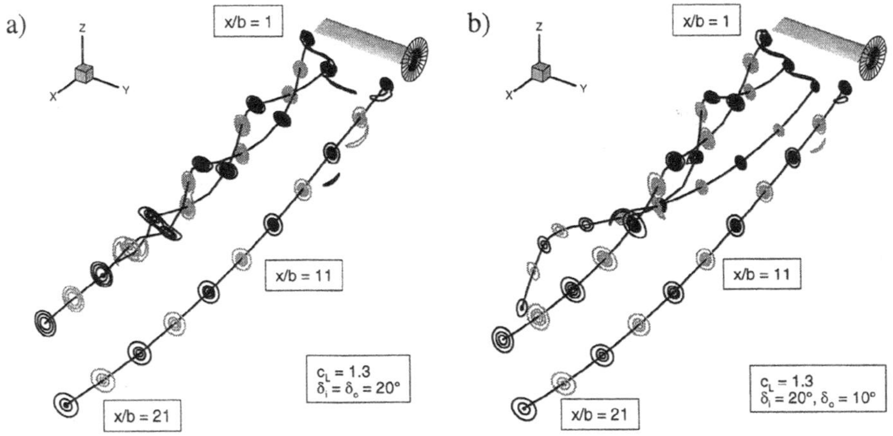

Figure 6. Far region of the wake computed by means of the 2D-Navier-Stokes method for a) $\delta_i = \delta_o = 20°$ and $\alpha = 2.143°$, b) $\delta_i = 20°$, $\delta_o = 10°$ and $\alpha = 4.079°$

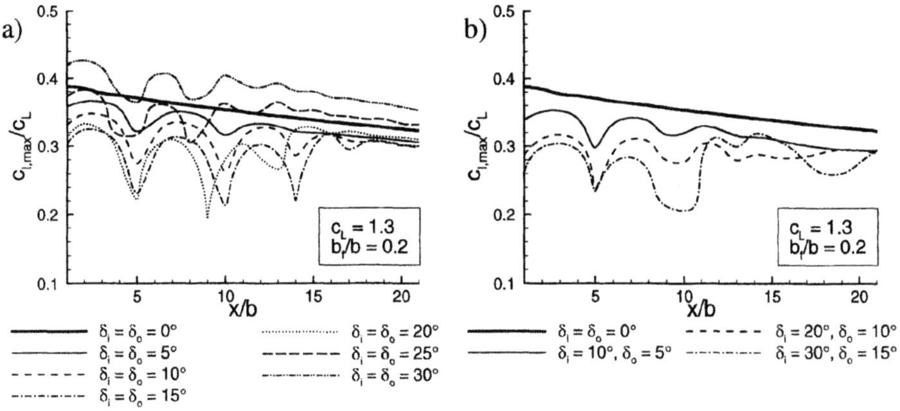

Figure 7. Maximum induced rolling moment $c_{l,max}$ induced on a following aircraft (wings horizontal) related to the lift coefficient c_L of the generating aircraft for a) flaps equally deployed, b) flaps not equally deployed versus distance to the trailing edge

Numerical Simulation of Wake-Vortex Encounters Using the Chimera-Technique

R. Heinrich
DLR Institute of Design-Aerodynamics,
Lilienthalplatz 7, 38108 Braunschweig
ralf.heinrich@dlr.de

1. Summary

Within the EU-project WAVENC the block structured DLR-FLOWer code has been applied to the inviscid simulation of a wake vortex encounter of a generic aircraft. The incoming vortices of the preceding aircraft are prescribed at the inflow boundary of the computational domain. This allows to capture a mutual interaction between aircraft and incoming vortices. In order to transport the vortices without too big numerical losses from the inflow boundary to the aircraft, very fine Cartesian grids, so „called vortex transport meshes", have been embedded into the grid structure. This is enabled by the Chimera technique implemented in the code. One main advantage of this strategy is, that no remeshing is needed to perform a parametric study of the aircrafts position relative to the incoming vortices. Global forces and moments as well as pressure distributions obtained for different positions of the aircraft in the vortex disturbed flowfield show a good agreement to experimental data.

2. Introduction

To avoid risks for an aircraft flying in the wake vortex field of a preceding aircraft, strict rules for minimum required separation distances have been devised. They are merely based on the weight of the aircraft and do not take into account details of the wake formation, the influence of specific meteorological conditions on the far wake evolution and the recovering capabilities of the aircraft behind. To find more realistic separation rules, the EU-project WAVENC (Wake Vortex Evolution and Wake Vortex Encounter) addresses the topics mentioned above. Experimental as well as numerical investigations have been carried out to get a better understanding not only of the vortex evolution behind an aircraft but also of the encounter of the aircraft flying next.

Within the EU-project WAVENC [1], a wake vortex encounter experiment has been performed in the large low speed wind tunnel DNW-LLF [2]. A small model named the SWIM-geometry has been placed downstream into the vortical flow-field of a preceding aircraft of more than double span. Forces, moments and pressure distributions have been measured at several model positions in the vortex disturbed flowfield. This experiment forms an excellent data base for the validation of simulation codes aiming at representing qualitatively and quantitatively the effects of wake vortex encounters.

Inviscid computations obtained with the DLR-FLOWer code are compared to these experiments, in order to test the capability of the numerical scheme for this kind of flow situation. The paper focuses on a strategy used to calculate the mutual interaction between the incoming vortices and the following aircraft, which is important, if a vortex trajectory is close to the aeroplane. This is shown by comparison to results of simpler methods [3] based on potential theory.

3. Numerical Method FLOWer

In the following, the numerical method used for the Euler simulations is briefly described with emphasis on the Chimera method implemented in FLOWer [4], since this technique has extensively been used within this work.

FLOWer solves the three-dimensional Reynolds-averaged full Navier-Stokes or Euler equations. The second order spatial discretization is based on a finite volume cell-vertex method. The fluxes are approximated using a central discretization operator. To avoid spurious oscillations, the well known scalar dissipation model of Jameson is implemented in FLOWer. A 5-stage Runge-Kutta scheme is used for the time-integration. The convergence can be accelerated using local time-stepping, implicit residual smoothing and a multigrid method.

3.1 Chimera Aspects

The so called Chimera overset grid approach has been implemented in FLOWer. It helps to open the door to the simulation of complex applications including, for example, bodies in relative motion. The approach involves the decomposition of the problem geometry in a number of geometrically simple overlapping component grids. Multiple-body applications, such as helicopter flow, are naturally treated in this way. The flow solution process is applied to topologically simple component grids. Body-fitted component grids are ideally suited to regions of thin shear flows such as boundary-layers, wakes, etc. In a Chimera-style overset grid approach, domain connectivity is achieved through interpolation of necessary intergrid boundary information from solutions in the overlap region of neighbouring grid systems. Examples of applications using the Chimera-technique option of FLOWer can be found in [5].

Following the idea of component grids enabled by the Chimera technique, a mesh around the SWIM geometry has been generated. Component meshes have been generated separately around the fuselage, the wings and the horizontal and vertical tails. The whole structure is embedded in a Cartesian background mesh, which is well suited for vortex dominated flows. Some details of the mesh are displayed in fig. 1. In regions of high gradients additional meshes can be embedded into the background mesh in order to resolve physical details like vortices more accurate, similar to adaptation techniques.

4. Idea of Vortex Transport Meshes

The incoming vortices shall be prescribed as an inflow boundary condition, to allow a mutual interaction between the vortices and the aircraft (lower left of fig. 2). One of the main problems arising is the convection of the vortex-induced disturbances from the inflow boundary to the SWIM geometry without incurring too big „numerical losses“. Standard block-structured meshes around similar configurations are not well suited to that kind of problem. This is caused by the big cell sizes near the far-field, usually some factors larger than the core radii of the incoming vortices. If the vortex cannot be properly represented at the inflow boundary, there is of course absolutely no chance at all to convect the vortex to the place of interest. A Cartesian type of mesh with a resolution adapted to the vortex core size would be ideally suited. Unfortunately it is quite a difficult task, to incorporate a Cartesian block into a usual multi-block mesh. Another difficulty arises if a parametric study of the model position relative to the vortices is of interest, just like in the WAVENC project; for each test case a new mesh has to be generated. The Chimera technique can help to bypass these problems: in contrary to standard block-structured grid topologies, overlapping meshes are allowed. This enables to embed Cartesian meshes in the block structure in the region of the vortices. The resolution of these vortex transport grids can be adjusted to the properties of the vortices, without changing the rest of the block topology. The same holds for a variation of the vortex position. The principal idea is graphically summarized in the left of fig. 3. For two test cases the inflow boundary is visualized including the vortex transport grids and the SWIM geometry. The question which still has to be answered is „what is a sufficient resolution of the vortex transport meshes in the region of the vortex cores?". This question is discussed in the following.

4.1 Grid Refinement Study for the Vortex Transport Mesh

To find out a proper resolution of the vortex transport meshes, a grid refinement study has been carried out for a single vortex transport mesh. Two different cases with different vortex core sizes were selected. Test I corresponds to the vortices measured during the experiment with a vortex core radius of 5 mm (= 10% of the root chord of the main wing of the SWIM geometry). A doubled value for the vortex core radius is used for test II. The vortices prescribed at the inflow boundary are described by analytical functions obtained by a least square fit [6] of the velocity field measured within WAVENC [2].

For a given vortex transport distance L (fig. 3), the number of cells $n=n_y=n_z$ near the vortex core region is changed from 24 to 120 in steps of 24 cells. The corresponding number of cells in the vortex core for test I is 2 to 10 in steps of 2 and 4 to 20 in steps of 4 cells for test II. The number of cells n_x in flow direction is constant. To analyse the mesh quality, the numerical solutions for the different resolutions are compared to the analytical solution in the wing tip plane.

In fig. 4 and fig. 5 results are shown for the pressure coefficient, the axial velocity u and the cross-flow velocity w in the region around the vortex core (wing tip plane). The largest errors of course can be found near the vortex core due to the high local gradients. Using a core radius of only 5mm, the maximum local difference between both analytical and numerical solutions becomes unacceptably high for the pressure distribution, if less than n=48 cells are used (4 cells in the vortex core radius). Using more than n=72 points (6 cells in the vortex core radius) there are still visible differences between both analytical and numerical solutions inside the vortex core. But the overall agreement of the curves becomes fairly good, especially for the important cross-flow velocity and the axial velocity which is visible in fig. 5. This is also expressed by the global errors plotted in fig. 4. The global error for the cross flow velocity is defined by

$$err_{w,\,global} = \frac{\int_0^{\infty} |w_{analytical}|\,dy - \int_0^{\infty} |w_{numerical}|\,dy}{\int_0^{\infty} |w_{analytical}|\,dy}\,100\%\,, \tag{1}$$

and for the pressure distribution in a similar manner. Using 96 or more cells (8 cells in the vortex core radius), the global error for the cross-flow velocity w becomes smaller than 2% for a core radius of 5 mm. This should be an acceptable value for the following encounter computations. The situation of course is improved, if a vortex core radius of double size is used (10 mm). On the one hand the local vortex resolution is increased by a factor of 2 and on the other the relative vortex transport distance L/r_{core} is reduced from 150 to 75. For the same resolution of the mesh the error is reduced to 0.5%.

5. Results for Wake Vortex Encounters

Computations have been run for five positions of the SWIM geometry in the flow-field for a Mach number of 0.176 and 0° angle of attack. The computations correspond to traverse 4 of the experiment described in [2]. An overview of the selected test cases is given in fig. 2. Global force and moment coefficients computed with FLOWer compared to the results of panel methods from NLR (PDAERO) and TsAGI (VORTPAN) [3] and to the experimental results are shown in fig. 6 (lift: c_L, side force: c_y, rolling moment: c_l). The squares correspond to the *method of vortex transport meshes* (FLOWer), the triangles and the '+' to the computations using panel methods. In general, the trends of the force and moment coefficients are well captured by all numerical codes. Especially the results of the rolling moment coefficient c_l are very close to the experiment, which is the most important coefficient from the pilot's point of view.

A large shift for the side force c_y relative to the experimental data is conspicuous. The agreement of the shape of the curves is good. This can be shown by moving the experimental data in such a way, that the values of c_y fit to the numerical results of case ii (y=0.0).

An interesting result is, that the simplified methods (panel methods PDAERO and VORTPAN), which are not able to take a mutual interaction between vortices and aeroplane into account, and the advanced Euler method (method of vortex transport meshes) are in fairly good agreement. So the influence of the aeroplane on the vortices seems to be negligible for most of the test cases considered, if global forces and moments are of main interest. Comparing the side forces between the *method of vortex transport meshes* and the panel methods, a good agreement for test cases i, ii, iii and v is visible. But for test case iv (y=-0.42m) there is a noticeable difference. Using the panel methods, the side force peak is overestimated, whereas the advanced method and the (shifted) experiment match. The reason for this discrepancy is, that for case iv one of the vortex trajectories is close to the *thick* fuselage (fig. 2). So there is a non-negligible influence of the aeroplane on the vortex trajectory. The deviation of the vortex track near the fuselage away from a straight line is large compared to the other test cases, where one of the vortices is close to a *thinner* wing. The errors due to the simplified assumptions of the panel methods result in a wrong pressure distribution on the fuselage, which mainly influences the side force. On the other hand it is remarkable, that the agreement to the experiment of the rolling moment coefficient for all numerical methods applied is very good, even for the critical test case iv. The reason is, that the influence of the incorrect computation of the pressure on the fuselage on the rolling moment is negligible. The rolling moment is mainly influenced by the pressure distribution near the main wing tips, which result in forces with large lever arms with respect to the rolling moment reference axis. The situation may change, if a more realistic aircraft is taken into account, including *bulky* high bypass engines. Then panel methods are not able to well predict the pressure in the area of the engines, if a vortex trajectory is close to one of the nacelles. And this will have a visible influence on the rolling moment coefficient.

The pressure distributions on the mid-span position of the port-side wing for all five test cases compared to the experimental data is shown in fig. 7. All in all the agreement between numerical results and the experimental data is good. Again it is remarkable, that the differences between the method of vortex transport meshes and the panel methods are very small (case ii, iv). But this may be due to the test cases selected. For all test cases, the distance between the location of the pressure holes and the vortices is relatively big. The situation would change, if one of the vortex tracks is close to the location of the experimental data. Then the mutual interaction between the wing and the vortices may play a significant role, like for the side force. That there are only small differences between Euler method and panel method can be explained by the small Mach-number combined with a 0° angle of attack. So compressibility effects can be neglected.

6. Summary and Conclusions

It has been shown, that the method of vortex transport meshes is well suited for the simulation of a wake vortex encounter situation. Compared to the experiment, global forces and moments tendencies as well as pressure distributions are well represented. The main advantages of this method are, that a mutual interaction of vortices and airplane is captured in the simulation and that a parametric study of the position of the aircraft relative to the vortices is possible without generating new meshes. The Euler results were additionally compared to those of panel methods, showing a good overall agreement. Noticeable errors due to the simplifying assumptions of the panel methods are visible for the side force, if one of the vortices is close to the *bulky fuselage*. In that case the influence of the aeroplane on the vortex trajectory cannot be neglected. The *overall* good agreement between panel methods and Euler method is due to negligible compressibility effects. This may change when switching to more realistic angles of attack for an aeroplane in landing or take-off situation. Then compressibility and viscous effects have to be taken into account for an accurate simulation of encounter situations.

7. References

[1] de Bruin, A. C. „WAVENC, Wake Vortex Evolution and Wake Vortex Encounter, Publishable Synthesis Report". WAVENC report no. R6 or NLR-TR-2000-079, 2000

[2] Hegen, G. H. „Wake encounter test in DNW wind tunnel - test number 98-1116". NLR-CR-98291 and WAVENC-TN10, 1998

[3] Brandsma, F. J. „Wake encounter computations for a simple configuration with the panel method PDAERO". NLR-TR-99441, WAVENC-TN11a, 1999

[4] Kroll, N. „National CFD-Project MEGAFLOW - Status-Report -". Notes on Numerical Fluid Mechanics, Volume 60, edited by H. Körner and R. Hilbig, Vieweg Verlag Braunschweig/Wiesbaden, 1997

[5] Heinrich, R.; Kalitzin, N. „Numerical Simulation of Three-Dimensional Flows using the Chimera-Technique". Notes on Numerical Fluid Mechanic, Volume 72, edited by Wolfgang Nitsche, Hans-Joachim Heinemann and Reinhard Hilbig, Vieweg Verlag Braunschweig/Wiesbaden, 1999

[6] Struijs, R.; Jonville, G.; Darraq, D. „Inviscid computation of the effect of wake vortices on a model airplane". WAVENC-TN16, 1999

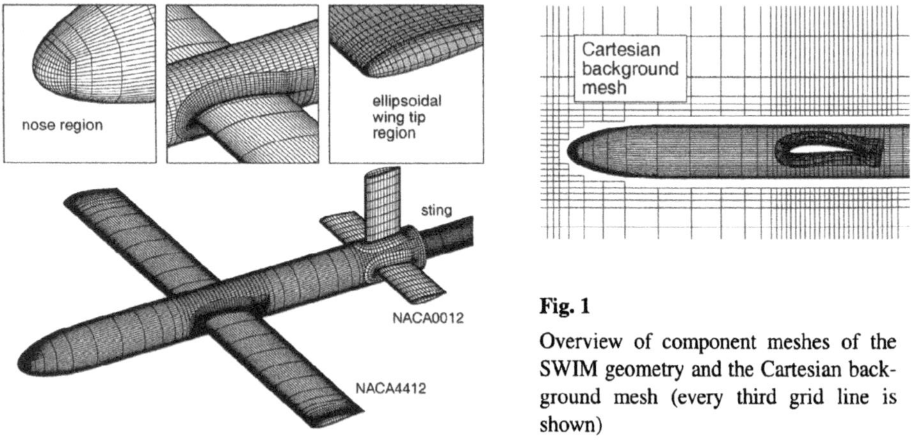

Fig. 1

Overview of component meshes of the SWIM geometry and the Cartesian background mesh (every third grid line is shown)

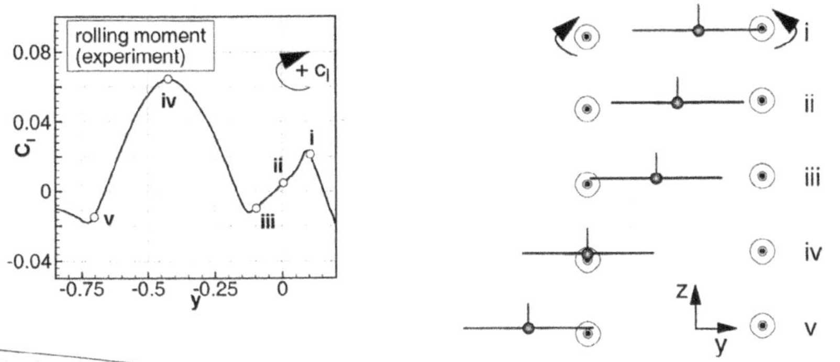

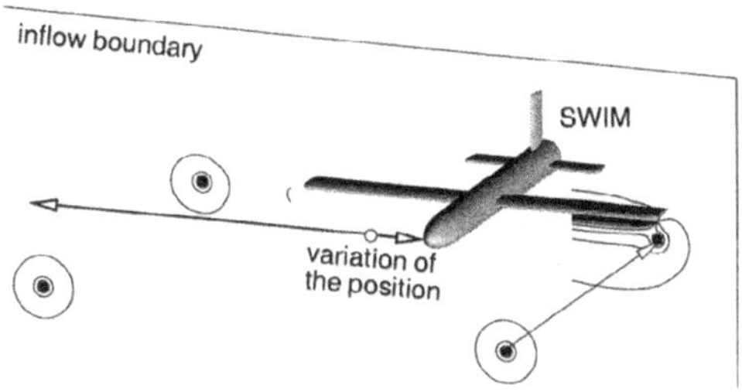

Fig. 2

Selected test cases for wake encounters of SWIM geometry (looking in flight direction)

Lower left: prescription of vortices as inflow boundary condition

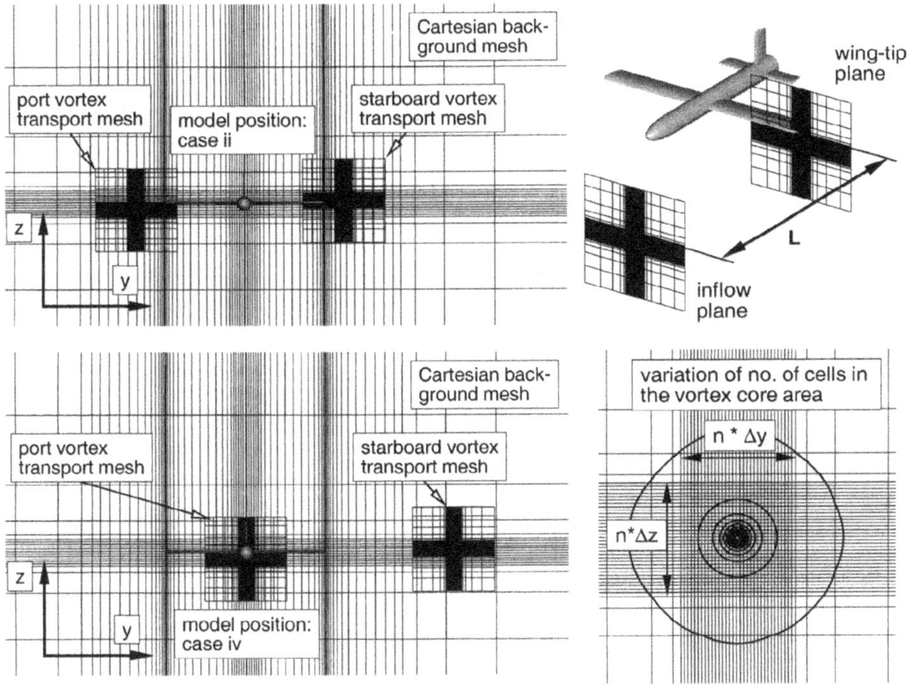

Fig. 3 Idea of vortex transport meshes (left)) and important values for the grid convergence study

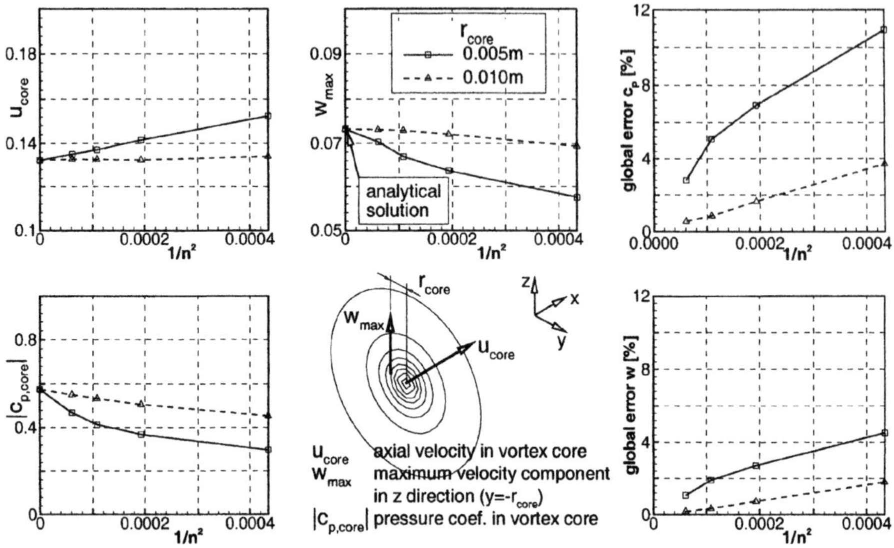

Fig. 4 Grid convergence (for axial velocity, the core pressure coefficient and maximum cross-flow velocity) and global error distribution (for the pressure coefficient and the maximum cross-flow velocity

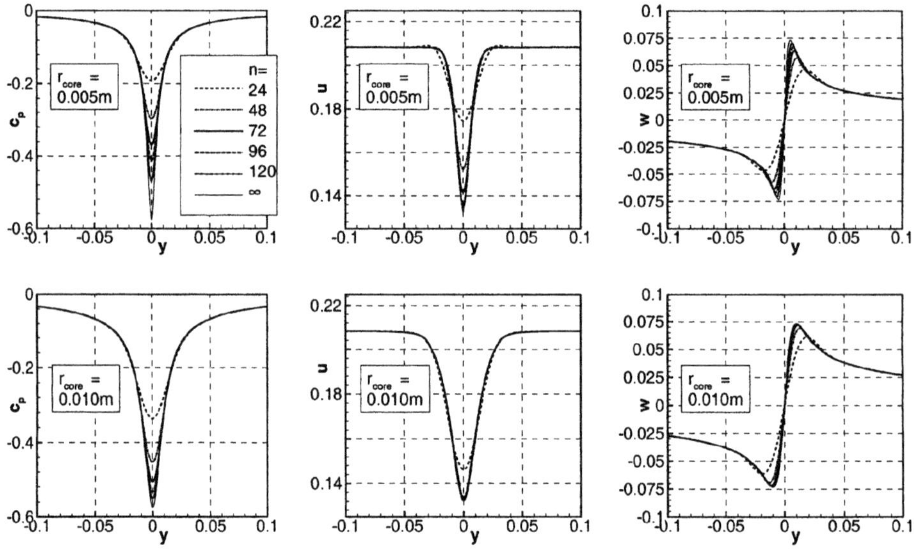

Fig. 5 Comparison of analytical (n=∞) and numerical solution in the wing tip plane for different mesh resolution (n=24, 48, 72, 96, 120) and two vortex core sizes on a line perpendicular to the vortex path

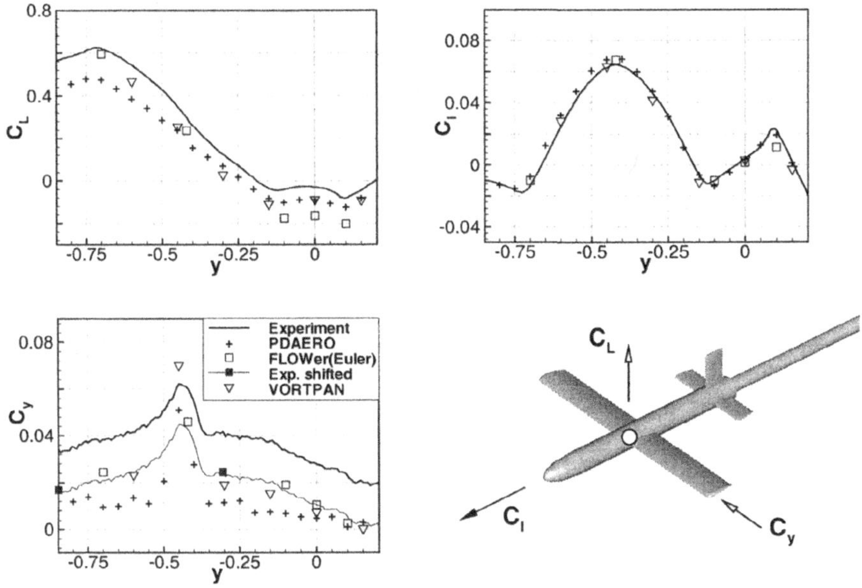

Fig. 6 Comparison of global force and moment coefficients of Euler- panel-methods and experiment

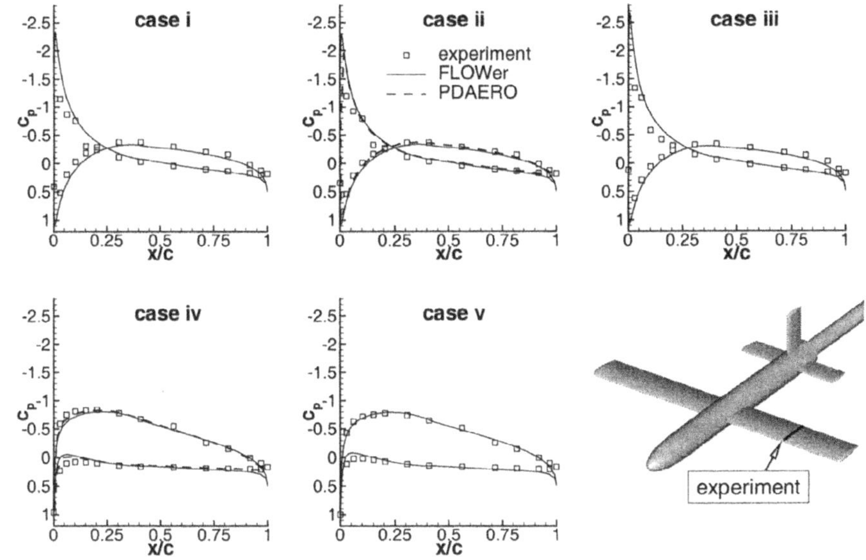

Fig. 7 Comparison of measured and calculated pressure coefficient distributions

Aspects of Low Speed Wind Tunnel Testing on an A400M Model with Propeller Simulation

M. HUHND[1], S. SCHMID-GÖLLER[2],

[1] EADS-Airbus, Germany, Experimental Aerodynamics Department
[2] EADS-Airbus, Germany, Product Aerodynamics Department
Hünefeldstr. 1-5, 28199 Bremen, Germany

Abstract

A400M is the future European military airlifter which is currently developed to fulfil the requirements of eight European Nations. The aircraft will be driven by four turboprop engines. The disc loading of the propellers is significantly higher than realised on former tactical transport aircraft like C130H or Transall C160. Due to the tactical and logistic tasks of this airplane the high and low speed propeller interference effects contribute essentially to the A/C sizing and configuration development. The following presentation deals with aspects of wind tunnel testing with propulsion simulation in the EADS Low Speed Wind Tunnel in Bremen. The simulation of realistic propeller effects is described. Investigations on an isolated propeller are illustrated and the main results of the test campaigns, focussing on propeller effects at high lift configuration are presented.

1 Introduction

Around 1995 the decision was made to use propellers instead of a typical Airbus turbofan configuration for the A400M. Due to the fact that wing mounted propellers have a significant effect on the aerodynamics of an airplane, the necessity of a realistic simulation of the whole slipstream effect of the propellers existed. The A400M (figure 1) will be driven by four modern turboprop engines with a high disc loading. Therefore large power effects especially on wing and high lift system were expected and various campaigns with the so called half model 'FLA5' have been carried out in the EADS Low Speed Wind Tunnel in Bremen.

2 Propeller Simulation on a scaled Wind Tunnel Model

For simulation of realistic power effects on a wind tunnel model it is essential to operate the scaled original propellers with the same thrust coefficient (same scaled thrust level) and the same advance ratio. The RPM at the propeller simulator have to be increased (compared to the real engine) due to the scaled diameter of the propeller. Therefore it is necessary to have an engine which is able to provide the required power at given RPM range. Additionally the size of this engine has to fit into the scaled nacelle geometry. In principle many different possibilities are conceivable. One is the use of electrical engines, but in our case the power density of such a solution was to low. Further alternatives are the use of hydraulic units or pressurized air driven axial turbines. For the half model 'FLA5.1' an air turbine unit has been developed and manufactured which was able to deliver the required power with a small turbine pressure ratio, in order to conduct the exhausted air through a return pipe system, see figure 2. This pipe system returned the expanded turbine air through the wing and out of the balanced model to the non balanced system, see figure 3. The high accuracy measurement of the actual thrust of the installed propellers is an important task because of the Thrust-Drag-Bookkeeping (see chapter 4.1) which delivers the aerodynamic data of the airplane. Therefore rotating balances have been designed which measure the thrust and the torque of the propellers. As usual on today's propeller driven aircraft on A400M all 4 propellers will have the same rotation sense. This results in an unsymmetrical flow situation. In order to simulate port and starboard wing flow with one single half model a second set of engines with opposite rotation sense were manufactured. Moreover, this allowed to study left and right wing slipstream effects separately.

3 Investigations of the Flow Field around a Propeller and Measurements of isolated Propeller Performance

3.1 Investigations of the Flow Field around the Propeller

At isolated propeller tests and during the measurements with the half model in the wind tunnel various investigations of the flow field around the propellers have been carried out. Measurements of the total pressure distribution, measurements with five-hole-probes and laser velocimeter measurements have been performed to get an impression of the flow field of the isolated propeller and the flow behavior on the wing with installed propellers at different thrust settings. Figure 4 shows as principle sketch the change of state of total pressure, static pressure and the axial velocity versus the rotating axis of the propeller. In figure 5 the real contraction of the propeller stream tube is visible by a laser light intersection. Figure 6 shows results of a velocity measurement. Here, additionally to the contraction of the stream tube, the swirl behind the propeller and the velocity distributions are presented. The swirl and axial velocity behind the propeller increase with larger thrust setting. If the slipstream with swirl interferes with the wing, a flow behavior as shown in figure 7 can be observed. In rear view the slipstream coming from a clockwise rotating propeller on a starboard wing is displaced to the wing tip above the wing and to the wing root below the wing. With these results of flow field measurements it was possible to support the interpretation of some unexpected results due to power effects. Therefore flow field measurements are inalienable tools for the understanding of such complex aerodynamics.

3.2 Measurements of the isolated Propeller Performance

To check the measurement of the rotating balance, to determine the isolated propeller performance and to investigate some other effects like a tilling propeller, a minimum body test on a sting in the wind tunnel has been carried out before starting with the half model measurements. In figure 8 the thrust measured by the rotating balance for various pitch angles of the propeller blades is shown versus RPM. The first measurement points with negative thrust were measured in wind milling conditions. Figure 9 shows the variation of thrust with various free stream velocities for one pitching angle. Usage of a non-dimensional diagram with the thrust coefficient

$$C_T = \frac{T_{\text{Propeller}}}{q_\infty \cdot A_{\text{Wing}}} \tag{1}$$

($T_{Propeller}$: Thrust, q_∞: dynamic pressure, A_{Wing}: reference wing area) versus propeller advance ratio

$$J = \frac{V_\infty}{n_{\text{Propeller}} \cdot D_{\text{Propeller}}} \tag{2}$$

(V_∞: free stream velocity, $n_{Propeller}$: RPM, $D_{Propeller}$: diameter of propeller)

accounts for main parameters influence and leads to a diagram in which all curves in figure 10 are joined to one curve for one pitch angle. Additionally to the thrust component, the torque has been measured directly by the rotating balance. Hence it was possible to calculate the shaft power, the power coefficient

$$C_P = \frac{P_{\text{Propeller}}}{q_\infty \cdot V_\infty \cdot A_{\text{Wing}}} \tag{3}$$

($P_{Propeller}$: shaft power, q_∞: dynamic pressure, V_∞: free stream velocity, A_{Wing}: reference wing area)

and finally the efficiency of the propeller which is defined as follows:

$$\eta = \frac{C_T}{C_P} \ . \tag{4}$$

Figure 11 shows the propeller efficiency for various pitch angles versus advance ratio. In this diagram the low speed range of advance ratio and thrust requirements with the corresponding efficiencies of the propeller are presented. Each pitch angle has its optimum efficiency at a specific advance ratio.

Figure 12 shows the effect of angle of attack on propeller performance. The propeller thrust increases with incidence and the value of the increments is relatively independent from the thrust level itself. Following explanation for this effect can be given: The downmoving blade of the propeller disc at angle of attack receives an increased incidence coupled with an increased approach velocity and the upmoving blade receives a decreased incidence coupled with a decreased approach velocity. Thus, with angle of attack, the increase of local thrust on the downmoving side of the propeller disc is larger than the decrease on the upmoving side. If the advance ratio is calculated using the axial component of the free stream velocity ($V_\infty \cdot \cos\alpha$), the main effect can be compensated in the thrust performance map. For the simulation of power settings of the propellers the main simulation parameters are blade pitch angle and advance ratio. During the wind tunnel tests with the model both parameters have been used to fix a specific power setting. The change of flow conditions behind the propeller at same thrust level but at different advance ratios have also been investigated during the isolated propeller test. One example is shown in figure 13: The velocity distributions for a thrust level of zero and different advance ratios versus the radius of the propeller blades. The velocity distributions are plotted as ratio to the free stream velocity of the wind tunnel. All curves shown in this figure have a mean velocity ratio of one and therefore a Net-Thrust level of zero, but a large variation of the distributions can be detected. The swirl distributions look quite similar than the velocity distributions. From this knowledge a power effect sensitivity to a variation of advance ratio at same thrust level was expected.

4 Low Speed Testing with Propeller Simulation on a Wind Tunnel Model

4.1 Thrust and Drag Bookkeeping Procedure

The wind tunnel model balance measures only total forces which act on model plus propeller forces. Therefore the investigation of propeller effects reveals the necessity of proper thrust determination and a Thrust-Drag-Bookkeeping allowing the breakdown of the aerodynamic effects. Furthermore it has to be defined which interference effects should be called "power effects". From figure 15 the basic equations can be derived. For example the drag coefficient can be calculated with

$$C_D = C_X + C_T \cdot \cos(\alpha + \sigma) - C_N \cdot \sin(\alpha + \sigma) \ . \tag{5}$$

(C_X: Value measured by WT-Balance in experimental system; C_T, C_N: Thrust and Normal propeller force coefficients in engine axis system; C_D: Aerodynamic drag coefficient in experimental system)

The other components of aerodynamic coefficients can be derived in the same way. Because of this bookkeeping it is important to determine the values of thrust with a very high accuracy. Example: The maximum thrust coefficient of one propeller amounts $C_T = 0{,}15$. If the accuracy of the thrust determination only amounts 1% of the maximum thrust level, the inaccuracy in thrust for one engine is $\Delta C_T = 0{,}0015$. For four engines the error amounts $\Delta C_T = 0{,}0060$. This means approximately 60 counts in turns of drag. The power effect of the propellers can be defined with the delta between power on and power off (blades off)

$$\Delta C_D = C_D - C_{D,POFF} = C_X + C_T \cdot \cos(\alpha + \sigma) - C_N \cdot \sin(\alpha + \sigma) - C_{D,POFF} \ . \tag{6}$$

For C_T and C_N the values directly measured by the rotating balance (apparent propeller condition) or the values calculated from the isolated performance map (isolated propeller condition) can be used. The differences between these two procedures are the installation effects created by changed propeller onset flow conditions, due to the presence of the aircraft.

4.2 Model Description and Test Arrangement

The A400M model 'FLA5.1' was a half model for development and aerodynamic data assessment investigations on the high lift system. It was designed with a scale of 1:17 to be tested in the LSWT-Bremen, see figure 14. The wing was equipped with a double slotted flap/tap system. On the wing nose region additionally to the clean leading edge slats or nose droop devices could be assembled. All wing elements were pressure plotted for load determination, for high lift wing development and for a better understanding of the power effects. As described before, the model was equipped with turboprop engine simulators with the original nacelle shape. The blades of the propellers could be disassembled for power off investigations. Two sets of propellers and turbines were available with different senses of rotation, to investigate the unsymmetrical behavior of the port and starboard wing. Various configurations have been measured in power off condition and with a matrix of thrust settings and advance ratios. In the analyses of the measurements the isolated propeller values determined by the isolated propeller performance map and the apparent propeller values measured by the rotating balance were used. Furthermore the data have been interpolated to constant apparent thrust conditions to eliminate small variations of the thrust adjustment between the polars. Additionally the measurements with both senses of rotation which simulated the port and starboard wing have been joined to one polar for the complete aircraft. Because of pitch angle, sense of rotation and advance ratio as additional parameters compared with turbo fan interference investigations, significantly more test days in the wind tunnel were necessary.

5 Slipstream Effects of Wing Mounted Propellers on High Lift Configurations

The following part of this paper illustrates the main power effects which were measured during the low speed investigations with the FLA5.1-Model.

5.1 Effect of Thrust Setting

Figure 16 shows the slipstream effect on lift for clean configuration and 2nd take-off configuration. The influence on the gradient of the $C_L(\alpha)$ polar and the increase in C_L is clearly visible. The maximum value of $C_{L,max}$ in landing configuratin with leading edge devices at high power setting was determined with $C_{L,max} > 4$. Principle effects of the slipstream on the wing are caused by axial flow acceleration and swirl which affect local flow velocity/Ma-number and incidence. Figure 17 shows the power effect on drag for the second take-off configuration. In figure 18 qualitative the local lift distribution versus the wing span for power off and for a high C_T-Level is shown which illustrates both, the effect of the overspeed and the effect of swirl.

5.2 Effect of Advance Ratio

As mentioned in chapter 3.2, the velocity and swirl distributions change at the same thrust level with the variation of advance ratio. These variations in the slipstream of the propellers produce slightly different flow conditions for the wing and therefore a variation of the power effect. Figure 19 shows polars with the same thrust setting and various advance ratios. The drag increases with decreased advance ratio (lower pitch angle).

5.3 Effect of Rotation Sense

With the two different propeller rotation senses it was possible to simulate the port and starboard wing with only one half model. Asymmetry of flow which creates different power effects on each half wing are from particular interest. The results are rolling and yawing moments which have to be trimmed if one sense of rotation is used for both sides of the wing. Figure 20 shows the power off polar and the polars with $C_T = 0$ for clockwise and counter clockwise rotation. A large asymmetry in $C_{L,max}$ and α_{max} can be detected which leads to asymmetric stall characteristics.

5.4 Effect on Wing Stall Characteristics

Figure 21 illustrates the stall characteristics detected by a flow visualization with tufts on the upper side of the wing for power off and power on conditions. The behavior of the local separation clearly shows the effect of the swirl. At the wing section of the upward rotating blade, the local angle of attack is increased and for the wing section of the downward rotating blade, the local angle of attack is decreased by the swirl within the slipstream. Measurements with power off conditions have shown that the inboard side of the wing/nacelle junction is the critical section where the wing separation starts. At the starboard wing, with higher thrust levels, this behavior is increased by the increase of the local angle of attack in this region. On the port wing, higher thrust levels decreases the local angle of attack in this region. This defuses the critical inboard side of the wing/nacelle section and the separation starts at the outboard engine on the outboard side where the local angle of attack is increased by the swirl. Only counter rotating propellers (low swirl) or different senses of rotation on each side of the wing are the principles to avoid this asymmetrical stall behavior.

5.5 Effect of Aircraft on the Propeller Performance

In chapter 3.2 which describes the isolated propeller investigations, the effects of velocity and incidence on the propeller performance are illustrated. When propellers are installed on the aircraft, the onset flow conditions change, due to the upwash of the aircraft. Hence at the locations of the propeller disc local velocity vector depends on high-lift configuration and angle of attack.

Figure 22 shows the power-off polar and the power-on polar for landing configuration. The power-on polars have been determined with isolated and apparent thrust (chapter 4.1). The difference of these two analyses can be defined as installation effects on the propellers. In figure 23 a RPM sweep at constant blade pitch angle in clean configuration and constant angle of attack is shown (no incidence of the propeller axis). Again the C_T was determined from the isolated performance map of the propeller and from the measured apparent thrust of the rotating balance. The change on propeller performance is clearly visible.

6. Conclusions

Investigations on power effects have shown that the slipstream of wing mounted propellers have a significant impact on the low speed aircraft characteristics. Especially the increments in lift and drag, the asymmetrical behavior of the port and starboard wing and the stall characteristics were very important results of the campaigns. The measured effects were larger than expected and today there are no reliable theoretical methods to fully predict these power effects. The installation effects between the isolated and apparent propeller performance show the importance of installed propeller force measurement. Additional flow field measurements have shown the complexity of the slipstream/wing interactions and were an important tool for the understanding of the phenomena. Large technical efforts were necessary to establish a realistic propulsion simulation with high data accuracy. Further tests are planned with an updated wind tunnel model which allows to be tested in high Reynolds number conditions.

Acknowledgement

The authors would like to thank all people who were involved in the FLA5-Campaigns. Special thanks to Mr. J. Rohde and his team of the Gasdynamic Test Facility, who were responsible for the development, manufacturing and operation of the whole propulsion systems, and for the flow field measurements. Special thanks to Mr. E. Graewe, Mr. B. Berdrow and their team, who were responsible for the development, manufacturing, calibration and operation of the rotating balances and for the flow field measurements. Also thanks to the Wind Tunnel and the Model Workshop teams.

Figure 1 A400M, Airdrop

Figure 2 Installation of engines on model
(1 air turbines, 2 wing, 3 turbine exhaused pipes)

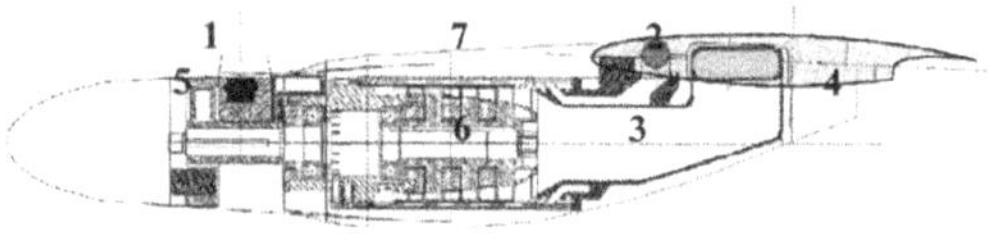

Figure 3 Propulsion simulator on model FLA5.1
(1 Propeller Disc, 2 Air Supply Pipe, 3 Exhaust Pipe, 4 Wing, 5 Rotating Balance, 6 Turbine, 7 Nacelle, Diameter of turb.: 75mm, Power: 40kW, RPM 15000min^{-1})

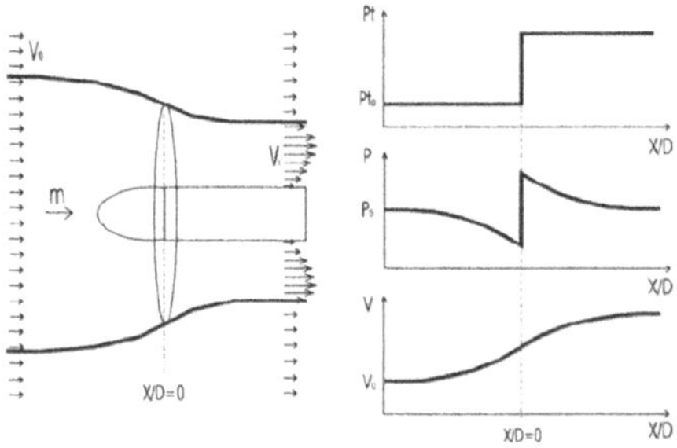

Figure 4 Change of mean values within the stream tube of a propeller

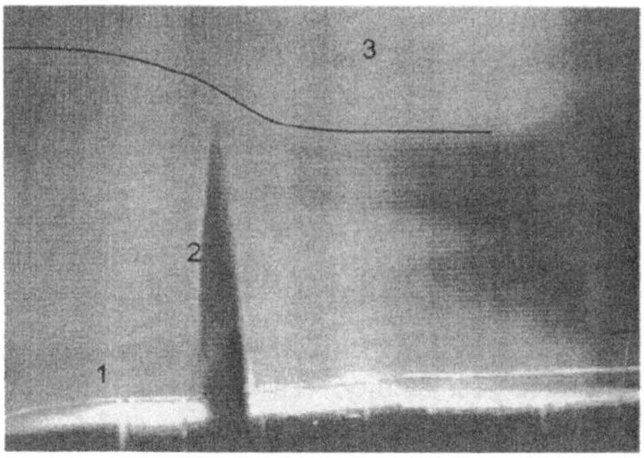

Figure 5 Contraction of stream tube
(1 Spinner, 2 Propeller, 3 Laser Light Intersection)

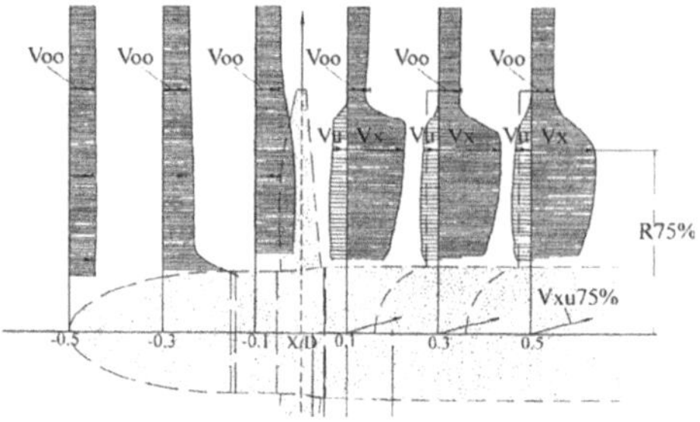

Figure 6 Measured axial velocity and swirl distributions around the propeller
(axial velocity: Vx, radial velocity: Vu)

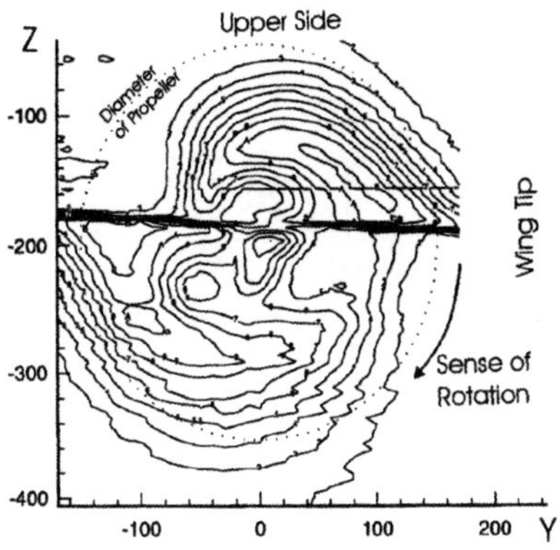

Figure 7 Propeller slipstream behind the clean wing (isobars of total pressure)

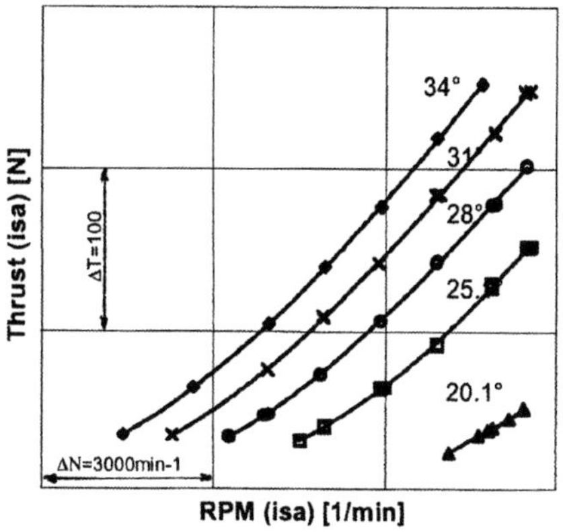

Figure 8 Thrust versus RPM for various blade pitch angles

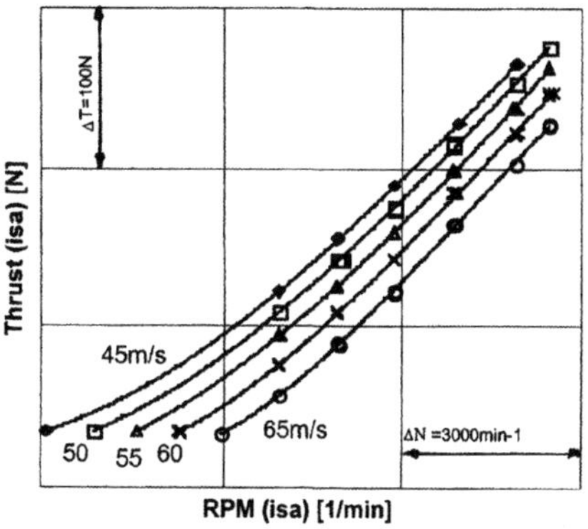

Figure 9 Thrust versus RPM for various free stream velocities ($ß_{75\%}$=const.)

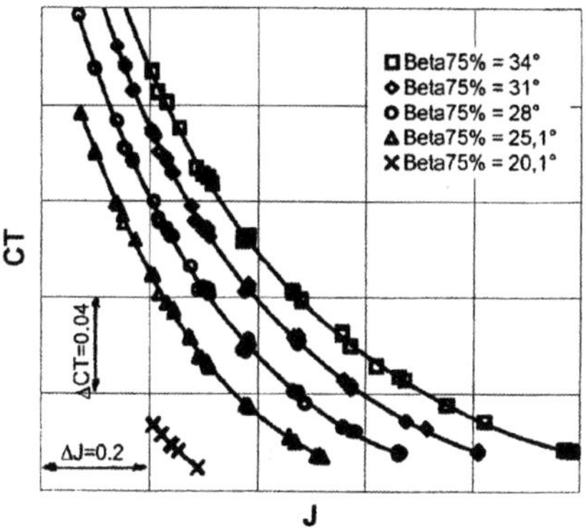

Figure 10 Thrust coefficient versus advance ratio for various pitch angles

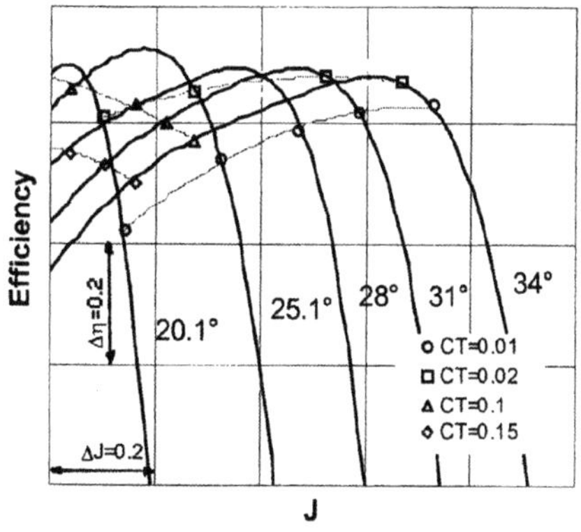

Figure 11 Propeller efficiency versus advance ratio for various pitch angles

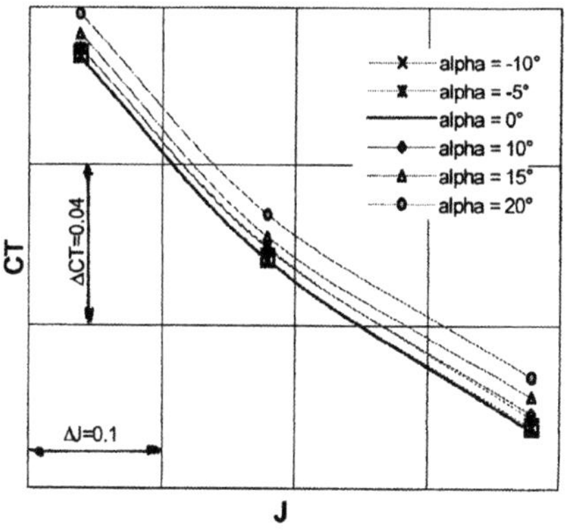

Figure 12 Effect of angle of attack on propeller performance

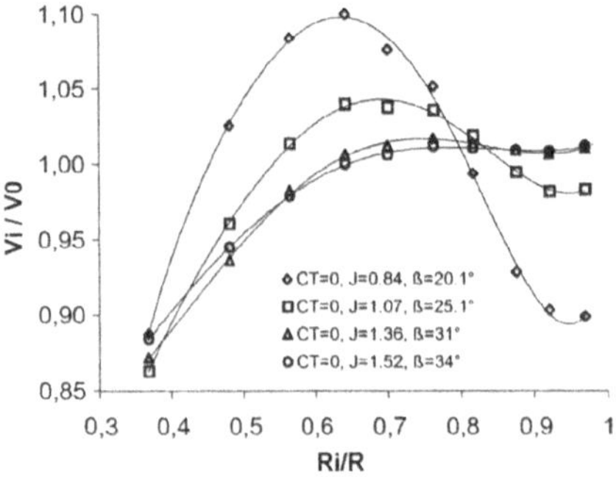

Figure 13 Velocity distributions versus blade radius for various J, CT = 0 = const.

Figure 14 Model FLA5.1 within the test section of the LSWT-Bremen

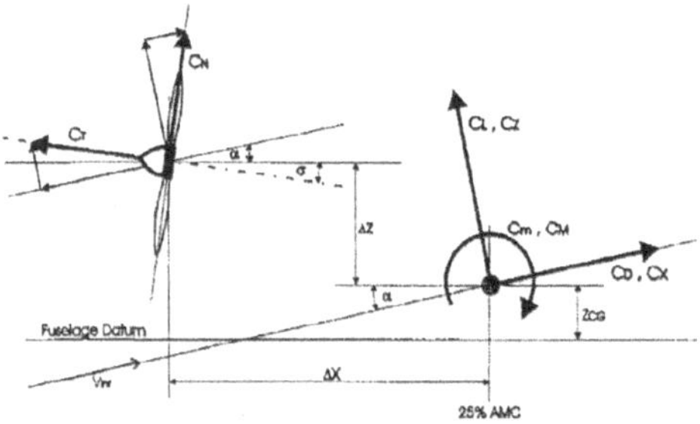

Figure 15 Scheme for Thrust-Drag-Bookkeeping procedure

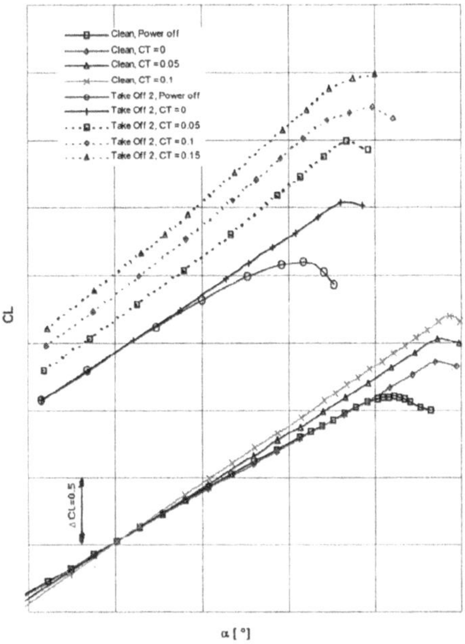

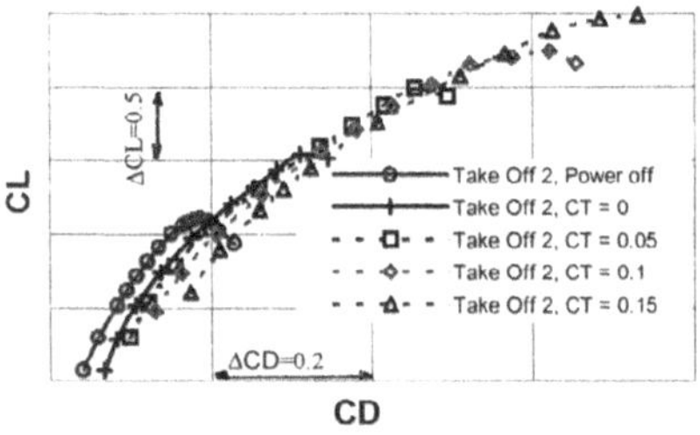

Figure 17 Power effect on drag (J=const.)

Figure 16 Power effect on lift (J=const.)

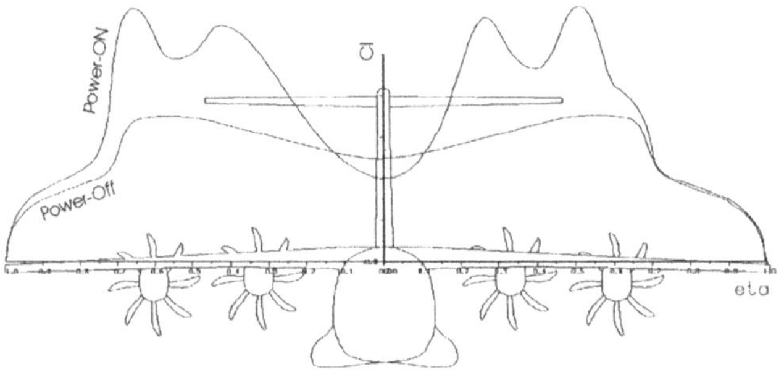

Figure 18 Typical power effect on local lift versus wing span (clockwise propeller rotation)

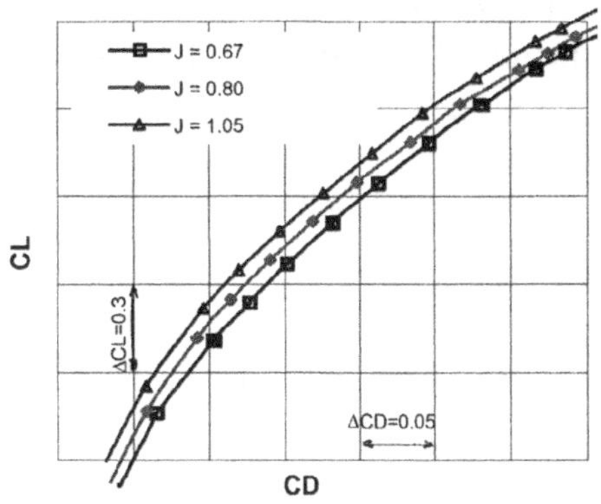

Figure 19 Effect of advance ratio at constant thrust level

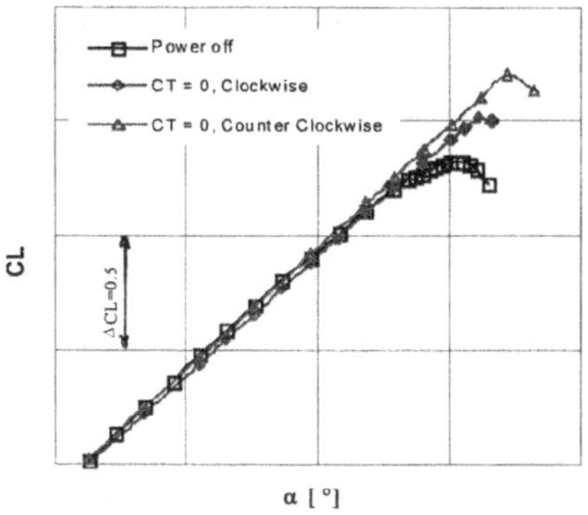

Figure 20 Effect of sense of rotation on CL, simulation of port and starboard wing

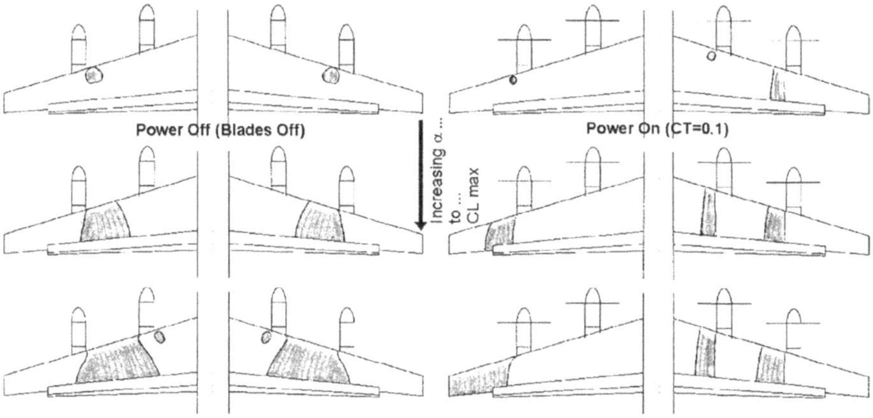

Figure 21 Local wing separation for power off and power on condition

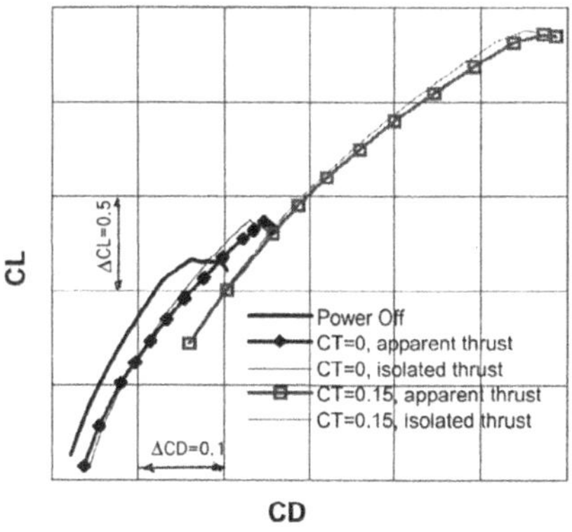

Figure 22 Thrust-Drag-Bookkeeping with isolated and apparent thrust condition

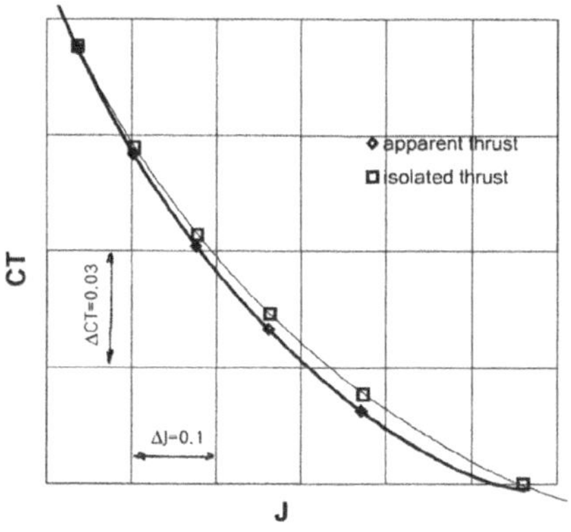

Figure 23 Effect of aircraft on propeller performance (RPM-Sweep, α=const., β=const.)

Coupling of DLR Navier-Stokes Solver FLOWer with an e^N-Database Method for Laminar-Turbulent Transition Prediction on Airfoils

A. Krumbein
German Aerospace Center (DLR), Institute of Design Aerodynamics
Lilienthalplatz 7, 38108 Braunschweig, Germany
e-mail: andreas.krumbein@dlr.de

Summary

This paper presents the strategy and application of the coupling of a Navier-Stokes solver with a transition prediction method. For industrial applications the coupling of the solver to an e^N-database method, based on local linear stability theory, results in a program system that is able to automatically handle transition prediction. The solver was coupled to a transition prediction module, consisting of a boundary layer method for swept, tapered wings and engineering transition prediction methods, a.) two algebraic, empirical transition criteria and b.) an e^N-database method for Tollmien-Schlichting waves. The sensitivity of the coupled system with regard to a change of system parameters was investigated. It could be shown that for some of the parameters the sensitivity does not exist and that it is very low for the others.

1 Introduction

The modeling of laminar-turbulent transition in Reynolds averaged Navier-Stokes solvers (RANS) is a necessary requirement for the computation of wings in aerospace industry. In many cases it is not possible to obtain quantitatively correct results, if the laminar-turbulent transition is not taken into account, e.g. the drag of the aircraft, or even qualitatively correct results such as suction peaks at the leading edge of wings at high angles of attack (take-off/landing condition).

A laminar-turbulent transition modeling consists of two major parts, the transition prediction and the transition prescription. The transition prediction determines the transition locations on the surface of the configuration. The transition prescription applies the determined transition locations in the flow solver and thus brings the information from the transition prediction into the solution process of the RANS equations.

In this paper, results obtained with the RANS flow solver FLOWer, [1], of the German Aerospace Center (DLR) are presented. FLOWer works with structured multiblock meshes and solves the RANS equations for 3-dimensional, compressible, steady and unsteady flows on body-fitted meshes of any topology. The space discretization is based upon the finite volume method and uses a cell-vertex approach and central differences to approximate the convective fluxes. The viscous fluxes are approximated by a node-centered-scheme using a compact cell and central differences. The discrete time integration is done by explicit hybrid multi-stage Runge-Kutta schemes using coefficients optimized with respect to numerical stability and smoothing properties and is accelerated by multigrid-techniques. In order to more efficiently solve low Mach number flows a pre-conditioning option is available. For turbulent calculations the algebraic Baldwin-Lomax model, the one-equation Spalart-Allmaras and the two equation k-ω transport models can be selected. The implementation of a coupling structure, which connects the flow solver to a transition prediction method and which is presented in this paper has been partly realized in the German research project MEGAFLOW, [2].

2 Transition Prediction

By 'pure' transition prescription the transition locations are imposed, [3-4]. Transition prediction is the next step of a transition modeling. Numerous transition prediction methods are available, ranging in cost and complexity from simple empirical transition criteria via local linear, non-local linear and non-local non-linear stability methods to direct numerical

simulations. Currently, the e^N-method based on local linear stability theory and the parallel flow assumption is used in aircraft industry for the design of laminar wings most frequently.
Transition prediction is an uncommon feature in most RANS solvers. A few evaluate empirical transition criteria or simplified analytical relations, some apply linear stability codes manually after having reached a preliminary solution of the RANS equations. The use of a transport equation approach is presented in [17].
The computation of flows with transition prediction in industrial applications using a RANS solver must be performed automatically and autonomously, i.e. without external interference by the user and without specific knowledge about the transition process itself (black box). For wings and tails of transport aircraft the transition locations can be determined during the run of the RANS solver using a transition prediction module coupled to the solver. For testing purposes, the transition prediction module was firstly provided with an empirical transition criterion, [5]. For industrial applications the coupling of the solver to an e^N-method, [6-7], provides a solution, that promises sufficiently high accuracy of the predicted transition locations. The use of an e^N-database method, [8], results in a program system that is able to automatically handle transition prediction. A weak coupling between a RANS solver and an e^N-database method, that are used separately one after the other with manual interference of the user, is performed in [7-8], [12] and [16]. In this paper, results of 2-dimensional configurations applying a strongly coupled program system, that is able to handle transition prediction without interruption of the solution process, are presented.

2.1 Coupling Structure

In order to compute transition on wings of transport aircraft FLOWer was coupled to a transition prediction module which is called after a certain number of iteration cycles, $k_{cyc.}$, of the RANS solution process. With the call of the module the solution process is interrupted and the module analyses the laminar boundary layers of previously specified components of the configuration, e.g. of an 2-dimensional airfoil or a wing section. The determined transition locations, $x_j^T(cycle = k_{cyc.})$ with $j = 1, \ldots, n_{loc.}$, $n_{loc.}$ being the number of transition points, are communicated to the RANS solver, which performs transition prescription applying an transition setting algorithm, [4], and continues the solution process of the RANS equations. In so doing, the determination of the transition locations becomes an iteration process itself. The structure of the approach is outlined graphically in figure 1.
The transition prediction module consists of a compressible boundary layer method for swept, tapered wings, [9], and engineering transition prediction methods, a.) two algebraic, empirical transition criteria, [4], and b.) an e^N-database method for Tollmien-Schlichting waves, [8].
At every call of the module the surface pressure, $c_p(cycle = k_{cyc.})$, along an airfoil computed by the RANS solver is used as input to the boundary layer calculation. The viscous data calculated by the boundary layer method are then subsequently analysed by the database method or an empirical transition criterion is evaluated at every surface point.
The application of a boundary layer method for the computation of all viscous data necessary for the transition prediction method ensures the high accuracy of the visous data required by the e^N-methods for the analysis of the laminar boundary layers. Thus, as shown in [7], the large number of grid points near the wall for a high resolution of the boundary layers, the adaptation of the Navier-Stokes grid in the laminar and turbulent boundary layer regions and the generation of a new adapted grid for the RANS solver after every step of the transition location iteration are avoided and the computational time can be massively reduced.

2.2 Transition Prediction Algorithm

a) The RANS solver is started as if a computation with prescribed transition locations should be performed with transition locations set far downstream on upper and lower side of the airfoil, e.g. at the trailing edge. The RANS solver now computes a fully laminar flow over the airfoil.

b) During the solution process of the RANS equations the laminar flow is checked for laminar separation. If laminar separation is detected, the separation point is used as approximation of the transition location and the computation is continued.

c) The RANS equations are iterated until the lift, or the lift coefficient c_l respectively, which can be represented as a function of the iteration cycles, $c_l = c_l(cycles)$, has become constant with respect to the iteration cycles.

d) The transition prediction module is called.

e) The determined transition locations $x_j^T(cycle = k_{cyc.})$ are underrelaxed, i.e. as new transition locations $\tilde{x}_j^T(cycle = k_{cyc.})$ coordinates located downstream of the coordinates $x_j^T(cycle = k_{cyc.})$ are used,

$$\tilde{x}^T(k_{cyc.}) = C_j^T(k_{cyc.})\, x_j^T(k_{cyc.}) \quad with \quad j = 1, \ldots, n_{loc.}, \tag{1}$$

with $C_j^T(k_{cyc.}) > 1$. Only after the last step of the transition location iteration $C_j^T(k_{cyc.}) = 1$ is applied. The underrelaxation of the determined transition locations prevents the case that at an unconverged stage during the transition location iteration transition coordinates are determined too far upstream which might not be shifted downstream again.

f) As convergence criterion $\Delta \tilde{x}_j^{T,l} < \varepsilon$ with $\Delta \tilde{x}_j^{T,l} = | \tilde{x}_j^T(k^l_{cyc.}) - \tilde{x}_j^T(k^{l-1}_{cyc.}) |$ is applied, l being the current iteration step. In the case that the criterion is satisfied, the iteration for x_j^T is finished, else the algorithm loops back to station b).

2.3 Empirical Criteria

Numerous empirical transition criteria are available in the literature, some of them taking into account pressure gradient effects, others include free-stream turbulence effects or both. All these criteria are correlations between boundary layer parameters at the location of transition onset, [5]. Two empirical criteria were tested. The first one is the criterion by Michel, [10], a correlation of Re_{δ_2}, the Reynolds number with respect to the momentum loss thickness δ_2, and Re_{x^T}, the Reynolds number with respect to the chordwise coordinate x^T, for two-dimensional flows over airfoils including pressure gradient effects. The pressure gradient is implicitly included in Re_{δ_2}. The second criterion by van Driest and Blumer, [11], includes pressure gradient as well as free-stream turbulence effects.

2.4 e^N-Database Method

The e^N-method used, [8], applies $N|_{x^T} = -\int_{x_0}^{x^T} \alpha_i \, dx$ as transition criterion. The limiting N-factor $N|_{x^T}$ at the transition location x^T must be determined experimentally and represents the total amplification of a perturbation of the mean flow of frequency f at the transition location, x_0, the x-coordinate of the point where this perturbation enters the unstable zone, and α_i, the local spatial amplification rate. α_i can be expressed as a function of the shape parameter H_{12}, Re_{δ_1}, the Reynolds number with respect to the displacement thickness δ_1, and the reduced frequency F, $\alpha_i = \alpha_i(H_{12}, Re_{\delta_1}, F)$. As described in [8] and [12], for 13 shape parameters H_{12}, which cover the range from highly accelerated up to separating flows, the boundary layers of the similarity solution using the approach of Falkner and Skan were generated. The growth of the boundary layer was simulated by varying Re_{δ_1}. For each boundary layer, stability computations were completed for a sufficient large range of excited frequencies. The results for

the amplification rates α_i were stored in a database. The stability computation for a real boundary layer, using the database method, is executed for a given frequency f in Hz in the following way: At each grid point on the airfoil the properties f, Re_{δ_1}, H_{12}, U_e, μ_e and ρ_e are known. Evaluating F from the value of f, α_i is obtained from $\alpha_i = \alpha_i(H_{12}, Re_{\delta_1}, F)$ via interpolation in the database. In [8] and [12] it was shown that the use of the database is a high quality approximation for the values resulting from a local linear stability code. In [7-8], [12] and [16] it was shown that the Falkner-Skan approach approximates all relevant properties of the 2-dimensional laminar boundary layer flow around airfoils in practice with high accuracy.
For the application of the e^N-method extensive experimental programs have been performed to determine the value of $N|_{x^T}$ which depends on the environmental disturbances in the flow. For free flight conditions $N|_{x^T} \approx 12$ was found for Tollmien-Schlichting waves. For wind tunnel flows, which are normally characterised by relatively high disturbances, one finds $N|_{x^T} \approx 6$.

3 Results

The transition prediction algorithm has been applied to the natural laminar airfoil of Somers, [15], using a 1 block mesh with 332 points in tangential direction and 72 in wall normal direction, as shown in figure 2. The aerodynamic parameters are Mach number $Ma = 0.1$ and Reynolds number $Re = 4\ 10^6$. In the computations the algebraic Baldwin-Lomax turbulence model and the k-ω transport equation model were used. First, the application of the transition prediction algorithm is illustrated for a single angle of attack, $\alpha = -6.2°$, in a computation with the Baldwin-Lomax model. The prediction of the transition locations was performed using the database method applying $N|_{x^T} = 11$. This value, that represents the environmental disturbances of the low turbulence wind tunnel where the measurements on the natural laminar airfoil of Somers have been carried out, was determined by Stock, [16], by comparison of the experimental data in [15] with results of linear stability calculations.
Figure 3 shows the iteration process of the transition locations. The initial transition locations were set at 75% of chord on upper and lower side of the airfoil at the beginning of the computation. The database was activated only when the RANS solution had been converged with temporarily fixed transition locations which were changed afterwards by the next call of the transition prediction module. It shows the convergence of the transition locations and the values of the force coefficients in the moments when the database was activated. The two curves of the transition locations start with the preset values from the start of the RANS solver. After 100 iteration cycles, for the first time the flow, which is computed as a laminar flow from the leading edge up to 75% of chord on upper and lower side, was checked for laminar separation. Laminar separation was found on the lower side at 2% of chord and this location is now used as approximation of the transition location on the lower side. After 250 cycles the database was activated for the first time. During the ongoing RANS solution only on the upper side excited frequencies reaching $N|_{x^T} = 11$ were found. All the transition locations determined directly by the database were set underrelaxed before the RANS solution continued. After its last call the value $x^T_{upp} = 0.4983$ was set without underrelaxation for a converged solution of the RANS iterations. Figure 4 shows the $c_l = c_l(c_d)$-polars of the experiment with free transition, [15], of fully turbulent computations and of computations with transition using the transition prediction algorithm and the database. For both turbulence models a strong improvement of the computational results is achieved. For all points, there is a clear tendency towards the experimental data. The drag deviations at high values of α ($\alpha \geq 6°$) and for the k-ω model in the range of $0° \leq \alpha \leq -2°$ still have to be investigated. Figure 5 compares the

experimentally determined transition locations, as given in [15], with the transition locations determined by the transition prediction algorithm using the database and the k-ω model. Indicated by filled, black symbols are the transition locations determined by the database. The hollow, black symbols mark the transition locations approximated by the x-coordinates of laminar separation, whose locations were detected by the laminar boundary layer method, [9]. The results of the computations using the Baldwin-Lomax model are very similar, [4]. Figure 6 shows parts of the $c_d(\alpha)$-polars – experimental, fully turbulent and with transition determined by the database – in a range of the angle of attack $-6° \leq \alpha \leq 10°$. Also shown are results of computations using the Baldwin-Lomax model in that the transition locations have been determined by the two empirical transition criteria. Criterion 1 (small circles) is the one given by Michel, [10], criterion 2 (small gradients) the one by van Driest and Blumer, [11]. In contrast to the application of the database, the transition prediction algorithm using the two empirical criteria resulted in a very short iteration process. While for the database the transition module was called about 4 to 6 times to reach a converged state of the transition prediction process, the empirical criteria detected unchanged values of the transition locations after 2 to 3 calls.

The convergence history of the coupled system in terms of the RANS solver (e.g. ρ-residual) and the $c_l(\alpha)$-polars are presented in [4].

For the coupled system using the database a sensitivity study with respect to a change of system parameters has been performed. The changed system parameters are the initial transition locations, $x_{0j}^T = x_j^T(cycle = 0.)$, the intervals of RANS cycles $k_{cyc.,ls}$ for the check for laminar separation, the intervals of RANS cycles $k_{cyc.,cl}$ for the call of the transition module and different underrelaxation factors $C_j^T(k_{cyc.})$ in equation (1). The studies were made for the case $\alpha = -6.2°$ with the Baldwin-Lomax model.

Figure 7 shows that there is no influence of different initial values for the transition locations x_{0j}^T on the computed transition locations and the iteration process. The computations, that have followed exactly the same strategy as described above and illustrated in figure 3 were performed with $x^T_{0,upp}/c = x^T_{0,low}/c = 0.65, 0.75, 0.85, 0.95, 1.00$. In figure 8 the missing dependence on the value of $k_{cyc.,ls}$ with $k_{cyc.,ls} = 100, 50, 10$ and $x^T_{0,upp}/c = x^T_{0,low}/c = 1.00$ with respect to the transition locations after the first call of the transition module is shown. A comparison of the c_l convergence histories, however, shows a stabilization of the RANS iteration the earlier the laminar separation is detected. Also a variation of $k_{cyc.,cl}$ with $k_{cyc.,cl} = 300, 250, 200, 150, 100, 50$ and $k_{cyc.,ls} = 100, 50, 10$ does not affect the values of the transition locations after the first call, as shown in figure 9, although in the cases for $k_{cyc.,cl} = 50$ and for the combination $k_{cyc.,cl} = 100$ and $k_{cyc.,ls} = 100$ the c_l convergence history is still in the transient phase and has high gradients. For $k_{cyc.,cl} = 100, 50$ and $k_{cyc.,ls} = 50$ the computations have been completed up to a converged state. The iteration ways are very slightly different, the results for $k_{cyc.,cl} = 50$ predict the transition location slightly upstream compared to the case $k_{cyc.,cl} = 100$, figure 10. Finally, differences can be found when lower values of the underrelaxation factors are applied. Figure 11 shows that the iteration converges faster and that the value of the converged transition location is lower than in the case of high underrelaxation factors.

4 Conclusions

For transition prediction the RANS solver FLOWer has been coupled with a transition prediction module consisting of a laminar boundary layer method and a transition prediction method. The coupling structure and the transition prediction algorithm were succesfully applied to a 2d flow over the natural laminar Somers airfoil. By the use of an e^N-database method for Tollmien-Schlichting waves the transition locations were predicted with excellent accuracy compared to all experimentally determined transition coordinates. The predictions were performed for computations using different turbulence models (algebraic Baldwin-Lomax model and k-ω transport equation model) and showed no dependence on the turbulence models. The application of empirical transition criteria in computations with the Baldwin-Lomax model did not yield satisfactory results and with respect to the accuracy of the prediction transition locations the database is superior to the empirical criteria. A sensitivity study for a variety of parameters of the coupled system using the database showed that the system does not depend at all on the change of the parameters or only very slightly. Different solutions were found only for different underrelaxation factors which result in transition points located too far upstream for too low factors.

The next steps for the future are the application of the presented transition prediction algorithm to a transonic testcase and the investigation of the stability of the coupled system to shock/boundary layer interaction and to a 2d multi-component high lift configuration. The computation of both testcases will be carried out using the values of $N|_{xT}$, that have been determined in corresponding experiments.

References

[1] FLOWer – Installation and User Handbook, Release 115, Doc.Nr. MEGAFLOW-1001, Institut für Entwurfsaerodynamik, Deutsches Zentrum für Luft- und Raumfahrt e.V., 1998

[2] Becker, K.; Kroll, N.; Rossow, C.C.; Thiele, F.: "The MEGAFLOW project", *Aerosp. Sci. Technol.* 4 (2000) 223-237

[3] Krumbein, A.: AVTAC Advanced Viscous Flow Simulation Tools for Complete Civil Aircraft Design - Transition Prescription and Prediction, Deliverable Task 3.2, AVTAC/DEL/DLR/D3.2C5, July 1999

[4] Krumbein, A., Stock, H. W.: "Laminar-turbulent Transition Modeling in Navier-Stokes Solvers using Engineering Methods", CD-ROM proceedings of the European Congress on Computational Methods in Applied Sciences and Engineering ECCOMAS 2000, Computational Fluid Dynamics, Barcelona, 2000

[5] Arnal, D: "Description and Prediction of Transition in Two-Dimensional, Incompressible Flow; in Special Course on Stability and Transition of Laminar Flow", *AGARD Report No.709*, Pages 2-1 - 2-71, 1984

[6] Radespiel, R.; Graage, K.; Brodersen, O.: "Transition Predictions Using Reynolds-Averaged Navier-Stokes and Linear Stability Analysis Methods", *AIAA 91-1641*, 1991

[7] Stock, H. W.; Haase, W.: "A Feasibility Study of e^N Transition Prediction in Navier-Stokes Methods for Airfoils", *AIAA J.*, **Vol.37**, pp.1187-1196, 1999

[8] Stock, H. W.; Degenhardt, E.: "A simplified e^N method for transition prediction in two-dimensional, incompressible boundary layers", *Z. Flugwiss. Weltraumforsch.*, **Vol.13**, pp.16-30, 1989

[9] Horton, H. P.; Stock, H. W.: "Computation of Compressible, Laminar Boundary Layers on Swept, Tapered Wings", *J. Aircraft*, **Vol.32**, pp.1402-1405, 1995

[10] Michel, R.: "Détermination du point du transition et calcul de la traînée des profils d'aile en incompressible", ONERA publication No. 58, 1952

[11] Blumer, C. B.; van Driest, E. R.: "Boundary layer transition: Free-stream turbulence and pressure gradient effects", *AIAA Journal*, Vol. 1, No. 6, 1963

[12] Stock, H. W.; Haase, W.: "Some Aspects of Linear Stability Calculations in Industrial Applications"; in Henkes, R. A. W. M.; van Ingen, J.L.: *Transitional Boundary Layers in Aeronautics*, pp.225-238, 1996

[13] Moir, I. R. M.: "Measurements on a Two-Dimensional Aerofoil with High-Lift Devices", *AGARD Report No.303*, Pages A2-1 - 2-12, 1994

[14] Wild, J. W.: "Direct Optimization of Multi-Element-Airfoils for High-Lift using Navier-Stokes Equations", Computational Fluid Dynamics '98, Proceedings of the Fourth European Computational Fluid Dynamics Conference, pp. 383, 1998

[15] Somers, D. A.: "Design and Experimental Results for a Natural-Laminar Flow Airfoil for General Aviation Applications", NASA Technical Paper 1861, Scientific and Technical Information

Branch, 1981
[16] Stock, H. W.: "Navier-Stokes Computations of Laminar Airfoils Using e^N Transition Prediction", DLR-Interner Bericht, DLR, German Aerospace Center, Braunschweig, Germany, Rept. IB 129-99/18, 1999
[17] Hassan, H. A., Warren, E. W.: "Alternative to the e^N Method for Determining Onset of Transition", *AIAA Journal*, Vol. 36, No. 1: Technical Notes, pp. 111-113, 1997

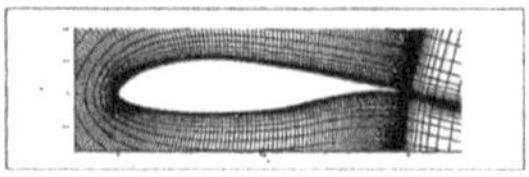

Figure 2: Natural laminar airfoil of Somers

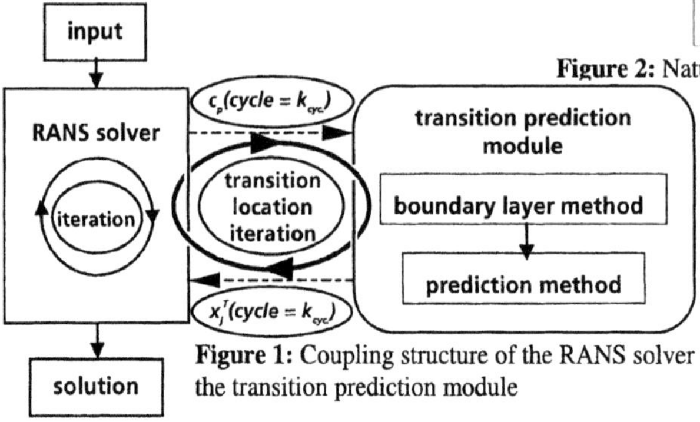

Figure 1: Coupling structure of the RANS solver and the transition prediction module

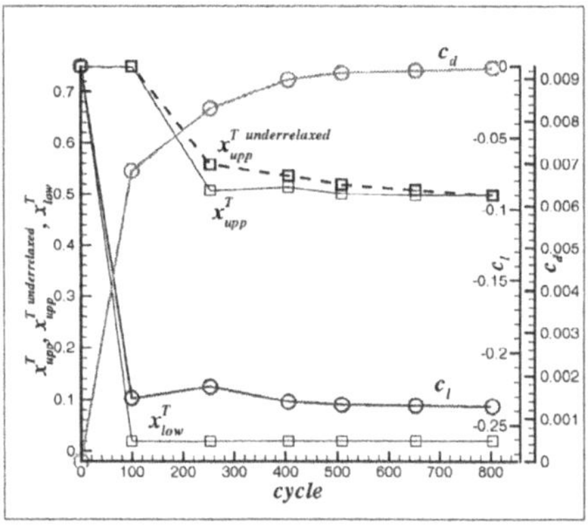

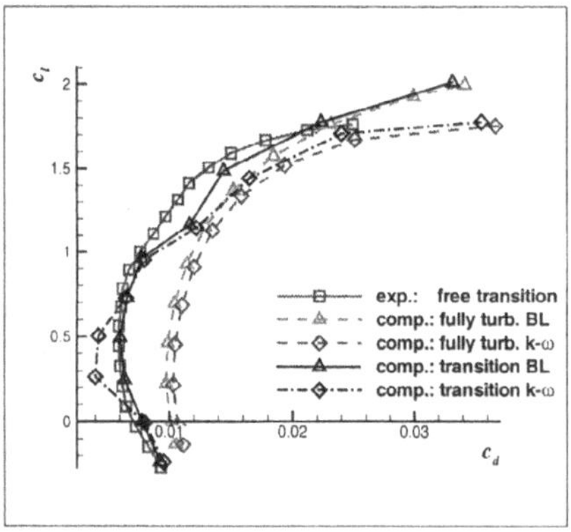

Figure 3: Convergence history of the iteration process of the transition locations
Figure 4: $c_l = c_l(c_d)$-polars of the Somers-airfoil

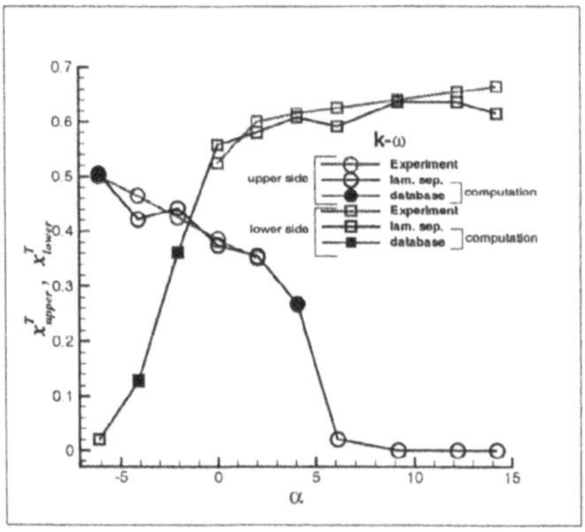

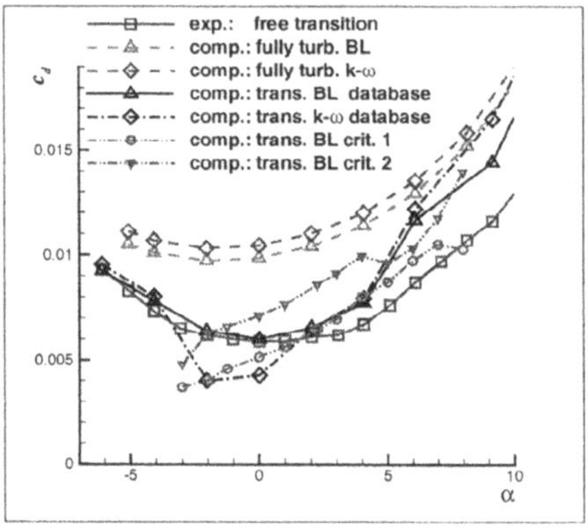

Figure 5: Experimental and computational transition locations of the Somers-airfoil (*k*-ω)
Figure 6: Comparison of results of the database and two empirical transition criteria, $c_d = c_d(\alpha)$

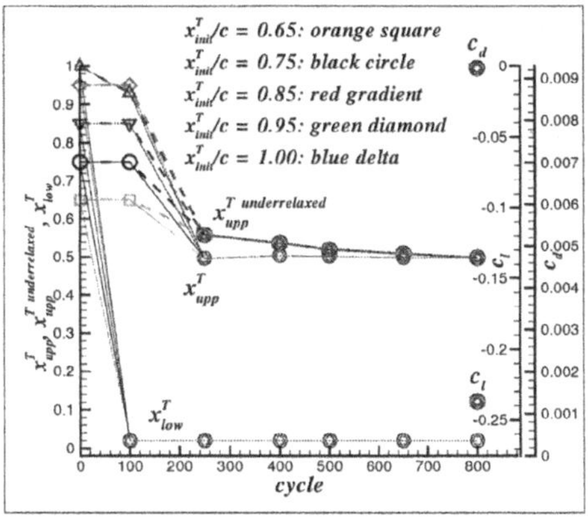

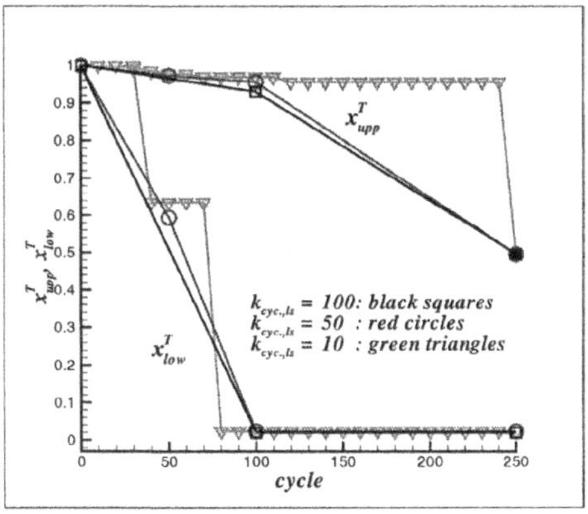

Figure 7: Influence of different initial transition locations on the iteration process
Figure 8: Influence of different intervals of laminar separation check on transition locations

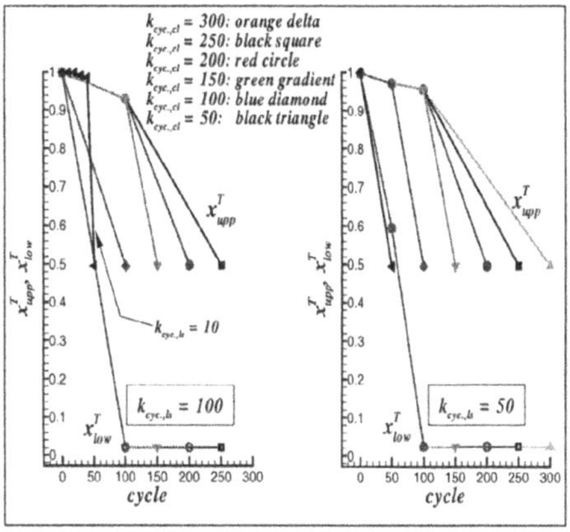

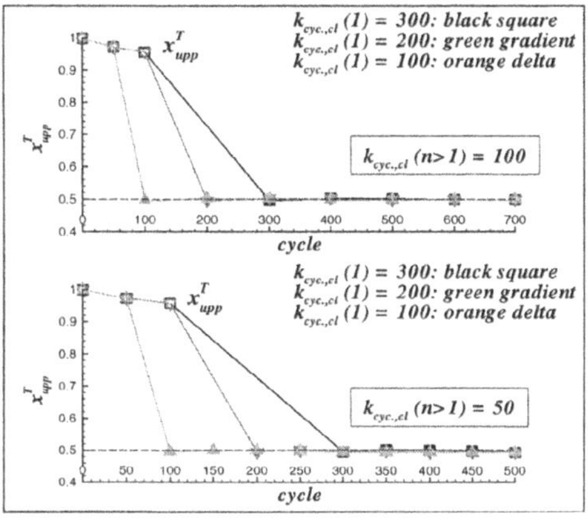

Figure 9: Influence of different intervals of 1st call of the transition module on transition points
Figure 10: Influence of different intervals of module call on transition points, converged state

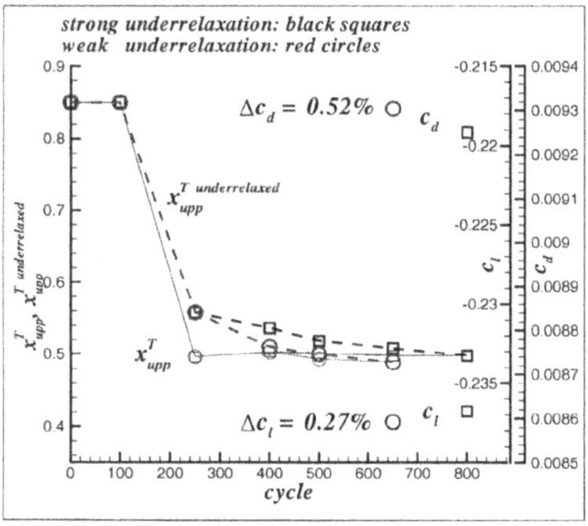

Figure 11: Influence of different values of underrelaxation factors

Low Aspect-Ratio Wings

Unsteady Measurements with the Continuously Rotating DLR-PSP-Model at the Transonic Wind Tunnel Göttingen (DNW-TWG)

M. BRUSE, K.-W. BOCK, S. TUSCHE, M. JACOBS
The German-Dutch Wind Tunnels DNW, Business-Unit GUK
Bunsenstr. 10, D-37073 Göttingen, Germany
internet: http://www.dnw.nlr.nl, email: dnw-guk@dlr.de

Summary

In the Transonic Wind Tunnel Göttingen (DNW-TWG) a test campaign was carried out to obtain experimental pressure data for a continuously rolling delta-wing-model. This campaign includes the design and manufacture of the model and the rolling apparatus for installation in the DNW-TWG. Since pressure data was required over the complete surface of the model, the "Pressure Sensitive Paint"-technique was used. Additionally, local unsteady pressure measurements with pressure sensors, and for the steady case, force and moment measurements were performed.

Introduction

The Western European Armament Group (WEAG) had established a common exercise to numerically simulate the flow over a delta wing, which rolls continuously around its axis of symmetry. To validate the numerical results, an experiment was required to provide data for transonic flow conditions. To deal with a well known model configuration a delta wing was chosen to be investigated. The model design and manufacture, and the wind tunnel test were carried out by the Business-Unit GUK (DNW) in co-operation with the DLR Institute of Fluid Mechanics, the Institute of Aeroelasticity and the Technical Service of the DLR in Göttingen and Braunschweig.

The DNW-TWG as part of the German-Dutch Wind Tunnels (DNW)

Since the beginning of 2000, twelve wind tunnels, formerly operated by the DLR (Germany) and NLR (Netherlands) have been amalgamated in the non-profit foundation German-Dutch Wind Tunnels (DNW). The main objective of the DNW is to provide a wide spectrum of wind tunnel facilities and simulation techniques to the customer. The DNW covers a wide range of Reynolds numbers (up to $Re=15x10^6$, reference length $0.1xS^{0.5}$, S cross-section) and Machnumbers (up to Ma=7).
The operation of the wind tunnels is organised in business-units (one for low-speed in the North East Polder/Braunschweig and two for transonic and supersonic speed (in Amsterdam and in Göttingen/Cologne). The Transonic Wind Tunnel Göttingen (DNW-TWG) is part of the Business-Unit GUK (Göttingen and Köln) and is located at the DLR centre in Göttingen. The DNW-TWG is a closed circuit wind tunnel for sub-, trans- and supersonic flow, equipped with either an adaptive, a perforated or a flexible Laval nozzle, as seen in **figure 1**.

Test set-up at DNW-TWG

For the test campaign described here, the wind tunnel was set up with the 'adaptive test section' and the '3D-support' to adjust the roll apparatus with the model. The Mach number was varied between 0.35 and 0.85 at a stagnation pressure from 0.4 to $1.0x10^5$Pa. For steady force and moment measurements, a six-component TASK balance was used. For the unsteady local pressure measurements, the model was equipped with Kulite® pressure sensors and the data were taken with the BU-GUK data acquisition system for unsteady measurements (IDa, max. sample frequency 50kHz per channel at 16 bit accuracy). In parallel, the Measurement Methods and Flow Analysis department of the Institute of Fluid Mechanics of the DLR in Göttingen carried out measurements with "Pressure Sensitive Paint" (PSP).

To generate a continuous rolling of the model at 10Hz a new apparatus had to be designed and manufactured. This was carried out by the Technical Services of the DLR Göttingen. In **figure 2** a sketch of this apparatus gives further details. Specifically, a hollow shaft, which guides the rotating shaft driven by the electrical motor, is mounted in the 10° crank (which already existed). A slip connector ring transforms the sensor signals from the rotating model to the non-rotating part of the sting. The angle of attack of the model is limited to the range -7° to +18°.

Investigations by the Institute of Aeroelasticity of the DLR Göttingen with forced oscillation had shown that the natural frequency of the rotating apparatus was close to the excitation frequency generated by the model rotation. To increase the natural frequency of the complete system, the hollow shaft was stabilized with two wires horizontally at the wind tunnel walls.

The DLR-PSP-model

The participants of the Common Exercise at the WEAG were looking for a model configuration, with which a vortex dominated flow is generated. A well known configuration is a generic delta wing with a sharp leading edge. With respect to the experiment, there are some more conditions to be satisfied. For the force and moment measurements with an internal balance, as well as for the instrumentation to use pressure sensors inserted in the model, something like a fuselage is needed. An important requirement for the PSP measurements is to avoid oscillation of the model. Thus the geometry of the model should be symmetrical. In addition there should be a 'smooth' surface with no chamfer to achieve good visibility of the complete surface and to avoid reflections, which could reduce the quality of the PSP technique. In the numeric investigation, the diameter of the sting is assumed to be the same as the diameter of the fuselage. At the rear part of the model a fairing is needed to get a smooth transition of the fuselage-wing combination to the circular sting. This fairing has to be separated from the delta-wing model to avoid influencing the force measurements from the sting, because the model and the fairing rotate with respect to the sting. A large diameter is advantageously for the hollow tube with an internally rotating shaft. Finally, the model configuration should be used in the DLR project AeroSUM (Aerodynamic Simulation of Unsteady Manoeuvres), which deals with transient rolling. For this project the model is equipped with flaps at the trailing edge of the wings.

To satisfy all the above mentioned requirements, a model design as shown in **figure 3** has been chosen: a delta-wing configuration with 65° swept leading edge, a length of 420mm, a span of 333mm and a diameter of the fuselage of 60mm. The location of the pressure holes were defined by the PSP team with respect to corrections and is shown on the photo in **figure 4** (left). On the photo on the right, the complete rolling apparatus with the model mounted at the 3D-support of the DNW-TWG is shown. For this photo the test section has been opened.

Results of the unsteady pressure measurements with four local pressure sensors.

To get a better idea about the orientation of the model (only one surface is PSP-coated) **figure 5** shows four typical roll angles for a positive angle of attack of the shaft. To describe the orientation of the model, we use the roll angle Φ and the angle of incidence i (the angle between the mean flow direction and the model axis).
In **figures 6** and **7** pressure, data for two cycles and in **figures 8** and **9,** for a half-cycle are shown.
In **figure 6** the model orientation in the wind tunnel is illustrated for selected roll-angles. **Figure 7** shows the unsteady pressure data for all sensors K1 – K4. The general behaviour of the four data curves is comparable. As expected, the amplitude at sensor K4 is the largest one, because this sensor is the one on the wing, far away from the axis of rotation.
The data in **figure 8** are taken for different sting configurations (a straight sting and the cranked sting with roll apparatus) at discrete roll angles of the model. The differences between the data curves are caused by the different interference caused by the size of the stings (one is straight and the other is cranked and equipped with a motor).
Figure 9 shows a typical data curve for the unsteady case in comparison to data for constant roll angle. In the unsteady case, the two kinks (marked with circles) are caused by unsteady effects. A significant data curve behaviour like this can also be found in other configurations.
Results of the PSP measurements and a comparison to the numerical data are reported in [1].

Conclusion

The success of this test demonstrates that the DNW can manage dynamic or unsteady measurements in co-operation with other institutions providing the design and manufacture of models and particular measurement techniques.

Acknowledgement(s)

The authors would like to thank the "Western European Armament Group" WEAG and the participants of the "Common Exercise V" for the co-operative work and the colleagues of the DLR institutions (the "Measurement Methods and Flow Analysis" department of the "Institute of Fluid Mechanics" and the "Institute of Aeroelasticity" of the DLR Göttingen and the "Technical Services in Göttingen and Braunschweig"), who were involved in the preparation and carrying out of this test campaign.

References

[1] K.A. Bütefisch. "Time resolved 2D measurement techniques for quantitative determination of flow fields and surface pressure distributions". In this volume.

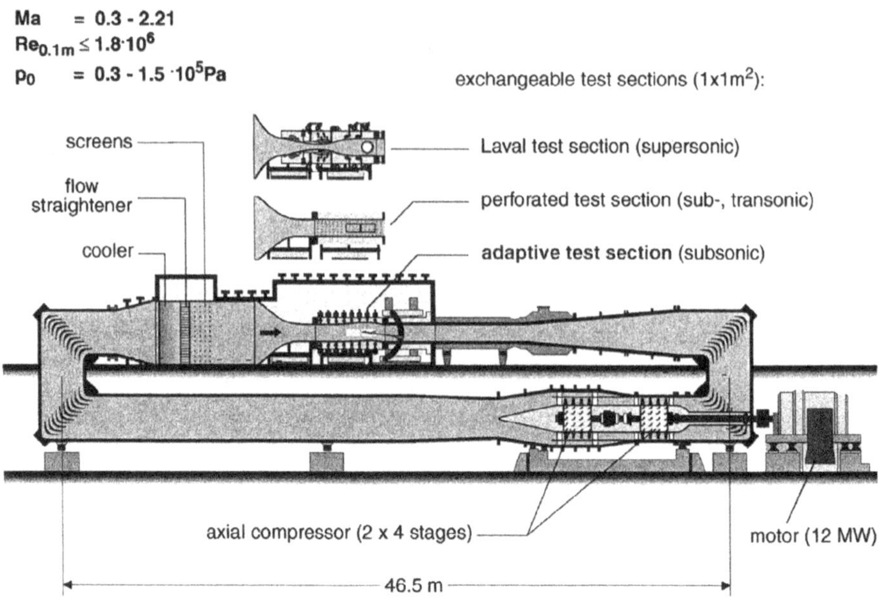

Figure 1 Scheme of the Transonic Wind Tunnel Göttingen (DNW-TWG).

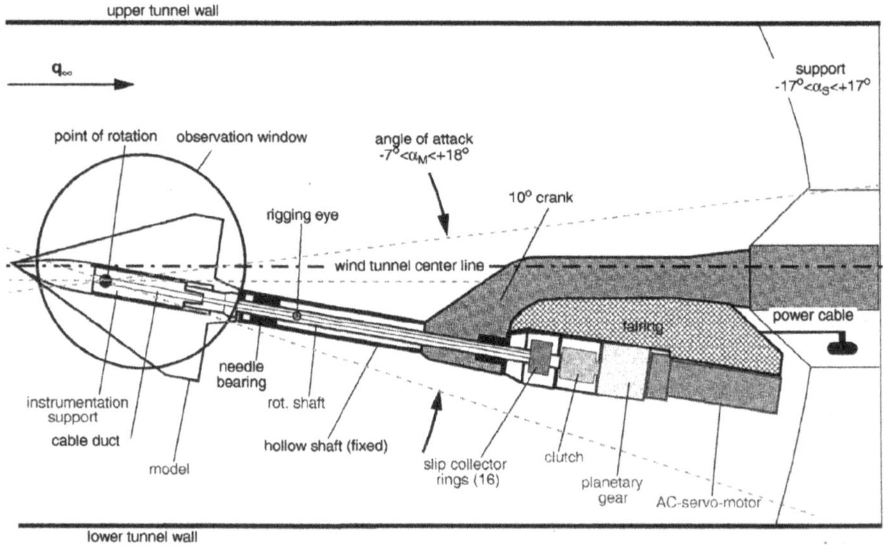

Figure 2 Sketch of the rolling apparatus with mechanical and electrical details.

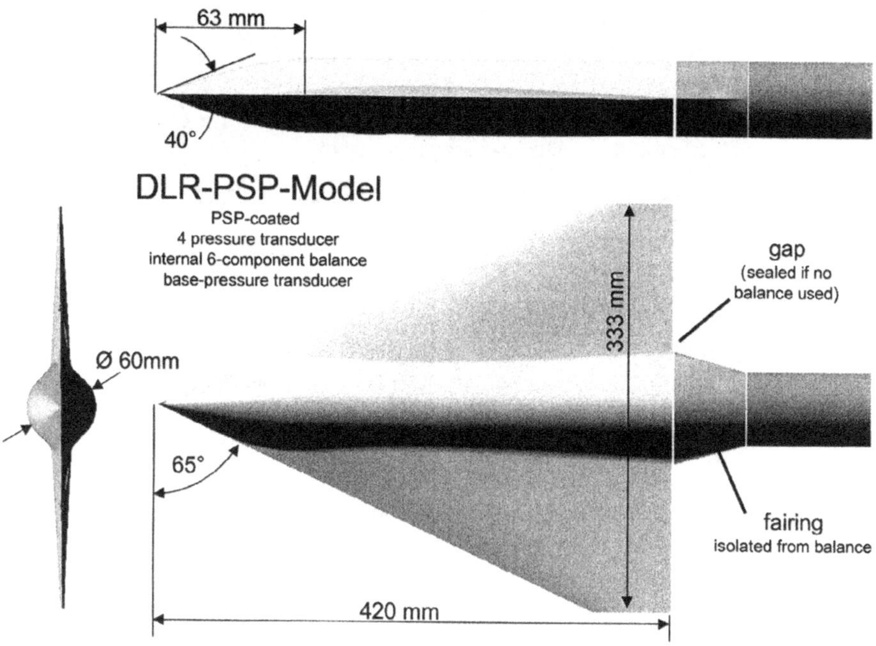

Figure 3 Sketch of the DLR-PSP-model with its dimensions.

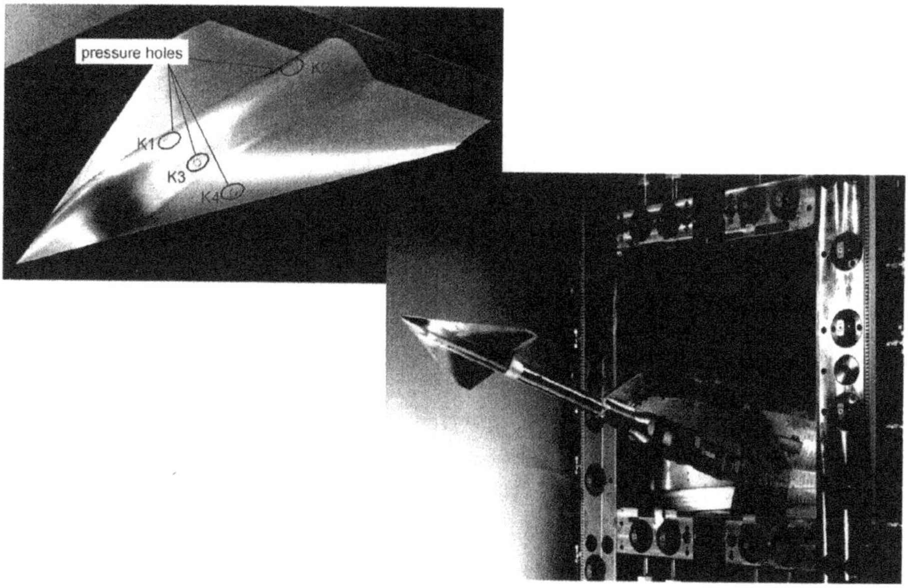

Figure 4 Photo of the DLR-PSP-model with pressure holes and mounted at the rolling-apparatus of the Transonic Wind Tunnel Göttingen (DNW-TWG)

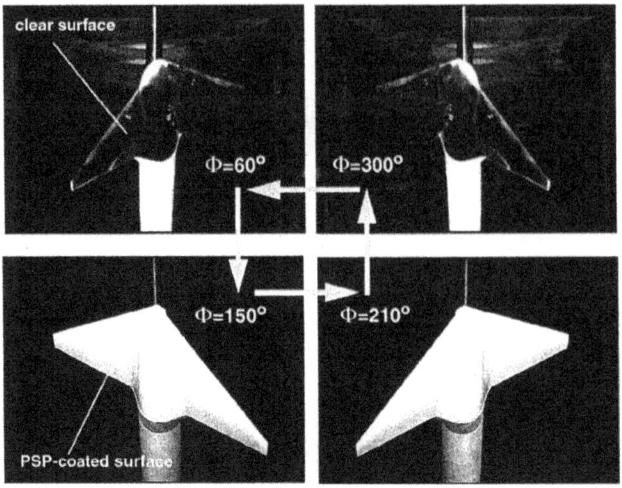

Figure 5 Photos of the DLR-PSP-model at four rolling angles Φ=60°,150°, 210°, 300° and an angle of incidence i=17°.

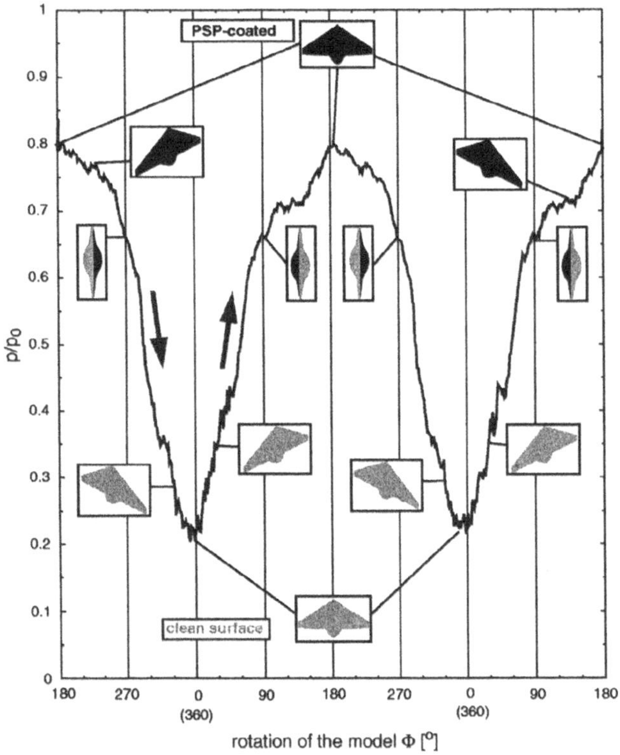

Figure 6 Data of pressure sensor K4 (wing) for two complete cycles of model rolling (Ma=0.8, p_0=40kPa, i=17°, rolling freq. 10Hz). The roll-angle of the model is illustrated.

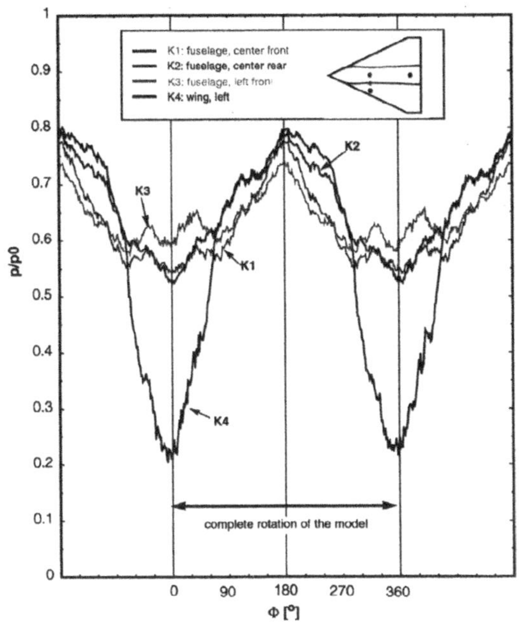

Figure 7 Comparison of the unsteady pressure data at the four pressure holes for two complete rotations of the model (Ma=0.8, p_0=40kPa, i=17°°, rolling freq. 10Hz).

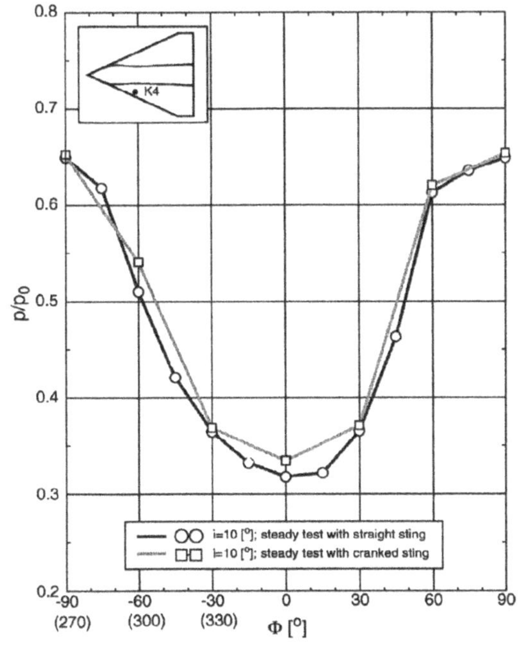

Figure 8 Comparison of pressure data for different stings (steady tests).

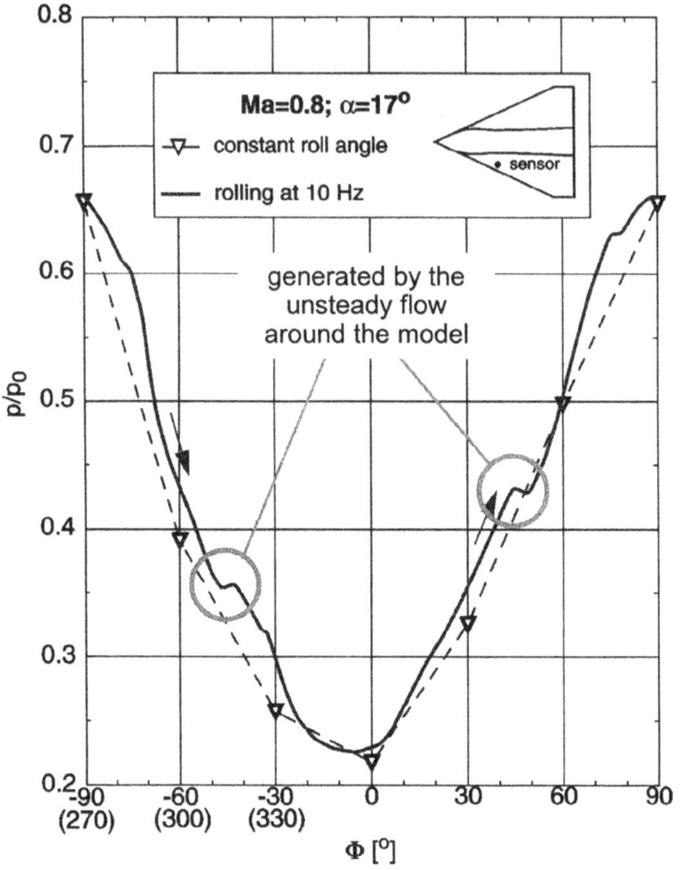

Figure 9 Comparison of steady and unsteady pressure data.

Hypersonic Flows

Turbulence Modeling for Reentry-Vehicles

C. Weber, R. Behr and C. Weiland
Astrium GmbH - Space Infrastructure
P.O. Box 801168, 81663 Munich, Germany

Summary

Several versions of the $k-\omega$ two-equation turbulence model are investigated for super- and hypersonic flows. Furthermore, in order to account for compressibility effects on turbulence, various compressibility corrections are considered in conjunction with the $k-\omega$ model. The models are applied to boundary layer flows up to a Mach number of 10, and to a hypersonic compression corner. Finally, the turbulent flow over the X-38 reentry vehicle at a Mach number of 6 and an angle of attack of 40° is investigated. Computed heat loads are compared with experimental data from the Langley Research Center.

1 Introduction

Within the framework of the national reentry technology program TETRA, the flow over NASA's X-38 crew return vehicle is extensively investigated using both windtunnel experiments and numerical simulations. The X-38 is a technology demonstrator for an emergency crew return vehicle for the International Space Station (ISS). After a de-orbiting maneuver, the X-38 will reenter into the atmosphere where an un-powered flight will follow until landing.

Since windtunnel testing is a time consuming and expensive task, and for certain flow conditions the complete set of similarity parameters cannot be simulated, Computational Fluid Dynamics (CFD) is nowadays extensively used for aeronautical design. In particular, one is interested in the thermal and mechanical loads at the nose of the vehicle and the body flap. The deflected body flap leads to boundary layer separation and shock/boundary layer interaction, similar to that occurring at a compression ramp, with very high pressure and heating rate peaks at the reattachment zone. Turbulence models play a crucial role for the correct prediction of these phenomena.

2 Turbulence Modeling

The governing equations are the Favre-averaged Navier-Stokes equations which can be found in any text book on fluid mechanics such as [1]. A perfect gas is assumed. The Reynolds stress tensor is modeled using the Boussinesq eddy viscosity assumption, and the turbulent heat fluxes are expressed by the standard temperature gradient approximation.

The Wilcox [2] $k-\omega$ two-equation model is used as baseline model. Compared to the $k-\varepsilon$ model, the $k-\omega$ model has the advantage that it does not need any damping functions near solid walls and simple Dirichlet boundary conditions can be specified at the wall. Furthermore, it has been shown that the behavior of the $k-\omega$ model is superior to that of the $k-\varepsilon$ model in the logarithmic part of the boundary layer [2].

The model has been extended to compressible flows by taking into account variable density effects through changes in the mean density, i.e. compressibility effects are accounted for only by the mean density field. Based on the Morkovin hypothesis, this approach might be justified for the Mach number range $M < 5$ for boundary layers. However, for larger Mach numbers compressibility effects on the turbulence structure cannot be neglected anymore.

Several corrections and modifications have been developed, among which the following are applied to the $k - \omega$ model in the present study. The total dissipation in the equation for k can be decomposed as the sum of solenoidal, ε_s, and dilatational, ε_d, components. The dilatational dissipation is then modeled as

$$\varepsilon_d = \xi^\star F(M_t)\, \varepsilon_s \tag{1}$$

where $M_t = \sqrt{2k}/c$ is the turbulent Mach number with c being the speed of sound. Three models are used which differ by the constant $\xi^\star$ and the function $F(M_t)$. The Wilcox [3] model uses

$$F(M_t) = (M_t^2 - M_{t0}^2)\, \mathcal{H}(M_t - M_{t0}) \ , \quad \xi^\star = 1.5 \ , \quad M_{t0} = 0.25 \tag{2}$$

where $\mathcal{H}$ is the Heaviside step function. The model of Sarkar *et al.* [4] is given by

$$F(M_t) = M_t^2 \ , \quad \xi^\star = 1. \ , \tag{3}$$

and the Zeman [5] model uses:

$$F(M_t) = \left(1 - e^{-\frac{1}{2}(\gamma+1)(M_t - M_{t0})^2/\Lambda^2}\right) \mathcal{H}(M_t - M_{t0}) \tag{4}$$

where $\xi^\star = 0.75$, $M_{t0} = 0.25\sqrt{2/(\gamma+1)}$, $\Lambda = 0.66$ and γ is the specific heat ratio. In the present study, the above values for M_{t0} and Λ are used which are those given by Zeman [5] for boundary layers, whereas $M_{t0} = 0.1\sqrt{2/(\gamma+1)}$ and $\Lambda = 0.60$ should be used for free shear flows.

Note that the compressible dissipation models have been mainly devised in order to correct the spreading rate deficiency for compressible mixing layers and that they should be regarded more as empirical corrections rather than true models for the compressible dissipation [1].

The pressure-dilatation appearing in the equation for the turbulence kinetic energy is approximated according to Sarkar [6] as

$$\overline{p'' \frac{\partial u_i'}{\partial x_i}} = \alpha_2 \tau_{ij} \frac{\partial \widetilde{u}_i}{\partial x_j} M_t + \alpha_3 \overline{\rho} \varepsilon_s M_t^2 \tag{5}$$

with the constants $\alpha_2 = 0.4$ and $\alpha_3 = 0.2$.

The turbulence length scale correction proposed by Vuong and Coakley [7] has been introduced in order to reduce the too high heat transfer rates predicted by two-equation turbulence models in the reattachment region of separated boundary layers. The turbulence length scale l is limited by an algebraic length scale as

$$l = \min\left(\kappa \beta^{\star -3/4} y, \sqrt{k}/(\beta^\star \omega)\right) \tag{6}$$

where $\beta^\star = 0.09$ and κ is the von Karman constant equal to 0.41. The eddy viscosity is then calculated using the limited length scale, $\mu_t = \alpha^\star \beta^\star\, \overline{\rho} \sqrt{k}\, l$.

In addition, three incompressible modifications to the $k - \omega$ model are investigated for high speed flows. The Wilcox [8] low Reynolds number version has been developed to yield a more "realistic description of transition" and a correct asymptotic behavior of certain turbulence quantities. It mainly consists of introducing a Reynolds number dependency into certain constants.

The Menter [9] SST correction ensures that Bradshaw's relation, $-\overline{u'v'}/k = const$, holds also in non-equilibrium boundary layers. It effectively reduces the eddy viscosity in non-equilibrium flows and therefore improves the ability of the model to predict separated flows.

Finally, a realizable $k-\omega$ model has been applied. Realizability can be defined as the condition that the normal Reynolds stresses must be positive and that the turbulent shear stresses must be bounded by the Schwartz inequality. In the present study, the general Moore and Moore [10] correction applied to the $k-\omega$ model is used

$$\alpha^{\star} = \min\left(1, \frac{1}{A_0' + A_s'(\overline{S}^2 + \overline{\Omega}^2)^{1/2}}\right) \tag{7}$$

where $\overline{S} = S/(\beta^{\star}\omega)$ and $\overline{\Omega} = \Omega/(\beta^{\star}\omega)$, S and Ω being the strain and vorticity invariants, respectively. The constants are those proposed by Thivet *et al.* [11]: $A_0' = 1 - A_s\sqrt{2\beta^{\star}}$, $A_s' = A_s\beta^{\star}$, $A_s = 1.77$ and $\beta^{\star} = 0.09$. Note that the constant A_0' is different to that given originally by Moore and Moore [10] in order to meet exactly the constraint for the logarithmic region.

3 Numerical Method

The Navier-Stokes equations are discretized using a cell-centered finite volume technique using structured multi-block meshes. The convective fluxes are discretized using the second-order upwind Total Variation Diminishing (TVD) scheme of Yee and Harten [12]. The viscous terms are centrally discretized. The convective terms in the turbulence equations are discretized using a first-order upwind discretization. In order to increase robustness and accelerate convergence, an implicit time-stepping procedure is used. The resulting linear system is solved by a Gauss-Seidel relaxation method.

At the wall a no-slip condition with either zero wall heat flux or constant wall temperature condition is applied. At the farfield, supersonic inflow or outflow conditions are imposed.

The code used is the Navier-Stokes flow solver DAVIS-VOL which has been developed at Astrium. It has been extensively applied to the analysis of hypersonic flows such as the flow over the Sänger two-stage spacecraft [13] or the Hermes reentry vehicle [14]. Recently, the code has been used to investigate the flow over the X-38 crew return vehicle [15].

4 Results

4.1 Flat Plate Boundary Layer

A mesh with 148×200 cells in streamwise and wall-normal direction has been used with different wall-normal distributions depending on the flow condition. Mesh convergence has been checked by doubling the number of mesh points.

The friction coefficient at a momentum thickness Reynolds number $Re_\theta = 9000$ for the different compressibility corrections applied to the standard $k-\omega$ model can be seen in Fig. 1 for the adiabatic wall. The Wilcox [3] dilatational dissipation model becomes active for Mach numbers larger than 4, approximately, and the skin friction falls slightly below that of the original model for higher Mach numbers. The model of Sarkar *et al.* [4] yields a too low skin friction for the entire Mach number range, whereas the compressible dissipation of Zeman [5] does not alter the skin friction profile. Note that the coefficients for the latter compressible dissipation model are those recommended by Zeman [5] for boundary layers, see section 2 for the values. However, as noted by Wilcox [3], this parameter setting leads to discrepancies in excess of 70% between computed and measured mixing layer spreading rate. Using the parameters for free shear flows, the skin friction is as low as that of the model of Sarkar *et al.* [4] (not shown here). The pressure-dilatation has only a small effect on the skin friction coefficient.

The cooled wall case is considered in Fig. 2 with regard to the compressible corrections. The effect of all corrections is higher than for the adiabatic wall, and also the Zeman [5] correction is now activated for large Mach numbers, although the effect on the skin friction coefficient is small.

4.2 Compression Corner

The flow parameters are: $M_\infty = 9.22$, $Re = 2.637 \times 10^7$, $T_\infty = 64.5K$, the wall temperature is $T_w = 295K$ and the ramp angle is 34° [16]. The mesh consists of 187 and 179 cells in streamwise and wall-normal direction, respectively. The first grid point above the wall is located such that $y^+ < 1$ in the reattachment zone. A finer mesh does not alter the results.

The pressure distribution at the compression corner is shown in Fig. 3 for the five different turbulence models. The $k - \omega$ and the low Reynolds number model yield identical results, both showing a very tiny separation bubble. The $k - \varepsilon$ model also fails to predict the flow separation. In contrast, both the SST and the realizable model predict a too large separation zone, and both models show a fairly similar behavior in the reattachment region. Though the SST model is not realizable, it relies on a limiter on the eddy viscosity which can be written in such a form that it belongs to the general class of limiters used to obtain realizable $k - \omega$ models. For both models, the large separation zone can be explained by the reduced eddy viscosity under rapid strain which facilitates flow separation.

The peak heating rate at the reattachment point is overestimated by all models as can be seen in Fig. 4. In particular, the Chien [17] $k - \varepsilon$ model shows an enormous excess in heat flux, a behavior that has also been shown by Coakley and Huang [18] for various other $k - \varepsilon$ models. Due to the larger separation, the SST and the realizable model yield larger heating rates than the $k - \omega$ model in the reattachment region. Interestingly, the $k - \varepsilon$ model yields the smallest heat transfer downstream the reattachment region.

The influence of the Wilcox [3] compressible dissipation is shown in Figs. 5 and 6 for the pressure and heat transfer, respectively. The compressible dissipation yields a separation zone that is about four times as large as that in the experiment. This is due to the fact that the boundary layer upstream the separation is less "turbulent", indicated by a momentum Reynolds number of $Re_\theta = 7386$ (at $x = -0.1024m$) compared to $Re_\theta = 9699$ and $Re_\theta = 9028$ for the original and the SST model, respectively. The pressure-dilatation model predicts a separation bubble that is about twice as large as the experimental one. In spite of the far too large separation, the compressible dissipation model yields a heat transfer at the impingement that is smaller than those obtained by the SST, realizable and pressure-dilatation models, and that has approximately the same magnitude as that of the original model which hardly predicts any separation.

The influence of the length scale limitation can also be seen in Figs. 5 and 6. The length scale limitation reduces mainly the maximum heat flux. Applied to the SST model, the heat flux peak is reduced by 26%.

4.3 Hypersonic Flow over the X-38 Crew Return Vehicle

The flow conditions are: Mach number $M = 6$, Reynolds number $Re = 3.571 \times 10^6$ and angle of attack $\alpha = 40°$. The body flap deflection angle is $\alpha_{bf} = 20°$. A constant wall temperature, $T_w = 300K$, is assumed. This flow condition leads to flow separation just upstream the body flap hinge line. Computed heat fluxes are compared with windtunnel data from the NASA Langley Research Center [19]. A phosphor thermography method was used to obtain the experimental heat fluxes. The mesh consists of 11 blocks with a total of 2.4×10^6 grid points including the wake behind the vehicle. The first grid point above the wall is located at $y^+ < 0.4$ for any location on the vehicle. Note that the body flap geometry corresponds to the "true" geometry including the cavity between the fuselage and the flap, and the gap between the two body flaps.

Although the freestream Mach number is $M = 6$, the Mach number at the boundary layer edge on the lower side of the vehicle is less than 2 due to the bow shock. Hence, the compressibility corrections have no influence on the windward boundary layer and only the $k - \omega$, the SST, and the realizable $k - \omega$ models are considered in the following. Although it is not active in the boundary layer, the Wilcox [3] dilatational dissipation is used for the computations in order to obtain the correct behavior of the models for free shear layers at large Mach numbers.

A comparison of the computed wall heat flux distribution on the windward side of the vehicle with experimental data is given in Fig. 7 for the standard $k-\omega$ model. Note that in the experiment transition from laminar to turbulent flow has been triggered, indicated by the dark blue spots at about $1/4$ of the vehicle length, whereas the simulations are fully turbulent. However, it can be assumed that at a distance far enough from transition the influence of the laminar region can be neglected. It can be seen that downstream the tripping device, the wall heat flux predicted by the $k - \omega$ model agrees well with the experiment. Upstream the separation, the heat flux is slightly too high compared with the experiment. On the body flap the heat transfer is correctly predicted, except near the outboard edge where the heat flux is somewhat too high. In particular, there is a small region in the corner between the hinge line and the outboard edge with relatively high heat loads.

The heat flux distribution predicted by the SST model is compared with experimental data in Fig. 8. With respect to the $k - \omega$ model, the heat flux is slightly lower and is, in particular near the separation region, in excellent agreement with experimental data. The heat transfer on the entire body flap is well predicted and there is no pronounced peak in the corner between hinge line and outboard edge.

Computed wall heat fluxes on the windward side of the X-38 are compared with experimental data along a $y = const$ (streamwise) and a $x = const$ (spanwise) cut in Figs. 9 and 10, respectively. The $k - \omega$ model predicts the wall heat flux in the streamwise cut slightly too high in the downstream and upstream regions of the body flap hinge line, see Fig. 9. The heat flux in the streamwise cut computed by the SST model is in excellent agreement with the experimental data, see Fig. 9. In particular, the heat flux after reattachment is well predicted along the entire body flap. The heat transfer upstream the body flap predicted by the realizable $k - \omega$ model is fairly close to that of the SST model. On the body flap itself the heat flux is too low, in particular in the vicinity of the reattachment.

The heat flux along the spanwise cut can be seen in Fig. 10. The $k - \omega$ model lies somewhat above the experimental data, and the local maximum at the corner of the fuselage at $y = 1.95m$ is overestimated. The spanwise heat flux predicted by the SST model is in very good agreement with the experimental data, particularly on the body flap. However, the local maxima at the curved corner of the fuselage are still too high, though they have been reduced compared to the $k - \omega$ model. The realizable $k - \omega$ model differs from the SST model mainly on the body flap, where it yields a smaller heat transfer. Note that the peaks at the flap edges at $y = 0.1m$ and $y = 1.4m$ are due to a relatively low resolution of the mesh around the 90° corner. However, these peaks are physical to a certain degree due to the flow around the rectangular corners, though they are not visible in the experimental data.

5 Conclusions

Considering zero pressure gradient boundary layers, the standard $k-\omega$ model gives good results concerning skin friction coefficient, for both adiabatic and cooled walls within the considered Mach number range $1.2 < M < 10$. The compressible dissipation models, which have been mainly developed to predict the correct spreading rate for free shear flows are not appropriate for boundary layer flows at very high Mach numbers, resulting in a too strong reduction of the wall shear stress.

The flow over the X-38 reentry vehicle at a freestream Mach number $M = 6$ has been computed and heat loads are compared with experimental data. The SST model is in very good agreement with the windtunnel data, whereas the standard $k - \omega$ model obtains heat fluxes that are somewhat larger. The realizable model is close to the SST model, except on the body flap where it yields a too low heat transfer, particularly in the reattachment zone. Due to the relatively small Mach number downstream the bow shock where $M < 2$, the compressibility corrections did not influence the boundary layer for this flow condition.

However, it is questionable if two-equation turbulence models based on the Boussinesq eddy viscosity assumption and a temperature gradient approximation for the turbulent heat fluxes are principally able to reliably predict non-equilibrium boundary layer flows involving shocks for very high Mach numbers. For the compression corner at $M = 9.22$, none of the various models or explicit compressibility terms predicted the correct size of the separation bubble. Furthermore, the heat transfer in the reattachment was strongly overestimated by all model versions.

For the X-38, though the geometry is much more complex, it is probably due to the relatively small Mach number at the boundary layer edge that compressibility effects are negligible and separation and heat fluxes are fairly well predicted by the standard turbulence models.

References

[1] Wilcox, D. C., *Turbulence Modeling for CFD*. DCW Industries, Inc., second edn. (1998).

[2] Wilcox, D. C., Reassessment of the Scale-Determining Equation for Advanced Turbulence models. *AIAA Journal*, **26(11)**, 1299–1310 (1988).

[3] Wilcox, D. C., Progress in hypersonic turbulence modelling. *AIAA Paper 91-1785* (1991).

[4] Sarkar, S., Erlebacher, G., Hussaini, M. Y. and Kreiss, H. O., The analysis and modelling of dilatational terms in compressible turbulence. *Journal of Fluid Mechanics*, 227 (1991).

[5] Zeman, O., A new model for super/hypersonic turbulent boundary layers. *AIAA Paper 93-0897* (1993).

[6] Sarkar, S., Modeling of the Pressure-Dilatation Correlation. *ICASE Report No. 91-42, 1991* (1991).

[7] Vuong, S. T. and Coakley, T. J., Modeling of Turbulence for Hypersonic flows with and without Separation. *AIAA Paper 87-0286* (1987).

[8] Wilcox, D. C., Simulation of Transition with a Two-Equation Turbulence Model. *AIAA Journal*, **32(2)**, 247–255 (1994).

[9] Menter, F. R., Two-Equation Eddy-Viscosity Turbulence Models for Engineering Applications. *AIAA Journal*, **32(8)**, 1598–1605 (1994).

[10] Moore, J. G. and Moore, J., Realizability in Two-Equation Turbulence Models. *AIAA Paper 99-3779* (1999).

[11] Thivet, F., Knight, D. D., Zheltovodov, A. A. and Maksimov, I. A., Some Insights in Turbulence Modeling for Crossing-Shock-Wave/Boundary-Layer Interactions. *AIAA Paper 2000-0131* (2000).

[12] Yee, H. C. and Harten, A., Implicit TVD schemes for Hyperbolic Conservation Laws in Curvilinear Coordinates. *AIAA Journal*, 25, 266–274 (1987).

[13] Schröder, W. and Hartmann, G., Analysis of Inviscid and Viscous Hypersonic Flows past a Two-Stage Spacecraft. *Journal of Spacecraft and Rockets*, **30(1)**, 8–13 (1993).

[14] Schröder, W. and Hartmann, G., Implicit Solutions of Three-Dimensional Viscous Hypersonic Flows. *Computers and Fluids*, **21(1)**, 109–132 (1992).

[15] Weber, C., Behr, R. and Weiland, C., Investigation of Hypersonic Turbulent Flow over the X-38 Crew Return Vehicle. *AIAA Paper 2000-2601* (2000).

[16] Coleman, G. and Stollery, J., Heat Transfer in Hypersonic Turbulent Separated Flow. *Imperial College Aero Report No. 72-05* (1972).

[17] Chien, K.-Y., Predictions of Channel and Boundary-Layer Flows with a Low-Reynolds-Number Turbulence Model. *AIAA Journal*, **20(1)**, 33–38 (1982).

[18] Coakley, T. J., Turbulence Modeling for High Speed Flows. *AIAA Paper 92-0436* (1992).

[19] Horvath, T. J., Berry, S. A. and Fitzgerald, S. M., X-38 Experimental Aerothermodynamics. *AIAA Paper 2000-2685* (2000).

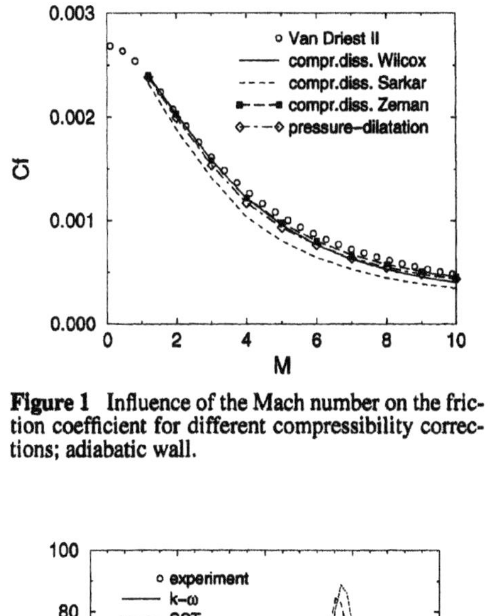

Figure 1 Influence of the Mach number on the friction coefficient for different compressibility corrections; adiabatic wall.

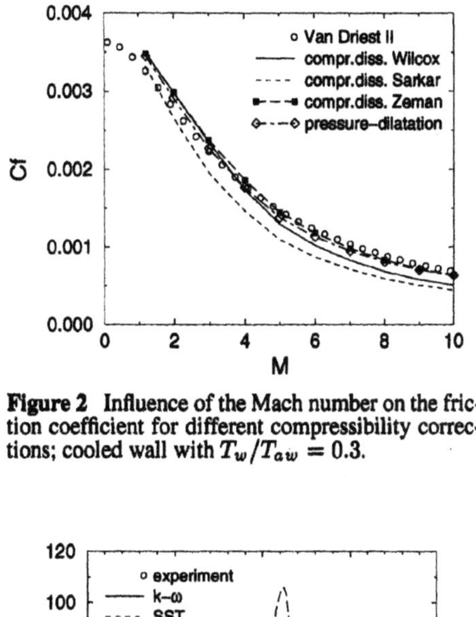

Figure 2 Influence of the Mach number on the friction coefficient for different compressibility corrections; cooled wall with $T_w/T_{aw} = 0.3$.

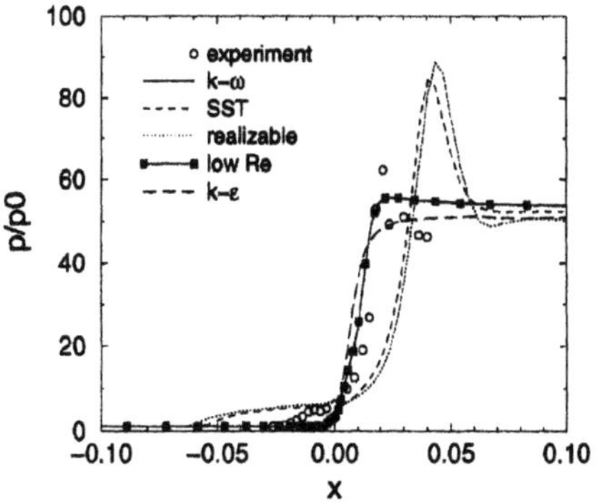

Figure 3 Influence of the turbulence model on the pressure at the compression corner.

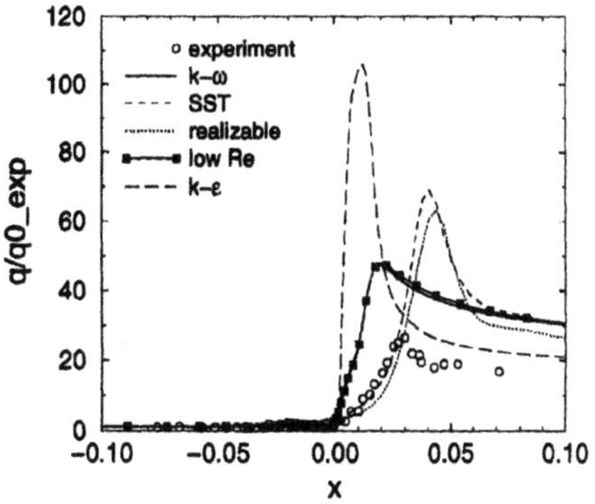

Figure 4 Influence of the turbulence model on the heat flux at the compression corner.

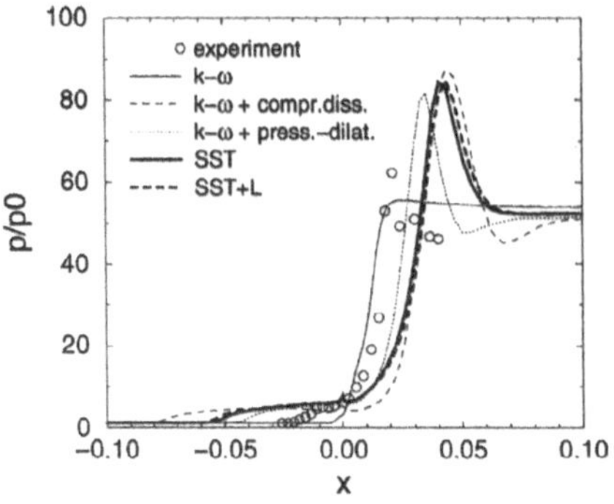

Figure 5 Influence of the compressible dissipation and pressure-dilatation model on the pressure at the compression corner.

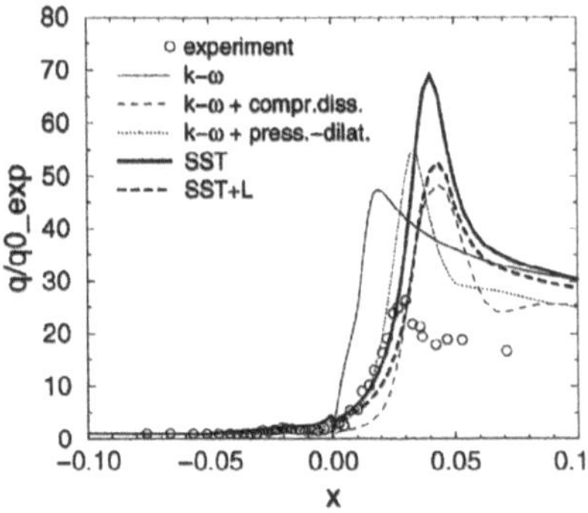

Figure 6 Influence of the compressible dissipation and pressure-dilatation model on the heat flux at the compression corner.

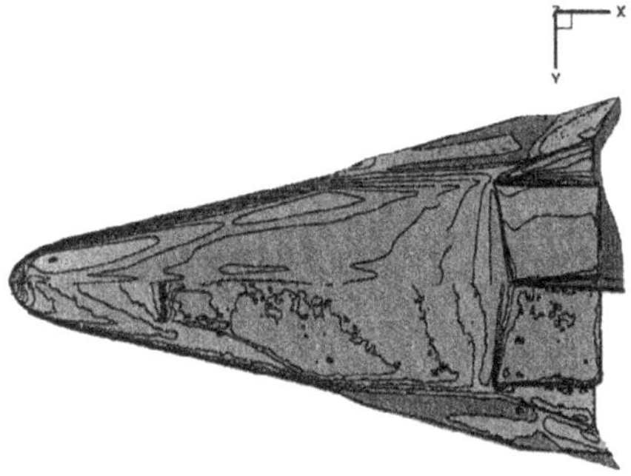

Figure 7 Comparison of the computed wall heat flux using the $k-\omega$ model (top) with experimental data (bottom); $y=0$ corresponds to the centerline.

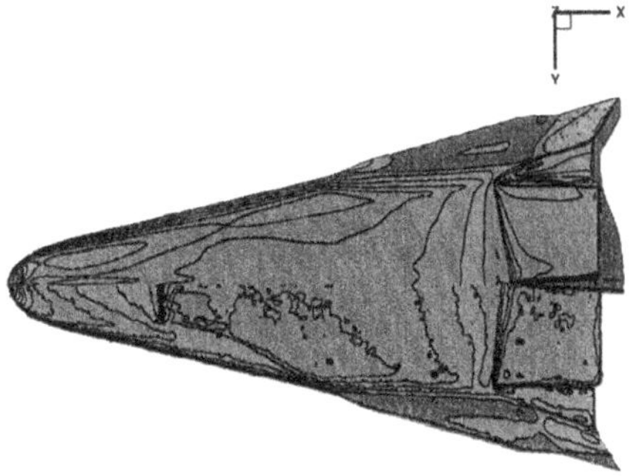

Figure 8 Comparison of the computed wall heat flux using the SST model (top) with experimental data (bottom); $y=0$ corresponds to the centerline.

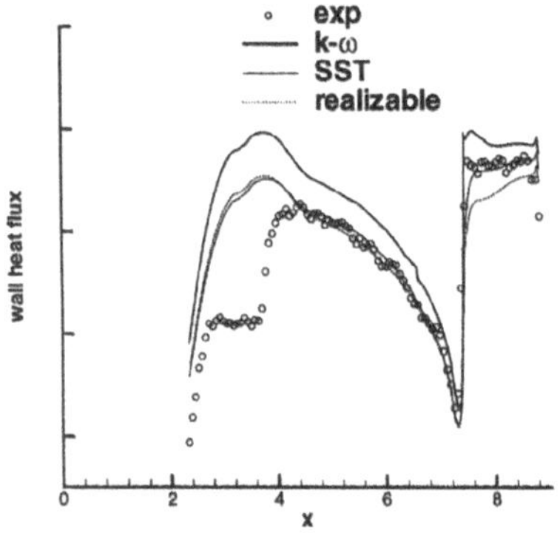

Figure 9 Wall heat fluxes in a cut $y = const$ approximately at the center of the body flap.

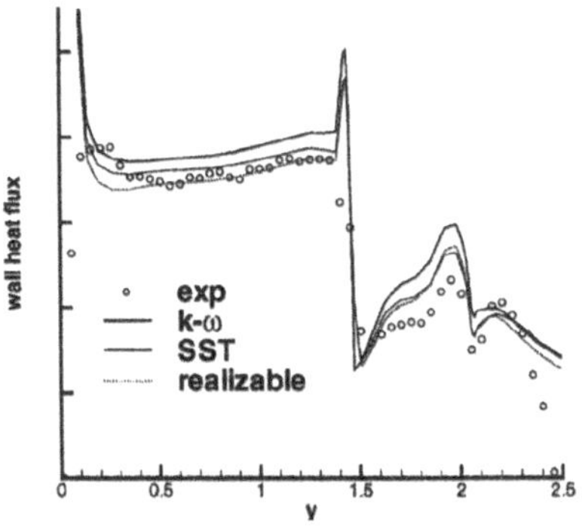

Figure 10 Wall heat fluxes in a cut $x = const$ approximately at 80% of the body flap length. Note: $y=0$ corresponds to the centerline.

Computation of the Two-Dimensional Flow in SERN Nozzles

Thomas Link and Wolfgang W. Koschel
Jet Propulsion Laboratory, Aachen University of Technology
Templergraben 55, D-52062 Aachen, Germany

Summary

For the design of air-breathing hypersonic vehicles the simulation of the flow inside the propulsion system is of essential interest. In order to allow accurate computation of the expanding flow of a hydrogen/air mixture at high temperatures, the chemical and vibrational nonequilibrium effects must be taken into account. In the present study the influence of thermal nonequilibrium on flows in SERN-type nozzles is numerically investigated. In the Finite Element code used for the flow simulation chemical reactions are modeled by finite rate chemistry and vibrational relaxations are modeled by the Landau-Teller theory. As test case a basic nozzle experiment is computed and the results are compared with measured data. In order to estimate the thermal nonequilibrium effects in true scaled nozzles the DLR/ONERA's JAPHAR nozzle flow is computed.

1 Introduction

Because of the deficiencies of ground based test facilities for high enthalpy flows, the design of airbreathing hypersonic spacecrafts requires efficient and accurate numerical methods. In the Collaborative Hypersonic Research Center (SFB 253) at the Aachen University of Technology different CFD codes are developed while at the same time experimental projects are conducted supplying a data base for validation of these codes. The presented project deals with the computation of the expanding flow in Single Expansion Ramp Nozzles (SERN) with respect to high temperature effects of a hydrogen/air burning gas mixture.

Due to the fact that inlet and exhaust momentum of an airbreathing propulsion system at hypersonic speed are of the same magnitude, the nozzle performance has a strong influence on the net thrust. In order to predict the nozzle flow field accurately, it is necessary to take the physical and chemical effects of high enthalpy expansion flows into account. At the beginning of the expansion, a considerable fraction of the combustion products is dissociated. The vibrational degrees of freedom of two and three atomic molecules are excited due to the high temperature. During the expansion in the nozzle, the velocity of the gas increases and the density decreases strongly. As a result, the gas molecules are not able to keep up the state of equilibrium because of an insufficient number of collisions. It is known that chemical nonequilibrium should be taken into account for nozzle computation [1, 2] but the importance of thermal nonequilibrium is weighted differently. While according to [3] thermal nonequilibrium is negligible, [1] estimates the thrust loss due to thermal nonequilibrium as of the same order of magnitude as the thrust loss due to chemical nonequilibrium. Since the experimental investigation of a nozzle flow conducted within a complementary project [4] of the SFB 253 showed a noticeable amount of thermal nonequilibrium, the code was extended to take thermal nonequilibrium effects into account.

2 Governing Equations

In this section the conservation equations for chemical and thermal nonequilibrium flows are considered. The equations are written in a non-dimensional form introducing the Reynolds number Re, the Mach number M, the Prandtl number Pr, and the Lewis number Le. To simplify the calculation of molecular diffusive fluxes a Lewis number of unity is assumed. The used symbols are given in table 1.

- continuity equation:

$$\frac{\partial \rho}{\partial t} + \nabla \cdot (\rho \mathbf{v}) = 0 \tag{2.1}$$

- species continuity equation of species k:

$$\frac{\partial(\rho Y_k)}{\partial t} + \nabla \cdot (\rho \mathbf{v} Y_k) + \frac{1}{RePr}\nabla \cdot (\frac{\lambda}{c_p}\nabla Y_k) = w_k \tag{2.2}$$

- momentum equation:

$$\frac{\partial(\rho \mathbf{v})}{\partial t} + \nabla \cdot (\rho \mathbf{v}\mathbf{v} + p\mathbf{I} - \sigma) = 0 \tag{2.3}$$

with:

$$\sigma_{ij} = \frac{\mu}{Re}[(\partial_j v_i + \partial_i v_j) - \frac{2}{3}(\nabla \cdot \mathbf{v})\delta_{ij}] \tag{2.4}$$

- energy equation:

$$\frac{\partial(\rho e_t)}{\partial t} + \nabla \cdot (\rho \mathbf{v} h_t - \sigma \cdot \mathbf{v}) + \frac{1}{RePr}\nabla \cdot \mathbf{q} = \sum_{k=1}^{N_{vsp}} -w_{vk} \tag{2.5}$$

with the heat flux $\mathbf{q}$:

$$\mathbf{q} = -\frac{1}{M^2(\kappa - 1)}\lambda \nabla T - \frac{\lambda}{c_p}\sum_{k=1}^{N_{sp}} h_k \nabla Y_k \tag{2.6}$$

- vibrational energy balance:

$$\frac{\partial(\rho Y_k e_{vk})}{\partial t} + \nabla \cdot (\rho Y_k \mathbf{v} e_{vk}) = w_{vk} \tag{2.7}$$

The total energy e_t includes the energy of motion of the gas, the energy of rotational degrees of freedom, and the heat of formation of the molecules

$$e_t = \frac{\mathbf{v}}{2} + e_{rot} + \sum_{k=1}^{N_{sp}} h^0_{f,k} Y_k \tag{2.8}$$

with

$$e_{rot} = \sum_{k=1}^{N_{sp}} TY_k c_{vrot,k}. \tag{2.9}$$

The vibrational energy e_v of an harmonic oscillator is according to Einstein's formula [5]

$$e_v = \frac{\Theta R}{e^{\Theta/T_v} - 1} \tag{2.10}$$

for molecules of two atoms, and

$$e_v = \frac{\Theta_1 R}{e^{\Theta_1/T_v} - 1} + \frac{\Theta_2 R}{e^{\Theta_2/T_v} - 1} + \frac{\Theta_3 R}{e^{\Theta_3/T_v} - 1} \tag{2.11}$$

for H_2O. Herein Θ_i are the characteristic vibrational temperatures and T_v is the vibrational temperature of the molecule or vibrational mode, respectively.

The equation system is closed by the thermal equation of state, the chemical source terms w_k and relaxation of vibrational energy w_{vk}, which are described in the next section. When a flow in thermal equilibrium is considered, the balance equation of vibrational energy is neglected and the contribution of vibrational energy is added to the total energy in equation (2.5).

Chemical Source Terms

The chemical source terms w_k are calculated by finite rate chemistry among the species H_2, O_2, H_2O, N_2, NO, OH, N, O, and H. This reduced reaction set was designed for expansion flows by Mühleck [6]. The reaction rate of an arbitrary reaction

$$\sum_{k=1}^{N_{sp}} \nu'_{k,r} B_k \underset{k_b}{\overset{k_f}{\rightleftharpoons}} \sum_{k=1}^{N_{sp}} \nu''_{k,r} B_k \quad r = 1,...,N_r \tag{2.12}$$

is given by

$$\dot{\omega}_r = k_{f,r} \prod_{k=1}^{N_{sp}} \frac{\rho Y_k}{M_k}^{\nu'_{k,r}} - k_{b,r} \prod_{k=1}^{N_{sp}} \frac{\rho Y_k}{M_k}^{\nu''_{k,r}} \tag{2.13}$$

herein the forward reaction rate $k_{f,r}$ and the backward reaction rate $k_{b,r}$ are computed according to Arrhenius's law

$$k_{f,r} = A_{f,r} \cdot T^{n_{f,r}} \cdot e^{\frac{-E_{f,r}}{\Re \cdot T}}, \tag{2.14}$$

$$k_{b,r} = A_{b,r} \cdot T^{n_{b,r}} \cdot e^{\frac{-E_{b,r}}{\Re \cdot T}}. \tag{2.15}$$

The parameters A, n, and E are given by the used finite rate chemistry model. The whole amount of the source term of a species is given by the summation of the source terms of every reaction, which involves this species:

$$w_k = \dot{\rho_k} = M_k \sum_{r=1}^{N_r} (\nu''_{k,r} - \nu'_{k,r}) \cdot \dot{\omega}_r. \tag{2.16}$$

Vibrational Relaxation

With regard to the computational effort that is necessary to describe the detailed relaxation mechanism a fairly simple model proposed by Pirumov [5] is used. In his publication, Pirumov gave a compilation of vibrational relaxation parameters for expansion flows of various combustion products obtained by experiments. The equations and parameters used in the following section are the processes of Pirumov's model concerned with combustion products of H_2 in air. The equations are valid for a Boltzmann distribution of the energy and molecules that can be treated as harmonic oscillators. The energy exchange between vibrational and translational energy according to the following processes:

$$
\begin{array}{llcl}
1.) & N_2 + M & \rightleftharpoons & {N_2}^* + M - 2330cm^{-1} \\
2.) & H_2O + M & \rightleftharpoons & H_2O^* + M - 1595cm^{-1} \\
3.) & H_2 + M & \rightleftharpoons & {H_2}^* + M - 4260cm^{-1} \\
4.) & NO + M & \rightleftharpoons & NO^* + M - 1876cm^{-1}
\end{array}
$$

is taken into account. The source terms of these processes are determined by the Landau-Teller theory

$$w_{vk} = (e_{vk}^0 - e_{vk})\Phi_i. \tag{2.17}$$

Herein the rate Φ of the i-th relaxation process is calculated by

$$\Phi_i = p \sum_{k=1}^{N_{vsp}} \frac{X_j}{\tau_{ij}} \tag{2.18}$$

where τ_{ij} is the reaction rate of the j-th species in the i-th relaxation process.

3 Numerical Method

Because of the chemical and vibrational source terms in the conservation equations, the equation system is stiff. A solution method capable for stiff equation systems is the semi-implicit algorithm proposed by Bussing [7]. This algorithm is implemented in a Finite Element code using unstructured hybrid grids and a Runge–Kutta time marching scheme [8]. The Finite Element code was comprehensively validated. Parameter studies, comparisons to analytical and experimental results, and details about grid generation and influence of grid refinement can be found in [9]. Further computation of experimental test cases were done in [10].

4 Results

As test case for the code a nozzle flow experimentally investigated by Weisgerber [4] is computed. The inlet conditions are a total temperature of 2376 K and a total pressure of 11 bar. Because of incomplete data at the nozzle inlet the composition of the mixture is assumed to be in chemical equilibrium. The following figures show the result of a Navier-Stokes computation of the nozzle for a thermal and chemical nonequilibrium flow. The nozzle shape and the computed Mach number distribution is presented in figure 1. The influence of the chemical reactions is shown in figure 2. As illustrated in the graph, the radical OH decreases strongly throughout the nozzle while H increases slightly. This behavior is often detected in such kind of nozzle flows [1]. The comparison between the measured and the computed wall pressures in figure 3 shows an excellent agreement. In figure 4 the measured and calculated profiles of rotational/translational and vibrational temperatures at the nozzle exit are shown. The difference between the rotational/translational and vibrational temperatures at the exit is remarkable because of the high water content in the gas. Water is known as very effective for de-excitation of the excited N_2 molecules. In figure 5 the measured and computed profiles of the velocity at the nozzle exit are plotted. Although the measured and computed values of rotational temperatures are in very good agreement the computed velocity is about 5% higher. Part of this difference is due to the particle slip of the employed L2F velocity measurements, but the remaining part is not explainable. In particular the lower velocity in the experiment is not confirmed by an energy balance for a streamline in the middle of the nozzle.

To evaluate the thermal nonequilibrium effects in flows throughout true scaled nozzles the nozzle flow of the JAPHAR configuration is computed. JAPHAR is a cooperation between DLR and ONERA where a hypersonic flight demonstrator with an airbreathing engine is investigated. The boundary conditions of the calculation are corresponding to an ambient flow at Mach 4.93. Figure 6 shows the computed Mach number and the computational domain. In order to take the ambient flow into account the computational domain is enlarged perpendicular to the upper and lower side of the afterbody. At these locations of the boundary air with a temperature of 296 K and a pressure of 0.0688 bar enters the computational domain. The inflow boundary of the nozzle is given by a stoichiometric hydrogen/air mixture of a temperature of 2604 K, and a pressure of 1.894 bar at Mach 1.12. The composition and the thermal degrees of freedom are treated as in equilibrium at the inflow boundaries. The effect of the thermal nonequilibrium is presented in figure 7. The graph shows the rotational and vibrational temperature of the N_2 molecules on a straight line with an angle of 10° to the nozzle axis. In figure 8 the largest difference between the two temperatures is shown in detail. The maximum difference is less than 20 K and consequently negligible in the thrust computation of the nozzle.

5 Conclusion

In this paper a flow solver for chemical and thermal nonequilibrium is presented. For validation of the code a basic nozzle experiment was computed and compared to measured results. With respect to the available experimental data, the computed and measured results are in good agreement. To evaluate the effect of thermal nonequilibrium in flows through nozzles in a real scale, the nozzle flow of the JAPHAR airbreathing engine was computed. It turned out that the maximum difference between the rotational and vibrational temperature of the N_2 molecule is on a negligible level. According to this result the amount of thermal nonequilibrium and the effect on the thrust reported in [1] seem to be overestimated.

Acknowledgment

The authors acknowledge the financial support of this work by the Deutsche Forschungsgemeinschaft (DFG) within the frame of the Sonderforschungsbereich 253 at RWTH Aachen.

References

[1] T. A. Grønland and J. L. Cambier. Sensitivity to physical modelling for nozzle/afterbody flowfields. AIAA-96-4547, 1996.

[2] J. J. Sangiovanni and T. J. Barber. Role of hydrogen/air chemistry in nozzle performance for a hypersonic propulsion system. *Journal of Propulsion and Power*, 1993.

[3] O. Rizkalla, W. Chinitz, and J. I. Erdos. Calculated chemical and vibrational nonequilibrium effects in hypersonic nozzles. *Journal of Propulsion and Power*, 1990.

[4] H. Weisgerber, E. Fischer, et al. Experimental analysis of the flow of exhaust gas in a hypersonic nozzle. AIAA-98-1600, 1998.

[5] U. G. Pirumov and G. S. Roslyakov. *Gas Flow in Nozzles*. Springer-Verlag, 1986.

[6] P. Mühleck. Numerische Untersuchung turbulenter reagierender Strömungen in Brennkammern und Schubdüsen von Hyperschall-Staustrahltriebwerken. DLR-FB 95-10, 1995.

[7] T. R. A. Bussing. *A Finite Volume Method for the Navier-Stokes Equations with Finite Rate Chemistry*. PhD thesis, Dept. of Aeronautics and Astronautics, Massachusetts Inst. of Technology, Cambridge, MA, 1985.

[8] W. Koschel, T. Link, and U. Fox. Computation of hypersonic propulsion system expansion flows considering hydrogen/air high temperature effects. AIAA-98-1601, April 1998.

[9] W. Rick. Adaptive Galerkin Finite Elemente Verfahren zur numerischen Strömungssimulation auf unstrukturierten Netzen. Dissertation, Institut für Strahlantriebe, RWTH Aachen, 1994.

[10] B. U. Reinartz. Wärmebelastung der Brennkammer eines Staustrahltriebwerks mit Überschallverbrennung. Dissertation, Institut für Strahlantriebe, RWTH Aachen, 2001.

Table 1: Symbols

c_p	specific heat at constant pressure	c_v	specific heat at constant volume
e_t	specific total energy	e_{rot}	specific rotational energy
e_v	specific vibrational energy	h_t	specific total enthalpy
h_t^0	heat of formation	$\mathbf{I}$	identity matrix
$k_{f,r}$	forward reaction rate of reaction r	$k_{b,r}$	backward reaction rate of reaction r
M_i	molar mass of species i	N_r	number of reactions
N_{sp}	number of species	N_{vsp}	number of species with vibrational relaxation
p	pressure	$\mathbf{q}$	heat flux
R	specific gas constant	$\Re$	universal gas constant
T	rotational temperature	T_v	vibrational temperature
$\mathbf{U}$	conserved variables	$\mathbf{v}$	velocity
x	Cartesian coordinate	X	molar fraction
Y	mass fraction	δ_{ij}	Kronecker delta
λ	thermal conductivity	μ	viscosity
ν',ν''	stoichiometric coefficients	$\dot{\omega}$	chemical production rate
ρ	density	σ	stress tensor
Θ	characteristic vibrational temperature	τ	relaxation time

Figures

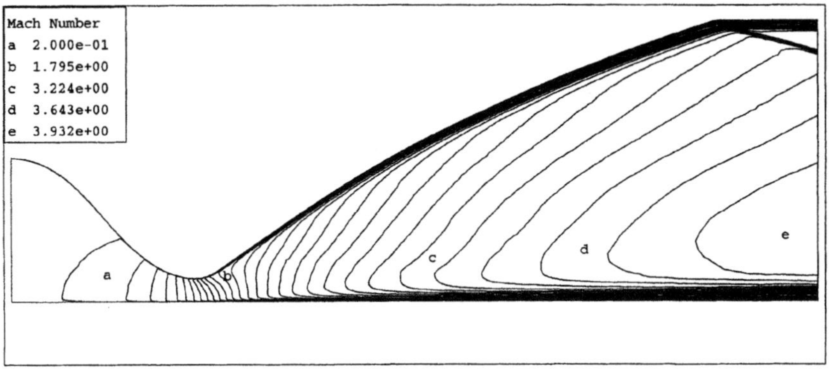

Figure 1: Computed Contour Plot of Mach Number

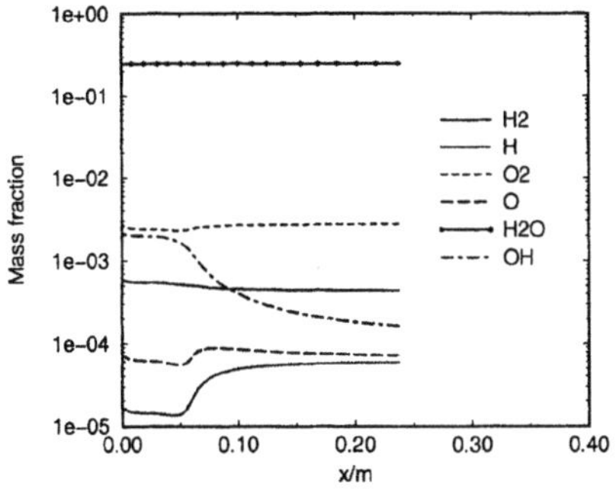

Figure 2: Computed Composition in the Middle of the Nozzle

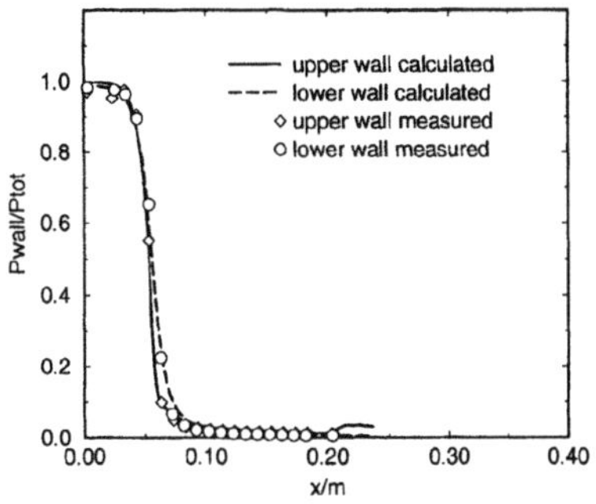

Figure 3: Computed and Measured Wall Pressures Versus Nozzle Length

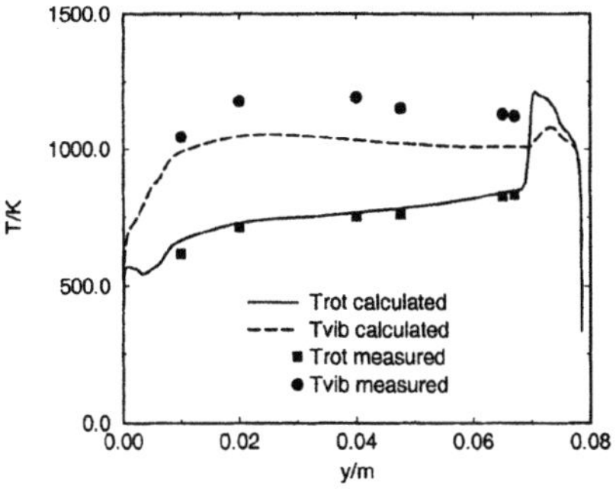

Figure 4: Temperature Profiles of N_2 at the Nozzle Exit

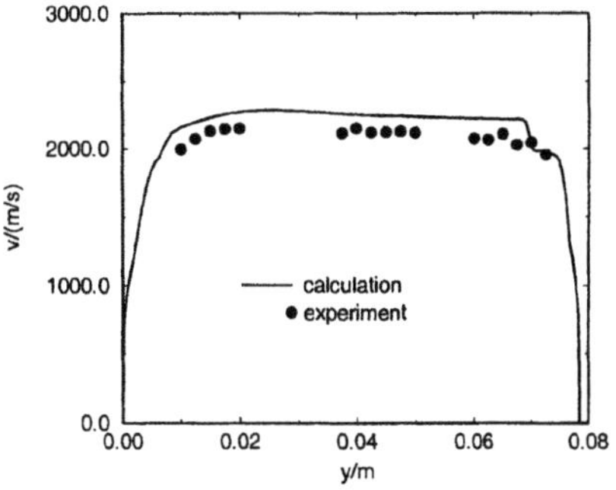

Figure 5: Velocity profile at the Nozzle Exit

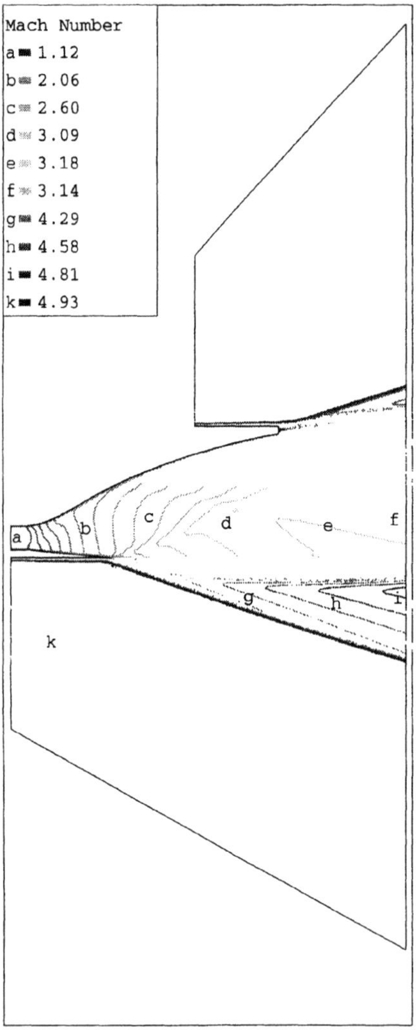

Figure 6: Computed Mach Number Distribution and Nozzle Shape of the JAPHAR nozzle

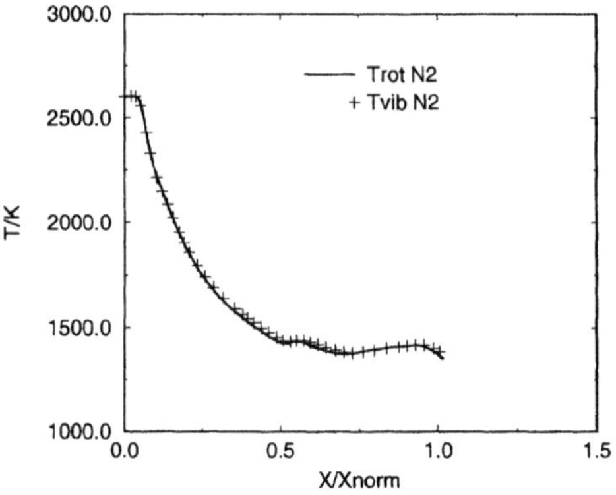

Figure 7: Computed Vibrational and Rotational Temperature of N_2 on a Straight Line with an Angle of 10° to the Nozzle Axis

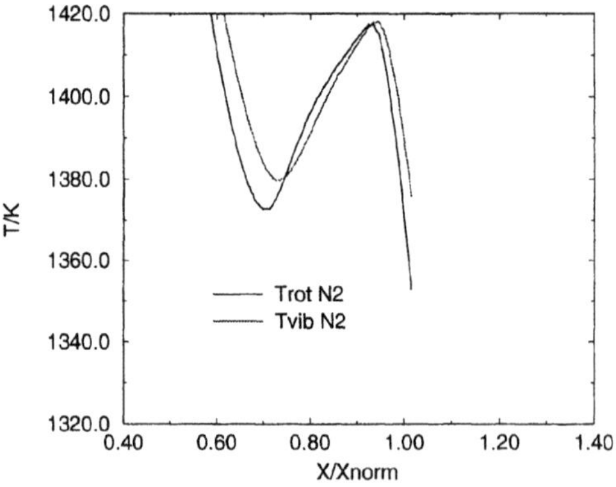

Figure 8: Largest Difference Between the Rotational and Translational Temperature

Experimental Investigation of a Hypersonic Inlet for the TSTO-Configuration ELAC

R. NEUBACHER, A. HENCKELS and T. GAWEHN
Deutsches Zentrum für Luft- und Raumfahrt,
Windkanäle Köln, Linder Höhe, D-51147 Köln, Germany

Summary

This paper describes viscous flow phenomena occurring on external compression ramps of a hypersonic RAM inlet for the two stage to orbit configuration ELAC. Main design parameters for the inlet ramps and side walls are the pressure and the thermal loads. In this study these heat loads were measured by infrared thermography on an inlet model in the H2K hypersonic facility at Mach 6. Additional information about the viscous ramp flow were gained by oil flow diagnostics, Schlieren optics and Pitot pressure measurements. The influence of model parts like side walls or cowl were examined. Separation regions were visualised with respect to the boundary layer state. The measured data may be used for validation of numerical tools. As the main result of the study recommendations for the improvement of the inlet geometry are given.

1 Introduction

After completion of the national Hypersonic Technology Program (HTP), the German Research Foundation (DFG) initiated three special collaborative research programs which aim at the development of two stage to orbit configurations. One of these programs, the SFB 253 "Fundamentals of Design of Aerospace Planes", is oriented towards the development of the reference configuration ELAC (*EL*liptic *A*erodynamic *C*onfiguration), propelled by six air breathing ramjet engines. This propulsion technology is an interesting option for hypersonic vehicles, because the oxidator needs not to be carried within the plane, but is taken from the atmosphere by an aerodynamic compression system (inlet) that provides the air required for the combustion.

The aerodynamic quality, i.e. the mass flow and the pressure recovery of the inlet should be as high as possible, because this essentially controls the efficiency of the whole ramjet. Thereby all inlet components have to fulfil high design requirements, particularly with regard to pressure and thermal loads. Therefore a lot of effort was spent in the optimization of ramjet inlets. In the past a number of research inlets were developed and tested at DLR in Cologne [1]. The object of the present study is the viscous flow field upstream the cowl shock at the ramps of the compression system. Here the thermal loads were measured on an inlet model by infrared thermography. Additional information about the ramp flow were gained by oil flow diagnostics, Schlieren optics and Pitot pressure measurements. The particular influence of the cowl shock was examined, too.

On the compression ramps of the inlet, boundary layer separation may be induced, due to the large positive pressure gradient created by the ramp shocks. In the literature criteria to predict boundary layer separation on ramps, but mostly without side wall influence, are found e.g. in [2, 3]. For the discussed inlet geometry, these criteria predict a laminar boundary layer separation upstream of the second ramp or a turbulent boundary layer separation upstream of the third ramp. The experimental verification of such predictions are of high relevance, because separation change shock structure drastically, leading to an off-design operation. For the present study separation regions were visualised with respect to the state of the boundary layer, in order to provide recommendations to improve the inlet geometry.

2 Experimental Techniques

The study was carried out at the hypersonic blow down facility H2K at DLR Cologne. Depending on the flow condition, the facility offers test times up to 30 seconds. To avoid condensation effects the flow upstream of the nozzle was heated by electrical heaters to 500K. Most of the runs were performed at Mach 6.0 using an axisymmetric Laval nozzle with 600 mm exit diameter. For the presented runs the Reynolds number was adjusted by the stagnation pressure. The inlet model was tested for unit Reynolds numbers of $5 \cdot 10^6\ m^{-1}$ and $18 \cdot 10^6\ m^{-1}$, which corresponds to Reynolds numbers $Re_L = 2.1 \cdot 10^6$ and $7.5 \cdot 10^6$ respectively for a overall model length of 420 mm.

The inlet model, scaled down to 1:21, features three compression ramps and geometrical dimensions as shown in Figure 1. It was mounted inside the plenum chamber of the facility behind the nozzle exit. To provide better optical access for the infrared thermography, the cowl of this inlet was removed a short distance downstream of the lip. The angle of the cowl lip to the incoming flow was set to 10° simulating an internal compression of about 20 percent. To avoid mass flow over the ramp sides, the inlet has walls which were equipped with glass plates for the optical access.

To reduce lateral heat transfer the compression ramps were machined out of PTFE (Teflon®), which were coated by a thin layer of black paint to ensure a high emissivity for infrared measurement. This permits the calculation of heat fluxes according to the one dimensional heat equation, i.e. for boundary conditions of a semi-infinite slab [4]. Thereby the temperature sensitivity of the thermal properties of PTFE requires to solve the nonlinear heat equation. The boundary condition at the wall surface was taken from the infrared measured temperature history, whereby the heat radiation between model and wind tunnel walls had to be taken into account. Due to aerodynamic forces and thermal loads at the leading edges, these parts of the ramps were machined out of metal. Because of thermal distortions the heat flux at these leading edges was not evaluated.

The infrared technique was supported by further measurement and visualization techniques: As far as possible, the position of shocks inside the inlet was visualized by Schlieren optics. Surface flow phenomena were shown by oil flow visualization images, which were generated using Silicon oils of different viscosities, mixed with fluorescent particles. To examine the flowfield in the external compression region of the inlet, a Pitot pressure rake with 15 probes of 0.6 mm inner diameter was used. A 45 PSID scanner module with 32 ports converted the measured pressures into electrical signals.

3 Results

3.1 Inlet flow for a natural boundary layer

At flight condition of Mach 7.5 the bow shock in front of the ELAC vehicle reduces the flow Mach number at the entrance of the inlet to Mach 6. To avoid that the thick boundary layer developing on the forebody of the vehicle enters the inlet, this boundary layer has to be diverted. In this case a new laminar boundary layer grows on the first compression ramp, becoming transitional about one meter behind the leading edge of the 4.25 m long first ramp of the real inlet. In the experiments the maximum achievable unit Reynolds number was $18 \cdot 10^6$ m^{-1}, about seven times smaller than in flight. A corresponding infrared image at tunnel condition visualizes turbulence wedges at the first compression ramp of the model (Figure 2). Furthermore, infrared and oil flow images prove that for this flow condition no boundary layer separation occurs.

The initiation of boundary layer transition on the first ramp is identified by three heat load profiles, which were taken parallel to the main flow direction and are added to Figure 2. For reasons of comparison a profile based on the relation of Eckert [5] for laminar flat plate heat transfer is plotted. The measured heat data between two turbulent wedges (Line 1) show a reasonable agreement to Eckert's relation, confirming a state of laminar boundary. The measured profile recorded inside such turbulence wedge (Line 2) shows enhanced convective heating and proves the initiation of a transition process [6]. On the second ramp both heat load profiles converge; i.e. the heat transfer is less effected by the presence of turbulent spots upstream on the first ramp. Figure 3 shows two heat load profiles taken in spanwise direction. They indicate that the heat peaks generated by turbulent wedges on the first ramp (Line 3) are smoothed on the second ramp (Line 4).

Directly downstream the second ramp shock Hung and Barnett's correlation for peak heating predicts a heat load increase by the factor of two for a laminar boundary layer and by a factor of fourteen for a turbulent boundary layer [7, 8]. At this location, a reliable data evaluation was not possible in the experiment due to heat conduction effects on second ramp's metal leading edge. But 23 mm downstream on the PTFE surface a Stanton number seven times higher than on the first ramp was found. From these facts it is concluded that the shock boundary layer interaction induced by the second ramp leads at least to a uniform transitional boundary layer with a heat transfer rate that is significantly larger than inside a laminar boundary layer.

3.2 Boundary layer tripping

For the real inlet boundary layer transition is initiated rather soon downstream of the first leading edge, as already mentioned above. To assure a geometrically more realistic transition location on the model inlet, tripping was necessary. Therefore a wire of 1 mm diameter was mounted in spanwise direction, 10 mm downstream of the first leading edge. To ensure its tripping effect, four runs, i.e. with and without trigger and at two different free stream Reynolds numbers were performed. The resulting heat load profiles on the first ramp are displayed in Figure 4.

At a low Reynolds number of $Re_L = 2.1 \cdot 10^6$ and without tripping the heat load is rather constant up to a position of 120 mm. Further downstream a decrease of the heat load denotes the beginning of a laminar boundary layer separation, as it was described for instance by Delery [6, 7]. This separation is visible in the oil flow pattern and the infrared image of Figure 5. In case of tripping, the heat load profile shows a rather strong increase along the first 120 mm to 160 mm as well as small fluctuations indicating the development of a transitional boundary layer. Further downstream, similar to the flat plate turbulent boundary layer, the heat load decreases. This result demonstrates that for an effective tripping the free stream Reynolds number has to be increased.

For the high Reynolds number without tripping, the boundary layer becomes transitional as already demonstrated in section 3.1. With tripping the heat load between x=80 and x=170 mm is significantly higher than without tripping, proving evidently an upstream shift of the transitional region of the boundary layer.

3.3 Turbulent boundary layer separation

Figure 6 displays an infrared image, showing the flow field close to the wall for the tripped boundary layer at high Reynolds number. Compared to the laminar boundary layer the turbulent one carries a larger energy and is less sensitive against pressure rise. In this case the separation occurs downstream on the second ramp and extends towards the side walls, generating a complex three dimensional corner flow field.

As reflected by the "free interaction theory" [7] the pressure rise in turbulent separation is by far more important than in laminar separation. The lip shock pressure rise induces a separation bubble with an embedded high temperature reverse flow. Thereby, particularly the third ramp is exposed to an enormous heat load. This is demonstrated by Figure 7 in comparison with the heat load measured for the same flow condition but with removed lip. From that it is noticed that a reduction of the internal compression decreases the heat load on the third as well as on the second ramp.

Further information about the corner flow field was gained by Pitot pressure measurements at a position of 300 mm downstream of the leading edge. Due to access problems these measurements were performed for the configuration without lip shock generator. The influence of the Reynolds number on the developing corner flow structure is demonstrated by the contour plots in Figure 8. These plots show the Pitot pressure normalized by the free stream total pressure, visualizing shocks developing at the inlet side walls. Above the nearly 6 mm thick boundary layer an homogeneous high pressure region is generated by the ramp shocks. The increase in Reynolds number from $Re_L = 2.1 \cdot 10^6$ to $7.5 \cdot 10^6$ reduces the boundary layer thickness only slightly.

Additionally, it was found, that the position and size of the separation bubble are very sensitive to slight changes of the flow conditions or to model modifications. For identical free stream conditions, the local energy level of an early tripped boundary layer is different compared to a natural undisturbed boundary layer with downstream transition. This may explain that in the case without tripping the transitional boundary layer does not separate at the second ramp.

3.4 Surface flow at yaw angles

A challenging topic is inlet off-design operation, for instance if the inlet is influenced by yawing flight conditions. Here the ramp flow is no longer parallel with the main flow direction which leads to reduced deflection angles at the ramps for the streamline. As mentioned before, the separation bubble is very sensitive against slight flow changes. Therefore one may expect that separation upstream of the third ramp would vanish in this case.

Within the present study, wind tunnel tests with the inlet at 1°, 3° and 5° angle of yaw were performed. Figure 9 shows the resulting oil flow and infrared image at 3° angle of yaw, indicating a separation bubble being shifted towards the windward side of the inlet wall. Additionally, a curvature in the surface streamline direction is detected on the first ramp. Figure 10 quantifies this curvature by presenting the surface velocity angle in spanwise direction some distance downstream of the leading edge and at the beginning of the second ramp. Here, on the windward side of the inlet, the surface flow was subjected to a redirection of nearly 10° due to pressure increase and hence boundary layer separation was initiated. Finally, Figure 9 visualizes the heating due to the reattaching boundary layer on the third ramp.

4 Conclusions

The experimental results pointed out that the boundary layer strongly influences the flow field on the compression ramps. In agreement with boundary layer predictions from literature, laminar boundary layer separation was confirmed upstream of the second ramp, while turbulent boundary layer separation was found upstream of the third ramp (Table 1). According to flight condition the most interesting test case of the inlet was at a Reynolds number of $7.5 \cdot 10^6$ with boundary layer tripping. Here, the formation of the separation bubble was found to be sensitive on changes of the flow condition or model geometry. With regard to side slip the inlet flow featured an unsymmetrical separation displaced towards the windward side wall.

By infrared thermography a high heat load occurring particularly at the third ramp was pointed out. Referring to the fact that the inlet was adjusted to the low speed flight condition by movable ramps, a design challenge became obvious. Based on this study, following recommendations may be given: The heat load resulting from boundary layer reattachment downstream of the third ramp should be reduced, either by an decrease of the third ramp angle and/or boundary layer bleed implementation [9]. To guarantee the same combustion pressures, the internal compression has to be increased, whereas choked inlet condition has to be avoided. Due to the boundary layer on the ramps the leading edge shock passes upstream of the lip, i.e. spillage occurs. Therefore the length of the first ramp should be reduced and ramp angles should be adjusted. Otherwise the position of the shock generator should be shifted upstream to avoid this spillage.

Acknowledgements

The presented investigations were carried out at the Institute for Fluid Mechanics, Wind Tunnel Division of the German Aerospace Center (DLR). This work is part of the DFG Sonderforschungsbereich 253 at the University of Aachen. The authors gratefully acknowledge the support by DFG.

References

[1] E.-O. Krohn und K. Triesch. "Untersuchungen an den ebenen Einläufen S01C und S01D im TMK". DLR Interner Bericht, IB-39113-95C01, 1995.

[2] D. A. Needham and J. L. Stollery. "Boundary-Layer Separation in Hypersonic Flow". AIAA-Paper No. 66-455, Juni 1966.

[3] W. H. Heiser and D. T. Pratt. "Hypersonic Airbreathing Propulsion". AIAA Education Series, J. S. Przemieniecki (ed.), Washington DC, 1994, pp. 240-242.

[4] U. Grigull and H. Sandner. "Wärmeleitung". Berlin, Springer-Verlag, 1979.

[5] E.R.G. Eckert. "Engineering Relations for Heat Transfer and Friction in High-Velocity Laminar and Turbulent Boundary-Layer Flow Over Surfaces With Constant Pressure and Temperature". ASME, Paper-No. 55-A-31, 1955.

[6] J. Delery. "Shock-Wave/Boundary-Layer Interactions in High-Mach-Number Flows". Experimental and Computational Capability, Improvement and Validation, AGARD-AR-319, 1996.

[7] J. Delery. "Shock/Shock and Shock-Wave/Boundary-Layer Interactions in Hypersonic Flows". Special Course on Aerothermodynamics of Hypersonic Vehicles, AGARD-R-761, 1988, pp. 9-16 to 9-23.

[8] F.T. Hung and D.O. Barnett. "Shockwave-Boundary Layer Interference Heating Analysis". 11th Aerospace Sciences Meeting, AIAA Paper No. 73-237, 1973.

[9] D. Schulte, A. Henckels and I. Schell. "Boundary Layer Bleed in Hypersonic Inlets". Notes on Numerical Fluid Mechanics, Vol. 60, 1996, pp. 296-303.

Table 1 Boundary layer state and separation

Re_L	Tripping	Begin of Transition	Position of separation	Boundary layer state
$2.1\cdot10^6$	No	---	Upstream 2nd ramp	Mainly Laminar
$7.5\cdot10^6$	No	1st ramp	No	Turbulent downstream 2nd ramp
$2.1\cdot10^6$	Yes	1st ramp	No	Transitional
$7.5\cdot10^6$	Yes	Downstream trigger	Upstream 3rd ramp	Transitional/ turbulent

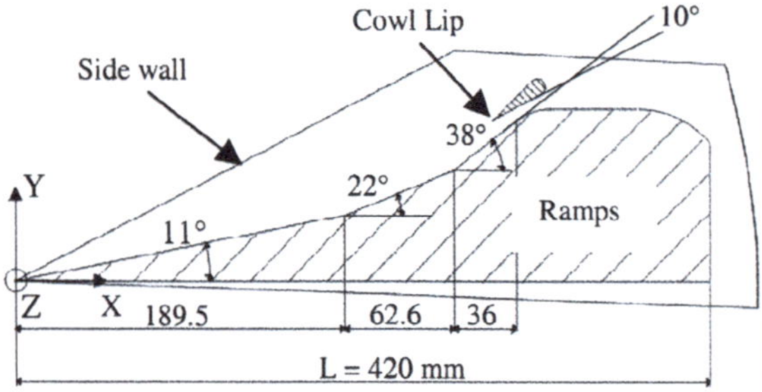

Figure 1 Sketch of inlet model with dimensions (left) and photo of inlet model (right)

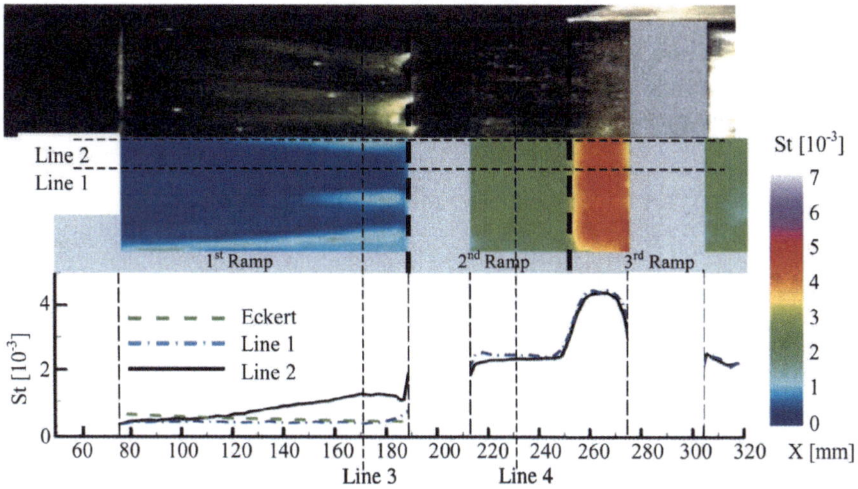

Figure 2 Development of turbulence on first ramp (M=6; Re_L=7.5·10^6; no tripping) and calculated (Eckert) heat load on the first ramp

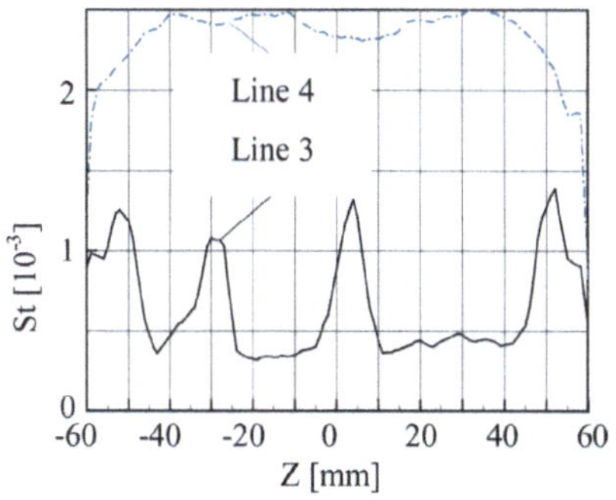

Figure 3 Spanwise Stanton number distribution for Figure 2

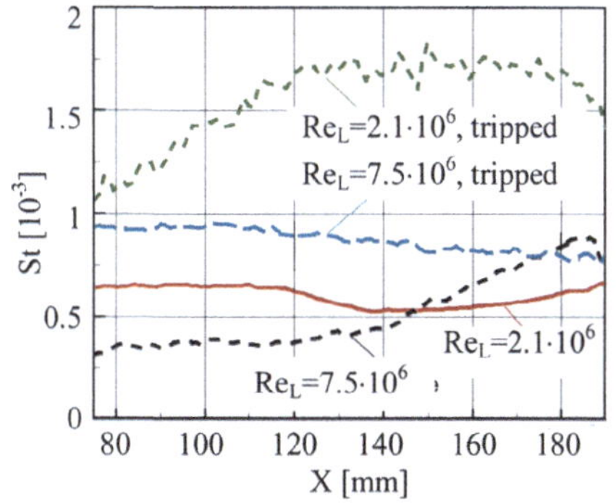

Figure 4 Stanton numbers on the first ramp for different boundary layer conditions

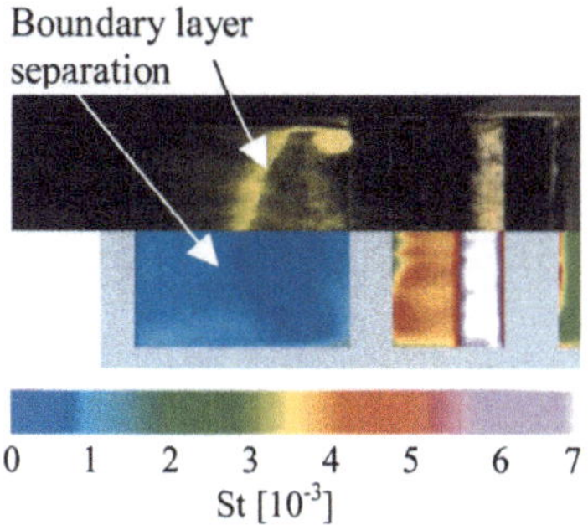

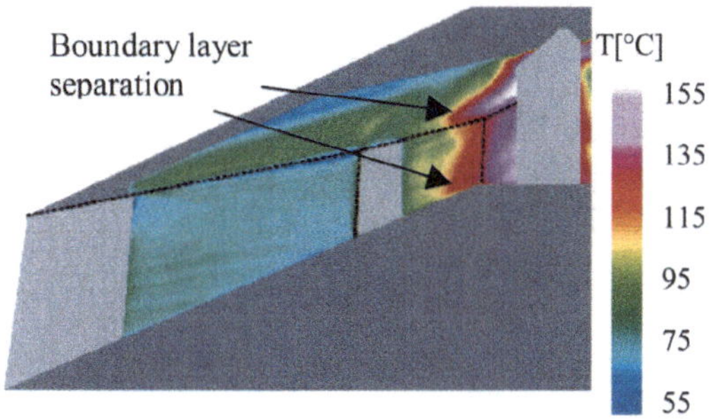

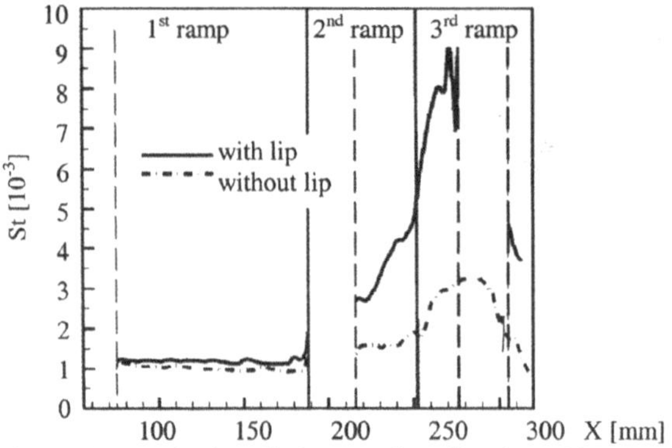

Figure 7 Heat load profile in mean flow direction with and without lip shock (M=6, $Re_L=7.5\cdot10^6$; tripped)

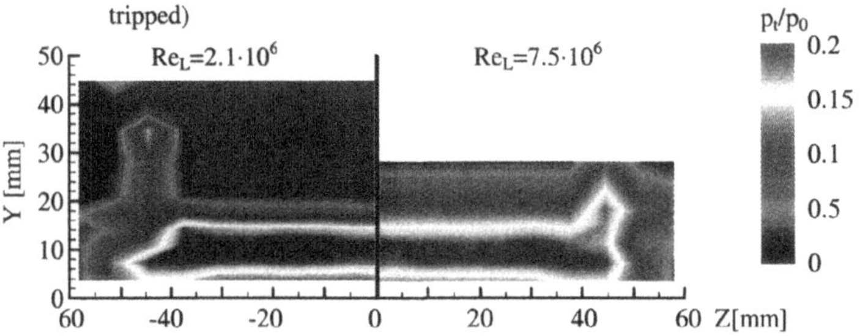

Figure 8 Pitot pressure contours at X = 300 mm for configuration without lip shock (M=6; tripped)

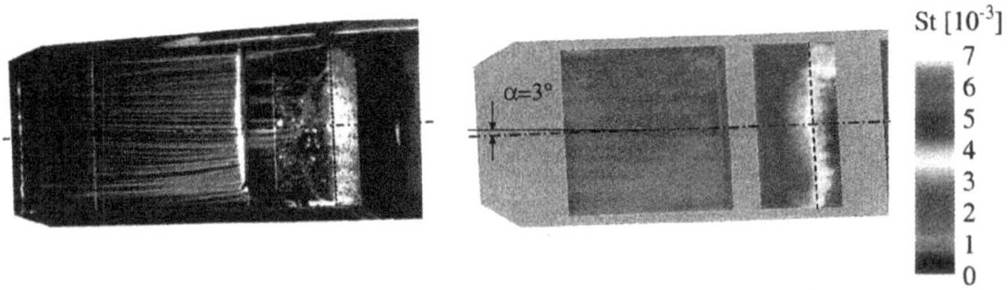

Figure 9 Oil flow and infrared image for 3° yaw angle (M=6; $Re_L=7.5\cdot10^6$, tripped)

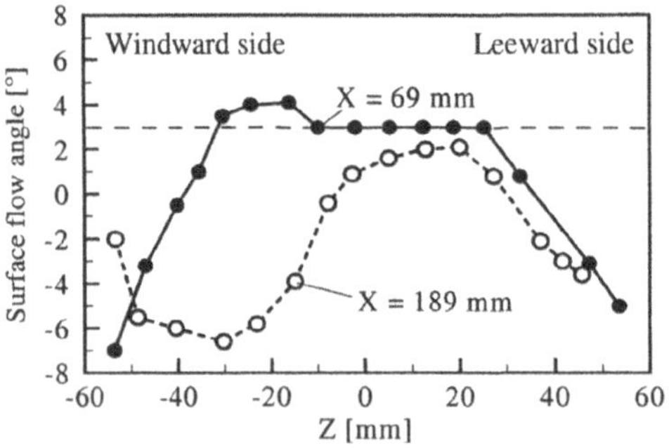

Figure 10 Surface flow angle as taken from the oil flow image Figure 9

Aerodynamic Analysis of the Dual-Mode Ramjet Vehicle JAPHAR

Thino Eggers

DLR, Institute of Design Aerodynamics, Lilienthalplatz 7, 38108 Braunschweig, Germany

Summary

Within the DLR - ONERA cooperation JAPHAR (**J**oint **A**irbreathing **P**ropulsion for **H**ypersonic **A**pplication **R**esearch), the feasibility of a dual-mode ramjet engine, combining subsonic and supersonic combustion, is investigated and a methodology for on-ground and in-flight performance demonstration is developed [1]. The studies are organised around the blue-print of the experimental vehicle JAPHAR (Fig. 1) which is considered to fly between M=4 and M=8 in order to cover both regimes, subsonic and supersonic combustion. The main objective of the vehicle is to provide realistic specifications for the different components of the propulsion system. It answers to one of the fundamental peculiarities of airbreathing propulsion, which is the high level of integration between vehicle and propulsion system requiring a strongly coupled design. This is important not only to establish the propulsive balance but also to obtain a realistic estimation of the component performances. The paper focusses on the aerodynamic analysis of the main design drivers of the vehicle which are the available thrust, longitudinal trim, longitudinal stability and the transition between the flight modes.

1. Introduction of the JAPHAR Vehicle

Preliminary studies for the definition of the JAPHAR vehicle discussed in [2] and [3] considered two concurrent approaches for an experimental vehicle satisfying the project specifications of JAPHAR. DLR considered a waverider vehicle while ONERA proposed a wing-body configuration. Both vehicles were consequently defined with the same propulsion system, which includes inlet, combustor and nozzle. This allowed to compare both concepts from the propulsive performance point of view as well as an investigation of their aerodynamic behaviour. The results of this study showed that the waverider allows a drastical improvement of the obtainable thrust but that it also revealed severe trim problems. As a consequence the excellent forebody characteristics of the waverider principle and the better volumetric efficiency and the smaller trim losses of a wing-body concept "V1" were combined by definition of the improved vehicle "JAPHAR". Some basic data of both vehicles are introduced in Fig. 2. They are the basis for the comparison of both configurations discussed in a later paragraph.

2. Assessment of the Aerodynamic Coefficients

Basis for the calculation of the aerodynamic coefficients was the DLR Shock-Expansion-Code HOTSOSE [4] and an appropriate trim program which also contains a bookkeeping procedure. This methodology allowed to take into account all forces acting on the vehicle and each component of the propulsion system inducing not only thrust but also lift, especially with view to the single-expansion ramp nozzle (SERN) that is used in the present case. The principle of the bookkeeping method is illustrated in Fig. 3. In order to obtain the total aerodynamic coefficients of the vehicle the coefficients of lift, drag and pitching moment had to be calculated for the outer surface of the vehicle (subcript: "surf."). The applied surface grid is shown in Fig. 4a. During these calculations the inlet area and the nozzle area (Fig. 3) were not taken into account. The contribution of the propulsion system then was introduced by addition of the aerodynamic coef-

ficients induced by the inlet, the combustor and the nozzle (subcripts: "inlet", "comb.", "nozzle", Fig. 3). Applying this method the complete thrust vector is contained within the aerodynamic data. The real entrance conditions of the propulsion system were obtained by 3D Navier-Stokes computations of both forebodies. These were performed assuming a turbulent boundary-layer in order to have a safety margin with view to the boundary layer thickness and the heatloads. The effect of initially laminar flow was not investigated within the forebody study. The entrance conditions were used, along with experimental results obtained for the inlet, to compute the combustor performances [2]. The SERN nozzle characteristics were obtained from preliminary Euler simulations. For the study discussed here, simple „1-D" computations were used to estimate the contributions of the propulsion system [3]. In further stages, one could use the results of 3D Navier-Stokes computations that will integrate, step by step, the real supply conditions of the upwind components. The pitching moments of the vehicle and its components were determined for the actually assumed center of gravity. Basis data for the trim calculations were the total aerodynamic coefficients C_L, C_D and C_m. A worst case estimation of the expected skin friction drag was performed based on the assumption of a turbulent flat plate as $C_{D,f} = c_f \cdot A_{wet}/A_{ref}$ [3]. The aerodynamic data cover the complete flight regime.
In order to reduce the uncertainties additional 2D-Euler calculations for the symmetry plane of the vehicle were performed with hybrid DLR TAU-Code [5]. Here, the isolated upper and lower profiles of the symmetry plane as well as the complete plane with all interactions inside the engine were considered. The $M_\infty=8$ result is illustrated in Fig. 5. Extensive grid adaptions, especially inside the engine, allow to resolve the complete internal and external shock system. The comparison of the results for the initial grid and the final adapted grid shows that the differences for C_L, C_D and C_m were always smaller than 3%.
The exemplary comparison of the $M_\infty=8$ TAU pressure coefficients and the aerodynamic coefficients for the isolated upper part of the symmetry plane with the corresponding HOTSOSE results shows an excellent agreement (Fig. 6). Thus, the expected accuracy of HOTSOSE for the case with propulsion is very high because the bookkeeping calculations only require the contribution of the outer vehicle surface.
Additionally, it was planned to consider also the case without propulsion. In order to allow this a second grid with additional blocks inside the engine was applied to estimate the pressure coefficients inside the engine and in the nozzle (Fig. 4b). Assuming that the effects of upper and lower engine wall together only yield a minor contribution to the total coefficients, the results were expected to be realistic. Unfortunately, this approach neglects the complex shock-shock-interactions inside the engine. In Fig. 7 the pressure coefficients of the internal part of the engine are shown for the case with interactions (complete symmetry plane). The comparison of TAU and HOTSOSE pressure distributions points to the fact that the shock-shock-interactions may not be neglected. The HOTSOSE result underlines the initial assumption, that the effects of upper and lower engine wall revoke each other. But the careful analysis of the TAU results shows that the internal part of the engine produces a substantial contribution to the total aerodynamic coefficients. This leads to the situation that the HOTSOSE lift coefficient is positive for the complete symmetry plane while the one of TAU is negative (Fig. 7). These results point out that the effect of the engine on the total aerodynamic coefficients of the vehicle may only be assessed by nose-to-tail calculations or force measurements.

3. Aerodynamic Analysis of the JAPHAR Vehicle

Within the aerodynamic analysis of the JAPHAR vehicle for the case with bookkeeping the present paper focusses on the effects of wing planform, wing airfoil and upper surface design on the propulsive efficiency, the trim results and the longitudinal stability. In the following the ONERA "V1" vehicle (Fig. 2 and Fig. 3) was considered as a reference because the aerodynamic study of [2] indicated that a wing-body combination is the most promising vehicle with view to flyability, trim and structural design. A comparison of the JAPHAR configuration and the "V1" vehicle is given in Fig. 2. It is obvious that the JAPHAR vehicle is more slender. Additionally, due to the shorter waverider forebody the length is reduced roughly by 0.5 m. As indicated by the smaller planform area this results in a higher wingload in order to allow angles of attack as high as possible. Fig. 8 compares the trim results of both configurations for q=80kPa and the complete considered Mach number range. The angles of attack obtained for the JAPHAR configuration only slightly differ from those of configuration "V1". On the upper surface of the forebody a compression surface had to be introduced in order to match the waverider forebody and the modified body of configuration "V1". This adaption partly compensates the very high lift of the lower waverider shape and allows to fly at angles of attack comparable with those of configuration "V1". The obtained thrust ratio is even higher than for the initial waverider vehicle [2]. Due to the fact that the forebody dominates the total lift, which is still higher than the one of the "V1" forebody, the flap deflections are increased in comparison to the reference vehicle. Another negative effect of the high forebody lift is the decreased stability margin. The objective of the given geometric parameter study was to indicate possibilities to reduce the flap deflections and to increase the longitudinal stability while preserving the high propulsive efficiency of JAPHAR. In order to reach this goal the neutral point of the configuration had to be positioned to a more backward position in order to be next to the point of neutral longitudinal stability.

One possibility to do this was to change the wing planform keeping the wing area constant (Fig. 9). This allowed to shift the center of pressure of the wing backward. In average it is possible to reduce the flap deflections by $\Delta\eta \approx 0.6°$ and to increase the stability margin by roughly 1% of the body length. The effect on angle of attack was negligible.

Another possibility to shift the center of pressure of the wing backward was to change its airfoil. This was performed based on wing C (Fig. 9) which gives the original wing load of JAPHAR. Five additional simple double wedge airfoils were considered in comparison to the originally used NACA 0004 (Fig. 10). These airfoils were not chosen to optimize the airfoil definition. The goal of the study was to estimate the effect which might be obtained by an adapted airfoil. Airfoils A and B allowed to assess the effect of the maximum thickness position t_{max}/l on the trim results. Airfoil C yields the effect of the upper surface. This was chosen parallel to the freestream keeping the thickness constant and allowed to produce the complete wing lift on the lower surface as it is proper for hypersonic flow conditions. A better balancing of the forebody lift was expected. Airfoils D and E allowed to investigate the effect of the maximum thickness position t_{max}/l and the thickness of the airfoil. The thickness itself was slightly increased from 4% to 6% for airfoil E. The influence of the airfoil on angle of attack was negligible for all cases. The position of the maximum thickness was of minor interest as long as positive pressures are produced along the upper surface (airfoil A and B). If the wing lift is produced on the lower surface (airfoil C) a noseward pitching moment is the result and in comparison to the original NACA airfoil a slight reduction of the flap deflection is obtained. The main effect was obtained by adaption of the maximum thickness position (Fig. 10). Airfoils D allows a reduction of $\Delta\eta \approx 1.5°\text{-}2°$ and an increase of the stability margin by 2% of the body length. An additional positive effect was obtained for increased thickness (airfoil E). Now the flap deflections were even smaller than those of the "V1" vehicle. Nevertheless, although the obtained flap deflections are promising, the configuration is still highly unstable ($-0.07 \geq (x_N - x_S)/l \geq -0.09$). Summarising these results it should be possible to reduce the flap deflections about $\Delta\eta \approx 3°$ only by

adaptation of the wing airfoil. The influence of the airfoils on angle of attack is very small because the total lift of the configuration is dominated by the forebody. This is also the reason why the stability margin cannot be sufficiently influenced by pure aerodynamic changes. In order to increase it to values of $(x_N-x_S)/l \geq -0.05$ the center of gravity of the configuration should be shifted to the region of $x_S/l \leq 0.59$.
A possibility to reduce the lift of the forebody and to produce more lift in the backward part of the body is to modify its upper surface. As shown in Fig. 11 this might be done by a simple thickening in the backward part of the body. As a result an additional compression in the forebody region reduces the forebody lift, and the expansion near the trailing edge of the vehicle gives an additional noseward pitching moment which allows to reduce the flap deflection about $\Delta\eta \approx 1°$. On the other hand, the disadvantage of the thickened upper surface is the decrease of the stability margin by roughly 1% in the case given here.
Based on this knowledge finally the "V1" and a modified JAPHAR vehicle were compared (Fig. 12). The considered JAPHAR configuration combines all positive effects found by the geometric adaptions discussed before. The figure shows that the flap deflections of the JAPHAR configuration may be reduced by $\Delta\eta \approx 2°$ in comparison to the values of the „V1" vehicle. The chosen modifications of the wing planform and the airfoil are sufficient to reduce the maximum flap deflection at $M_\infty=4$ from roughly $\eta \approx 7°$ to a value of $\eta < 4°$ which is even more promising than the reference configuration. The results were obtained at thrust ratios which are between 30% and 110% higher than those of the "V1" vehicle. The obtained thrust ratios never fall below 2.2. In comparison to the „V1" vehicle the longitudinal stability is reduced but if it is assumed that a center of gravity at $x_S/l=0.59$ may be realised the stability margin could be increased to values of $(x_N-x_S)/l \geq -0.07$. Further gains could be obtained by adaptation of the wing area and its planform.

4. Conclusions

The dual-mode ramjet vehicle JAPHAR was analysed by application of the hybrid DLR TAU-Code and the Shock-Expansion-Method HOTSOSE. The comparison of both methods points to the fact that the vehicle without engine effects can only be investigated by complete nose-to-tail-calculations or force measurements. In case that only the pressure distribution of the outer vehicle surface is required, the HOTSOSE results show a high accuracy. Here, the contribution of the engine is considered using a bookkeeping method.
The analysis of geometric adaptions of wing planform, wing airfoil and a modified upper surface showed that the required flap deflections are very small over the complete considered Mach number range if all positive effects are combined in a slightly adapted definition of the JAPHAR vehicle. The ratio between available thrust and required thrust is always higher than 2 and provides an appreciable safety margin with view to the uncertainties of the propulsion dataset.
Remaining problem is the small stability margin of $(x_N-x_S)/l \leq -0.05$. The total aerodynamic coefficients are dominated by the forebody and the low wing lift does not allow to increase the stability margin sufficiently. The only efficient way to increase the stability margin seems to be an adaptation of the center of gravity position.

References

[1] Ph. Novelli, W. Koschel: "JAPHAR: A Joint ONERA-DLR Research Project on High Speed Airbreathing Propulsion". XIV ISABE Congress, Paper IS-093, Florence, 1999.

[2] Th. Eggers, Ph. Novelli: "Design Studies for a Mach 8 Dual Mode Ramjet Flight Test Vehicle". AIAA Paper 99-4877, 1999.

[3] Th. Eggers: "Studies for the Design of a Dual-Mode Ramjet Flight Test Vehicle". DLR-IB 129-2000/1, 2000.

[4] U. Reisch, Y. Anseaume: "Validation of the Approximate Calculation Procedure HOTSOSE for Aerodynamic and Thermal Loads in Hypersonic Flow with Existing Experimental and Numerical Results". DLR-FB 98-23 (1998).

[5] M. Galle: "Ein Verfahren zur numerischen Simulation kompressibler, reibungsbehafteter Strömungen auf hybriden Netzen". DLR-FB 99-04 (1999).

Figures

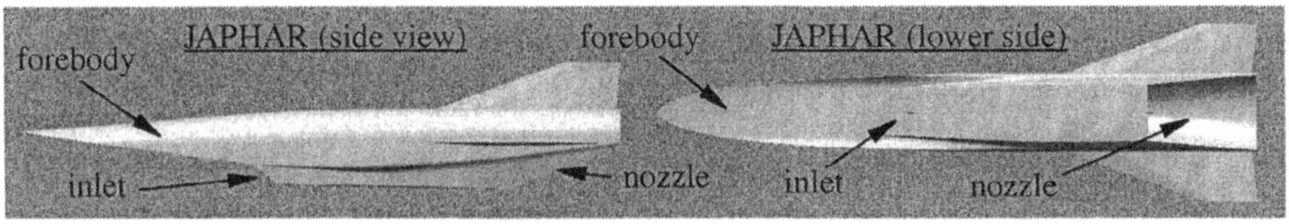

Fig. 1 The experimental vehicle „JAPHAR"

parameter	unit	„V1"	JAPHAR
l_{ref}	[m]	11.50	10.95
A_{ref}	[m^2]	17.55	15.63
volume	[m^3]	13.86	10.16
A_{wet}	[m^2]	53.56	48.97
span	[m]	3.18	3.09
aspect ratio	[-]	0.58	0.61

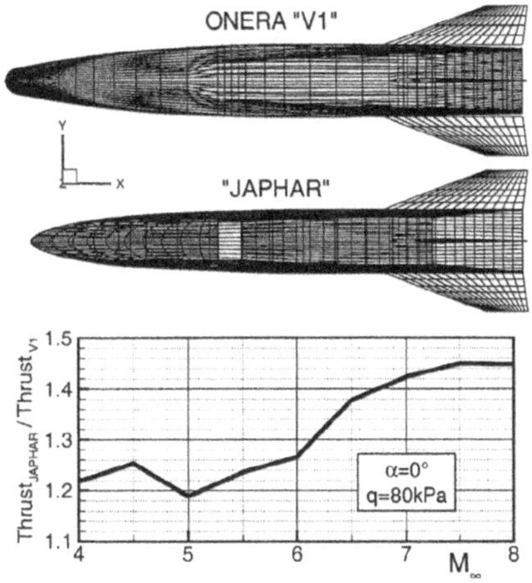

Fig. 2 Introduction of the considered vehicles

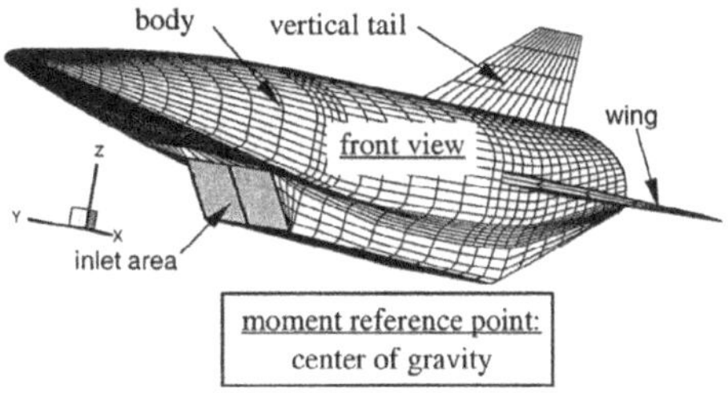

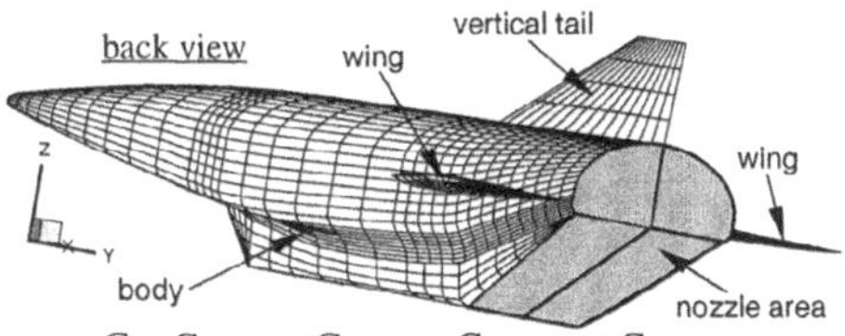

$C_L = C_{L,\ surf.} + C_{L,\ inlet} + C_{L,\ comb.} + C_{L,\ nozzle}$

$C_D = C_{D,\ surf.} + C_{D,\ inlet} + C_{D,\ comb.} + C_{D,\ nozzle}$

$C_m = C_{m,\ surf.} + C_{m,\ inlet} + C_{m,\ comb.} + C_{m,\ nozzle}$

Fig. 3 Principle of the bookkeeping procedure

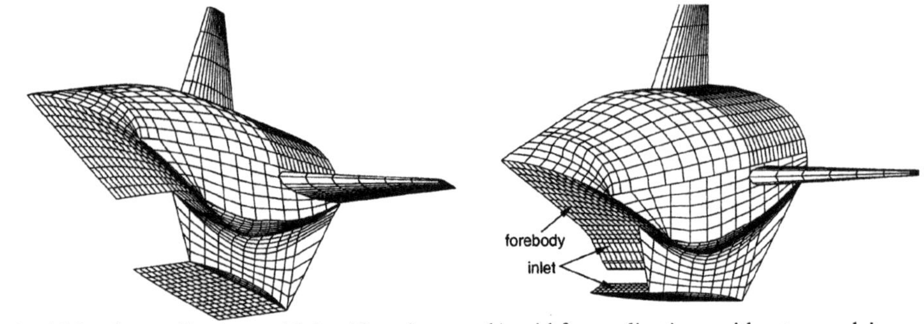

a) grid for the applications with bookkeeping b) grid for applications without propulsion

Fig. 4 HOTSOSE surface grids

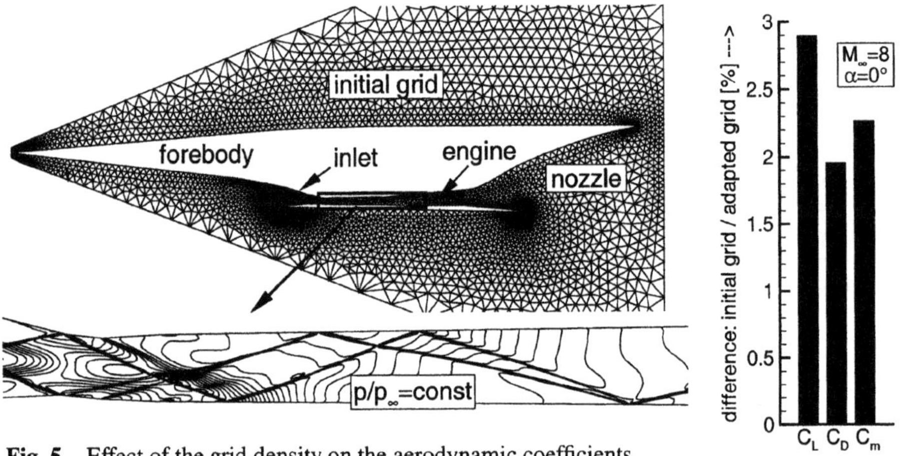

Fig. 5 Effect of the grid density on the aerodynamic coefficients

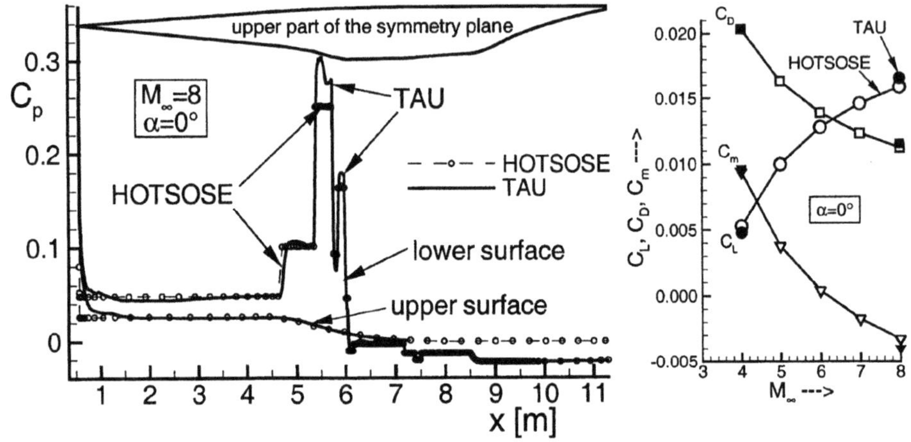

Fig. 6 Aerodynamic coefficients for the isolated upper part of the symmetry plane

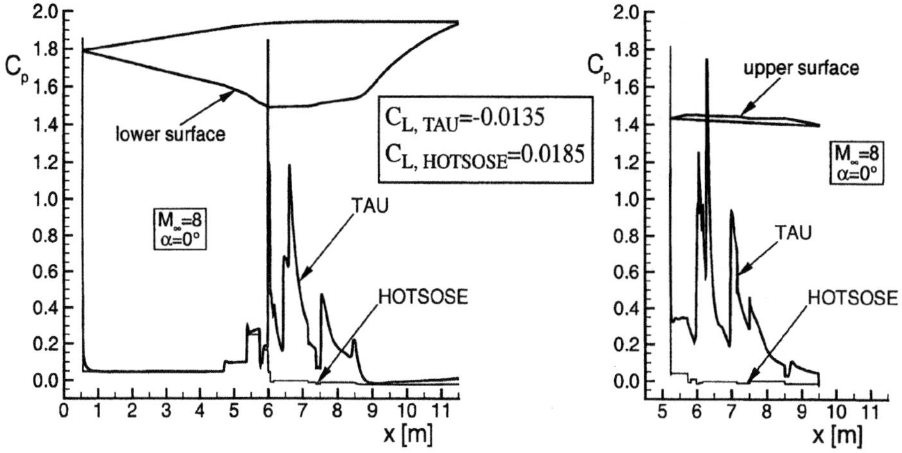

Fig. 7 Influence of the internal shock-shock-interactions on the pressure coefficients

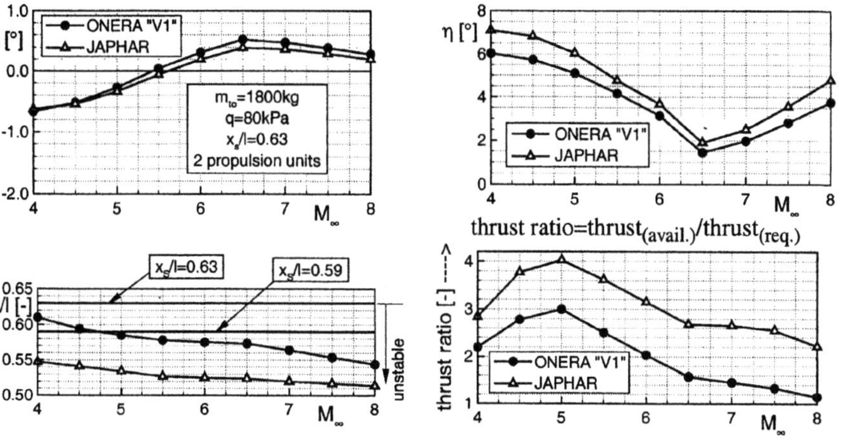

Fig. 8 Comparison of the trim results for configuration ONERA "V1" and JAPHAR

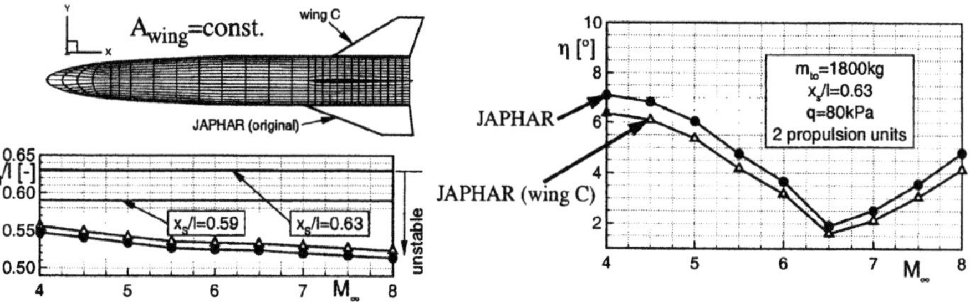

Fig. 9 Influence of the wing planform on the trim results of JAPHAR

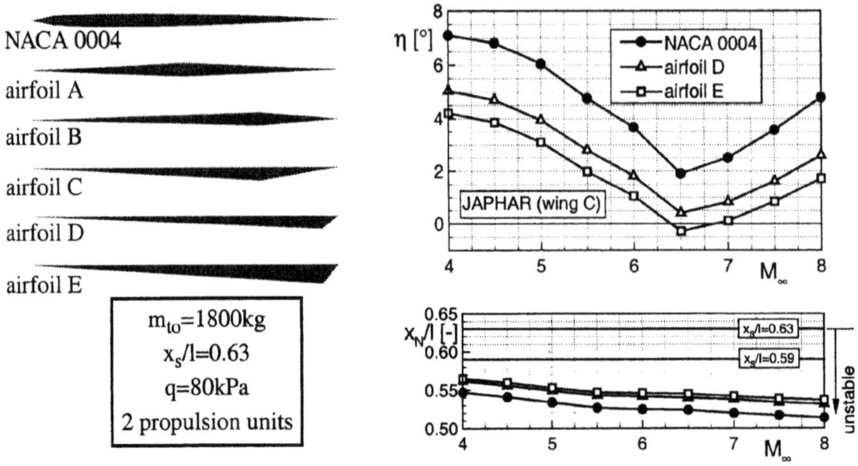

Fig. 10 Influence of the airfoil on the trim results of JAPHAR

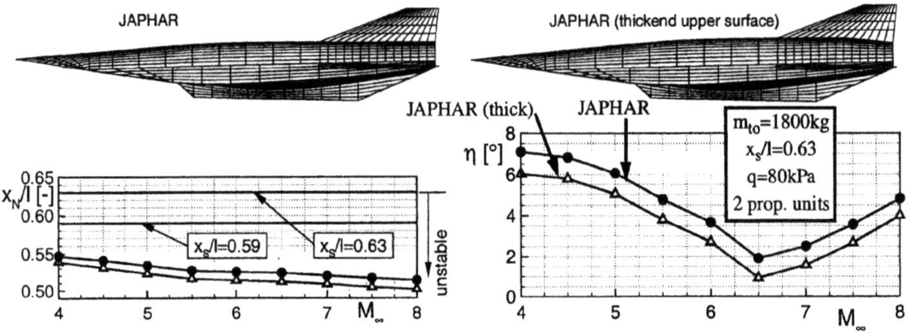

Fig. 11 Influence of the upper surface of the body on the trim results of JAPHAR

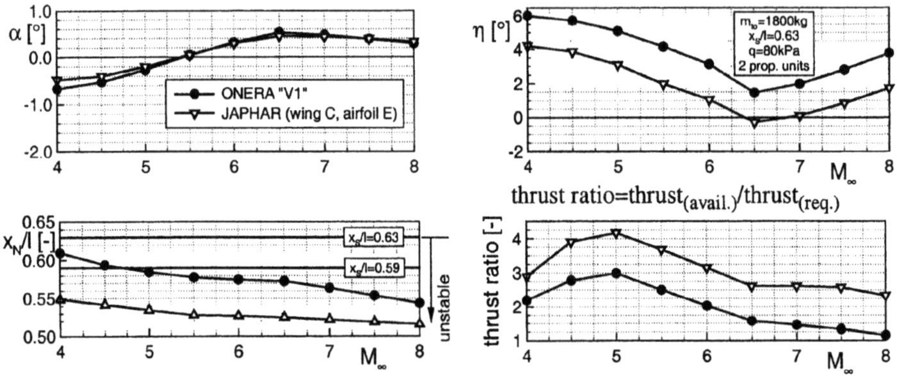

Fig. 12 Coefficients of final JAPHAR version compared to V1 design

Study of Shock-Wave/Boundary-Layer Interaction Flowfield on a 24° Ramp Model

S.B. VERMA (Humboldt Fellow) and G. KOPPENWALLNER(Head)
HTG, Hyperschall Technologie Göttingen, Max-Planck-Strasse 19
37191 - Katlenburg-Lindau, Germany

Summary

An experimental investigation was carried out to study the unsteadiness related to shock-wave/boundary-layer interaction flowfield using a two-dimensional 24° compression ramp model in a 8.94 Mach flow with air as test gas. All the tests were carried out at a charge tube pressure of 50 bar and T_0=773 K, corresponding to a unit Reynolds number of 2.078×10^5 m^{-1}. Off-surface analysis of the complete interactive flowfield is made using a Laser Schlieren system that measures time-dependent voltage signals representative of the fluctuating density gradient field. An array of 16 very-high-response photodiodes arranged on a 1 mm pitch were used for multichannel measurements. Analysis shows the shock-legs close to the model surface to vary in intensity and exhibit axial movement in accordance to the expansion and contraction movement of the separation bubble.

1 Introduction

It has been known for quite sometime now that all shock-wave/boundary-layer interactions exhibit some degree of unsteadiness [1, 2] that can generate severe surface pressure fluctuations, which under the influence of aerothermal effects, can pose significant problems to the structural integrity of the aircraft. Due to their inherent complexity, these interactions are only partially understood and hence, experimental studies are still necessary to gain better understanding of the flowfield physics. Most of the previous surface pressure studies reveal the external shock wave to be stationary while the shock foot in the viscous interaction region may undergo large-scale motion [3]. However, the structure of the interaction flowfield has been comparatively neglected or poorly understood. Therefore, an attempt is made to study this behavior at HTG.

The unsteadiness of the shock wave [1, 2, 4], particularly when the flow is separated, is one of the little known characteristics in shock-wave/boundary-layer interaction. Kistler [1] was perhaps the first to make fairly detailed high-frequency measurements in an attempt to characterize such unsteadiness providing valuable quantitative data for flow over forward-facing steps in supersonic boundary layers. Later studies focussed on the fluctuating wall pressure in the interaction regions of compression-ramp flowfields. Their results, although obtained from different flow geometries, exhibited statistical properties that were quite similar to those reported by Kistler [1]. Similar results have been obtained in 3-D flows, notably in those induced by blunt fins and cylindrical protuberances, at transonic and supersonic speeds [2]. In these cases, high-speed photography and wall pressure fluctuation measurements have shown that large variations occur in the position and structure of the separation-shock-wave system. At hypersonic speed (Mach 7) in an interaction generated by a flare mounted concentrically on an cone-ogive cylinder model, large-scale unsteadiness of the separation onset and reattachment regions has been observed [5].

The goal of this study is to further expand our knowledge of ramp-induced shock-wave/boundary-layer interactions and to be able to relate surface and flowfield features so as to gain insight of the interaction behaviour as a whole. In order to study the off-surface flowfield, an instrumentation was developed at HTG using Schlieren principle with Laser as the light source. The present method depends on a narrow laser sheet that is more accessible and can be used to analyze the fluctuating density gradient field. Previous studies were carried out on hyperboloid [6] and HAC [7] models using this instrumentation which helped reveal useful details of the interaction phenomena. The present study was carried out in a Mach 8.94 ramp-induced shock-wave/boundary-layer interaction (SWBLI) with air as test gas. The SWBLI flowfield was analysed using conventional Schlieren system and Laser Schlieren system.

2 Experimental Techniques

2.1 Wind-Tunnel and Test Model

The experiments were performed in the hypersonic wind tunnel at HTG (Hypersonic Technology Göttingen). HTG designed a completely new concept based on the Ludwieg-tube principle implementing state-of-art technologies [8]. This wind tunnel is an intermittent blowdown type with a test-section diameter of 250 mm and a variable Mach number capability over the range of Mach 6-11 with maximum run-time of 100 ms. Figure 1 shows the schematic diagram of the existing hypersonic facility at HTG.

The test-section Mach number for the present experiments was $8.94 \pm 4\%$. A pitot-tube rake was installed below the test model but sufficiently high enough to be close to the model. Caution was exercised, however, to keep the leading edge of the rake tubes ahead of the oblique shock from the leading edge of the model. The test-section Mach number was calculated using the stagnation pressure measured by the pitot-tube closest to the model and later on using the Rayleigh Pitot-tube formula. The study was carried for charge tube pressure of 50 bar$\pm 1\%$ which results in a test-section stagnation pressure of 42.1 bar$\pm 2\%$. To avoid condensation of the relieved flow in the nozzle, the gas column which is ejected during the shot is heated to 773 K$\pm 0.2\%$. The corresponding freestream Reynolds number per unit length was $2.078 \times 10^5\,\mathrm{m}^{-1}$.

A two-dimensional compression ramp model with corner angle of 24° was used. Figure 2 shows a schematic of the model. The model was mounted on a support-pipe along the test-section centerline. There was a 3 cm gap on either side of the model for passage of the sidewall boundary layers. No side fences were used since the study of the fluctuating density field close to the model surface required an unobstructed view of the interactive flowfield. This may induce 3-dimensional effects. But it may be pointed out that even in 2-dimensional cases, 3-dimensional flow exists.

A MICROLINK 3000 data acquisition system was used to record the measured data. In the present state, up to 32 channels can be recorded at a speed of up to 40 kHz. Data stored on the hard disk are available for post processing or graphical output. Piezoresistive pressure gauges have been calibrated and installed to measure the flow in the charge tube, test section and vacuum tank.

2.2 Laser Schlieren System

Figure 3 shows the schematic of the Laser Schlieren set-up used. The principle of operation of a Laser Schlieren system is the same as that of an ordinary Schlieren system except that in this voltage signals are obtained in accordance with the density gradient fluctuations in the flowfield. This arrangement is free from structural vibrations as it is mounted on a table completely isolated from the wind-tunnel.

For this purpose a parallel sheet of laser light (a 15 W diode-laser) is passed through the test section and made to fall on an array of 16 photodiodes (each with an active area of 0.66 mm^2 arranged on a 1 mm pitch in common cathode configuration) placed on the other side of the test section. Figure 4 shows the experimental procedure followed for flowfield analysis using this technique. It may be pointed out that the tunnel wall boundary-layer equally influences each photodiode measurement, irrespective of its position in freestream/interactive field, as shown in Fig. 4. When the laser sheet touches an oscillatory shock, the diffracted light appears periodically, and a time trace obtained from the array of photodiode shows unsteady voltage signals. The arrangement gives an idea of the existing density fluctuations in the flow and hence, an insight to the fluctuating nature of flow. Since the response time of the photodiode is extremely small (4 ns), extreme high-frequency shock oscillations can be detected using this optical technique. In parallel mode, the complete array was kept 0.6$\pm$2% mm above the model surface. This, however, initially required very careful alignment of the laser sheet itself. All the data were sampled at 5 kHz.

The Laser Schlieren technique seems to be more promising because the Schlieren signals are more directly related to a property of the flow, i.e., density gradient[9]. This condition implies that the Laser Schlieren system gives voltage signals representative of the fluctuating density flowfield integrated across the span of the flow. The instrument, however, was not calibrated against a known density gradient.

The primary advantage of the current method over the quantitative Schlieren or Shadowgraph system lies in its simplicity and accessibility. The best use of this technique, perhaps, is in measuring the unsteady shock motion. Measurements were made, on the basis of optical Schlieren recordings, at locations wherever necessary. The interaction flowfield was scanned parallel to model surface, to investigate the movement of the separation shock.

3 Results and Discussions

3.1 Flow Visualization

Figures 5(a)-(b) show the Schlieren pictures of the shock-wave boundary-layer interaction flowfield on the ramp model with air as test gas at two consecutive time intervals. Due to limitations of the recording camera, photo frames with only 40 ms time gap could be captured. The first frame, Fig. 5(a), shows a completely established interactive flowfield with a separation bubble along with separation and reattachment shocks. The next time frame, Fig. 5(b), shows a considerable growth in the size of separation bubble with forward movement of separation shock and a backward movement of reattachment shock. This expansion and contraction movement of the separation bubble is thought to be the principle mechanism [2] responsible for the unsteady nature of SWBLI flowfield. The random movement of the bubble causes the separation shock to oscillate and since there exist large pressure gradients in the vicinity of the separation, such a motion will cause high fluctuating pressure levels to be generated. As such the phenomenon is of much concern and needs to be investigated in detail. Though the Schlieren pictures give an overall idea of the interaction phenomena, it does not shed light on the time-dependent features which are crucial in understanding the oscillating phenomenon. For this purpose, voltage signals from Laser Schlieren system were acquired and analyzed.

3.2 Signal Analysis

Figure 6 shows the simultaneously acquired off-surface voltage signals in the vicinity of separation and reattachment points, respectively. These locations were selected in accordance with the Schlieren recordings and with the help of a scale mounted on the test-section window. The Laser sheet is accordingly targeted perpendicular or parallel to freestream direction as per need. In the present plots, the Laser sheet was targeted perpendicular to freestream direction with the lowest photodiode, in the array, 0.6±2% mm above the model surface. Two distinct features can be observed from these signals, i.e., a high-amplitude low-frequency part and a very low-amplitude high-frequency part. The low-frequency part seems to be the outcome of the movement of the separation bubble while the high-frequency part represents the general unsteadiness of the flowfield which is a combination of the disturbances in the incoming boundary layer and those triggered by the interaction process.

These two characteristics are observed to change near separation and reattachment points. For both these locations, the density signals from two neighbouring photodiodes experiencing a fluctuating density gradient follow opposite trends almost to the extent of mirror imaging, Fig.6 (i),(b)-(c) and Fig.6 (ii),(a)-(b). In this case the fluctuating density gradient is across the separated shear layer, see Fig. 4, which oscillates in accordance with the expansion and contraction movement of the separation bubble. The height of the photodiodes capturing this feature is different at the two locations under study depending upon the height of the shear layer from the model surface at these locations. However, only low-frequency oscillations behave in this manner; high-frequency changes in density signal do not necessarily follow this trend. Away from the fluctuating density gradient, the signal shows no trace of low-frequency oscillation indicating an undisturbed flow away from the interactive field, Fig.6 (i),(a) and Fig.6 (ii),(c).

A careful observation of these signals also shows that in the vicinity of separation point, (i) only one low-frequency oscillation can be detected while two can be seen near reattachment and (ii) the amplitude of high-frequency fluctuations increases considerably near reattachment relative to that near separation. These evidences suggest that considerable discrepancies exist in the flowfield characteristics near separation and reattachment points although they share a common cause. Though the type of separation is laminar(due to the low Reynolds number), the large oscillations of the separation bubble observed in these experiments could possibly be due to the transition of boundary layer from laminar to turbulent occuring at separation.

3.3 Fluctuating Density Field

3.3.1 Near Separation

Figures 7(a)-(c) show the streamwise density gradient profiles at three time instances in the vicinity of separation region. For investigating the shock-leg motion, the laser sheet is targeted parallel to the freestream direction close to the model surface, as shown in Fig. 4. It can be seen that the arrangement can capture very subtle changes in density gradient field. The plot shows undisturbed flow upstream of the separation point. However, near the separation point, a jump in signal is observed in the form of three peaks representing three shock-legs.

Further, these shock-legs are observed to fluctuate both in intensity and axial distance as a function of time. Figures 8(a) and 8(b) show the variation in intensity and axial distance of each shock-leg as a function of time. It is discernable that the first shock-leg does not change in intensity with time whereas the second leg shows a slight change in intensity with no axial movement. The third shock-leg, the strongest of all, however shows considerable variation in intensity and axial movement over a period of time. This suggests that (i) the shear layer in the vicinity of separation is curved resulting in a series of compression waves the strength of which increases in the downstream direction and, (ii) only the strongest shock-leg exhibits axial oscillation.

3.3.2 Near Reattachment

Figure 9 shows the time-dependent density gradient field near the reattachment region. Ahead of the reattachment point and behind the plate-ramp junction, the signal shows approximately undisturbed flow conditions. Near reattachment point the signal jumps considerably forming two distinct peaks which vary both in intensity and axial location with time. The first peak represents the shear layer and the second peak represents the reattachment shock. A careful examination of Figures 8(b) and 9 shows that the peaks in density gradient signals show variations in intensity or axial distance only in a particular time interval. This seems to indicate that the fluctuations near separation and reattachment regions are caused by the expansion and contraction motion of the separation bubble.

3.4 Power Spectrum

Figure 10 shows the power spectrum of the voltage signals from the photodiodes closest to the model surface throughout the interaction. Ahead of separation, the amplitude of flow unsteadiness close to the surface is low(approx. in e-08 range), Fig. 10(a), whereas immediately behind separation, Fig. 10(b), a considerable increase in amplitude over the entire frequency range can be seen(approx. e-06). Further downstream and probably under the separation bubble, Fig. 10(c), the flow unsteadiness grows still more up(approx. e-05) to the plate-ramp junction, where it reaches a maximum(approx. e-04) in the entire frequency range, Fig. 10(d). However at this point the high amplitude unsteadiness persists more in the low frequency range. Near reattachment point, Fig. 10(e), the flow unsteadiness shows a drastic increase in the entire frequency range while behind the reattachment point the spectrum shows a considerable decrease in the amplitude of fluctuations(approx. in e-06 range), Fig. 10(f). The above results indicate that the flowfield fluctuations occur both in the low and high frequency range with stronger contribution from high-frequency fluctuations inside the separated region.

4 Conclusions

An experimental study has been carried out to study the off-surface flowfield characteristics related to shock-wave/boundary-layer interaction near the separation and reattachment point on a 24° ramp model. For this purpose, Laser Schlieren technique was used which helped reveal important results that can be summarized as follows.

Considerable discrepancies seem to exist in the flowfield characteristics near separation and reattachment points. Flowfield fluctuations are observed to increase and decrease in amplitude at various location throughout the interaction. Shock-legs exist at the separation point, due to curving of the shear layer, the strength of which increases in the downstream direction. Further, the strongest shock-leg exhibits axial movement.

The peaks in the density gradient signal near separation and reattachment show variations in intensity and axial location during the same time interval indicating a common cause of fluctuation. Both high-frequency and low-frequency fluctuations seem to dominate the interactive flowfield with stronger contribution from high-frequency fluctuations inside the separated region.

References

[1] A.L. Kistler. Fluctuating wall pressure under a separated supersonic flow. *Journal of the Acoustical Society of America*, 36(No.3):pp. 543–550, 1964.

[2] D.S. Dolling and M.T. Murphy. Unsteadiness of the separation shock wave structure in a supersonic compression ramp field. *AIAA Journal*, Vol. 21(No. 12):pp. 1628–1634, 1983.

[3] B.T. Gibson and D.S. Dolling. Exploratory study of wall pressure fluctuations in a mach 5 sharp fin-induced turbulent interaction. *AIAA Journal*, 30(9):2188–2195, 1992.

[4] D.J. Maull. Hypersonic flow over axially symmetric spiked bodies. *Journal of Fluid Mechanics*, Vol.8(Part 4):pp. 584–592, 1969.

[5] C.C. Horstman and F.K. Owen. New diagnostic technique for the study of turbulent boundary layer separation. *AIAA Journal*, Vol.12(No.10):pp. 1436–1438, 1974.

[6] S.B. Verma and G. Koppenwallner. Experimental study of shock unsteadiness on a hyperboloid model using laser schlieren. *9th AG-STAB Workshop*, pages 156–157, Nov. 9-11, 1999, DLR Standort, Gottingen, Germany.

[7] S.B. Verma and G. Koppenwallner. Study of flow unsteadiness in a hypersonic swbli flowfield on a hac model. *39th AIAA Aerospace Sciences Meeting and Exhibit*, AIAA Paper No-2001-0566, Session APA-9, January 2001, Neveda, USA.

[8] Koppenwallner G. Friehmelt, H. and R. Muller-Eigner. Calibration and first results of redesigned ludwieg expansion tube. *AIAA/DGLR 5th International Aerospace Planes and Hypersonics Technologies Conference*, AIAA-93-5001, 30 Nov.-3 Dec.,Munich, Germany.

[9] B.H. Funk and K.D. Johnston. Laser schlieren cross-beam measurements in a supersonic jet shear layer. *AIAA Journal*, Vol.8(No.11):pp. 2074–2075, 1970.

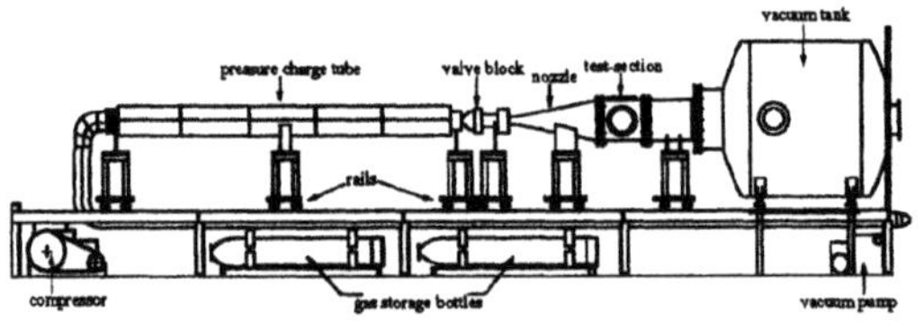

Figure 1: Schematic diagram of the hypersonic wind tunnel facility at HTG

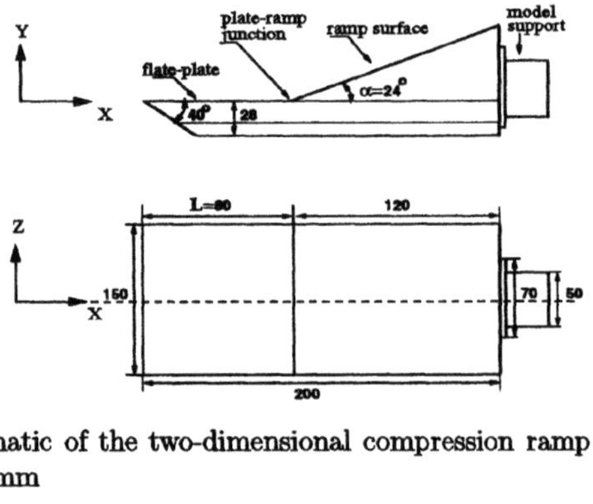

Figure 2: Schematic of the two-dimensional compression ramp model details; all dimensions are in mm

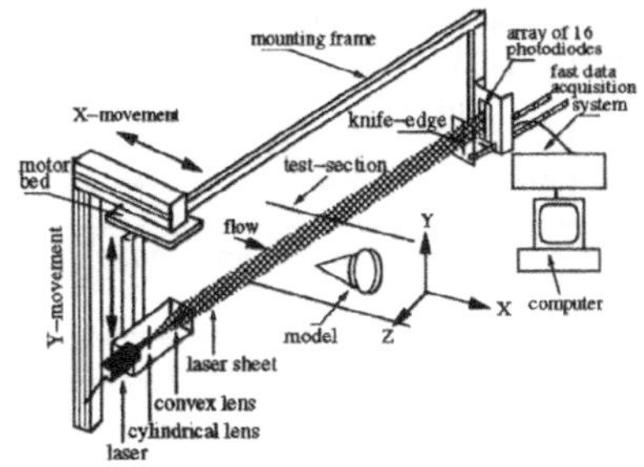

Figure 3: Schematic of the Laser Schlieren setup

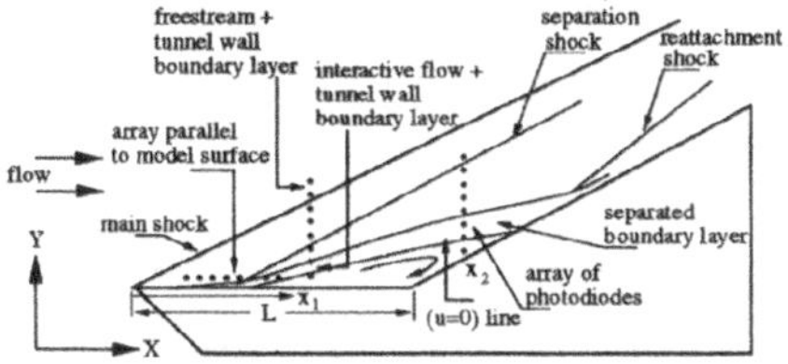

Figure 4: Experimental procedure followed for flowfield analysis with Laser Schlieren system

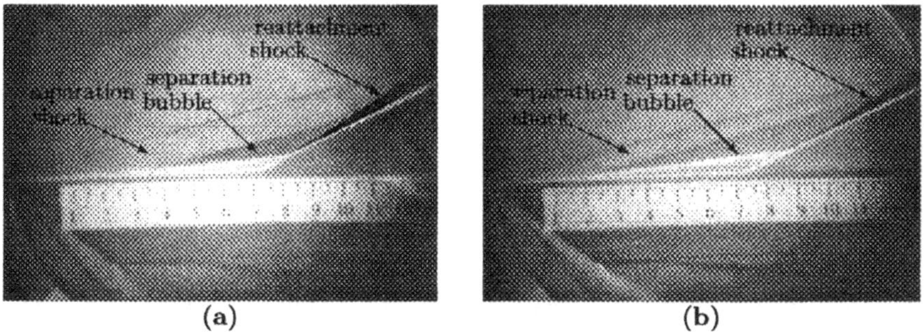

Figure 5: Schlieren pictures of the 24° ramp model at two time intervals with air as test gas, flow is from left to right; (a) time t_1 (b) time t_2 (40 ms later)

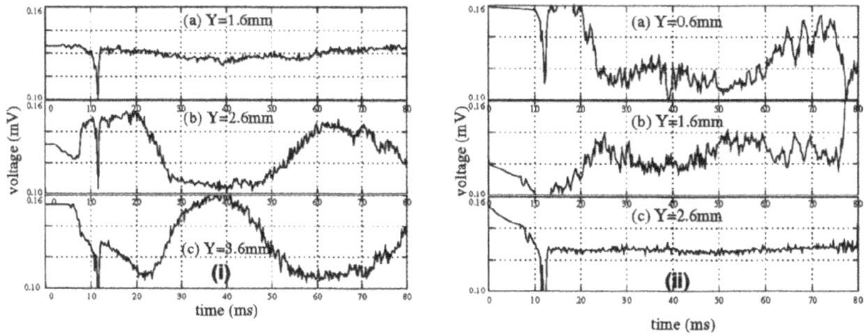

Figure 6: Simultaneously acquired voltage signals from photodiode array placed normal to freestream direction (i) behind separation point (X/L=0.34) and, (ii) ahead of reattachment point (X/L=1.18)

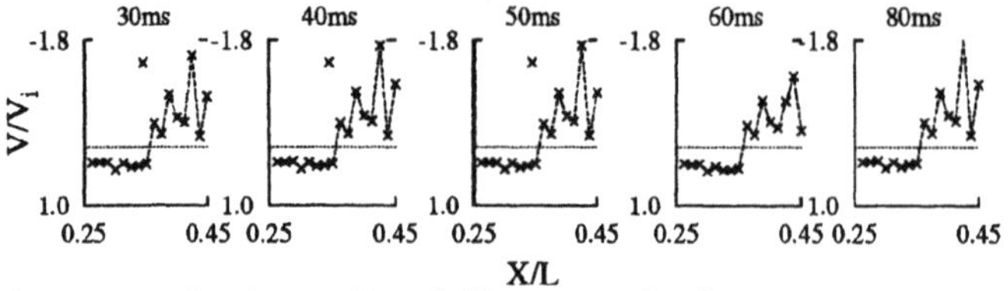

Figure 7: Streamwise density gradient field at 6 time levels near separation region; V/V_i is the signal voltage, V, non-dimensionalised by its initial value, V_i, at the beginning of flow

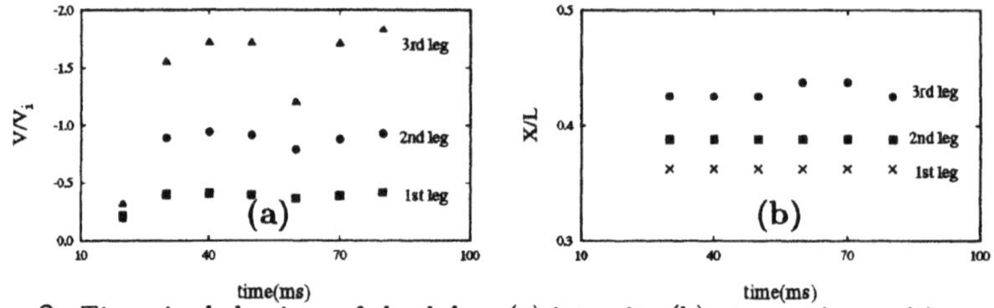

Figure 8: Timewise behaviour of shock legs (a) intensity (b) streamwise position

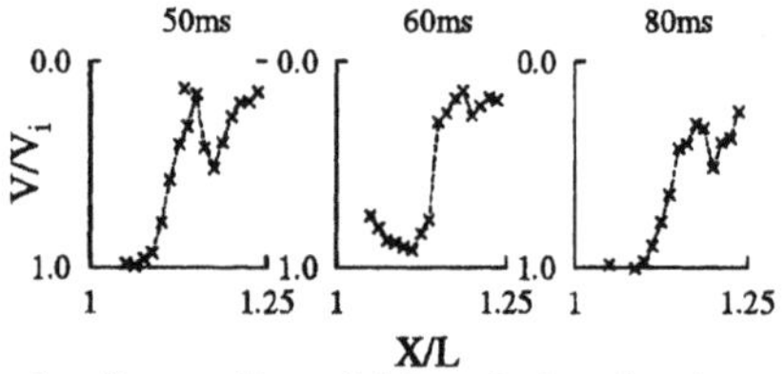

Figure 9: Streamwise density gradient field at 3 time levels near reattachment region; V/V_i is the signal voltage, V, non-dimensionalised by its initial value, V_i, at the beginning of flow; laser sheet parallel to ramp surface

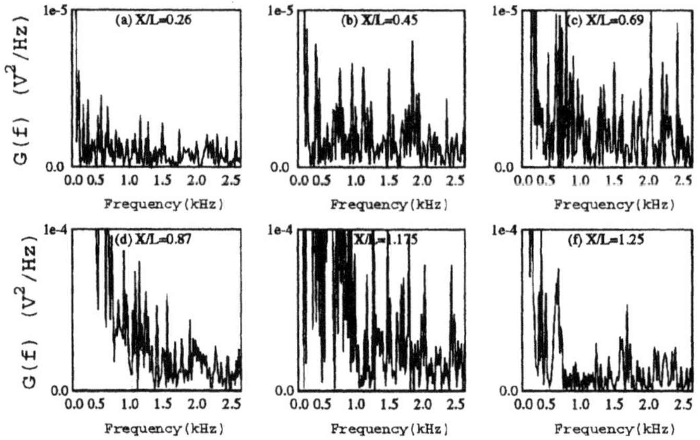

Figure 10: Power spectrum of voltage signals from photodiodes closest to ramp surface at different axial locations

Numerical Investigations of Small-Amplitude Disturbances in a Laminar Boundary Layer with Impinging Shock Waves

A. Pagella, U. Rist, S. Wagner
Institut für Aerodynamik und Gasdynamik, Universität Stuttgart
Pfaffenwaldring 21, 70550 Stuttgart, Germany

Summary

Two-dimensional shock/boundary-layer interactions on a laminar boundary layer over a flat plate at cold super- and hypersonic conditions have been computed in direct numerical simulations based on the complete Navier-Stokes equations. Comparison with both experimental and numerical data has shown good agreement for $Ma = 2$. For $Ma = 4.8$ and different shock strengths the steady flow-field is discussed and then stimulated with small perturbations. The perturbations are amplified already upstream of the shock in accordance to pressure and temperature gradients. Behind the point, where the shock hits the boundary layer, amplification reduces. Increasing shock angle causes separation and leads to higher amplification of the perturbations as well as larger non-parallel effects. To validate this observation, a laminar boundary layer without impinging shock is compared with linear stability theory and non-parallel effects are clearly identified.

1 Introduction

Transition to turbulence in high-speed boundary layers is not yet fully understood because of its higher complexity compared to subsonic flows. However, starting with the pioneering work of Mack on compressible linear stability theory (cf. [1]), the use of parabolized stability equations [2, 3], direct numerical simulations [4] and experimental work (e.g. [5, 6]) focusing on generic flows over flat plates or cones, considerable progress has been achieved within the past several years (cf. [7]). When additional features, such as shock induced separation by an oblique shock come into play, it can no longer be expected that the parallel-flow assumption typically used in linear stability theory is justified and that investigations other than DNS or experiments yield further insight.

Based on earlier successful investigations by Eißler & Bestek [8] on flat-plate boundary layer transition in zero-pressure gradient boundary layers the present work aims at continuing that work for cases where an impinging shock generates a separation bubble in a laminar boundary layer. As a first step towards DNS of transition in shock-induced separation bubbles, the method developed by Eißler is validated for computations of flow fields including shocks by comparisons with literature and the influence of oblique shocks with different strengths on a small-amplitude Tollmien-Schlichting wave is then investigated. Since it turned out that non-parallel effects are considerably larger than in subsonic flows, the original case of Eißler at M=4.8 had to be reconsidered in order to assess the magnitude of these effects.

2 Numerical method

In order to numerically solve the two-dimensional conservative formulation of the Navier-Stokes equations, spatial streamwise (x) and wall-normal (y) derivatives are approximated by split-type compact finite differences of at least 4th-order accuracy [9]. Dissipative terms in y-direction are discretized by 4th-order-accurate central differences. For the time-evolution, a standard 4-step Runge-Kutta scheme is used (for more details and earlier validations see [10], [11]). Due to the existence of very large gradients caused by the shock, finite difference schemes tend to yield oscillations. To suppress these oscillations, an implicit filter of 4th-order accuracy [12] is implemented in x-direction.

The basic setup of the problem is shown in figure 1 by a visualization of the resulting density gradients (computed Schlieren image) for the case with $Ma = 4.8$ and a shock angle of $\sigma = 13.2^o$. At the inflow boundary, which is indicated in figure 1 as line $A - B$, solutions of the compressible boundary-layer equations are held constant, while the flow variables at the outflow boundary (figure 1, $C - D$) are calculated under neglection of second derivatives in streamwise direction. For the conditions at the wall (figure 1, $A-D$), temperature can be chosen as a constant value or distributed adiabatically. The perturbation of the flowfield origins at the wall. Variations of $(\rho v)'$ (prime denotes disturbance quantities) simulates blowing and suction. The perturbation follows a sine-function multiplied with a Gaussian:

$$f_{\rho v}(\varsigma,t) = \hat{a} * \sin(\omega t) * \sin(n\varsigma) * e^{-b\varsigma^2}, -2\pi \leq \varsigma \leq 2\pi, \quad (1)$$

where n and the length of the disturbance strip is chosen to exactly excite one single wave length, which can be found by linear stability theory. Flow variables at the freestream boundary ($B - C$) are calculated by means of a characteristic condition [10]. The shock is prescribed at the freestream boundary using Rankine-Hugoniot relations to calculate flow variables behind the shock. Hence, flow variables are held constant around the shock along the upper boundary. This explains why a small steady disturbance enters through the upper boundary until it is swept away in downstream direction (cf. figure 1). In the plots within this paper, the local Reynolds number is $R_x = \sqrt{x * \rho_\infty u_\infty / \mu_\infty}$. The perturbed flowfield is excited with the disturbance frequency $F = (2\pi f \mu_\infty)/(u_\infty^2 \rho_\infty)$. The pressure of the flow field p and the disturbance amplitude of the wall pressure $|p'_w|$ is non-dimensionalized by $\rho_\infty u_\infty^2$. For purpose of comparison with validation data, wall pressure p_w in figure 2 is normalized by the wall pressure upstream of the area of influence of the impinging shock. Flow velocities u and v are normalized with free-stream velocity u_∞ and T with the free-stream temperature T_∞. For profiles of flow variables and amplitude distributions a similarity coordinate $\eta = y * 10^5 / R_x$ is used.

3 Results

3.1 Unperturbed laminar boundary layer

Although the focus of our work is on unsteady disturbances at $Ma = 4.8$, comparisons with [13] and [14] for $Ma = 2$ with freestream temperature $T_\infty = 320K$, total pressure in freestream $p_t = 0.1995bar$, shock angle $\sigma = 32.58^o$ and adiabatic wall will be presented here, because suitable steady data for validating the unperturbed flow-field at higher Ma-numbers was not available.

In figure 2, δ_0^* is the displacement thickness at location x_0 where the shock would hit the plate in the inviscid case. The difference to the experimental results regarding the length of the separation bubble which can be identified by its negative skin friction might be caused by the Stanton probe used in the experiments [13] which contacted the wall and thus disturbed the boundary layer. Wall pressure agrees well with both [13] and [14], so do velocity profiles (not presented here).

We now turn to simulations for $Ma = 4.8$ and increasing shock strengths, i.e. $\sigma = 12.5^o$ and $\sigma = 13.2^o$. The other parameters of the flow are $T_\infty = 55.4K$ and $p_t = 4bar$. Wall temperature is held constant at $T_w = 270K$. A typical grid for these simulations consists of 801x151 grid-points. Some interesting details of the flow field can be seen in the computed Schlieren image in Figure 1. The oblique shock, which impinges on the boundary layer turns into a normal shock when approaching the sonic line in the boundary layer. The initial reflection consists of an expansion fan emanating from the impingement location of the shock on the sonic line. It turns into downstream direction as it leaves the boundary layer. Behind the expansion, a recompression is observed which becomes the reflected shock in the far field.

Here, inviscid shock impingement is at $R_{x0} = 1400$. For the case with $\sigma = 13.2^o$ presented in figure 3, the shock angle is strong enough to force separation, which can be identified as an area of negative skin friction while for $\sigma = 12.5^o$, it is not yet sufficient to cause the boundary layer to separate.

For $R_x = 1330$ in figure 4 the different wall-normal profiles for velocity v and pressure p show significant quantitative differences, especially when passing the impinging shock. For the stronger shock these effects are more pronounced, but profiles also differ for u. The same can be seen for $R_x = 1440$ in figure 4. Note the different behaviour of v and p compared to $R_x = 1330$.

Let us now take a closer look into the flow field and what should be expected in terms of stability behaviour. For this, let us draw our attention once again on the computed Schlieren image of the case with $\sigma = 13.2^o$ in figure 1. The pressure gradient which results from the impinging shock causes the boundary layer to thicken. Upstreamward of the shock, the fluid motion is delayed and heated by the compression waves, which result from the deflection of the flow due to this thickening. In the first mode instability region (see [1]), delayed and heated flows are known to cause increasing amplification rates. In contrast, acceleration and cooling yields reduced amplification. This is the case downstream of the shock, where an expansion fan is located. Further downstream a set of compression waves can be identified, which are due to the re-deflection of the flow. This is more pronounced, the more the boundary layer thickens which is directly influenced by the strength of the impinging shock.

3.2 Perturbed laminar boundary layer

For the cases presented in this section we discuss the disturbance waves introduced by activation of the disturbance strip at the wall with $n = 13.5$ (see equation 1). The disturbance strip ranges from $R_x = 477$ to $R_x = 970$. Figure 5 shows fourier transforms of wall pressure disturbances over one perturbation cycle for the different cases. The disturbance frequency was chosen close to the highest amplification upstream of the impinging shock for a two-dimensional first mode instability according to [1].

Up to $R_x = 977$, the disturbance strip can be identified. Compared to [10] our disturbance strip extends over a much larger distance in streamwise direction. This is a possible means to excite a single Tollmien-Schlichting instability mode more precisely than in [10], where two modes were excited with the same disturbance frequency. For $\sigma = 12.5^o$ it can be seen that the influence of the shock causes the Tollmien-Schlichting wave to rise, which results in an increased amplification rate. After reaching a peak value, the amplitude is decreasing, yielding negative amplification, later increasing again to a positive amplification rate. Figure 6 compares the pressure and the temperature at a location parallel to the wall near the sonic line at the impingement point of the shock, as well as the wall-pressure amplitudes in the same sector. It can be seen that the local amplitude maxima's location corresponds well with the location of the highest temperature peak and a local peak in the pressure distribution which is caused by the impinging shock. This is in accordance with the general behaviour of a heated and delayed flow, which are both known to cause amplification in the region of first mode instability, as stated before.

Linear stability theory did not deliver satisfying results for the amplification curves shown in figure 5. For the case without shock we could show non-parallel effects, which rendered some differences to linear theory (to be explained later in this section). With the shock, there is a deflection of the flow caused by the shock itself. The magnitude of velocity normal to the wall thus reaches larger amounts (see figure 4) which we interpret as a higher degree of non-parallelism. Turning to the stronger shock ($\sigma = 13.2^o$), the area of upstream influence is bigger than before and the wall-pressure disturbances rise to a ten times higher level. Downstream the shock, areas of rise and decrease can be observed. This behaviour is not yet clear and has to be investigated further.

Figure 7 gives normalized amplitude distributions for u, v and p at location $R_x = 1330$. Shapes are almost congruent for tangential velocity u when comparing the case without impinging shock wave and $\sigma = 12.5^o$. But amplitude distributions do differ significantly for the normal velocity v and pressure p. This is the case for $\sigma = 13.2^o$, as well. However, differences for u can be observed here, too. It is especially interesting to note here that the secondary amplitude maxima which are further away from the wall than the primary ones are decreasing with increasing shock strength. This lies in accordance with the observations of non-parallel effects and a stronger amplification at closer distance to the wall.

Figure 8 shows amplification rates of second wall-normal maxima of the disturbance amplitudes for a boundary-layer flow at $Ma = 4.8$ without impinging shock, which are located in a farther position normal to the wall than first maxima. For comparison, results of linear theory and simulations of an idealized, parallel boundary layer profile at discrete locations R_x are shown. Amplification rate is $\alpha_i = -(\partial ln(A(x)/A_0))/(\partial x)$, where $A(x)/A_0$ is the amplitude ratio of the concerning flow variable, respectively, and $x = R_x^2 * 10^{-5}$. For the given frequency, our integration domain covers both the first and second mode instability domains [1]. The second instability mode starts to grow at $R_x \approx 1500$. In linear theory both instability regions are separated by a shallow region with $\alpha_i > 0$. However, in DNS this is not the case, due to large non-parallel effects, which were not documented before.

Linear theory (see [1]) gives one single amplification rate for each flow variable everywhere in the boundary layer. To prove that the differences in our simulations are really caused by non-parallelism (linear stability theory assumes parallel flow), simulations were carried out using an idealized, parallel boundary layer at discrete locations R_x. These results agree well with linear theory.

Conclusions and future research

Shock/boundary layer interaction imposes adverse and favourable local pressure and temperature gradients on the boundary layer apparently leading to considerable amplification of small-amplitude Tollmien-Schlichting waves. Non-parallel effects become increasingly large with growing shock strength. Future research aims at explaining the influence of three-dimensional and large-amplitude disturbances on the transition process as well as other unsteady phenomena in the flow-field.

Acknowledgement

The authors would like to thank the DFG for supporting this research within SFB259.

References

[1] Mack, L., Boundary layer stability theory. *Jet Propulsion Laboratory, Pasadena, Tech. Rep.*, 900-277 (1969).

[2] Hein, S., Bertolotti, F., Simen, M., Hanifi, A. and Henningson, D., Linear nonlocal instability analysis – the linear NOLOT code –. *DLR-IB*, 223-94 A56 (1994).

[3] Pruett, C., A comparison of PSE and DNS for high-speed boundary-layer flows. *Transitional and Turbulent Compressible Flows, L.D. Kral and T.A. Zang (eds.)*, FED Vol. 151, ASME, New York, 57–67 (1993).

[4] Eißler, W. and Bestek, H., Direct numerical simulation of transition in Mach 4.8 boundary layers at flight conditions. *Engineering Turbulence Modelling and Experiments*, 3, 611–620 (1996).

[5] Kosinov, A., Maslov, A. and Shevelkov, S., Experiments on the stability of supersonic boundary layers. *J. Fluid Mech.*, 219, 621–633 (1990).

[6] Stetson, K. and Kimmel, R., On hypersonic boundary-layer stability. *AIAA Paper*, No. 92-073 (1992).

[7] Saric, W., Reshotko, E. and Arnal, D., Hypersonic Laminar-Turbulent Transition. *AGARD*, AR-319 (1998).

[8] Eißler, W. and Bestek, H., Spatial numerical simulations of linear and weakly nonlinear wave instabilities in supersonic boundary layers. *Theoret. Comput. Fluid Dynamics*, 8, 219–235 (1996).

[9] Kloker, M., A Robust High-Resolution Split-Type Compact FD Scheme for Spatial Direct Numerical Simulation of Boundary-Layer Transition. *Applied Scientific Research*, 59, 353–377 (1998).

[10] Eißler, W., *Numerische Untersuchungen zum laminar-turbulenten Strömungsumschlag in Überschallgrenzschichten*. Dissertation, Universität Stuttgart (1995).

[11] Fezer, A. and Kloker, M., Transition Process in Mach 6.8 Boundary Layers at Varying Temperature Conditions Investigated by Spatial Direct Numerical Simulation. *Notes on Numerical Fluid Mechanics, Vieweg*, 72, 138–145 (1999).

[12] Lele, S., Compact Finite Difference Schemes with Spectral-like Resolution. *J. Comp. Phys.*, 103, 16–42 (1992).

[13] Hakkinen, R., Greber, J., Trilling, L. and Abarbanel, S., The interaction of an oblique shock wave with a laminar boundary layer. *NASA MEMO*, 2-18-59W (1959).

[14] Katzer, E., On the lengthscales of laminar shock/boundary-layer interaction. *J. Fluid Mech.*, 206, 477–496 (1989).

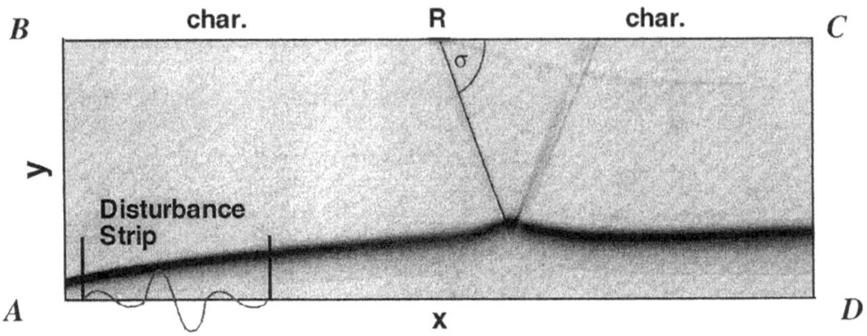

Figure 1 Integration domain (square $A - B - C - D$), disturbance strip and computed Schlieren image ($| \partial\rho/\partial y |$) for $Ma = 4.8$, $\sigma = 13.2^o$, $R_{x0} = 1397$.

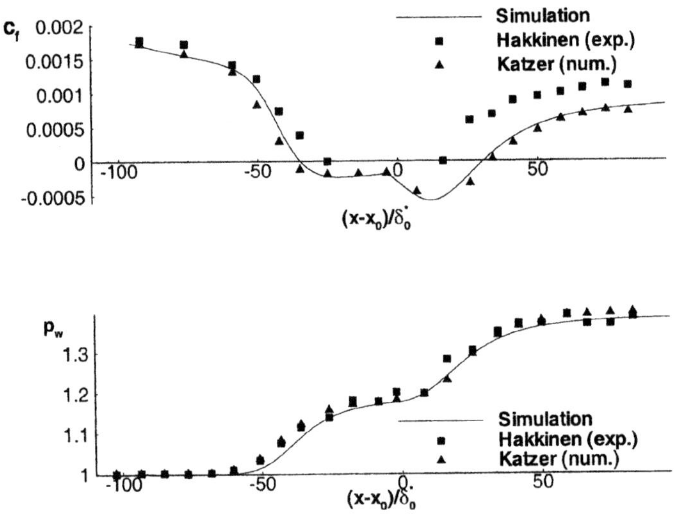

Figure 2 Comparison of skin friction (upper image) and wall pressure (lower image) at $Ma = 2$. $R_{x_0} = 548$.

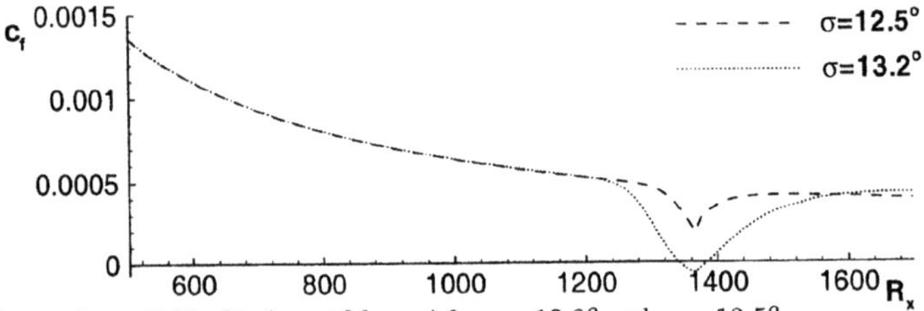

Figure 3 Comparison of skin friction at $Ma = 4.8$, $\sigma = 13.2^o$ and $\sigma = 12.5^o$.

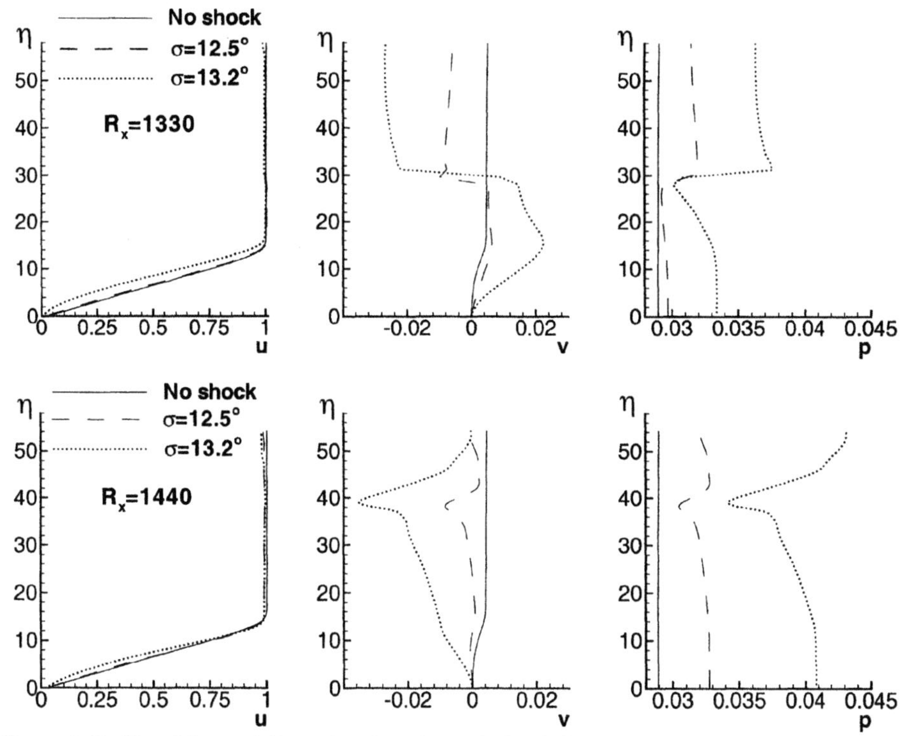

Figure 4 Profiles of flow variables at locations $R_x = 1330$ and $R_x = 1440$.

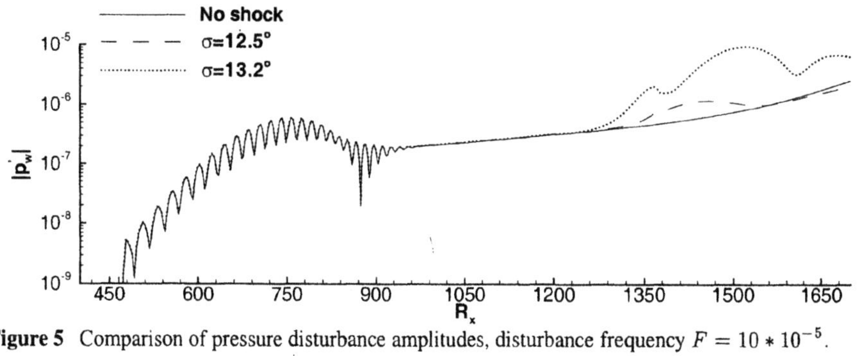

Figure 5 Comparison of pressure disturbance amplitudes, disturbance frequency $F = 10 * 10^{-5}$.

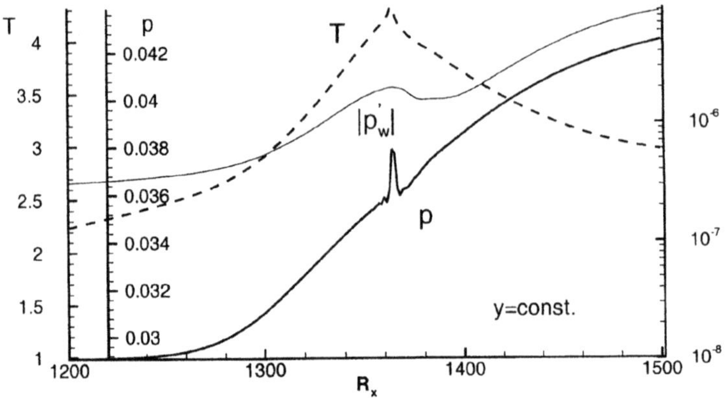

Figure 6 Temperature T, pressure p parallel to the wall near sonic-line at impinging shock as well as wall-pressure amplitudes in a sector of the flow-field for $\sigma = 13.2^o$.

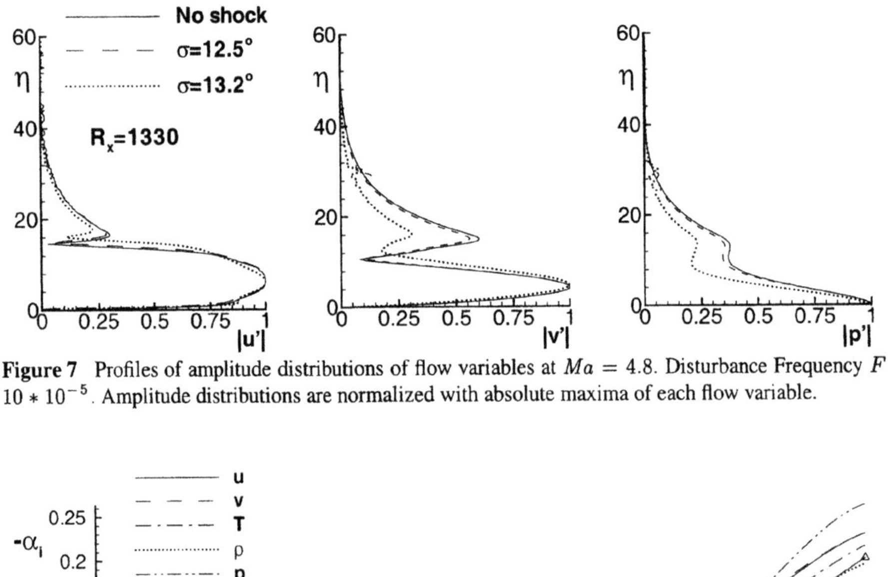

Figure 7 Profiles of amplitude distributions of flow variables at $Ma = 4.8$. Disturbance Frequency $F = 10 * 10^{-5}$. Amplitude distributions are normalized with absolute maxima of each flow variable.

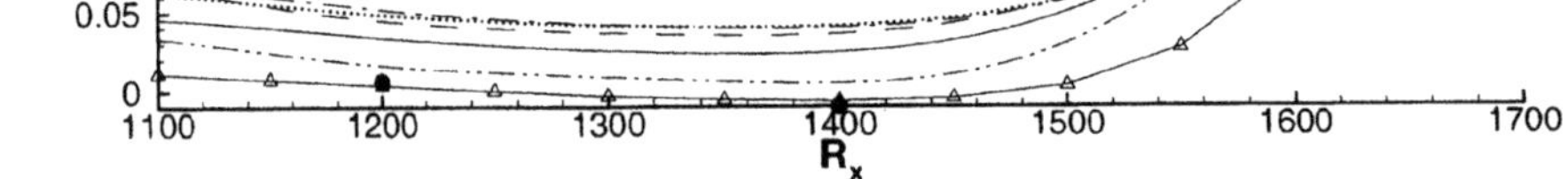

Figure 8 Amplification rates of relevant flow-field variables at $Ma = 4.8$ along with results from linear stability theory and simulation of idealized boundary layer. No impinging shock.

Surface Temperature Effects on Shock-Wave Boundary-Layer Interaction of Ramp Flows

M. Bleilebens, H. Olivier
Shock Wave Laboratory, RWTH Aachen
Templergraben 55, 52056 Aachen, Germany
e-mail: martin@swl.rwth-aachen.de, olivier@swl.rwth-aachen.de

Summary

Experiments in the shock tunnel TH2 have been conducted using heated ramp models with deflection angles of 15° and 24°. Surface temperatures of $700\,K$ were reached with electrical resistance heating and $1000\,K$ with radiation heating. The experiments showed an increase in separation length for increasing surface temperature for laminar flow. By suitable scaling, agreement can be achieved between the experimental results and theoretical approaches describing the separation length scaled with the boundary-layer thickness. The results show the independence of the ratio of separation length to boundary-layer thickness at separation on the wall-to-total-temperature ratio T_W/T_0 proposed by Stollery and Needham. The boundary-layer thickness was determined with pitot-rake measurements on cold surfaces.

1 Introduction

The experiments conducted at the Shock Wave Laboratory (SWL) within the Special Research Center (SFB 253) "Fundamentals of the Design of Aerospace Planes" have to provide aerodynamic coefficients, pressure distributions and heat loads on the two stages of the reference configuration (ELAC/EOS) at high free-stream Mach numbers ($M_\infty \approx 8$). Attention was focused on the effects of free-stream unit Reynolds number and surface temperature on the shock-wave boundary-layer interaction on control surfaces and the external oblique-shock compression of the air-breathing first stage. In the experiments these aerodynamic problems have been simplified by the more fundamental ideal ramp flow. In order to study the influence of high wall temperatures on different flow phenomena, experiments have been conducted with two heated ramp models. The influence of the wall-to-total-temperature ratio on, e.g., separation length is obvious and can be explained by semi-empirical models.

2 Models and test conditions

The two-surface radiation equation

$$Q = \sigma\varepsilon AF(T_W^4 - T_E^4) \tag{1}$$

shows that with increasing temperature T_W of the model surfaces the radiation losses due to the temperature difference to the environment T_E (test section wall temperature) increase with the power of four. σ is the Stefan-Boltzmann constant, ε represents the effective emissivity of the two surfaces. The geometric situation is defined by surface area A and form factor F. The models used are heated by internal heating elements.

This is necessary because of the rough environment during and after the experiment. A heating device installed into the test section outside the model has to be protected against the flow. Heating prior to mounting the model into the test chamber is not advisable, because the chamber has to be evacuated before the experiment. Due to the relatively long evacuation time the model surfaces would cool down and therefore high temperatures could not be reached with this method. For this reason the model has to be heated already mounted inside the test section.

Keeping this in mind two heatable models have been tested in the shock tunnel TH2 with internal heating systems. A 24°-ramp model is equipped with electrical resistance heating elements. The aerodynamic surfaces consist of high temperature resistant steel. Total length of the model is $410\,mm$ with a width of $200\,mm$. Because the heated parts of the model are of a relatively big mass, the heating time is about 1 hour to achieve surface temperatures of about $T_{W,max} \approx 700\,K$. The model is equipped with 21 thermocouples to measure the temperature distribution during heating and heat fluxes during experiment.

The second model with a ramp angle of 15° was provided by former DaimlerChrysler Aerospace GmbH. The model width of $190\,mm$ and total length of $265\,mm$ provide a somewhat better two-dimensional flow. The aerodynamic surfaces consist of quartz-glass plates which are coated with a high-temperature resistant paint. The heating is performed by infrared radiators mounted inside the model. The main part of the radiation is expected to pass the glass, which has a transmissivity of $\tau \approx 0.7$. Only the thin coating is heated up to $T_{W,max} \approx 1000\,K$ within 2 minutes. This model was used in a work package of the FESTIP-program [1].

The geometrical requirements to assure a two-dimensional ramp flow were treated in [2]. In general, the model width W has to be bigger than the length of the ramp-model forebody L and ramp length R must be bigger than L. Ratios of $W/L \approx 1$ and $R/L \approx 1$ were found to result in a good approximation of a two-dimensional ramp flow. To give an idea of the influence of the aspect ratio, in [2] the measured pressure coefficients showed a maximum deviation of +5% for $W/L = 1.16$ and of +20% for $W/L = 0.33$ compared to the ideal reference model for 10° ramp angle and $M_\infty = 4$. Therefore, the 15°-ramp model with $W/L = 1.81$ and $R/L = 1.55$ provides a better approximation of a two-dimensional flow than the 24°-model with $W/L = 0.61$ and $R/L = 0.37$.

The experiments have been performed in the Aachen shock tunnel TH2 [3]. Table 1 gives the reservoir and free-stream conditions of the experiments performed.

3 Theoretical modeling of shock/boundary-layer interaction

Many theoretical investigations have been done regarding the dependence of flow fields on control surfaces as function of Reynolds number Re_∞, Mach number M_∞, wall-to-total-temperature ratio T_W/T_0 and peak pressure ratio p_{pk}/p_{ref}. The influence of these parameters on upstream influence or the separation length of a ramp flow, boundary-layer thickness δ, displacement thickness $\delta^\star$, laminar-turbulent transition and heat fluxes on the surfaces is summarized in [4].

As one visible feature of ramp flows the separation length L_{sep} can be characterized as

$$L_{sep} = f(Re_{L_{sep}},\ M_\infty,\ \delta^\star,\ \alpha,\ \frac{T_W}{T_0}) \tag{2}$$

where $Re_{L_{sep}}$ is the Reynolds number at separation position and α is the ramp angle. It was found by Needham and Stollery [5], that for laminar flow the influence of the temperature ratio T_W/T_0 should disappear, if the separation length is normalized by the boundary-layer thickness or its displacement thickness

$$\frac{L_{sep}}{\delta^\star} = f(Re_{L_{sep}},\ M_\infty,\ \alpha). \tag{3}$$

This is an interesting result which, in conjunction with heated model testing, may be useful to explain the influence of increased surface temperatures on this special flow behaviour. There exists a large variety of theoretical approaches to describe the dependencies in Eq. 3. Comparing numerical results with triple deck theory and experimental values, Katzer [6] found the length of the separated region L_{sep} for supersonic adiabatic flow to be given by

$$\frac{L_{sep}}{\delta^\star_{x0}} = K \cdot \frac{p_3 - p_{inc}}{p_1} \cdot \frac{\sqrt{Re_{x0}/C}}{M_\infty^3}, \tag{4}$$

where p_1 is the pressure just upstream the separation and p_3 the maximum pressure at reattachment (K = 4.4). The displacement thickness $\delta^\star_{x0}$ is determined at the position of the ramp hinge line x_0, C is the Chapman-Rubesin factor. The ramp pressure for incipient separation p_{inc} can be calculated by free interaction theory [6]

$$p_{inc} = p_1 \cdot \{1 + 0.5\gamma M_\infty^2 \cdot 1.85\sqrt{2} \cdot [\frac{c_{f0}}{(M_\infty^2 - 1)^{0.5}}]^{0.5}\}. \tag{5}$$

c_{f0} is the local wall shear stress of an undisturbed laminar boundary layer on a flat plate determined at the position of incipient separation.

For the boundary-layer thickness on a flat plate, which for hypersonic flow is nearly equivalent to the displacement thickness, the following formula is given in [7]

$$\frac{\delta}{x}\sqrt{Re_x} = 1.721 \cdot \{2.39 + \frac{T_W}{T_\infty} + 0.097\sqrt{Pr}(\gamma - 1)M_\infty^2\}. \tag{6}$$

The boundary-layer thickness according to Eq. (6) is used to scale the separation length.

4 Results

With the heated ramp models experiments have been performed for wall temperatures varying from 300 K to 1000 K. The temperature distribution and thermal characteristics of the model surfaces were proved inside the test section before doing experiments with the shock tunnel. Figure 2 shows the results of temperature measurements on the centerline of both models at four different levels of mean surface temperature up to about 700 K with electrical and at three different levels up to about 800 K with radiation heating elements. For comparison the axial coordinate was made dimensionless with the length of the ramp-model forebody L. One temperature distribution approximated for numerical simulation is added [8]. For the 24°-ramp model the temperature distribution was determined by thermocouples whereas infrared-thermography was used for the 15°-ramp model.

The deviation of the surface temperature was found to be $\pm 25\,K$ with the resistance heating elements (24°-ramp model, steel surfaces) and $\pm 50\,K$ with the radiation heating system (15°-ramp model, quartz glass surfaces). Using radiation heating up to now higher temperatures have been reached. Local temperature maxima at five positions above the radiators are visible. Both models need a solid front wall to protect sensors and heaters against the hypersonic flow. Because of this heat sink, temperatures of the front part of both models are lower than for the central part.

It is not possible to run the heating systems during the experiments because of vibrational loads which would cause a damage of the heating units. The maximum cooling rates of the surfaces were determined to be $2 \cdot 10^{-4}\ K/ms$ using steel surfaces and $5 \cdot 10^{-2}\ K/ms$ for coated glass surfaces. Compared to the surface heating rate during the experiments of about $5\,K/ms$ the cooling due to radiation and heat conduction is negligible.

The length of the separation region is determined from colour Schlieren photographs taken during the run. The interaction of the separation shock with the forebody boundary layer is clearly visible (Fig. 1). On the pictures the separation shock has a finite thickness. The intersections of the upstream and downstream edge of the separation shock with the boundary-layer edge give two positions l_1 and l_2 on the plate. The average value of these two is taken as separation position. The reattachment position is determined the same way by averaging the impact locations of the shear-layer edges on the ramp. For the 15°-ramp model on all Schlieren pictures transition phenomena of the separated boundary layer could not be detected.

Figure 3 shows the separation position L_S and reattachment position L_R as function of the temperature ratio T_W/T_0 for both ramp models. The temperature ratio T_W/T_0 is determined with the measured mean surface temperature. A nearly linear increase of the size of the separation bubble with temperature ratio is found. For comparison numerical results [8] for the 15°-ramp are added. For the computations the wall temperature distribution was approximated from the measured temperature distribution avoiding the local temperature maxima (see Fig. 2). The experiments and numerical results yield a quite good agreement showing that the local maxima of surface temperature have no crucial impact on the measured separation length.

To check the statement of Needham and Stollery [5], the boundary-layer thickness must be known. It was measured on a flat plate for a cold model surface ($T_W = 300\,K$) to check whether Eq. (6) is applicable for the flow conditions used in the shock tunnel TH2. As example Fig. 4 shows pitot pressure distributions within the boundary layer using two different unit Reynolds numbers of $6 \cdot 10^6\ 1/m$ and $1.2 \cdot 10^6\ 1/m$, respectively. The measurements are normalized with the pitot pressure of the flow outside the boundary layer. The distance, where the pitot pressure ratio reaches the value one, defines the boundary-layer edge. Figure 5 shows the results in comparison with Eq. (6). For cold model surfaces ($T_W = 300\,K$) a good agreement is achieved with the theoretical values. Based on this result Eq. (6) was also used to calculate the boundary-layer thickness δ for the heated surface experiments. This value is used to normalize the separation length $L_{sep} = L_S + L_R$ of Fig. 3. The result of this is given in Fig. 6.

Within the experimental uncertainty at least for the 15°-model the ratio L_{sep}/δ shows an independent behaviour from the temperature ratio T_W/T_0. This confirms the hypothesis of Needham and Stollery. That means, the growing of the separation bubble with increased surface temperature is caused by the thickening of the undisturbed boundary layer. This leads to a reduced skin friction, which for the same ramp angle causes a larger extent of the separation bubble in the upstream direction. Equation (4) given by Katzer to calculate the length of the separation region was derived for supersonic adiabatic flow with $M_{\infty,max} = 3.4$. Using Katzer's original formula for the present case of a hypersonic nonadiabatic flow with reference conditions of an undisturbed flow at the ramp hinge line (x_0, Re_{x_0}, δ_{x0} and c_{f0}), no realistic result could be achieved. But scaling Eq. 4 with the values at the separation position (x_1, Re_{x_1}, δ_{x_1}), the theoretical results achieved with the 15°-ramp model are in good agreement with the experiments, if the factor $K = 4.4$ in Katzer's formula is replaced by $K = 3.9$. The pressure p_3 was determined by oblique-shock theory and the angle of the separating stream line from the Schlieren pictures. With the reference conditions chosen at separation ($x = x_1$) and $K = 4.4$ the ratio L_{sep}/δ is somewhat bigger. These values are indicated by the solid line in Fig. 6.

For the 24°-ramp model a larger deviation between the experimental and theoretical results occurs, which can be explained by three-dimensional flow phenomena due to the relatively long flat plate as forebody. Transitional effects in the separated boundary layer due to the relatively large ramp angle can be another reason. This also explains, why for the 24°-ramp model the ratio L_{sep}/δ in Fig. 6 shows not a constant value as for the 15°-ramp model.

5 Conclusions

A semi-empirical approach to estimate the separation length on ramp models has been compared to measurements with heated ramp models. For constant stagnation temperature T_0 the temperature ratio T_W/T_0 was varied from 0.1 up to 0.6 for the highest surface temperature of $1000\,K$. Within these limits for 15° and 24° ramp angle the separation length increased with growing surface temperature, but the ratio of separation length to boundary-layer thickness at separation was proven to be independent of the wall-to-total-temperature ratio. Comparison with an approach proposed by Katzer to calculate this ratio showed that it is only applicable modifying the reference condition. With this modification his approach shows the same independence on wall-to-total-temperature ratio as Needham and Stollery proposed. The experimental results suggest that within certain limits a homogeneous temperature distribution is not absolutely necessary to achieve agreement with theoretical methods and numerical simulations.

Acknowledgements

The authors thank Prof. Charbonnier and Michael Playez for their help doing the infrared-measurements and the DaimlerChrysler Aerospace AG for providing one heatable ramp model. This investigation was funded by the Deutsche Forschungsgemeinschaft within the Special Research Center (SFB 253) "Fundamentals of the Design of Aerospace Planes" at the RWTH Aachen and by ESA within the FESTIP-program.

References

[1] H. Olivier; M. Bleilebens. Final report: Hot experimental testing. Technical Report ESA WP5510, Shock Wave Laboratory, RWTH, 1999.

[2] J.E. Lewis; T. Kubota; T. Lees. Experimental investigation of supersonic laminar, two-dimensional boundary-layer separation in a compression corner with and without cooling. *AIAA Journal*, 6:7–14, 1968.

[3] H. Olivier; H. Grönig. Instrumentation techniques of the Aachen shock tunnel TH2. In 16th *Int. Congress on Instrumentation in Aerospace Simulation Facilities*, 1995.

[4] J. Délery. Shock/shock and shock-wave/boundary layer interactions in hypersonic flows. AGARD Report 761, NATO, 1989.

[5] D.A. Needham; J.L. Stollery. Hypersonic studies of incipient separation and separated flow. AGARD Conference Proceedings Vol. 4, NATO, 1966.

[6] E. Katzer. On the lengthscales of laminar shock/boundary-layer interaction. *Journal of Fluid Mechanics*, 206:477–496, 1989.

[7] N.R. Mudford; S.G. Mallinson; S.L. Gai. High enthalpy, hypersonic compression corner flow. In 33rd *Aerospace Sciences Meeting and Exhibit*, 1995. AIAA Paper 95-0155.

[8] A. Henze; W. Schröder. On the influence of thermal boundary conditions on shock boundary-layer interaction. In *Deutscher Luft- und Raumfahrtkongress 2000*, 2000. Jahrbuch 2000.

Table 1: Calibrated test condition used for heated ramp model testing

Condition	p_0 [MPa]	T_0 [K]	M_∞ [-]	Re_∞ [$10^6 m^{-1}$]	T_∞ [K]	u_∞ [ms^{-1}]	p_∞ [kPa]	ρ_∞ [$10^{-2} kgm^{-3}$]
I	7	1500	7.7	4.2	125	1745	0.76	2.1

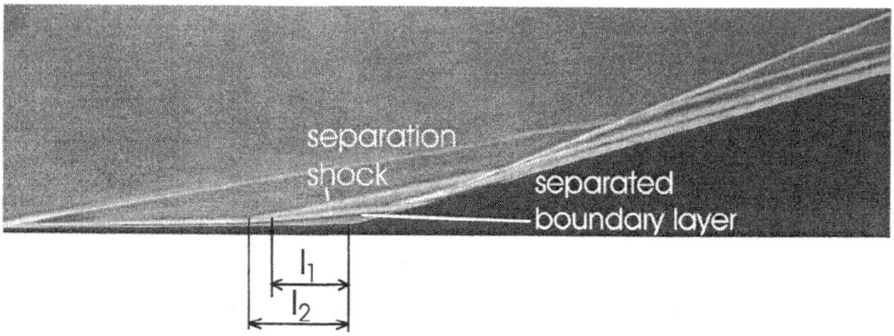

Figure 1: Example of a schlieren picture showing all features of the ramp flow (15° ramp).

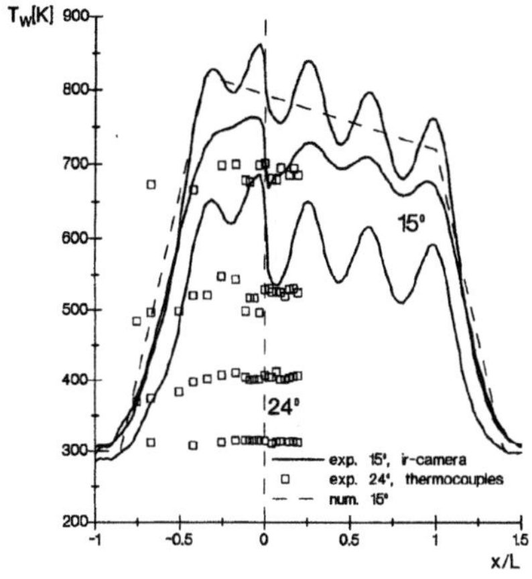

Figure 2: Temperature distribution on the centerline of the heated ramp models using different heating power levels.

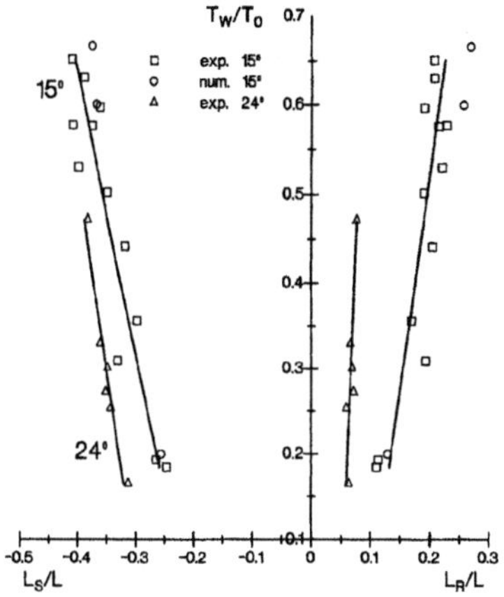

Figure 3: Separation and reattachment positions as function of wall-to-total-temperature ratio.

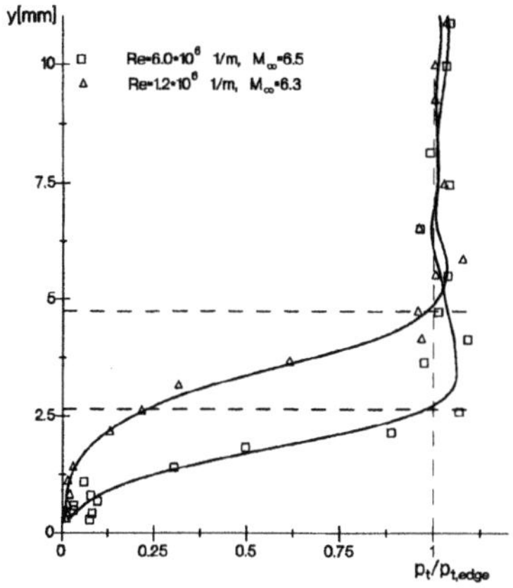

Figure 4: Examples of boundary layer pitot pressure distributions for two unit Reynolds numbers.

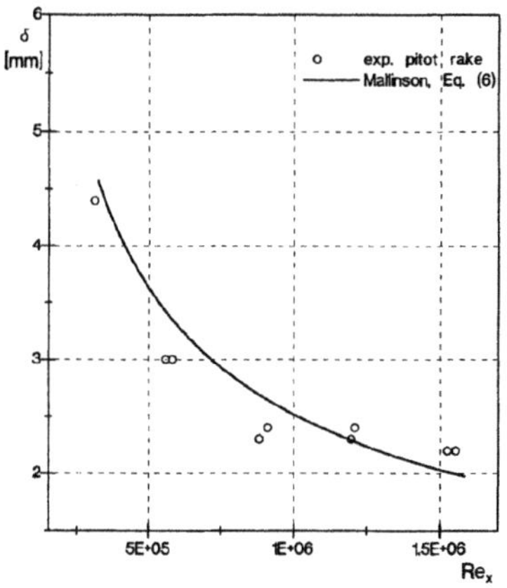

Figure 5: Comparison of pitot rake measurements with equation (6).

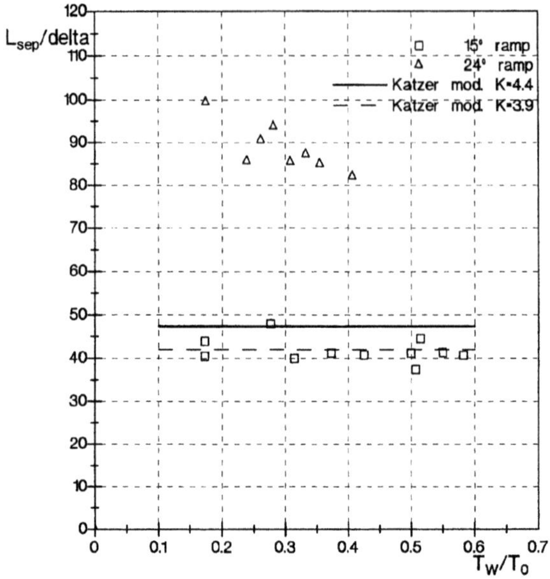

Figure 6: Normalized separation length for 15° and 24° ramp models.

Fluid-Structure-Coupling

Preliminary Results of Numerical Simulations of Elastic Panels in Transonic Flows

J. Hurka and J. Ballmann
Lehr- und Forschungsgebietg für Mechanik
Templergraben 64
RWTH Aachen, D-52062 Aachen, Germany

Summary

We consider the two-dimensional (2D) panel flutter problem, i.e., a 2D flow over an one-dimensional (1D) plate, with the objective to study the influence of structural nonlinearities and shear deformation on the aeroelastic solution. Preliminary results of numerical simulations of linear and geometrically nonlinear classical plates in transonic Euler flows are presented. We apply a loose fluid structure coupling scheme where we use the DLR FLOWer code to compute the flow field and a modal Galerkin method to compute the plate deformation. An algebraic grid deformation algorithm to adjust the fluid grid to the deforming panel surface is presented.

1 Introduction

We consider the aeroelastic problem of a rectangular plate of infinite span with hinged leading and trailing edges in 2D flow, see Fig. 1. The problem is solved by numerical integration of the corresponding dynamic equations in the time domain. In the transonic and supersonic flow regime, this system may exhibit aeroelastic instabilities such as divergence or flutter. Different linear and nonlinear aerodynamic and structural theories have been employed in the literature to address the problem at hand, see, e.g., [5], [6] and references therein. Davis and Bendiksen [3], [2] presented numerical simulations of panel flutter and divergence which they obtained by using nonlinear classical plate theory and Eulerian aerodynamics.

One goal of our research project is to investigate the influence of the mathematical model used to describe the dynamics of the structure on the solution of the aeroelastic problem at hand. Linear and geometrically nonlinear plate models are considered. Geometrically nonlinear effects in plate dynamics become important when the plate deflection reaches the order of the plate thickness, as it may occur for coupled fluid plate dynamics beyond the aeroelastic stability boundary, see, e.g., [6]. Furthermore, we wish to study the influence of shear deformation. Classical plate models, which do not account for shear deformation, yield parabolic sets of partial differential equations (PDEs) which allow for infinite (and hence, unphysical) information propagation. Reissner-Mindlin (RM) type plate models, which incorporate shear deformation, on the other hand, result in hyperbolic sets of PDEs which exhibit finite information propagation.

While the implementation of a finite element method to solve the nonlinear classical and RM type plate equations is on its way, preliminary results of coupled aeroelastic simulations of linear and geometrically nonlinear classical plates are presented here. In Sects. 2-4 we give a brief discussion of the methods used to solve the problem at hand. Results of our simulations are presented in Sect. 5, while in Sect. 6 we summarize.

2 Structural method

Within the framework of classical plate theory, the nonlinear dynamic equation governing the plate lateral deflection w in 1D may be written as

$$D\frac{\partial^4 w}{\partial x^4} - (N_{x_0} + N_x)\frac{\partial^2 w}{\partial x^2} - \alpha\frac{\rho h^3}{12}\frac{\partial^4 w}{\partial t^2 \partial x^2} + \rho h\frac{\partial^2 w}{\partial t^2} + \Delta p = 0, \tag{2.1}$$

where N_{x_0} and

$$N_x = \frac{Eh}{2c}\int_0^c (\partial w/\partial x)^2\, dx. \tag{2.2}$$

denote in-plane loads due to externally applied forces and lateral deflection, respectively, while $\Delta p = p - p_0$, D, E and ρ denote the aerodynamic pressure loading on the panel, the plate's bending stiffness, Young's modulus and mass density, respectively.. The case $\alpha = 1$ corresponds to the inclusion of rotary inertia, while $\alpha = 0$ means its neglection. In what follows, we will take $p_0 = p_\infty$ and $N_{x_0} = \alpha = 0$, and (2.1) reduces to a 1D version of von Karman's plate equation. The nonlinearity is introduced into Eq. 2.1 through the assumption that the plate's leading and trailing edges are immovable.

A modal Galerkin method is employed to solve Eq. (2.1). We expand w in terms of the eigenmodes of the linear Kirchhoff plate

$$w(x,t) = \sum_n q_n(t)\sin(n\pi x/c) \tag{2.3}$$

whereby we restrict ourselves the boundary conditions of hinged support. Inserting Eq. (2.3) into (2.1), multiplying by $\sin(r\pi x/c)$ and integrating the resulting equation between the limits 0 and c yields the following coupled set of nonlinear ordinary differential equations (ODEs) for the modal amplitudes q_r, see, e.g., [6],

$$\rho h\ddot{q}_r = \left[-D\left(\frac{r\pi}{c}\right)^4 - \left(\frac{r\pi}{c}\right)^2\left\{\frac{Eh}{4}\sum_k\left(\frac{r\pi}{c}\right)^2 q_k^2\right\}\right] q_r - 2p_r, \tag{2.4}$$

where

$$p_r = \frac{1}{c}\int_0^c \Delta p(x)\sin(r\pi x/c)\, dx. \tag{2.5}$$

We assume that the pressure on the panel surface varies linearly between the discrete pressure values computed by the flow solver and use the trapezoidal rule to evaluate the integral in Eq. (2.5). In order to compute the time evolution of the modal amplitudes q_r, Eq. (2.4) is rewritten as a system of first order ODEs. Explicit time integration is performed by a standard 4-step Runge-Kutta (RK) method, see, e.g., [8]. Furthermore, we implemented Jameson's second order accurate implicit dual time stepping method [9] which employs the RK method to achieve convergence during each implicit time step. For the results presented in Sec. 5, the implicit Jameson method was used.

3 Fluid method

3.1 Fluid solver

We use the DLR FLOWer code to compute the flow field over the deforming panel. FLOWer is a block structured finite volume solver to the three-dimensional unsteady Euler and Reynolds averaged Navier Stokes equations in a deforming mesh formulation. A central space discretization with artificial viscosity is employed. Explicit Runge-Kutta time stepping and several convergence accelerating methods such as multigrid may be used for the calculation of steady flow fields. Jameson's implicit dual time stepping method is used for time accurate calculations of unsteady flows. For a detailed discussion of the FLOWer code, see [1] and [10].

3.2 Grid deformation algorithm

We implemented an algebraic grid deformation algorithm, which is a modification of the method described in [11] and [7], to adjust the fluid grid to the deforming plate surface during each time step. The algorithm may be sketched as follows.

1. For each grid line leaving the panel surface, we perform a translation and rigid body rotation such that the new grid lines are normal to the deformed panel surface.

2. We calculate the minimal distance of points, where adjacent new grid lines intersect, to the panel surface, and call it a_0. Finally, we take $a_0 = \min(a_0, \epsilon)$, where ϵ is some prescribed threshold value.

3. We add the displacement of the grid line foot point on the panel surface, weighted by some function g, to the points of the old grid line. The weight function is an exponential of the form $g = \exp\{-(\kappa a/a_0)^2\}$, where a is the arc length along the grid line and κ is some numerical factor depending on the grid dimensions. We choose κ such that g is close to unity for $a \ll a_0$, i.e., near the panel surface, and $g \approx 0$ for $a \geq a_0$.

4. Finally, we interpolate between the rotated grid lines, taken from step 2, and the modified old grid lines from step 3 by use of a weight function f, such that the final grid lines leave the deformed panel surface in a nearly perpendicular fashion while far away from the panel the old and new grid points are identical. The weight function f is of the same generic type as g with a slightly different numerical value for κ. Note that the parameter a_0 is recalculated in each time step to avoid intersection of grid lines.

For grid lines leaving the fixed wall in the neighborhood of the panel end points, in step 2 of our algorithm, we perform rotations appropriate to the panel end points weighted by some smoothly decaying function. Again, this is done to avoid grid line intersection. Fig. 2 shows a deformed mesh configuration which was obtained by the algorithm described.

4 Fluid structure coupling scheme

We apply a conventional loose fluid structure coupling scheme which allows for a time lag of one integration time step between fluid and structure. A schematic diagram of the coupling algorithm is presented in Fig. 1. Given the numerical solutions U_n and W_n for structure and fluid, respectively, the structural displacements are used to update the fluid boundary and mesh. In a second step, the fluid solution is advanced to time level $(n+1)$. The fluid pressure acting on the structure is passed to the structural solver. Finally, the structural solution is advanced in time to level $(n+1)$.

5 Results

We performed numerical simulations of 2D transonic Euler flows over 1D linear and geometrically nonlinear classical flat panels with hinged ends. The flow field was computed on a 208 by 72 structured grid, shown in Fig. 2. The panel end points are located at $x = 0.0$ and $x = 1.0$ on the $z = 0.0$ boundary of the computational domain . We have 80 grid cells on the panel surface. We took the distance of the first grid line parallel to the panel to be $\Delta z = 10^{-3}$. We refined the grid lines orthogonal to the panel towards its leading and trailing end points, such that at the panel end points we have a grid line spacing of $\Delta x = 10^{-3}$. Dual time stepping in combination with four level multigrid acceleration were used to obtain the flow field.

For the computation of the structural solution, 20 modes were included into the analysis. For the aeroelastic cases considered here, we found the number of excited modes to be well below 10. For the simulation of linear panels we set $N_x = 0$ in Eq. (2.1) as for simply supported ends which are free to move axially.

Following Davis & Bendiksen [3], we initiate our simulations by setting the flow variables equal to the far field values throughout the fluid mesh. We take the panel to be in an undeformed state with an initial velocity that would lead to a small amplitude first mode oscillation for a linear panel in vacuum. We specify the initial velocity such that this amplitude would reach approximately 20 per cent of the plate thickness.

In Figs. 3 and 4 we compare results of simulations of nonlinear panel divergence and flutter obtained by the present numerical method to results given by Davis & Bendiksen in [3], [2]. Davis & Bendiksen apply the finite element method and explicit 5-stage RK time stepping to solve the fluid and structure equations. In their fluid structure coupling scheme, information between fluid and structure is exchanged in each RK substep. The diagrams in Fig. 3 depict the plate deflection and plate surface Mach number at divergence of an aluminium panel with thickness ratio $h/c = 0.004$ at sea level and Mach number $M = 0.9$. The left diagram is taken from [3], the right diagram displays our results. While we find two details to differ quantitatively, the overall agreement between the two solutions is quite good. In contrast to Davis and Bendiksen, our solution shows tiny supersonic flow regions at the leading and trailing edge of the panel. We attribute this to our higher fluid grid resolution near the panel end points. In addition, we find a smaller plate midpoint deflection as Davis and Bendiksen (about 75 % of theirs). This may be due to the way they discretized the nonlinear stretching term N_x in [3], see their own remark on this matter in [2]. The diagrams in Fig. 4 display the deflection of three points on the panel surface versus dimensionless time for the transonic flutter of a double hinged aluminium panel at sea level ($M = 0.95$ and $h/c = 0.002$). The left diagram is taken from Bendiksen and Davis [2], the right diagram shows our results. Good agreement between the two methods is found for this case.

In addition to the simulations of nonlinear panels, we performed calculations of linear panels. The absence of the nonlinear stretching term in the plate equation can lead to large plate deflections for flow conditions beyond the stability boundary. In some cases, the plate deflection became so large that fluid grid lines near the panel surface intersected and the calculation broke down. In Figs. 5 and 6 we present results of a simulation of a linear panel with $h/c = 0.005$ at 11 km above sea level and $M = 0.85$. While for a nonlinear panel we find stationary divergence with a relatively low amplitude ($\sim 0.5\ h$) first mode deflection for these flow parameters, in the case of a linear panel our simulation predicts oscillatory plate motion around a mean amplitude of approximately 22 h, see Fig. 5. Figure 6 shows the Mach number distribution over the oscillating panel for two instants corresponding to minimal (Fig. 6 left, $t = 185.0$) and maximal (Fig. 6 right, $t = 195.0$) deflection during the panel oscillation. For $t = 185.0$, the panel deflection is such that the shock wave terminating the supersonic region on the panel surface is located near the trailing edge (at $x = 0.93$), i.e., a relatively large portion of the panel is exposed to a high speed flow and thus experiences lift that the linear panel bending stiffness cannot compensate. Consequently, the amplitude of the panel deflection grows with time. This in turn causes the shock to move towards the leading edge of the plate, whereby the total lift of the panel decreases again, until for $t = 195.0$ the shock is located at $x = 0.79$ and the motion of the panel is reversed by the structural forces. In the course of the downward motion the shock wave moves towards the trailing edge again and the cycle is repeated. Note that our Euler solution predicts the existence of a vortex near the trailing edge of the plate. We find that the size of vortex region grows as the panel moves upwards, while its size decreases during the downward motion of the panel, see Fig. 6.

6 Conclusions and outlook

We applied a loose fluid structure coupling scheme to perform numerical simulations of 2D transonic inviscid flows over linear and geometrically nonlinear 1D classical plates. The DLR FLOWer code and a modal Galerkin method have been used to compute the flow field and the plate deformation. An algebraic grid deformation algorithm was presented, which is capable to adjust the fluid grid to moderately large plate deformations. Two simulations of nonlinear panels were presented, where divergence and limit cycle flutter occur. The results of these simulations are in reasonable agreement with results given by Davis & Bendiksen in [3] and [2]. In the case of a simulation of a linear panel, we found panel vibrations around a high amplitude mean deflection.

Simulations of linear and nonlinear RM type plates will be performed as soon as the implementation of an appropriate finite element method is completed. First calculations of viscous flows over a deforming panel have been performed and further viscous flow simulations will follow.

References

[1] P. Aumann, W. Bartelheimer, H. Bleecke, B. Eisfeld, J. Lieser, R. Heinrich, N. Kroll, M. Kuntz, E. Monsen, J. Raddatz, U. Reisch, B. Roll: FLOWer installation and user handbook, Release 115, 1998

[2] O.O. Bendiksen, G.A. Davis: Nonlinear traveling wave flutter of panels in transonic flow, AIAA Paper 95-1496, 1995

[3] G.A. Davis, O.O. Bendiksen: Transonic panel flutter, AIAA Paper 93-1476, 1993

[4] E.H. Dowell: Nonlinear oscillations of a fluttering plate, AIAA Journal, Vol. 4, No. 7, 1966, pp. 1267-1275

[5] E.H. Dowell: Panel flutter: A review of the Aeroelastic stability of plates and shells, AIAA Journal, Vol. 8, No. 3, 1970, pp. 385-399

[6] E.H. Dowell: Aeroelasticity of plates and shells, Noordhoff International Publishing, Leyden, 1975

[7] G. Britten, J. Ballmann: Navier-Stokes-based direct numerical simulation, to appear in: ICAS 2000 Conference Proceedings

[8] C. Hirsch: Numerical computation of internal and external flows, Volume 1, John Wiley & Sons, New York, 1988

[9] A. Jameson: Time dependent calculations using multigrid, with applications to unsteady flows past airfoils and wings, AIAA Paper 91-1596, 1991

[10] N. Kroll, C.-C. Rossow, K. Becker, F. Thiele: The MEGAFLOW project, Aerospace Science and Technology 4, 2000, pp. 223-237

[11] M. Werle: Gittergenerierung für die Strömungssimulation von elastischen Auftriebskörpern, Diplomarbeit, RWTH-Aachen, 1997

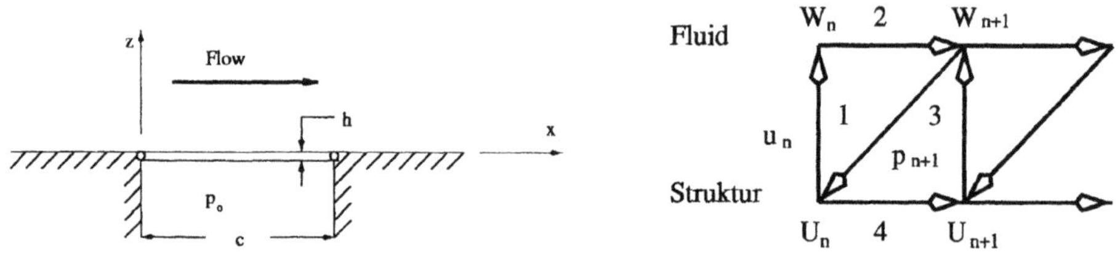

Figure 1 **Left**: Geometry of the considered doubled hinged plate fluid problem. h: plate thickness, c: plate chord length, p_0: pressure below the plate. **Right**:Schematic diagram of the used fluid structure coupling algorithm.

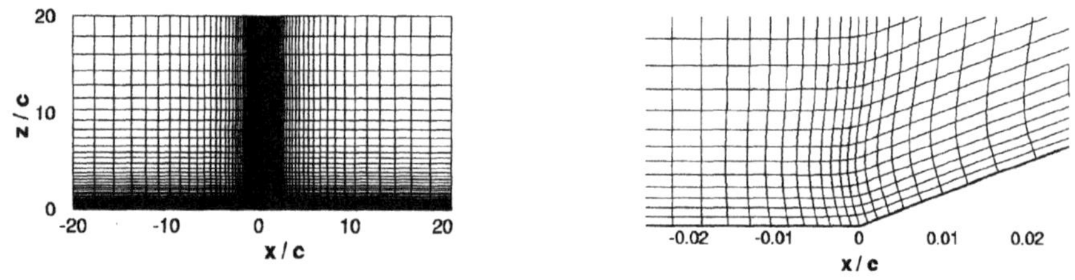

Figure 2 Fluid computational grid. **Left**: Display of far field boundaries. **Right**: Deformed fluid grid near the panel leading edge (taken from the simulation presented in Figs. 5 and 6).

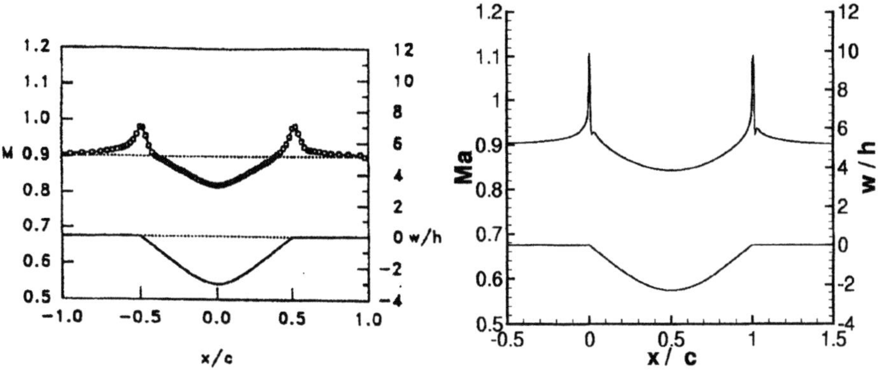

Figure 3 Plate deflection and Mach number distribution on the plate surface at divergence. Double hinged aluminium plate at sea level, $M = 0.9$, $h/c = 0.004$ (M: Mach number). Euler flow – (nonlinear) von Karman plate. **Left:** Results of Davis & Bendiksen, taken from [3]. **Right:** Results of the present method.

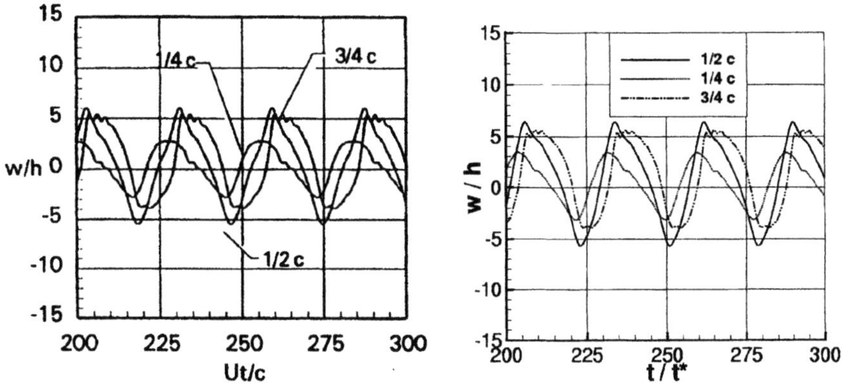

Figure 4 Transonic limit cycle flutter of a hinged aluminium plate at sea level, $M = 0.95$, $h/c = 0.002$. Time history of the displacement of three points on the plate's surface: $1/4c$, $1/2c$, $3/4c$. Euler flow – von Karman plate. **Left:** Results of Bendiksen & Davis [2]. **Right:** Results of the present method.

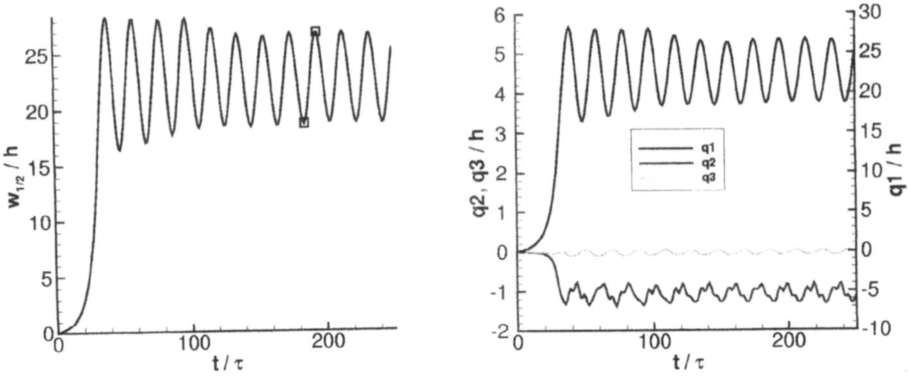

Figure 5 Simply supported aluminium plate at 11 km above seal level, $M = 0.85$, $h/c = 0.005$. Euler flow – (linear) Kirchhoff plate. **Left**: Time history of plate midpoint displacement. **Right**: Time history of the first three modal amplitudes q_1-q_3 . The right y-scale refers to the q_1, the left one to q_2 and q_2.

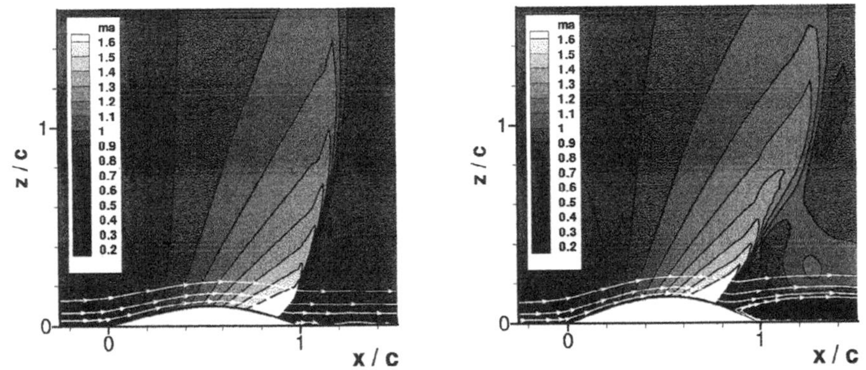

Figure 6 Mach number distribution over the oscillating panel of Fig. 5 at two times $t = 185.0$ (**left**) and $t = 195.0$ (**right**) (labeled in Fig. 5 by the square symbols). The white lines indicate tangents to the instantaneous velocity field near the panel surface.

Fundamentals of Fluid Flow

Numerical Simulation of Laminar Separation-Bubble Control

U. Rist, K. Augustin, S. Wagner
Universität Stuttgart, Institut für Aerodynamik und Gasdynamik
Pfaffenwaldring 21, D-70550 Stuttgart, Germany

Summary

In the present paper an active control mechanism for the control of laminar separation bubbles on airfoils is investigated by means of direct numerical simulation and linear stability theory. Boundary layer instabilities excited by periodic oscillations are utilized to control the size and length of the separation bubble and to make it finally disappear when desired. Unlike traditional vortex generators a sensor-actuator system based on this method will be adaptive to the respective flow conditions and will cause no additional undesired drag.

Introduction

Operated in a low Reynolds number regime the drag characteristics of laminar airfoils at "off-design" conditions can be considerably deteriorated by *laminar separation bubbles (LSB)*. The LSBs form due to the inability of the laminar boundary layer to overcome the stronger adverse pressure gradient caused for example by a higher angle of attack or a deflected trailing edge flap. Such an effect is shown in figure 1 by the polars c_L/c_D of a laminar airfoil at two different Reynolds numbers Re.

At $Re = 0.7 \cdot 10^6$ (low speed) with an extended flap the laminar boundary layer separates from the lower side of the airfoil just upstream of the flap. Laminar-turbulent transition occurs subsequently and the turbulent boundary layer reattaches at the rear end of the airfoil, thus forming a laminar separation bubble. The reattachment results from the increased momentum transfer towards the wall within the turbulent boundary layer. This laminar separation bubble can be removed by mounting of vortex generators upstream of the bubble which force the boundary layer into turbulence and therefore reduce the drag coefficient c_D accordingly (dashed line with diamonds). In contrast to the low speed case the properties of the airfoil at high speed ($Re = 2.5 \cdot 10^6$) and no extended flap are vice versa. The laminar boundary layer is able to overcome the pressure gradient and is not separating (solid line with squares). Vortex generators in this case produce undesired additional drag (dashed line with squares) by shortening the length of the laminar boundary on the lower side.

To obtain an adaptive vortex generator with respect to the present flow conditions we suggest to use controlled excitation of boundary layer perturbations upstream of the bubble. These disturbance waves become amplified by instability mechanisms within the laminar boundary layer and in the separation bubble itself which leads to an upstream shift of the transition and the now turbulent boundary layer does not separate. As a constraint, the upstream shifting should be achieved by a minimum of disturbance amplitude A_0 respectively of disturbance energy introduced into the flow. But the necessary amplitude A_0 can only be minimized by choosing the proper wavelength λ_x, frequency β and of course the location and length of the disturbance strip exciting sinusoidal disturbance waves.

An indicator for the effectiveness of this control is the location of transition within the LSB as shown in figure 2 where "S", "T" and "R" mark separation, transition and reattachment, respectively. The physical investigation of the laminar separation bubble phenomenon by means of direct numerical simulation (DNS) and linear stability theory (LST) is used to estimate a control method capable to avoid separation in the "off-design" case without deteriorating the design properties of the airfoil.

Based on the considered method an actuator-sensor concept will be designed, where the LSB is detected by sensors and an upstream actuator driven by a controller (figure 3) excites the boundary layer disturbances.

Numerical method

The implemented 3D DNS method [1] solves the complete Navier-Stokes equations for incompressible flow in a vorticity-velocity formulation

$$\frac{\partial \underline{\omega}}{\partial t} - \mathrm{rot}\,(\underline{v} \times \underline{\omega}) = \frac{1}{Re}\,\Delta \underline{\omega} \qquad (1)$$

$$\text{with} \quad \underline{v} = (u,v,w) \quad \text{and} \quad \underline{\omega} = (\omega_x,\omega_y,\omega_z)$$

within a rectangular integration domain (A-B-C-D in figure 2) over a flat plate.

All spatial coordinates are non-dimensionalized by a reference length $\tilde{L}$ and velocities by the free-stream velocity $\tilde{U}_\infty$, where $\tilde{}$ denotes dimensional variables. This leads, together with the kinematic viscosity $\tilde{\nu}$, to the following definitions for the vorticity vector $\underline{\omega}$ and the frequency β

$$\underline{\omega} = -\,\mathrm{rot}\,\underline{v}\,, \quad \beta = \frac{2\pi \tilde{f}\,\tilde{\nu}}{\tilde{U}_\infty} \cdot 10^5\,, \quad \mathrm{Re} = \frac{\tilde{U}_\infty \tilde{L}}{\tilde{\nu}}. \qquad (2)$$

Fourth-order accurate finite differences are used on the equidistant numerical grid in streamwise (x-) and wall-normal (y-) direction whereas a spectral Fourier ansatz in spanwise (z-) direction is applied. The explicit time integration is realized by a fourth order, four step Runge-Kutta scheme. Once the three vorticity components are obtained, three Poisson equations for the remaining velocities u, v and w have to be computed. Due to the spectral ansatz in spanwise direction the Poisson equations for the u and v velocity reduce to ordinary differential equations (ODE) and only the v-equation has to be solved iteratively by a line relaxation method accelerated by a multigrid scheme [7].

At the inflow boundary steady Falkner-Skan or, at zero pressure gradient, Blasius velocity profiles are prescribed. In a relaminarization zone [2] upstream of the outflow boundary the unsteady vorticity components are damped to steady state values to avoid non-physical reflections and as a consequence the unsteady velocity components $\underline{v}'$ vanish exponentially. At the wall the no-slip condition is applied except for a disturbance strip where 2D and 3D perturbations can be introduced into the flow by periodic suction and blowing.

To take into account the strong boundary layer displacement effects of the LSB a viscous-inviscid interaction model is implemented into the multigrid algorithm for the v-velocity component. At run time a velocity distribution which allows for the typical boundary layer displacement effects of a LSB evolves from the initially prescribed potential flow. For a detailed summary of the interaction model and boundary conditions refer to [3]. Figure 4 shows the initial potential distribution of the u-velocity (dashed line) and the resultant distribution considering the effects of the LSB (solid line).

Once the DNS has yielded a reasonable mean flow the development of excited boundary layer perturbations can be studied with less effort by using the linear stability theory (LST) [8]. However, the LST can provide correct results in the linear regime of the disturbance development only, i.e. upstream of transition ("T") and cannot cover variations of the mean flow due to the strong influence of different disturbance amplitudes A_0 on the size of the LSB. Applying the LST the spatial amplification rate α_i of discrete fluctuations specified by their streamwise wavenumber α_r and disturbance frequency β in

$$v(x,y,t) = \Re\left\{\hat{v}(y)\; e^{i(\alpha x - \beta t)}\right\} \qquad (3)$$

$$\text{with} \quad \alpha = \alpha_r + i\,\alpha_i$$

can be determined by solving the Orr-Sommerfeld equation. Values of $\alpha_i < 0$ mean amplification and $\alpha_i > 0$ damping of the disturbance amplitude. The disturbance amplitude $A(x)$ can thereon be obtained by integrating the amplification rate α_i in streamwise x-direction and multiplying these values with an initial disturbance amplitude A_0.

Control of LSB by boundary layer disturbances

The present quantitative investigations are performed for a pressure-induced LSB in a flat-plate boundary layer where a Blasius boundary layer solution with $Re_{\delta_1} = 1722$ is prescribed at inflow and a streamwise velocity deceleration by 10% is prescribed at the free stream boundary (cf. figure 4). The other parameters of the flow are Re=100000 (for $x = 1$), $\Delta x = 0.008376$, $\Delta y = 0.002296$, and $\Delta t = 0.002513$. This case is comparable to the "midchord-bubble" considered in greater detail in [6].

Comparing the amplification curves $u'_{max}(y)/U_\infty$ for the fundamental disturbance mode with $\beta = 5.0$ in figure 5 yields an excellent agreement between LST and DNS. This is illustrated in the $A_v = 10^{-6}$ case (dash-dotted and dash-dotted line with squares), where A_v is the amplitude of the wall-normal velocity disturbance within the disturbance strip at the wall. Both curves evidently coincide which shows that the LST is able to predict the disturbance development in flows with LSB once the flow field has been computed.

The amplification curves in figure 5 clearly show an upstream shift of the point of amplitude saturation with increasing A_v which coincides with transition. It can be expected that this upstream shift will reduce the size of the LSB.

This influence on the size of the bubble is also present in the local mean velocity profiles of the streamwise velocity u at certain x-positions. In figure 6 the velocity profiles at the point of separation a) and at the point of reattachment b) of the case $A_v = 10^{-4}$ (with the smallest LSB) are shown for comparison with the other considered cases. The reference case without disturbance excitation shows the greatest differences. The boundary layer has separated upstream and is still separated in b) where a fully developed separation profile with reverse flow ($u < 0$) close to the wall is visible. All other cases show the same properties at these positions but with higher disturbance levels the size and the strength of the region of reverse flow becomes smaller and smaller and together with it the overall size of the LSB reduces.

This strong influence of even very small disturbances ($A_v \ll 1$) on the shape and the size of the time averaged separation bubble is displayed in figure 7. On the left hand side by streamlines including the separation line (emphasized) versus the streamwise coordinate x and on the right hand side by the mean skin friction distribution ω_{zw} for the five cases $A_v = 0$ to $A_v = 10^{-4}$ already considered in the preceding figures.

The large separation bubble which contains a large recirculation vortex at its rear end is reduced in size. Only a very small bubble is visible at a disturbance level of $A_v = 10^{-4}$ and no recirculation vortex. Moreover, the strong displacement of the boundary layer by the large separation bubble, which would affect the pressure distribution of an airfoil and cause additional drag, almost vanishes with an increased disturbance level. Both the separation and reattachment point are shifted leading to a reduction of the bubble in size. The reattachment point is shifted upstream by the growing influence of the disturbance excitation and the thereby earlier transition. In addition to that the separation point moves further downstream towards the reattachment because of upstream effects of the transition location on the wall-pressure (cf. [6]). In other words, the bubble becomes shallower with increased forcing and exerts less displacement on the potential flow.

The LSB can be detected by the negative values of the time averaged (in this case over two disturbance cycles) skin-friction distribution ω_{zw} in the right-hand-side figures. The streamwise length of the bubble is marked by arrows. The initially only small negative values of ω_{zw} turn into strong skin friction at the rear end of the bubble with a small region of positive values in between. This region is the footprint of the recirculation vortex at the end of the bubble whose strength weakens with increasing disturbance amplitude. High positive values of the skin friction ω_{zw} downstream of the bubble are caused by the large-amplitude (saturated) disturbances that mimic a turbulent boundary layer in the present computations.

Therefore, the skin friction can be used to sense the occurrence of a LSB. Different shear stress sensors for this purpose are currently under development, for example in [5], but not yet available.

Based on the present results a controlling mechanism (cf. figure 3) consisting of an oscillating piezoceramics actuator, which has already been built and tested [4], will be derived. For the present design the oscillation frequency signal for the actuator is provided by a signal generator and only the amplitude A_v will be controlled. An estimation of suitable dimensional disturbance frequencies $\tilde{f}$ for maximum amplification is quoted in table 1 with respect to different free-stream velocities $\tilde{U}_\infty$.

The already tested actuator showed the capability of a maximum frequency of more than 1 kHz and a maximum amplitude of the surface deformation of several μm. The controller will read the signal from the sensor and determine the size of the LSB by averaging the signal over a certain period of time, e.g. ten disturbance periods. The controller is necessary to avoid excessive disturbance amplitudes, to react on nonlinearities in the control loop, and to enable the mechanism to respond to changing flow conditions where a LSB might not be present.

Conclusion

A boundary layer flow with laminar separation bubble has been ivestigated by means of linear stability theory and direct numerical simulation. Hereon a method has been suggested that utilizes the instability mechanisms of the flow to influence the size of the separation bubble. An amplification of properly chosen disturbances can be observed up to an amplitude where an evident change in the size of the bubble results. Further simulations are expected to lead to a sensor-actuator system based on wall shear stress sensors and a piezoceramics driven actuator. This system will be capable of generating the desired disturbances if activated when a separation bubble is about to occur. Otherwise it will not generate any additional drag.

Acknowledgments

The financial support of the Deutsche Forschungsgemeinschaft (DFG) via the Sonderforschungsbereich 409 "Adaptive Strukturen im Flugzeug- und Leichtbau" at the Universität Stuttgart is gratefully acknowledged.

References

[1] Markus Kloker. Direkte numerische Simulation des laminar-turbulenten Strömungsumschlags in einer stark verzögerten Grenzschicht. Dissertation, Universität Stuttgart, 1993.

[2] Markus Kloker, Uwe Konzelmann and Hermann F. Fasel. Outflow boundary conditions for spatial Navier-Stokes simulations of transition boundary layers. *AIAA Journal*, 31(4):620–628, April 1993.

[3] Ulrich Maucher, Ulrich Rist and Siegfried Wagner. Refined interaction method for direct numerical simulation of transition in separation bubbles. *AIAA Journal*, 38(8):1385–1393, 2000.

[4] Jörg Müller. Konzeptstudie für ein aktives Oberflächenfeld zur Verhinderung laminarer Strömungsablösung. Studienarbeit, Bereich Flugzeugentwurf, Fakultät Luft- und Raumfahrttechnik, Universität Stuttgart, September 1999.

[5] Tilmann von Papen and Ha Duong Ngo. Entwicklung eines mikromechanischen Oberflächenzauns für den Einsatz in Rückströmgebieten. In Martin Schober and Jan Schulz, editors, *Doktorandenkolloquium des SFB 557 "Beeinflussung komplexer turbulenter Strömungen"*. Hermann Föttinger Institut, TU Berlin, Oktober 1999.

[6] Ulrich Rist. Zur Instabilität und Transition in laminaren Ablöseblasen. Habilitationsschrift, Universität Stuttgart, 1998. Shaker Verlag, 1999.

[7] Ulrich Rist and Hermann F. Fasel. Direct numerical simulation of controlled transition in a flat-plate boundary layer. *J. Fluid Mech.*, 298:211–248, 1995.

[8] Hermann Schlichting. *Boundary Layer Theory*, chapter Origin of Turbulence I, pages 459–483. McGraw-Hill, 1979.

Tables

Table 1 Estimation of required dimensional frequencies $\tilde{f}$ with respect to the free stream velocity $\tilde{U}_\infty$.

$\tilde{U}_\infty$	$\frac{m}{s}$	20	30	40	50
$\tilde{f}$	Hz	425	955	1700	2650

Figures

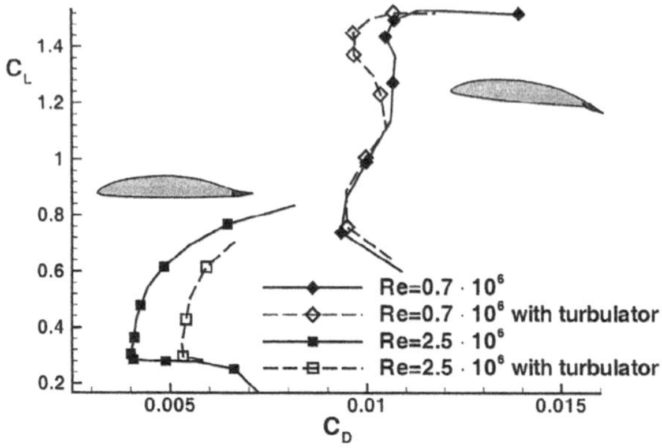

Figure 1 Lift vs. drag coefficients at two different Reynolds numbers with and without vortex generators on the lower side of a laminar airfoil.

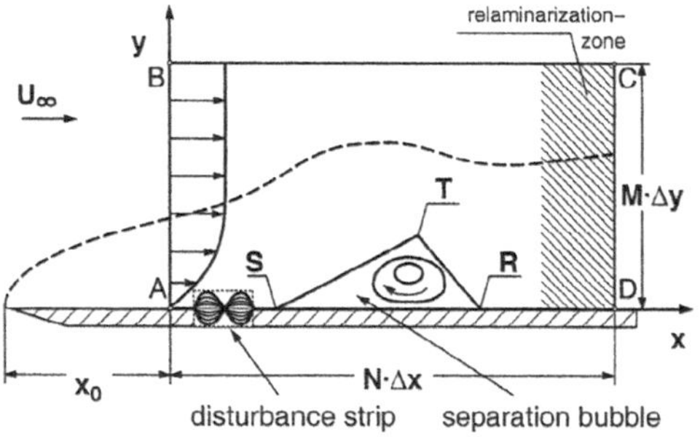

Figure 2 Schematic overview of the integration domain of the DNS with LSB.

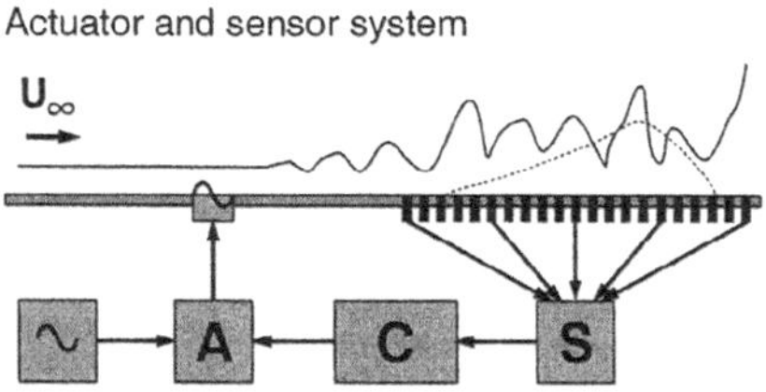

Figure 3 Sensor (S)-actuator (A) concept with controller (C) and signal generator (~).

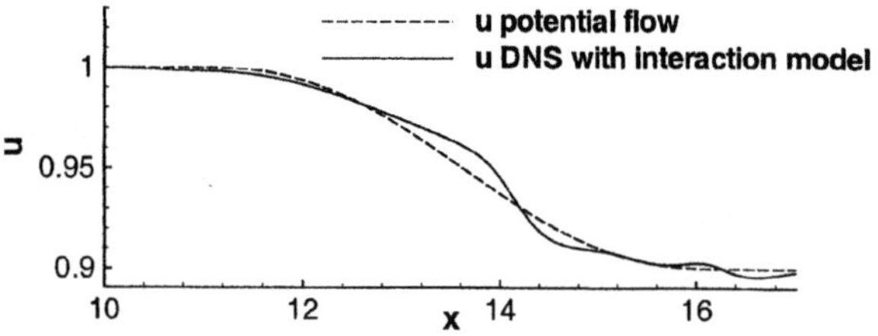

Figure 4 u-velocity distribution at the free-stream boundary.

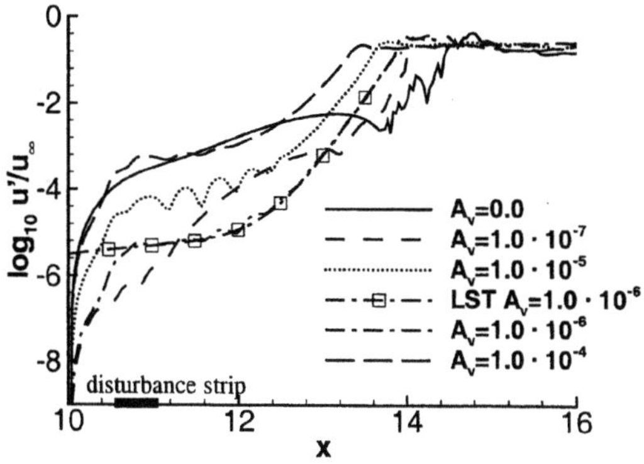

Figure 5 Amplification curves as a result of DNS and LST at different levels A_v of excited boundary layer perturbations with frequency $\beta = 5$.

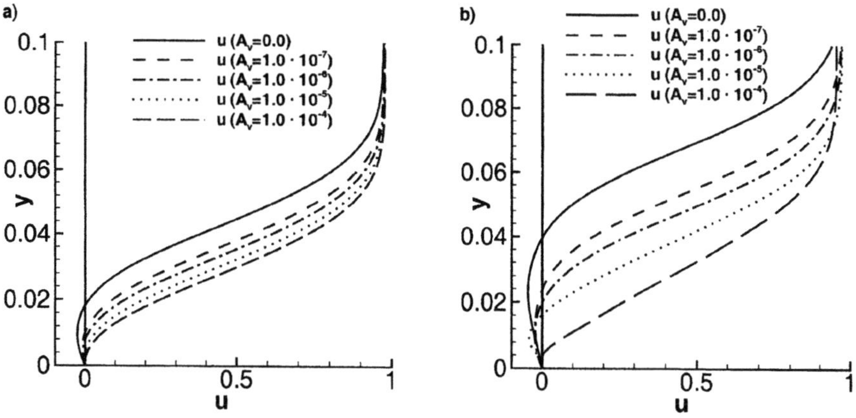

Figure 6 Wall normal distribution of the streamwise time-averaged velocity u at a) separation of the $A_v = 10^{-4}$ case and b) reattachment of the $A_v = 10^{-4}$ case.

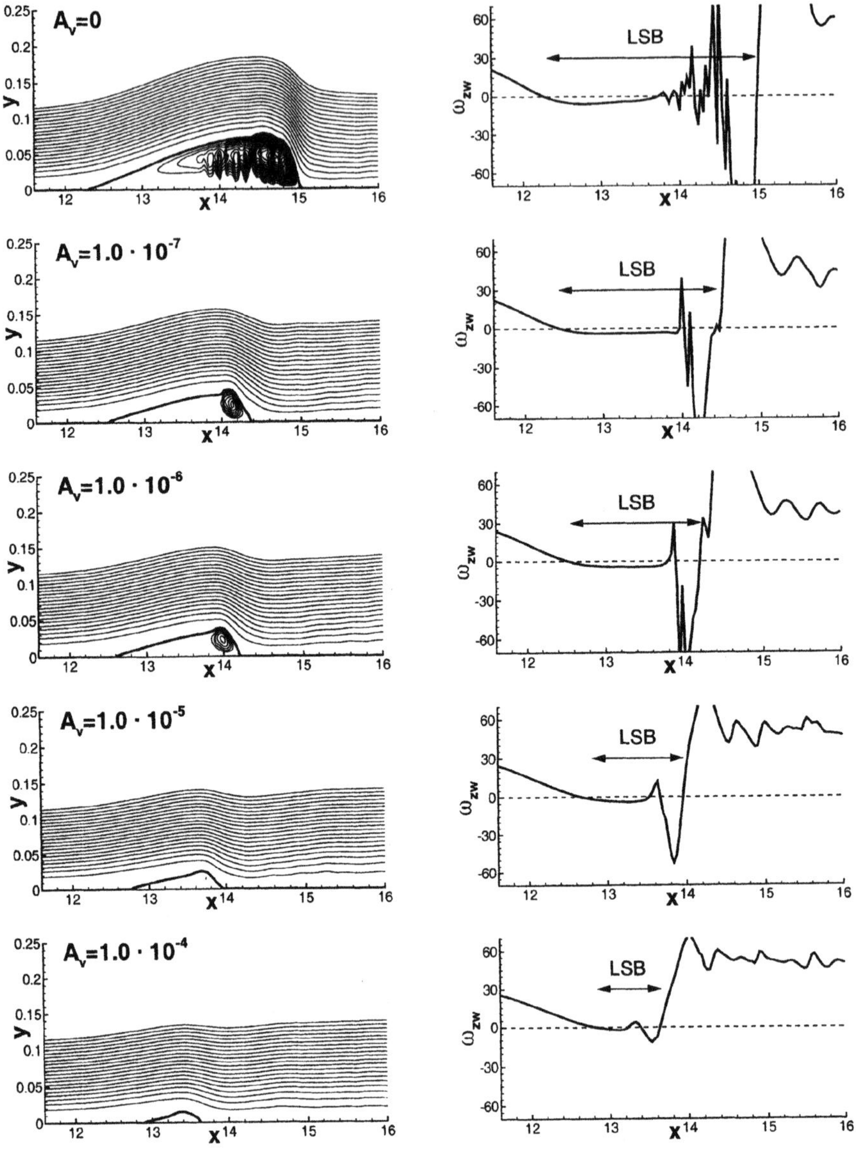

Figure 7 Streamlines and separation line of the time-averaged flow with separation bubble and corresponding skin friction distribution ω_{zw}.

On the Structure of Turbulence in the Transition Cylinder Wake

M. BREDE and A. LEDER
Universität Rostock, Institut für Maritime Systeme und Strömungstechnik,
Albert-Einstein-Straße 2, 18051 Rostock, Germany

Summary

In the present paper the near wake of a circular cylinder is investigated experimentally at a cylinder diameter based Reynolds number of 20.000. Using 3-D Laser-Doppler anemometry velocity data have been acquired on a grid perpendicular to the cylinder to determine the distribution of the Reynolds stresses in the near wake from 0.5 to 3 cylinder diameters downstream. A trigger mechanism detecting the phase of the primary vortex shedding allowed to separate the measured data in 16 phase ensemble averaged data sets. From the series of phase frames the development of the Reynolds stress distribution in the shedding process was monitored.

1 Introduction

In the wake of a circular cylinder periodic vortex shedding occurs for Reynolds numbers larger than 45. For low Reynolds numbers up to about 160 these primary vortices are laminar, the sequence formed by these laminar vortices is the well known Kármán vortex street. Above Re = 160 secondary instabilities lead to a complex 3-D vortex structure including streamwise secondary vortices (Brede, Eckelmann, and Rockwell 1996). Additionally the separated shear layer in the near wake shows the amplification of Kelvin-Helmholtz waves (Bloor 1964) which results in the formation of vortices in the shear layer for Reynolds numbers of approx. 1000 and larger (Unal and Rockwell 1988, Prasad and Williamson 1997). This vortex system in the shear layer is being absorbed by the vortex formation of the primary vortices in a yet unknown process. The resulting complex vortex structure then shows a significant amount of turbulence (see measurements of Cantwell and Coles 1983 with hot wires and Leder 1991 with 2-D LDA for circular cylinder, Lyn, Einav, Rodi, and Park 1995 with 2-D LDA for square cylinder), so that the instability of the shear layer and the process of interference between the shear-layer vortices and the primary vortices can be regarded as a keystone to the transition to turbulence.

Turbulence production in the wake of bluff bodies has major significance for the industrial application of fluid dynamics and the fundamental understanding of the turbulence process itself. Unfortunately the complex 3-D vortex structures which appear in the transitional wake of bluff bodies as a result of the interaction of several flow instabilities, are difficult to investigate experimentally and numerically. Therefore, the wake of a circular cylinder has been chosen as a model case which allows to analyse the typical transition scenario in detail. In the following sections the results from 3-D LDA measurements in the cylinder wake at a Reynolds number of 20.000 are presented. In this investigation it has been possible for the first time to determine all Reynolds stresses in the cylinder wake from phase ensemble averaged 3-D LDA data.

2 Experimental Setup

All measurements were performed in the open test section of a closed return wind tunnel with a cross section area of 0.65 x 0.65 m and a turbulence level of 0.5%. A cylinder of 0.02 m diameter and 0.58 m length was chosen to obtain an aspect ratio of 29 in accordance with recommendations given by Norberg (1994). Endplates were used as described by Eisenlohr and Eckelmann (1990) to cut off end-effects and to avoid any interference between the cylinder wake and the test section boundaries. The cylinder wake was observed at several flow speeds, for the presentation here a flow speed of 15 m/s was chosen, resulting in a Reynolds number of 20.000 and a Strouhal number of 0.191, both based on the cylinder diameter.

Investigating the development of the turbulence in the unsteady wake requires the simultaneous measurement of all three velocity components to determine all Reynolds normal and shear stresses. Additionally, the phase of the primary shedding cycle has to be recorded to allow phase averaging of the Reynolds stresses coupled to the primary shedding process.

For the present experiment these tasks are being performed by a 3-D Laser-Doppler-Anemometer and a triggering mechanism driven by a hot wire probe in the wake of the circular cylinder. The 3-D LDA consists of a 500 mW Argon-Ion laser providing two colors for a 2-D LDA-probe and a 100 mW frequency doubled NdYAG laser providing a third color for a 1-D probe measuring the third velocity component. The complete experimental setup is displayed in figure 1. In this configuration the Laser-Doppler anemometer allows the acquisition of coincident 3-D velocity data with an average data rate of about 800 Hz. Simultaneously, the trigger mechanism detects the beginning of the primary shedding cycle from the signal of the hot wire probe and produces electronically a TTL pulse which is recorded by the LDA data processor in addition to the velocity data. This allows to calculate the phase angle for the following sequence of velocity data samples (bursts) up to the next TTL pulse. This method is described in detail in Leder and Geropp (1988).

Data are being acquired on a measurement grid of 46x21 nodes on the plane orthogonal to the cylinder. Given a spacing of 0.1 D the complete measurement grid covers an area of 4.6 D x 2.1 D. To focus more towards the near vicinity of the cylinder, in the following only the area of 2.5 D x 2.1 D next to the cylinder is being displayed. On each node velocity and phase data have been sampled for 100 seconds, resulting in a total data set of about 8 GBytes size. These data sets are then processed using custom made software to compute all turbulent flow properties including the Reynolds stresses and triplecorrelations.

3 Phase Triggered 3-D LDA Measurements

In figures 2 to 5 phase averaged data from a 3-D LDA measurement series is presented which is covering one cycle of separation of a pair of primary vortices. In each figure eight panels are being displayed representing eight phase steps of the full motion cycle with a timestep of Δt=0.83 ms. In figure 2 the velocity vector (u,v) is depicted, figure 3 shows the Reynolds stress $\langle u'^2 \rangle$, figure 4 $\langle v'^2 \rangle$, and finally figure 5 $\langle u'v' \rangle$.Due to the limited space of this presentation the display of $\langle w'^2 \rangle$ was omitted as well as the Reynolds shear stresses including the w-component which show almost no intensity above the noise level

First, the general distribution of the Reynolds stresses in the wake can be inferred from the figures. The Reynolds normal stress $\langle u'^2 \rangle$ (figure 3) shows a peak intensity in the regions of strong velocity gradient $\partial u/\partial y$, particularly in the separated shear layer close to the cylinder. The distribution of $\langle v'^2 \rangle$ (figure 4) is characterized by one extended peak area which matches the formation region of the primary vortex. The Reynolds shear stress $\langle u'v' \rangle$ in figure 5 shows the extremal peaks in the regions of strong velocity gradient $\partial u/\partial y$ but also at areas of a strong change of flow direction (see also figure 2). The peak area here stretches more from the cylinder along the separated shear layer to the vortex formation region of the primary vortex (frame 9, figure 5). Close to the cylinder two minor peaks are visible in several phases which coincide with vortices in the region of reverse flow.

The sequence of phase frames of the Reynolds stresses now allows to analyze the turbulence distribution during the separation process. The Reynolds stress $\langle v'^2 \rangle$ is clearly periodical in terms of the intensity modulation coinciding with the growth and separation of the primary vortex from the cylinder. A maximum is reached with the moment of separation of the primary vortex (frames 5 and 13, figure 4). After the separation the peak of high $\langle v'^2 \rangle$ intensity is moving downstream with the vortex. The intensity in that area hereby decreases with the movement downstream.

The Reynolds stresses $\langle u'^2 \rangle$ and $\langle u'v' \rangle$ seem to be bound to the separated flow region or vortex formation region up to an x/d of 2.5. The peaks of $\langle u'^2 \rangle$ shows a considerable oscillation in the cross stream direction. It appears as if phases with high intensities of $\langle u'^2 \rangle$ are synchronized with the movement of the peak area towards the center of the wake. For the Reynolds stress $\langle u'v' \rangle$ the same process is more clearly visible: In the sequence from frame 5 to frame 13 in figure 5 the region of negative $\langle u'v' \rangle$ starts to grow in *upstream* direction until it is amplified there (frame 9, figure 5, see isoline $\langle u'v' \rangle = -0.05$). Afterwards it is moved back downstream towards the vortex formation region (frame 11, figure 5) where it unifies with the main peak (frame 13, figure 5) which reaches the maximum negative value after the process in frame 1 (figure 5).

4 Conclusions

The experimental results indicate, that there is a complex interaction process between the separated shear layer and the primary vortex which is being formed in the near wake. First, the distribution of $\langle u'^2 \rangle$ shows that the maximum appears in the shear layer close to the cylinder. This peak area is also modulated periodically during the shedding cycle.

Second, from the phase frames of the distribution of $\langle u'v' \rangle$ the mechanism of interaction can be obtained. It appears as if there is an upstream growth of the intensity of the Reynolds stress $\langle u'v' \rangle$ starting from the vortex formation region and extending into the separated shear layer. In the shear layer the instability is amplified and the intensity of $\langle u'v' \rangle$ is increasing. Finally, together with the general movement of the wake, this turbulent area is moved back into the vortex formation region and merges with the primary vortex.

Third, in the phase frames of $\langle v'^2 \rangle$ the evolution of the primary vortex shedding can be monitored. After reaching a maximum value in $\langle v'^2 \rangle$ the primary vortex separates from the vortex formation region. The separated vortex shows also a local maximum of $\langle v'^2 \rangle$ but the intensity is decreasing with the distance to the cylinder. In opposition to the other Reynolds stresses intensity of <v'²> is convectively transported out of the formation region with the primary vortices. Further experiments will allow to study the three main processes described in the previous paragraph in greater detail. Also the influence of the Reynolds number will be investigated.

Acknowledgements

The authors would like to thank the DFG for supporting the project "unstable separated shear layers" with financial funding.

References

[1] S. Bloor 1964: The transition of turbulence in the wake of a circular cylinder, J. Fluid Mech. 19, 290-304

[2] M. Brede, H. Eckelmann and D. Rockwell 1996: On secondary vortices in the cylinder wake, Phys. Fluids 8, 2117-2124

[3] B.J. Cantwell and D. Coles 1983: An experimental study of entrainment and transport in the turbulent near wake of a circular cylinder, J. Fluid Mech. 135, 321-374

[4] H.Eisenlohr and H. Eckelmann 1989: Vortex splitting and its consequences in the vortex street wake of cylinders at low Reynolds numbers, Phys. Fluids A 1, 189-192

[5] A. Leder 1991: Dynamics of fluid mixing in separated flows, Phys. Fluids A 3, 1741-1748

[6] A. Leder and D. Geropp 1988: Phase-averaged LDA measurements in turbulent separated flows; in: Durăo, D.F.G. (Editor): Proceedings of the fourth International Symposium on Applications of Laser-Anemometry to Fluid Mechanics, Lissabon, Portugal

[7] D.A. Lyn, S. Einav, W. Rodi and J.-H. Park 1995: A laser-Doppler velocimetry study of ensemble-averaged characteristics of the turbulent near wake of a square cylinder, J. Fluid Mech. 304, 285-319

[8] C. Norberg, 1994: An experimental inverstigation of the flow around a circular cylinder: influence of aspect ratio, J.Fluid Mech. 258, 287-316

[9] A. Prasad and C.H.K. Williamson 1997: The instability of the shear layer separating from a bluff body, J.Fluid Mech. 333, 375-402

[10] M.F. Unal and D. Rockwell 1988: On vortex formation from a cylinder. Part 1. The initial instability, J. Fluid Mech. 190, 491-512

[11] H.-Q.Zhang, U. Fey, B.R. Noack, M. König und H. Eckelmann 1995: On the transition of the cylinder wake, Phys. Fluids 7, 779-795

Figure 1 Experimental setup of the 3-D LDA-system and the cylinder configuration

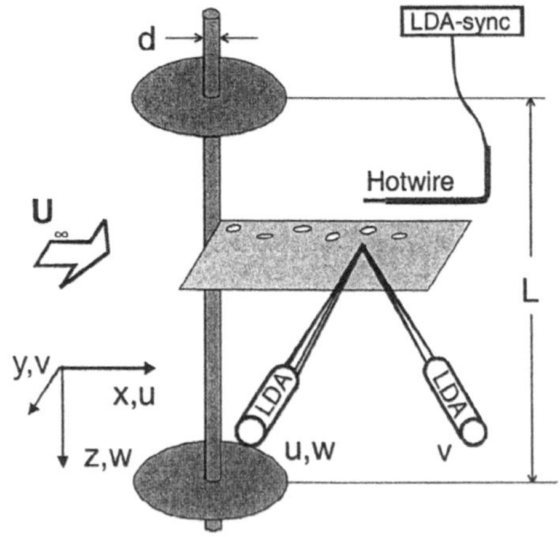

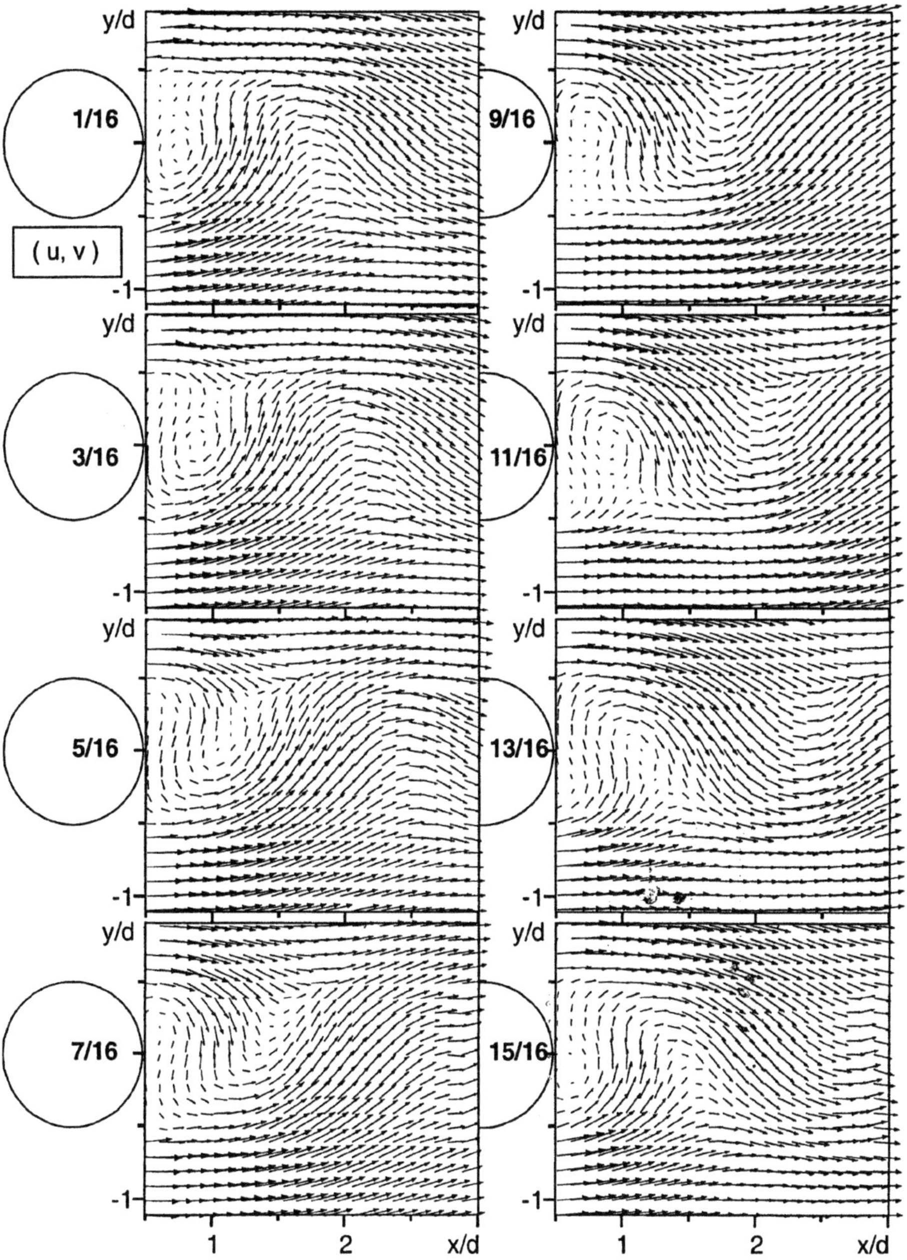

Figure 2 Velocity vector (u,v) in the cylinder wake, 8 phase frames covering one primary vortex shedding period, Re = 20.000

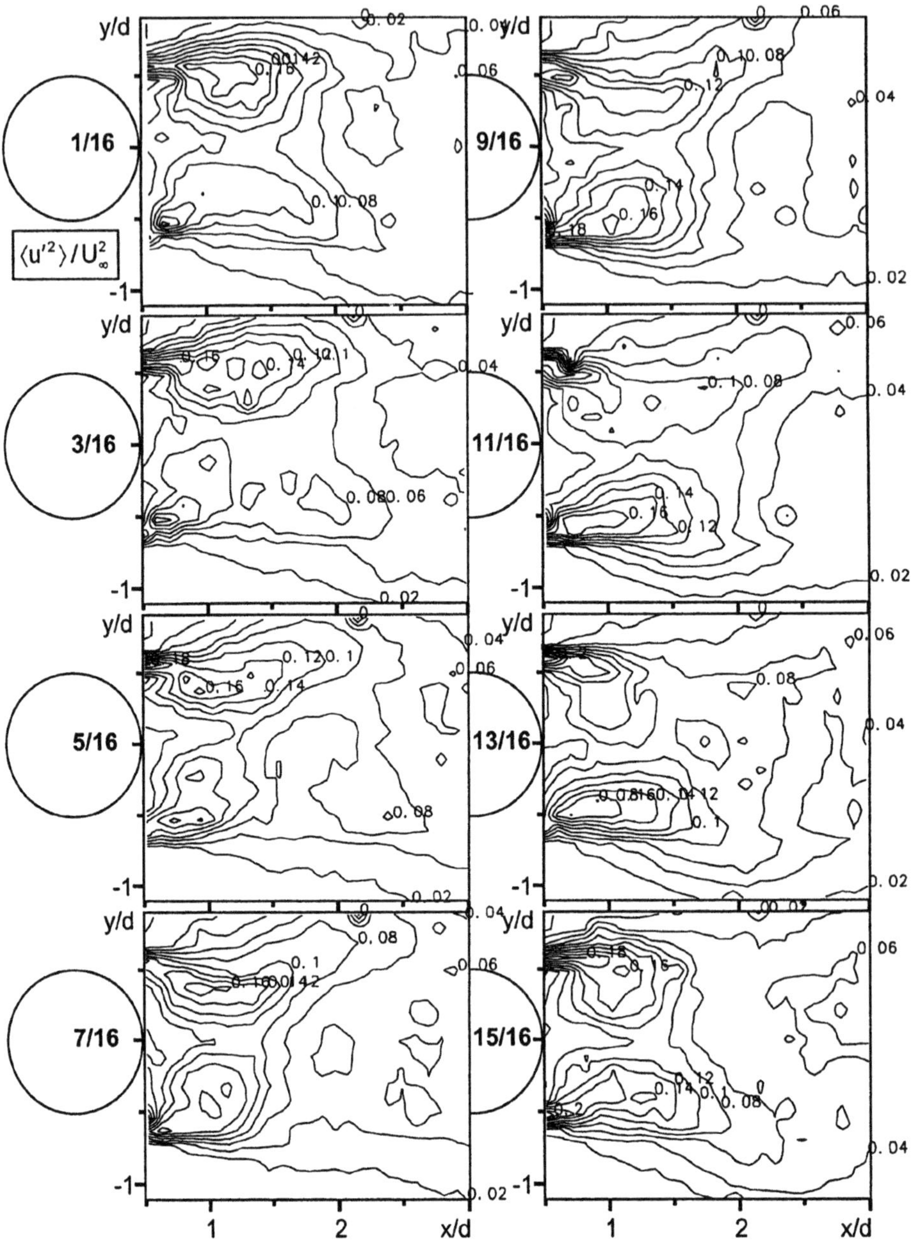

Figure 3 Distribution of normalized Reynolds stress $<u'^2>/U_\infty^2$ in the cylinder wake, 8 phase frames covering one primary vortex shedding period, Re = 20.000

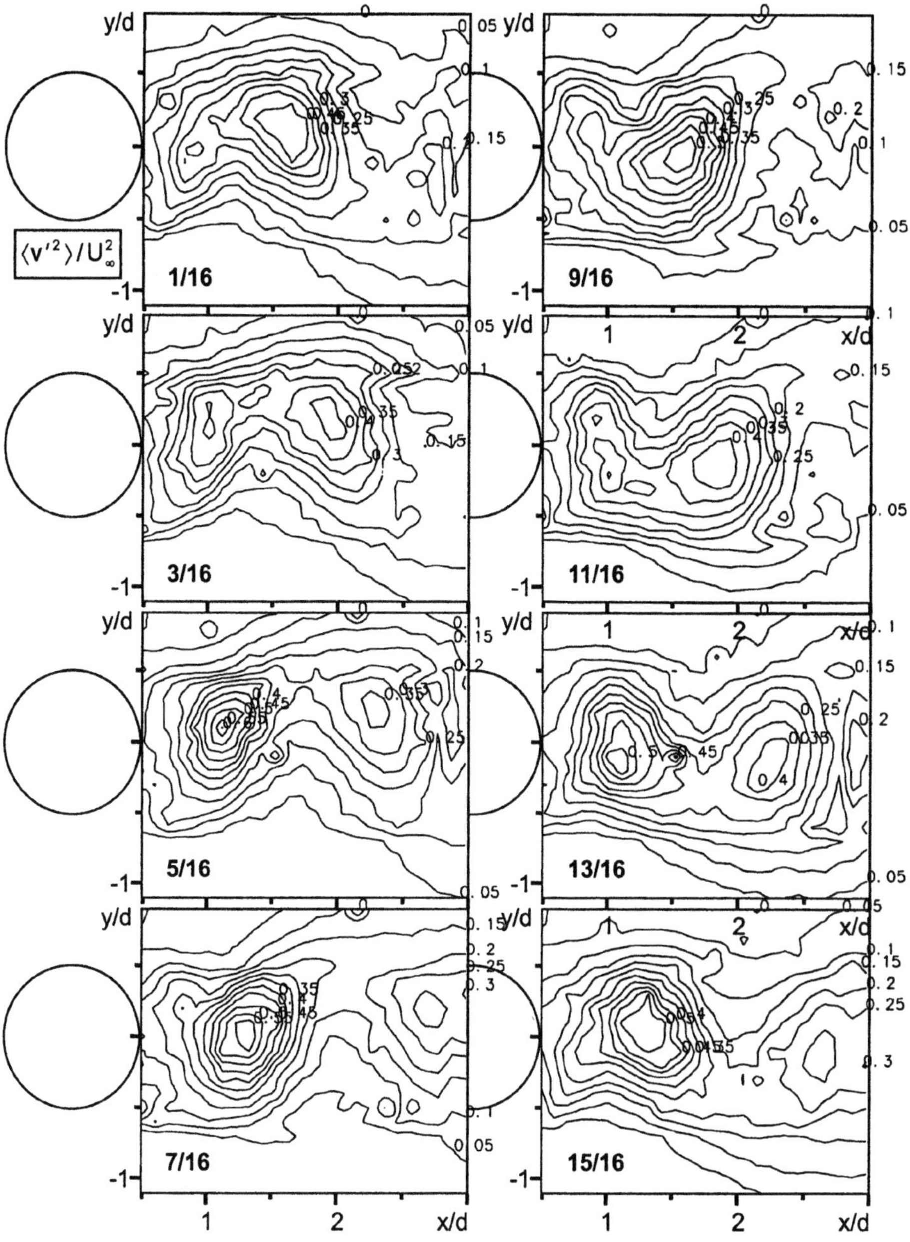

Figure 4 As figure 3, here with the normalized Reynolds stress $\langle v'^2\rangle/U_\infty^2$

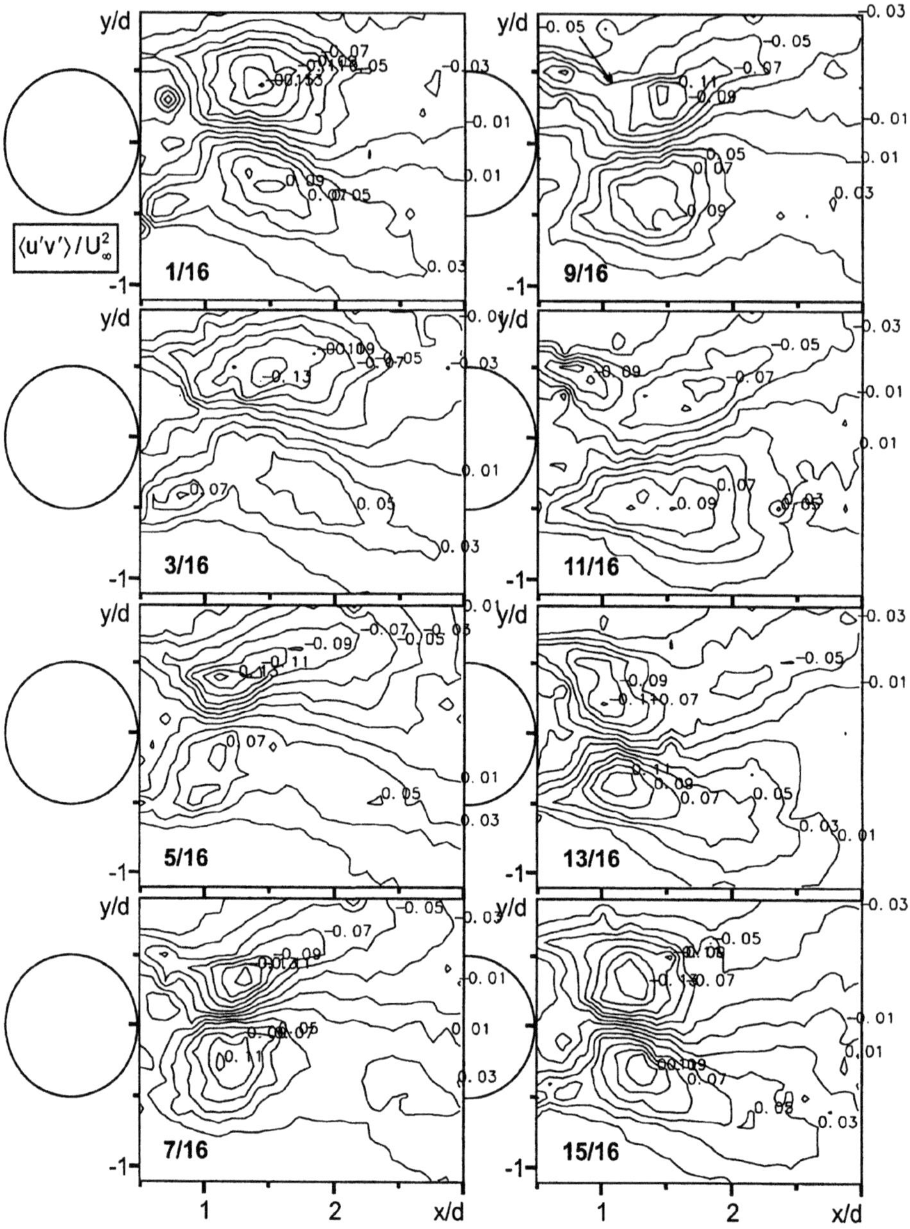

Figure 5 As figure 3, here with the normalized Reynolds stress <u'v'>/U_∞^2

Transition and Fluidmechanical Modelling

Investigation of Subgrid-Scale Models in LES of Turbulent Flows with Separation

S. Schmidt, F. Thiele
Hermann–Föttinger–Institut für Strömungsmechanik, Technische Universität Berlin
Müller–Breslau–Straße 8, D-10623 Berlin, Germany

Summary

The method of Large-Eddy Simulation (LES) has provided a closer insight into the structures of turbulent flows at Reynolds numbers up to $Re = 10^5$. When applying LES to more complex flows of industrial relevance at higher Reynolds numbers the subgrid-scale modeling is of much larger importance than for lower Reynolds number configurations. This study aims to perform a critical analysis of the performance of subgrid-scale models at Reynolds numbers of up to $Re = 10^6$ for the complex flow around a NACA 4412 airfoil at 12 degrees angle of attack. It turns out that the most commonly used subgrid-scale models (i.e. Smagorinsky Model and Dynamic Models) display deficiencies in capturing the dominating near-wall motion which highly affects flows with separation. While these drawbacks only have a minor effect in flows at rather low Reynolds numbers, the observed model shortcomings lead to unsatisfactory results in high Reynolds number flows.

1 Introduction

1.1 Large-Eddy Simulation

In LES, only the large scales are explicitly resolved by the numerical grid while the smaller ones are represented by a subgrid-scale model. The motivation for this approach is that the large-scale vortices are dominated by geometrical constraints and boundary conditions. Due to turbulent transport phenomena, these vortices pass their kinetic energy on to smaller vortices. In this process the orientation of the initial vortices gets lost during this energy cascade. Therefore, the small-scale turbulence is expected to be isotropic without any preferred orientation and should consequently be much easier to prescribe than the whole spectrum of turbulence.

1.2 Subgrid-Scale Models

The most commonly used model is the Smagorinsky Model (SM) [1] which is based on a simple mixing-length approach. It includes one parameter (C_S) that has to be fixed according to the flow problem and is usually set to $C_S = 0.1$ for most turbulent flows. Under strong shear and near rigid walls, however, this parameter has to be reduced by a van Driest function [2] to damp out the eddy viscosity. This model is numerically stable since it allows only for positive eddy viscosity. Therefore, in many flows it is too dissipative because it cannot account for backscatter of kinetic energy from smaller to larger scales.
In contrast, the Dynamic Smagorinsky Model (DSM) [3] extracts this parameter from the filtered velocity field and evaluates a local, time-dependent value $C(x,y,z,t)$. This model is capable of predicting backscatter and adapts itself to the local turbulence structure. Despite the fact that it avoids the parameter problem, the procedure suffers from numerical difficulties possibly destabilising the calculation and leading to divergence. However, in the presence of homogeneous directions, these shortcomings can be suppressed by averaging the parameter in these directions.

The more elaborate Dynamic One-Equation Model (DOEM) [4] employs the turbulent subgrid kinetic energy to avoid these numerical difficulties while retaining the key features of the dynamic approach. This model remains unconditionally stable also in complex geometries without any homogeneous directions and is in this sense superior to the standard model. The favourable stability properties as compared to the dynamic model usually outweigh the higher computational costs of solving the additional transport equation.

Dynamic models are known for being flexible and very sensitive to the local turbulent structure and in many testcases outperform the Smagorinsky Model. This is at least true for low Reynolds number flows up to $Re = 10^5$. In this regard, the question arises whether this fact also holds for high Reynolds number flows beyond $Re = 10^6$. Often, the overall quality of the flow prediction is very sensitive to the correct resolution of the turbulent near-wall structures, which are becoming smaller with increasing Reynolds number, but have a large influence of the flow dynamics. Although highly stretched grids or zonal grid approaches are employed to resolve the most important flow structures, there still remain underresolved regions in the computational domain. Due to this insufficient spatial resolution, more unresolved motion has to be captured by the subgrid-scale model, which becomes an even more important task in high Reynolds number flows. As all subgrid-scale models described above are based on a rather simple mixing-length approach, this study aims at working out if this type of model can be successfully applied to high Reynolds numbers flows or, beside some minor model differences, might give an indication of a general model failure of current subgrid-scale models.

2 Numerical Scheme

The flow solver ELAN3D [5] is based on an fully-implicit pressure-based Finite-Volume Navier-Stokes procedure applying a cell-centred discretization on semi-structured grids. It uses an 'all-speed' pressure-correction scheme and a generalised Rhie and Chow interpolation. The code is second-order accurate in space and time and uses multi-block algorithms. Several gradient-based HOC schemes are available. Furthermore, the solver has been parallelised using domain decomposition, where the performance has been quantified to roughly 60 MFLOPs per PE on a CRAY T3E-900.

3 Low Reynolds Number Flows

3.1 Pipe Flow

In turbulent flows of Reynolds numbers beneath $Re = 10^5$, especially with homogeneous mean flow directions (e.g. channels, pipes), only smaller model differences are to be expected. The reason for this is that the model parameter is only a function of the wall distance and therefore can be represented by either a constant parameter with wall damping or be obtained by the dynamic procedure with planewise averaging. A typical example is the turbulent pipe flow of $Re_\tau = 360$, which can be used to demonstrate the similarity of all models. The size of the domain is set to $L \times R \times \Theta = 10R \times R \times 2\pi R$ and was discretized by a semi-structured hybrid H/O-type grid with a different streamwise resolution in the core and in the wall region ($y^+ = 0 - 63.64$). The spatial resolution in the wall region is chosen to $\Delta z \times \Delta r \times \Delta\Theta = 0.1R \times 0.025R \times 0.02R$ with wall-normal distance of $\Delta r^+ = 1.3$ over the entire domain. The time-averaged streamwise velocity $\overline{w}$ (Figure 1) is in good agreement with numerical DNS data by Eggels et al. [6] and experimental LDA data by den Toonder [7]. Small model differences can be seen in the buffer layer, whereas at the centreline all velocity profiles coincide. The time-averaged streamwise Reynolds stress $\overline{w'w'}$

(Figure 2) reveals that all models behave similarly under uniform shear in the pipe flow case. Regarding the lack of numerical stability of the DSM, the OEDM will be used in more complex flows without homogeneous directions.

3.2 Flow around Bluff Bodies

Turbulent bluff-body flows have been intensively investigated with LES for more than five years. These configurations provide a good test basis to investigate the LES' capability to capture certain complex flow phenomena as well as strong transient motion, which is known as being a major problem for Reynolds-Averaged Methods (RANS). Besides influences of the numerical scheme and the computational grid, one key issue to resolve the correct flow physics is the subgrid-scale modeling.

Square Cylinder

The flow around a square cylinder at $Re = 22{,}000$ forces the flow into a geometry-induced separation at the leading edge of the cylinder resulting in a periodic von Kármán vortex street in the wake. This configuration was experimentally investigated by Lyn et al. [8]. The domain $X \times Y \times Z = 20D \times 14D \times 4D$ was discretized by a block-structured grid consisting of 32,304 × 32 (xy × z) cells with a local cell spacing of Δx,$\Delta y = 0.03 - 0.125$ and $\Delta z = 0.125$ satisfying a wall-normal distance of $\delta = 0.064$ around the body. While some of the global flow parameters like the Strouhal number range can be quite easily captured (Table 1), the model influences the resulting mean recirculation length. This property is coupled with the vortex dynamics in a way that a larger separation bubble reduces the Strouhal number while a shorter bubble leads to a higher shedding frequency. The mean centreline velocity $\overline{u}$ behind the cylinder is given in Figure 3 for both models. Giving almost the same size of the recirculation zone, the Smagorinsky Model achieves a satisfactory agreement with experimental data while the Dynamic One-Equation Model overpredicts the velocity relaxation in the wake downstream of the location $x = 2.0$. The total resolved turbulent kinetic energy $\overline{k}$ (Figure 4) contains all resolved transient motion including the vortex shedding. It turns out that both models predict nearly the same energy level which compares satisfactorily with Lyn's data [8]. As a result, it can be assumed that the model difference has only a minor effect on the overall flow prediction, inducing only a slight change in the mean flow parameters.

Circular Cylinder

In contrast, the circular cylinder has no fixed separation point and features pressure induced separation, which besides other parameters of influence, is strongly affected by the subgrid-scale model. At $Re = 3{,}900$, the oncoming laminar flow separates just before the apex, undergoes transition in the separated shear layer and forms a turbulent wake. For this simulation a similar block-structured grid was used with spatial resolution in the cross-sectional area (x-y) of $\Delta x = 0.00625 - 0.19$, $\Delta y = 0.00625 - 0.46$ and $\Delta z = 0.098$ in the spanwise direction (z) keeping a wall-normal distance of $\delta = 0.00625$. The subgrid-scale model has to be able to capture all these effects to predict the correct separation point and flow development. In case of the Smagorinsky Model, the constant model parameter is disadvantageous to the flow prediction because it is not able to distinguish between different kinds of flow regions. In laminar flow, the model should have no influence, however, the Smagorinsky Model overestimates the dissipation rate, leading to a stronger removal of turbulent kinetic energy, thus resulting in a longer mean recirculation zone in the wake, which compares quite reasonable to the experimental data from Lourenco & Shih [9] and SM results from Fröhlich et al. [11]. The dynamic model, shows an opposite behaviour as it keeps more turbulent energy in the flow, leading to an enhanced momentum exchange in the wake, which results in an earlier relaxation and a shorter separation bubble, as can be seen from Figure 5 in comparison with experimental data. The total kinetic energy (Figure 6), however, shows a deviation for both models. The Smagorinsky Model underpredicts the kinetic energy level of Fröhlich showing that other numerical parameters, such as grid resolution and the size of the computational domain might be of the same influence to the solution. As explained, the dynamic model overshoots the energy level, which is the result of an underpredicted dissipation in terms of the eddy viscosity. Beside this, both models fail to capture the stress anisotropy correctly, leading to this different behaviour. Since the Strouhal number (Table 2) of the Smagorinsky Model simulation is significantly lower as compared to the dynamic model and the experimental value, it turns out that the overreduction of kinetic energy in case of the SM slows down the whole vortex dynamics, leading to a larger recirculation zone and resulting in a shift of the energy peak downstream.

4 High Reynolds Number Flows

At higher Reynolds numbers, the observed model differences are becoming even larger. As already mentioned, this is due to the fact that the small-scale turbulence contains a significantly higher amount of kinetic energy of the turbulent spectrum, thus requiring a more accurate prescription of the subgrid-scale motion. To clarify the influence of the Reynolds number on the subgrid-scale model, the flow around the circular cylinder at Re = 140,000 is investigated.

4.1 Flow around Bluff Bodies

This configuration allows a direct comparison with the low Reynolds number case (Re = 3,900) described above. Using the same box-shaped computational domain ($X = 19D$, $Y = 14D$) and a spanwise length of $Z = D$ with a spatial resolution of ($\Delta x = 0.0001 - 0.0035$, $\Delta y = 0.0001 - 0.0094$, $\Delta z = 0.015625$), the Smagorinsky Model builds up two different separation bubbles on both sides of the leeside cylinder walls whereas the dynamic model predicts one large recirculation zone on the centreline. Therefore, only the latter obtains a mean backflow component on the centreline. The size of the separation bubble ($l_r = 0.40$) agrees only qualitatively with the experimental value of $l_r = 0.50$ (Figure 7).
In a circular domain ($R = 15D$) with a wider spanwise length ($Z = \pi D$), the results show a totally different behaviour. This domain was chosen to allow for a detailed comparison with other numerical data obtained by Breuer [12]. In this case the dynamic model yields too large a recirculation zone ($l_r = 1.188$) while the Smagorinsky Model ($l_r = 0.624$) remains closer to the experiment, however, the latter also overpredicts the experimental value (Figure 8). It appears that the size of the computational domain is of at least the same importance to flows with pressure-induced separation as the subgrid-scale model. In order to separate and quantify all effects, it is advisable to carry out further studies based on one numerical scheme, see also Breuer [12].

4.2 NACA 4412 Airfoil Flow

The application of LES to the flow around a NACA 4412 airfoil is even more challenging for LES than bluff-body flows because this configuration represents an entry testcase towards real-life flow problems in which RANS usually give quite reasonable results with only a fraction of computational effort. At an angle of attack of $\alpha = 12^o$, this airfoil operates at nearly maximum lift and the flow remains attached almost over the entire airfoil until it separates at 82% chord and forms a closed separation bubble. The experiments carried out by Wadcock [13] and Hastings &

Williams [14] at Reynolds numbers of $Re = 1.64 \times 10^6$ and $Re = 4.2 \times 10^6$ respectively, reveal that the flow reaches a steady state without strong transient motion or massive vortex shedding. The task of capturing the correct flow development from the leading edge, where a wall-mounted strip forces the flow to transition, downstream to the trailing edge highly depends on the adequate spatial resolution of the boundary layer. The airfoil was discretized using a C-shaped semi-structured grid by 4×10^6 cells enforcing a wall-normal distance along the upper airfoil surface between $\Delta y+ = 0.33 - 3.41$. The far-field boundary was located four chord length away and the wing span of Z=5% c was resolved by 48 cells.
The pressure distribution (Figure 9) is showing a reasonable overall agreement with the experimental data over the whole profile despite some distinct differences at the trailing edge. In this region (Figure 10), the Smagorinsky Model nearly reaches the correct pressure plateau, indicating a separation bubble of almost the size determined in the experiments. The dynamic model, however, fails to get the correct pressure distribution, showing only a small tendency to capture the separation at all. This is confirmed by the time-averaged streamlines at the trailing edge as computed with both models shown in Figure 11. The reason for this opposite model behaviour is again the underprediciton of the eddy viscosity in case of the dynamic model, which vanishes at the trailing edge, leaving a surplus of kinetic energy within the boundary layer and keeping the flow attached. In contrast, the Smagorinsky Model drives the flow towards separation because of a too strong boundary-layer growth downstream of the transition strip (Figure 12). It is evident that both models fail to predict the main flow features; either the momentum thickness or the separation bubble size is captured correctly. It turns out that in case of attached flows with only a small transient motion and almost no vortex shedding, LES results highly depend of the quality of the correct near-wall modelling.
Summarizing, it can be noted that at high Reynolds numbers, satisfactory LES results can only be obtained for transient flows with substantial contents of kinetic energy within that motion. As a consequence, attached flows have to be accurately resolved to capture the prevailing near-wall motion which affects the overall flow. This necessitates large numbers of grid points and makes LES rather unfeasible for complex aerodynamic flows with current subgrid-scale models.

5 Conclusions

At low Reynolds numbers, both the Smagorinsky Model and the Dynamic One-Equation Model are able to predict unsteady flows with reasonable success. As DOEM and DSM show nearly the same behaviour in the fully turbulent pipe flow, the DOEM is applied to flows without homogeneity in the mean flow direction. In bluff-body flows which are characterized by large-scale vortex shedding, the models are in case of a fixed separation point (e.g. square cylinder), capable of capturing all relevant flow features as well as vortex dynamics around the body and in the wake. In contrast, the flows around the circular cylinder at different Reynolds numbers show that the overall results are fairly sensitive to the subgrid-scale model and other numerical influences such as the size of the computational domain. While these differences have only a small effect at low Reynolds numbers, they can be significant in high Reynolds number flows.
LES keeps being attractive for transient flows with vortex shedding, but gives rather poor results when applied to aerodynamic configurations in which the flow remains attached or only builds up a small recirculation zone. This becomes obvious from the simulation of a NACA 4412 airfoil at 12° angle of attack. Due to insufficient spatial resolution at the leading edge, the subgrid-scale model has to capture a larger portion of the unresolved motion. Both models fail to predict the correct development of the dominant turbulent structures on the suction side of the airfoil which is vital to obtain the correct size of the separation bubble at the trailing edge.

Currently, the main problem in LES remains the near-wall modelling, which turns out to be a major obstacle to apply this technique to general turbulent flows at high Reynolds numbers. These shortcomings can be alleviated by increasing the spatial resolution in the vicinity of the wall, however, this keeps the main problem - the performance of the subgrid-scale model - untouched. Future studies should therefore concentrate on the improvement of current models or the development of accurate wall functions.

6 Acknowledgement

The authors gratefully acknowledge the DFG (Deutsche Forschungsgemeinschaft) for the financial support (SFB 557). The calculations were carried out at the Konrad-Zuse-Zentrum Berlin (ZIB) on a Cray-T3E.

References

[1] Smagorinsky, J., General circulation experiments with the primitive equations. *Monthly Weather Review*, 91, 99–164 (1963).

[2] Van Driest, E., On turbulent flow near a wall. *Journal Aero. Science*, 23, 1007–1011 (1956).

[3] Germano, M., Piomelli, U., Moin, P. and Cabot, W. H., A dynamic subgrid-scale eddy viscosity model. *Physics of Fluids*, **A3(7)**, 1760–1765 (1991).

[4] Davidson, L., Large Eddy Simulation: A Dynamic One-Equation Subgrid Model for Three-Dimensional Recirculation Flow. In: *11th Symposium on Turbulent shear Flows*, pp. 26–1–26–6, Grenoble, France (1997).

[5] Xue, L., *Entwicklung eines effizienten parallelen Lösungsalgorithmus zur dreidimensionalen Simulation komplexer turbulenter Strömungen*. Dissertation, Technische Universität Berlin (1998).

[6] Eggels, J., Unger, F., Weiss, M., Westerweel, J., Adrian, R., Friedrich, R. and Nieuwstadt, F., Fully developed turbulent pipe flow: a comparision between direct numerical simulation and experiment. *J. Fluid Mech.*, 268, 175–209 (1994).

[7] Toonder, J. d., *Drag reduction by polymer additives in a turbulent pipe flow: laboratory and numerical experiments*. PhD-Thesis, Delft University of Technology (1995).

[8] Lyn, D., Einav, S., Rodi, W. and Park, J.-H., A Laser-Doppler Velocimetry Study of Ensemble Averaged Characteristics of the turbulent Near Wake of a Square Cylinder. *J. Fluid Mech.*, 304, 285–319 (1995).

[9] Lourenco, L. and Shih, C., Characteristics of the plane turbulent near wake of a circular cylinder. A particle image velocimetry study (1993), (Daten aus Beaudan & Moin, 1994).

[10] Beaudan, B. and Moin, P., Numerical experiments on the flow past a circular cylinder at sub-critical reynolds-number. Report TF-62, Center for Turbulence Research (1994).

[11] Fröhlich, J., Rodi, W., Kessler, P., Parpais, S., Bertoglio, J. and Laurence, D., Large Eddy Simulation of Flow around Circular Cylinders on Structured and Unstructured Grids. In: E. H. Hirschel (ed.), *Notes on Numerical Fluid Mechanics, Numerical Flow Simulation I*, vol. 66, pp. 319–338. Vieweg Verlag (1998).

[12] Breuer, M., A Challenging Test Case for Large Eddy Simulation: High Reynolds Number Circular Cylinder. In: S. Banerjee and J. Eaton (eds.), *Turbulence and Shear Flow Phenomena – 1*, pp. 735–740. Begell House Inc., New York (1999).

[13] Wadcock, A., Investigations of low-speed turbulent separated flow around airfoils. Contractor Report 177450, NASA (1987).

[14] Hastings, R. and Williams, B., Studies of the flow field near a NACA 4412 aerofoil at nearly maximum lift. *Aeronautical Journal*, 91, 29–44 (1987).

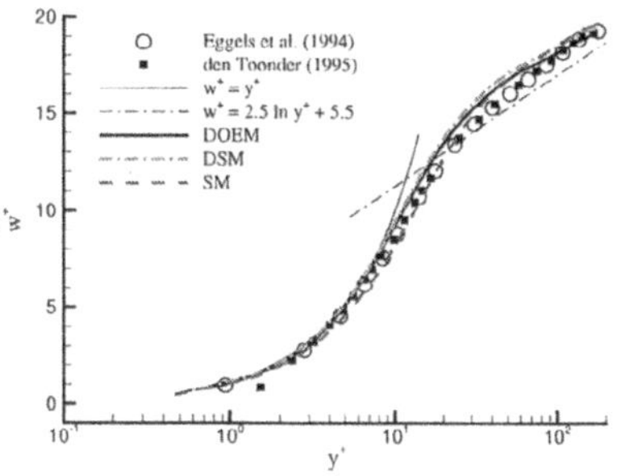

Figure 1 Turbulent pipe flow (Re=5,300): profile of the mean velocity $\overline{w}$

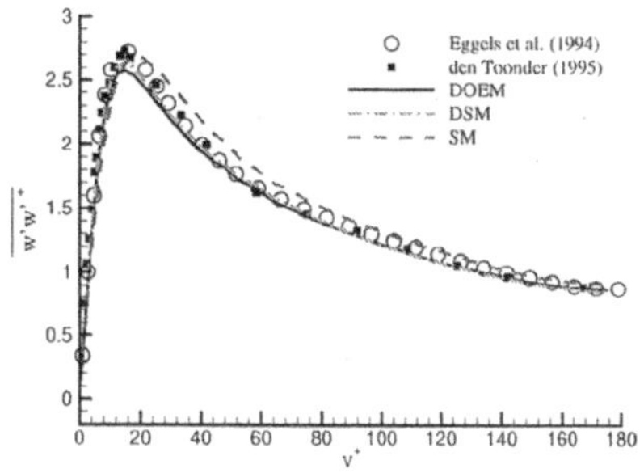

Figure 2 Turbulent pipe flow (Re=5,300): profile of the normal stress $\overline{w'w'}$

Table 1 Square cylinder flow (Re=22,000): Global Flow Parameters

	$\overline{c}_d$	$\overline{c}_{bp}$	$\overline{l}_r$	St
Exp.	2.1	–	–	0.132
DOEM	2.38	-1.69	0.76	0.137
SM	2.41	-1.69	0.81	0.134

Table 2 Circular cylinder flow (Re=3,900): Global Flow Parameters

	$\overline{c}_d$	$\overline{c}_{bp}$	$\overline{l}_r$	St
Exp.	0.99	-0.88	1.33	0.215
Fröhlich [11]	1.08	-1.03	1.09	0.216
DOEM	1.40	-1.46	0.89	0.210
SM	1.16	-1.21	1.11	0.233

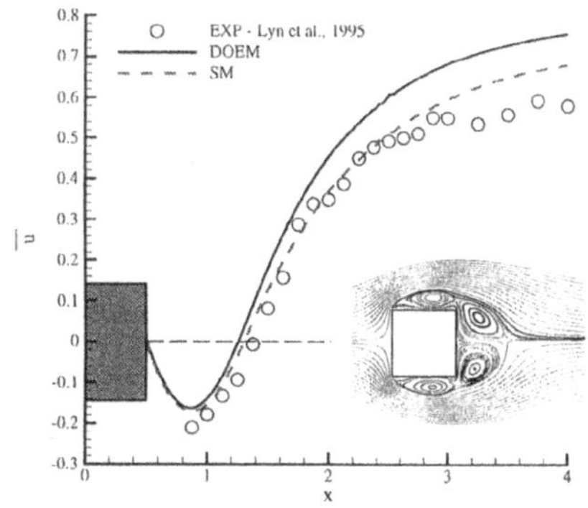

Figure 3 Square cylinder flow (Re=22,000): mean centreline velocity $\overline{u}$

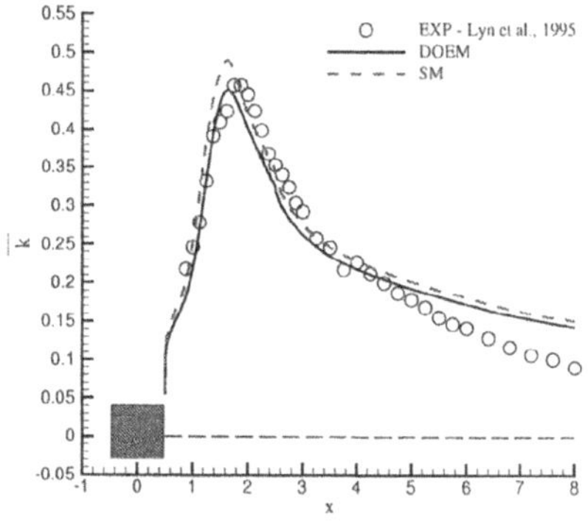

Figure 4 Square cylinder flow (Re=22,000): total resolved kinetic energy $\overline{k}$

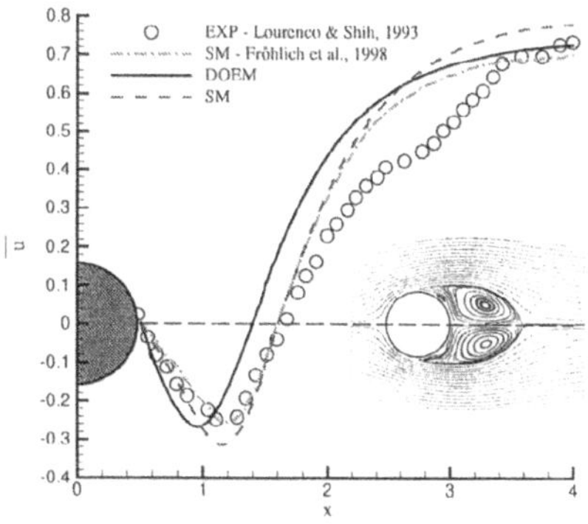

Figure 5 Circular cylinder flow (Re=3,900): mean centreline velocity $\overline{u}$

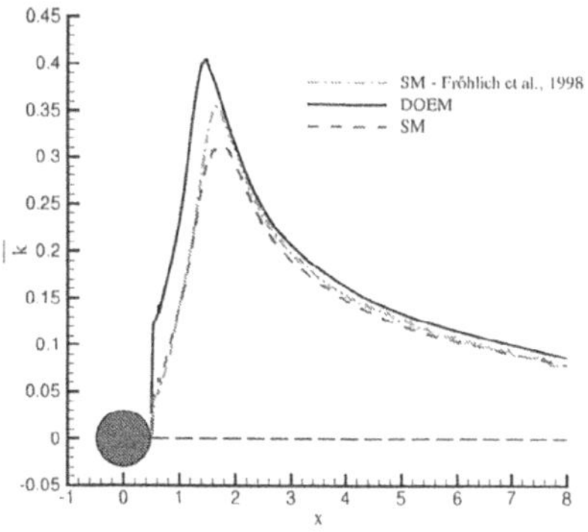

Figure 6 Circular cylinder flow (Re=3,900): total resolved kinetic energy $\overline{k}$

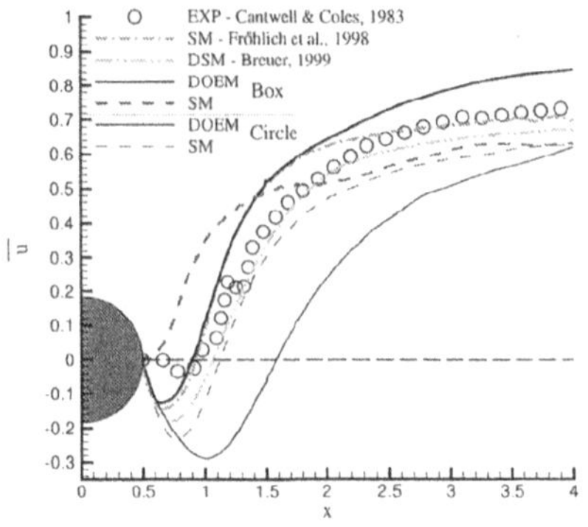

Figure 7 Circular cylinder flow (Re=140,000): mean centreline velocity $\overline{u}$

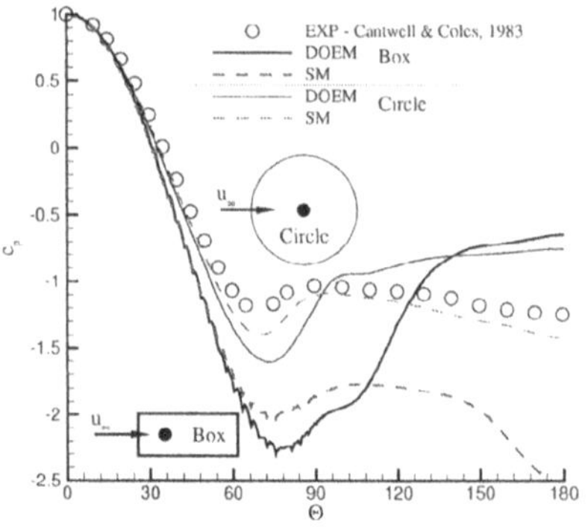

Figure 8 Circular cylinder flow (Re=140,000): pressure distribution $\overline{c}_p$

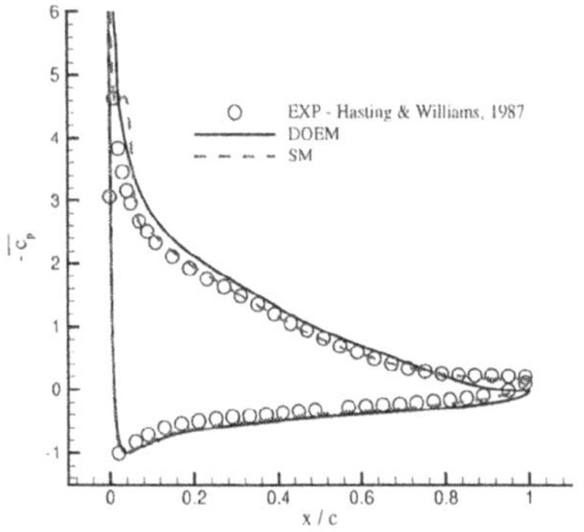

Figure 9 Airfoil Flow (Re=1.64×10^6): pressure distribution over the entire airfoil

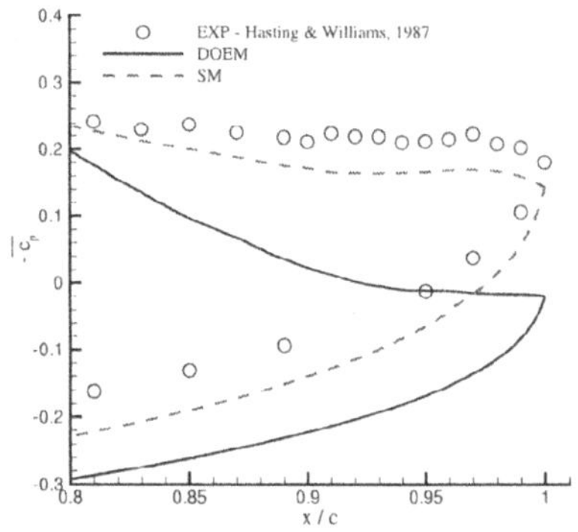

Figure 10 Airfoil Flow (Re=1.64×10^6): pressure distribution at the trailing edge

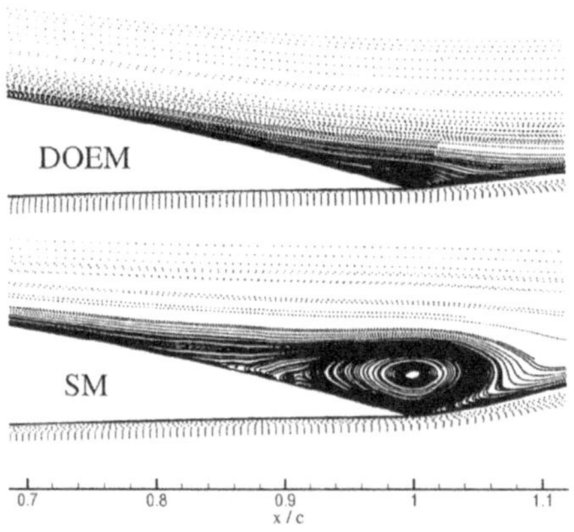

Figure 11 Airfoil Flow (Re=1.64×10^6): streamlines at the trailing edge

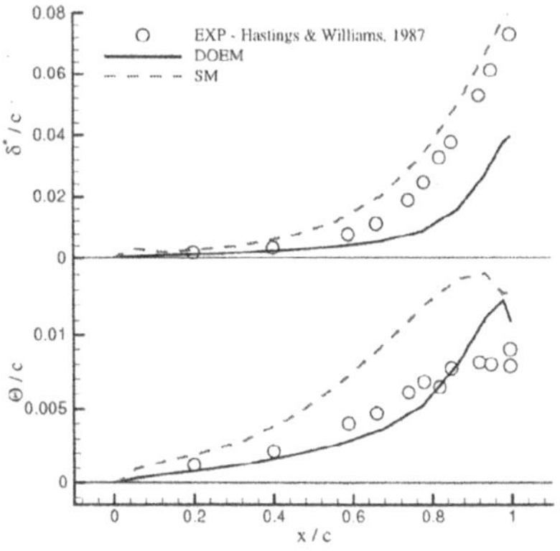

Figure 12 Airfoil Flow (Re=1.64×10^6): boundary-layer parameters on the suction side

Experimental and Numerical Investigations on Transition in a Laminar Separation Bubble

M. Lang, O. Marxen, U. Rist, S. Wagner
Universität Stuttgart, Institut für Aerodynamik und Gasdynamik (IAG)
Pfaffenwaldring 21, 70550 Stuttgart, Germany

Summary

A laminar boundary layer separates in a region of adverse pressure gradient on a flat plate, undergoes transition, and finally the turbulent boundary layer reattaches. Laminar-turbulent transition within this laminar separation bubble (LSB) is investigated by means of measurements with a Laser-Doppler-Anemometer (LDA), flow visualization in water and direct numerical simulation (DNS). The role of unsteady disturbances with and without controlled spanwise variation in the occuring mechanism of transition are examined in detail.

1 Introduction

The laminar water tunnel at the IAG was especially designed to study transition in Blasius boundary layers and in separation bubbles [1]. Because of the large scales in water, this water tunnel is particularly suitable for flow visualization. Wiegand et al. [2] and Kruse & Wagner [3] confirmed the high quality of the tunnel flow. Recent investigations on laminar separation include an experimental study of a LSB excited by an impulsive point-source disturbance [4]. Transition in a LSB employing DNS was examined in [5] with controlled forcing and in [6] without forcing. Despite these efforts to study LSB's, no detailed comparison of numerical results with experimental data obtained in a LSB has been made up to now, in particular for unsteady phenomena. In the present work it will be shown that by mutual comparison of results from calculations and measurements, mechanisms leading to the breakdown into turbulence in existing experimental facilities can be identified in a way that is not possible when applying only one of these methods.

2 Experimental and Numerical Methods

2.1 Experimental Set-Up

A flat plate with an elliptical nose is mounted in the test section of the laminar water tunnel. Optical access is permitted from all sides of the test section. A screen at the end of the flat plate leads to a pressure difference between the upper and the lower side of the flat plate. A gap between plate and sidewalls allows fluid transport beneath the plate. This prevents the formation of an unstable corner boundary layer, therefore extending the range of laminar flow on the plate. At a free stream velocity of 125 $\frac{mm}{s}$ a shape factor of $H_{12} = 2.6 \pm 1\%$ was measured.

To generate a pressure induced laminar separation bubble, a displacement body is positioned in the test section above the plate (Fig. 1). After a short region of favourable pressure gradient the adverse pressure gradient is applied. The boundary layer developing on the displacement body is sucked off at two suction stripes in streamwise direction to prevent separation and transition on the displacement body itself.

A 2-D Tollmien-Schlichting wave is generated by an oscillating wire which is positioned before the displacement body in the region of favourable pressure gradient. The frequency of this disturbance is $f_0 = 1.1\ Hz$. This is also the frequency which is most amplified according to linear stability theory. In the undisturbed case the separation bubble oscillates with the same frequency but not with a constant amplitude as in the case with disturbance input. The oscillating wire generates one trigger signal each period which is taken as time reference for the LDA-measurements. Therefore, the phases, wavelengths and phase velocities can be determined.

To investigate the development of a 3-D disturbance input and to fix appearing vortex structures in spanwise direction, thin ($1\ mm$) metal plates (spacers) are placed regularly underneath the wire. Interaction of the spacers with the oscillating wire results in a 3-D wave combination with typical peak-valley structure, where the boundary layer thickens in the peak plane compared to the valley plane [3, 7]. In the present measurements with 3-D disturbance input the spanwise wavelength was set to $\lambda_z = 58\ mm$, so that regularly appearing vortex structures could be seen in the transition region. For that purpose, visualizations using a hydrogen bubble technique and fluorescent colors in combination with Laser light sheets were carried out (Fig. 2).

The applied 2-D LDA measures the velocity components $\hat{U}(t)$ in mean flow direction and the vertical component perpendicular to the flat plate $\hat{V}(t)$. All velocities are non-dimensionalized by a velocity $\hat{U}_\infty$, taken in the narrowest section ($\hat{x} = 0\ mm$) at a wall-normal distance $\hat{y} = 50\ mm$, where ^ denotes dimensional variables.

In order to achieve near-wall measurements with the probe aligned almost perpendicular to the side walls, the lower beam can be shifted to the optical axis of the probe. However, this halves the beam intersection angle for the vertical velocity component $\hat{V}$ thus causing a higher noise level compared to the streamwise component $\hat{U}$. The lowest turbulence level measurable with the LDA is 0.8% for the $\hat{U}$ velocity [3]. In contrast to hot-film anemometry, a LDA is a non-intrusive measurement technique which provides the possibility to identify velocities with their value and direction. In the transition region of the shear layer with its strong bidirectional velocity fluctuations, this is an important benefit. Thus, the dividing streamline can be determined directly from the measured velocity profiles.

2.2 Numerical Method and Calculation Parameters

Spatial direct numerical simulation (DNS) of incompressible flow is used to compute the pressure induced LSB described in the previous section. In the numerical method, fourth-order accurate finite differences are used for downstream and wall-normal discretization, while a spectral ansatz is applied in spanwise direction. A fourth-order Runge-Kutta scheme is used for time integration. Upstream of the outflow boundary a buffer domain smoothly returns the flow to a steady laminar state. The streamwise pressure gradient is imposed by prescribing a potential velocity distribution for the streamwise velocity u_p at a constant distance y from the wall, while the displacement effect of the boundary layer is captured by a boundary layer interaction model. The numerical method applied in this study is described in detail in [8]. A small-amplitude 2-D Tollmien-Schlichting (TS) wave is forced upstream of the LSB. Additional 3-D disturbances are excited to trigger the break-up of the separated shear layer into regions of small-scale turbulence.

Parameters of the flow are chosen to match the experimental setup introduced before. The length of the computational box in streamwise direction is $\hat{x} \approx 800\ mm$, with the inflow placed at the position of the narrowest section (Fig. 1). The height of the box is $\hat{y} = 50\ mm$ in wall-normal direction, corresponding to approximately 12.5 times the displacement thickness δ_1 of the incoming boundary layer. In spanwise direction a single spacer wavelength fits into the domain ($\hat{z} = \hat{\lambda}_z = 58\ mm$). The resolution is 185 and 1778 grid points in wall-normal and streamwise direction, respectively. One TS wave is discretized with approximately 120 grid points. In spanwise direction 27 Fourier modes are computed. The resolution was confirmed to be sufficient by a calculation with increased number of grid points (approximately by a factor of 1.5 in each direction) yielding the same results in the first part of the LSB, which is discussed in this paper.

The velocity profile at the inflow boundary was obtained from a preceeding 2-D DNS of a strongly accelerated boundary layer (starting from a Blasius solution well upstream of the displacement body in the experiment) with the favorable pressure gradient chosen to match the experimental values up to the narrowest section, since the measured profile could not be represented by an analytical (Falkner-Skan) solution ($H_{12} \approx 1.6$ at the inflow). As can be seen in Fig. 3, the u-velocity profile from measurements and computations at $\hat{x} = 0\ mm$ are in excellent agreement. Hence, this profile is expected to be an appropriate inflow condition for the subsequent DNS of the LSB.

The potential velocity distribution u_p is chosen according to measurements where separation was suppressed by artificially causing transition prior to the laminar separation point (Fig. 3). It serves as initial condition at the free-stream boundary in the DNS and finally the mean edge velocity distribution u_e almost matches the experimental measurements with separation bubble. Since this velocity condition at the upper boundary is not fixed, but rather comes out of the boundary-layer interaction model, it can already be considered a result of the DNS. Good agreement of DNS data with experiments (Fig. 3) therefore gives a first prove of the comparability of the separation bubbles observed in the experiment and in the DNS. Furthermore, it should be emphasized that the actual velocity prescribed at the upper boundary is unsteady and the results presented in the diagram are time-averaged.

2.3 Data Analysis

In order to obtain the spanwise wavenumbers of the 3-D disturbance input, velocity profiles were measured in streamwise and spanwise direction. The resolution in span was 16 points per λ_z and the record length of all measured points was about 30 wire cycles. The time-signals are cut to multiples of the wire cycle time to evaluate the entire spectrum of the measured data.

For an exact determination of the amplitudes and phases of the oscillations in the flow a phase averaging technique with respect to the oscillating wire is used [3, 9]. A Fourier transformation in time followed by a transformation in spanwise direction yields the amplitudes $a_{h,k}$ and phases $\Phi_{h,k}$ of the disturbance [7].

$$
\begin{aligned}
u'(x,y,z,t) &= \sum_{h=1}^{H} a_h(x,y,z) \sin(h2\pi f_0 t + \phi_h(x,y,z)) \\
&= \sum_{h=1}^{H} \sum_{k=-K}^{K} a_{h,k}(x,y) \sin(h2\pi f_0 t - k\frac{2\pi}{\lambda_z} z + \phi_{h,k}(x,y)),
\end{aligned}
$$

The indices h and k represent the wave-number coefficients in time and spanwise direction and specify the (h,k)-modes of this double Fourier transformation. This method can be used until strong non-periodicities occur in the reattachment region of the laminar separation bubble.

In DNS, data from four TS cycles are Fourier analyzed in time using a Hanning window function to suppress the influence of low frequency drifting of the bubble. This drift would otherwise contaminate the Fourier-analyzed signals. The low level of subharmonic disturbances resulting from this way of data processing confirms the clear distinction of drifting and TS frequency, supporting the validity of this procedure.

3 Results

In the present results the Reynolds number based on the displacement thickness at the separation line is about $Re_{\delta_1} = 900$. The extent of the transitional separation bubble of about 250 mm is given by the mean dividing streamline with the net mass flux set to zero.

$$\int_{z=0}^{z_T} \bar{u}(z)dz = 0$$

First of all, general properties of the LSB are discussed using base flow variables. Here, base flow means averaged quantities in time and spanwise direction. Note that the LSB is a highly unsteady phenomenon, and the base flow can never be observed instantaneously at any timestep.

In Fig. 4 grayscale contours of the streamwise velocity are shown. The darkest color shows the reversed flow region bounded by the line of vanishing streamwise velocity ($u = 0$). Good agreement between DNS results and measurements reveals that a separation bubble of approximately the same size is formed in both cases. At the point of laminar separation S the dividing streamline (dash-dotted) moves away from the wall until the flow undergoes transition (around location T). It ends at point R, where the flow reattaches in the mean as a turbulent boundary layer. The mean recirculation vortex is illustrated by a closed streamline in the rear part of the separation bubble. In Fig. 5 a comparison of some boundary layer quantities is shown. Boundary layer thickness δ as well as displacement thickness δ_1 considerably increase along the separation bubble, while the momentum thickness δ_2 starts to grow not until the onset of transition. As expected from the streamwise velocity, results from experiment and DNS again correspond very well. Since the LSB is of unsteady character as emphasized before, it appears inevitable to compare time-dependent results before claiming that the same physics are captured in the numerical and experimental investigations. This shall be done in the next paragraph for 2-D quantities first.

The first section of the LSB is dominated by a primary convective instability of the 2-D TS wave (see [10]). The same holds for the size of the bubble.Therefore, the development of the 2-D TS wave appears to be very important and is examined in detail. Fig. 6 shows the amplification curve for the streamwise velocity. It can be seen that the TS wave (1,0) is strongly amplified in the region of adverse pressure gradient. Experimental and numerical results perfectly match from $x = 230$ mm onwards even shortly beyond saturation. Good agreement with linear stability theory (LST) confirms the primary convective nature of the disturbance. Calculation and experiment also predict the same amplitude, growth rate and saturation level for the non-linearly generated higher harmonic disturbance (2,0). The deviation of the amplitude development between experimental and DNS results up to $x = 230$ mm is probably due to a different kind of disturbance visible in the experiment. This disturbance is less damped than the TS wave in the accelerated region, therefore starting with higher initial amplitude into the region of adverse pressure gradient, until the TS mode catches up and its amplitude outnumbers all other perturbations. This is supported by the fact that the disturbance observed in the experiment halves its wavelength at the position of the sharp bend in the amplification curve. Disturbances in DNS are generated at a streamwise position where the TS mode is already the most amplified disturbance and therefore correspond to linear stability theory from the beginning.

Wall-normal distributions of amplitudes and phases for the streamwise velocity u and wall-normal velocity v are shown in Fig. 7 at a location in the middle of the region of linear amplification. In the lower part of the base flow profile the reversed flow is clearly visible. The detached shear layer corresponds to the part of maximum streamwise velocity gradient around $y = 13\ mm$. It is at this distance from the wall where the disturbance amplitude of the 2-D TS wave reaches maximum values for the u-velocity.

In the experiment, a strong steady disturbance with half the spacer wavelength (0,2) can be observed (Fig. 8). From good agreement with LST for the development of the 2-D TS wave as described in the previous section it can be concluded that the spanwise modulation of the base flow does not exert any influence on 2-D instability characteristics of the flow field. In contrast, the growth of spanwise modulated perturbations of fundamental frequency is decisively affected by the presence of this steady mode. Strong growth of these disturbances cannot be explained by linear theory, which would give considerably lower growth rates, nor by a secondary (convective [11] or temporal [10]) instability, since it already sets in well before the TS wave has gained a sufficient amplitude. DNS calculations in which a similar steady perturbation was excited (see Fig. 8 and amplitude distribution Fig. 9) revealed that the growth rate of mode (1,2) can be explained by non-linear generation, though the amplitude level in the DNS and the experiment are different. Interaction of disturbances (0,2) and (1,0) produces a mode (1,2) that exhibits the same growth rate as the one observed in the measurements (Fig. 8). Agreement of amplitude and phase distributions between experiment and DNS is reasonable (Fig. 9), albeit deviation from symmetry can be seen in the measurements, which are not included in the DNS method in this study. Despite the importance of 3-D disturbances for the breakdown to turbulence, the dominance of the 2-D fundamental perturbation even far downstream of reattachment is remarkable [10].

4 Conclusions

A detailed comparison between measurements and numerical calculations of the velocity field in a transitional separation bubble showed very good agreement for time-averaged and 2-D Fourier analyzed quantities. Transition in a separation bubble under consideration here is driven by convective primary amplification of 2-D TS waves, mainly determining the size and position of the bubble. This mechanism could be clearly identified from the existing numerical and experimental data. The same holds for the formation of spanwise variations of the incoming 2-D flow field, which can be attributed to large steady perturbations and non-linearly generated time-dependent disturbances. The influence of the amplitude of these disturbances, which differed between experiment and DNS remains open to further investigation.

Acknowledgements: The financial support of this research by the Deutsche Forschungsgemeinschaft DFG under grant Wa 424/19-1 is gratefully acknowledged.

References

[1] Strunz, M. and Speth, J. F., A new laminar water tunnel to study the transition process in a Blasius boundary layer and in a separation bubble and a new tool for industrial aerodynamics and hydrodynamic research. *AGARD CP-413*, pp. 25–1–25–5 (1987).

[2] Wiegand, T., *Experimentelle Untersuchungen zum laminar-turbulenten Transitionsprozess eines Wellenzuges in einer Plattengrenzschicht*. Dissertation, Universität Stuttgart (1996).

[3] Kruse, M. and Wagner, S., LDA Measurements of Laminar-Turbulent Transition in a Flat-Plate Boundary layer. 8th International Symposium on Applications of Laser Techniques to Fluid Mechanics, 8.–11. July, Lisbon, Portugal (1996).

[4] Watmuff, J., Evolution of a wave packetinto vortex loops in a laminar separation bubble. *J. Fluid Mech.*, 397, 119–169 (1999).

[5] Alam, M. and Sandham, N. D., Direct numerical simulation of 'short' laminar separation bubbles with turbulent reattachment. *J. Fluid Mech.*, 410, 1–28 (2000).

[6] Spalart, P. and Strelets, M. K., Mechanisms of transition and heat transfer in a separation bubbles with turbulent reattachment. *J. Fluid Mech.*, 403, 329–349 (2000).

[7] Kruse, M., *Einsatz der Laser-Doppler-Anemometrie zur Untersuchung des laminar-turbulenten Grenzschichtumschlags an der ebenen Platte*. Dissertation, Universität Stuttgart (1997).

[8] Maucher, U., Rist, U. and Wagner, S., Refined Interaction Method for Direct Numerical Simulation of Transition in Separation Bubbles. *AIAA Journal.*, **38(8)**, 1385–1393 (2000).

[9] Lang, M., Marxen, O., Rist, U., Wagner, S. and Würz, W., LDA-Messungen zur Transition in einer laminaren Ablöseblase. In: *Lasermethoden in der Strömungmeßtechnik*. 8. Fachtagung der GALA, 12.–14. Sep. 2000, Freising-Weihenstephan, FRG, Shaker Verlag, Aachen (2000).

[10] Maucher, U., Rist, U. and Wagner, S., Transitional Structures in a Laminar Separation Bubble. In: W. Nitsche, H. Heinemann and R. Hilbig (eds.), *Notes on Numerical Fluid Mechanics*, vol. 72, pp. 307–314. Vieweg Verlag, Wiesbaden (1999), 11th Stab Symposium 98, Berlin.

[11] Rist, U., *Zur Instabilität und Transition in laminaren Ablöseblasen*. Habilitation, Universität Stuttgart, Shaker Verlag, Aachen (1998).

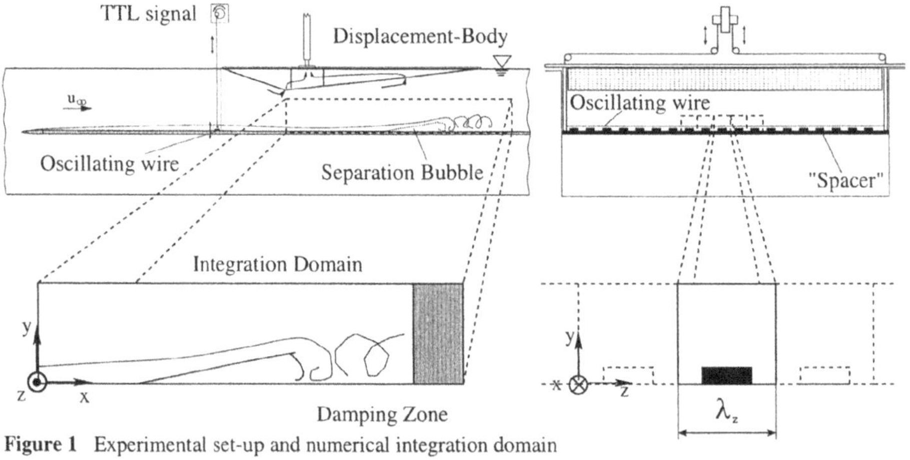

Figure 1 Experimental set-up and numerical integration domain

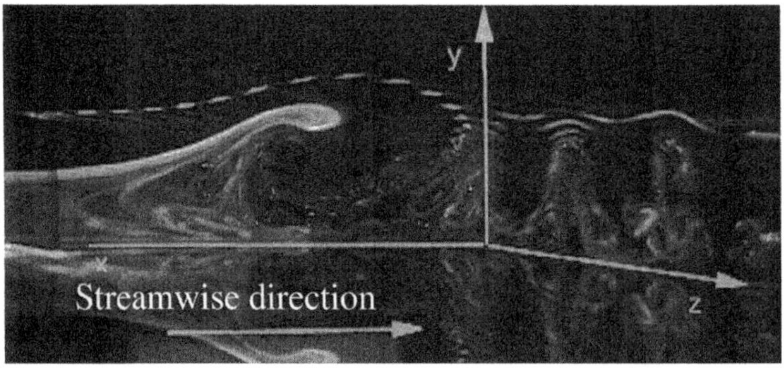

Figure 2 Flow visualization of the transition region. 2 simultaneous Laser light sheets in xy-plane (Peak-Position) and yz-plane (x=330 mm), respectively.

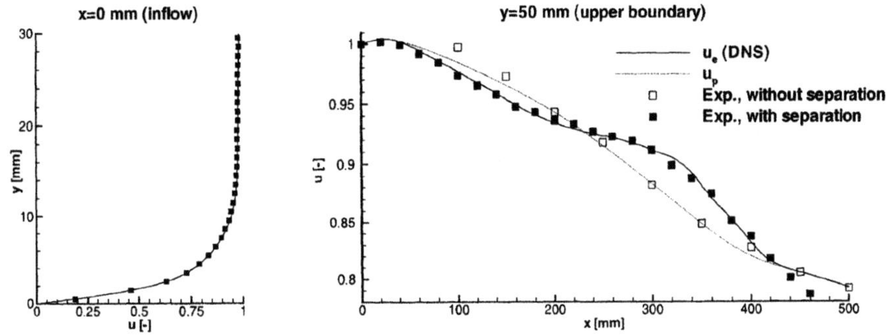

Figure 3 Streamwise mean velocity at $\hat{x} = 0\ mm$ (left) and $\hat{y} = 50\ mm$ (right). DNS results u_e (solid line), potential velocity u_p prescribed in DNS (dotted line); Measurements with (filled symbols) and without separation (open symbols).

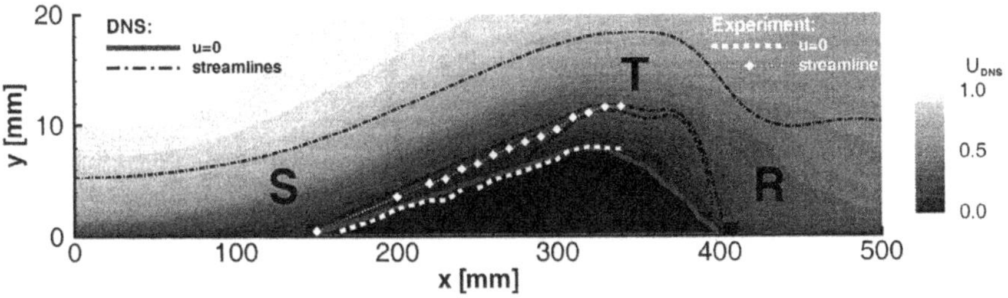

Figure 4 Contours of streamwise mean velocity (averaged in time and span) together with lines $u = 0\ mm$ and streamlines.

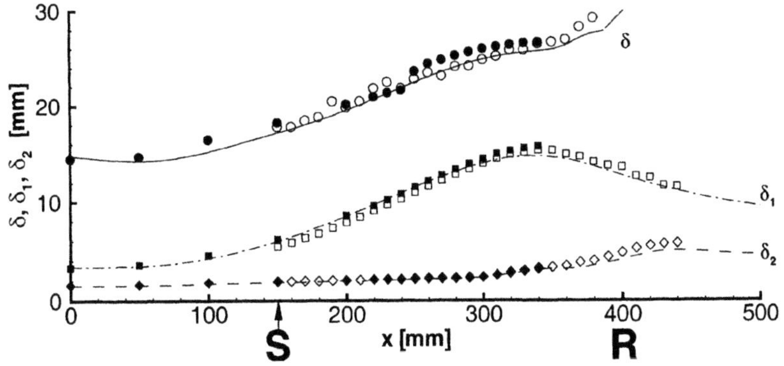

Figure 5 Time-averaged boundary layer quantities in the vicinity of a LSB. Results from spanwise-averaged DNS (lines) and measurements with spacer and spanwise-averaging (filled symbols) and without spacer at a single spanwise position (open symbols).

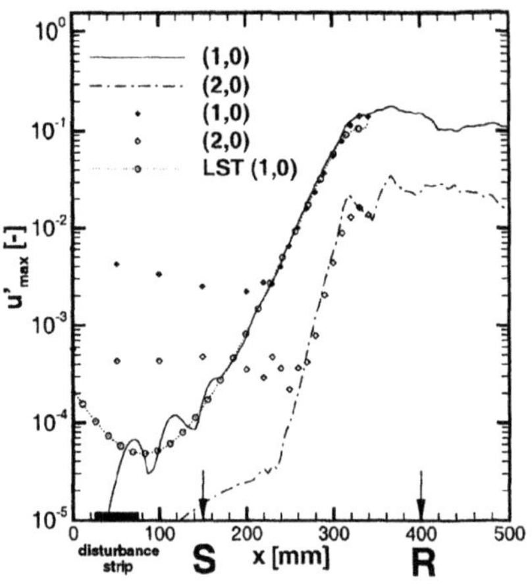

Figure 6 Amplification of the maximum velocity fluctuation u'_{max}. Results from DNS (lines), measurements (symbols) and LST.

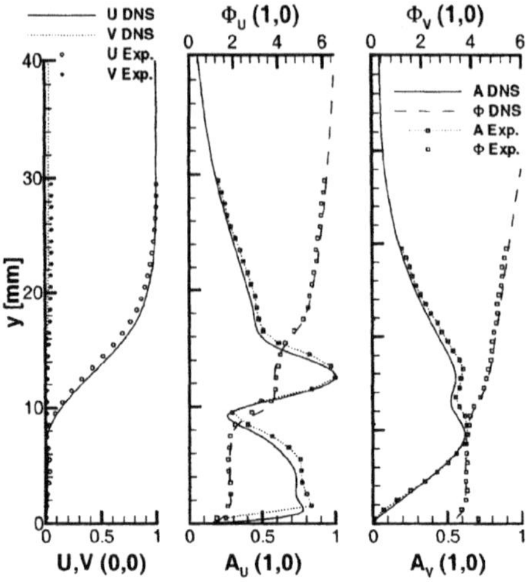

Figure 7 Normalized 2-D amplitude and phase distributions. Results from DNS (lines) and measurements (symbols) at $x = 290\ mm$.

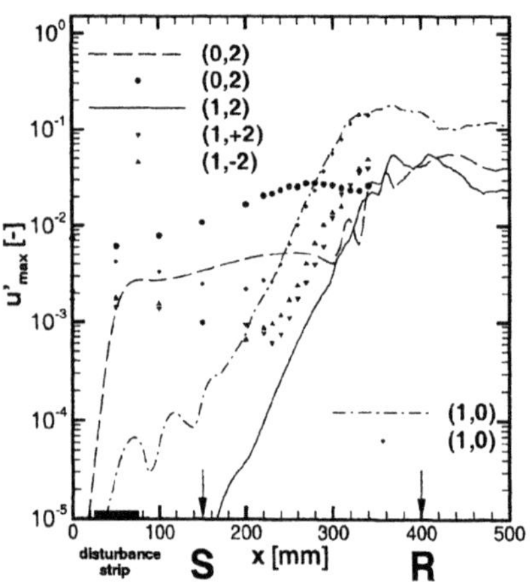

Figure 8 Amplification of the maximum velocity fluctuation u'_{max}. Results from DNS (lines) and measurements (symbols).

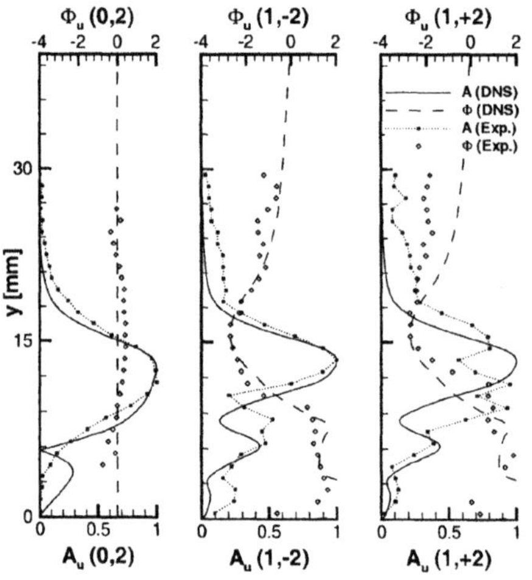

Figure 9 Normalized 3-D amplitude and phase distributions. Results from DNS (lines) and measurements (symbols) at $x = 290\ mm$.

Boundary-layer Separation Computed by Second-Moment Closure Models

S. Jakirlić[1], I. Hadžić[2], A. Djugum[1] and C. Tropea[1]

[1] Fachgebiet für Strömungslehre und Aerodynamik, Technische Universität Darmstadt, Petersenstr. 30, 64287 Darmstadt, Germany

[2] Arbeitsbereich Fluiddynamik und Schiffstheorie, Technische Universität Hamburg-Harburg, Lämmersieth 90, 22305 Hamburg, Germany

Summary

An analysis of computational results of the turbulent boundary layer separation obtained by second-moment closure (SMC) models was performed. Several flows separating due to the effect of an imposed adverse pressure gradient, ranging from the separation on the plane and inclined flat plates to the separation on a curved surface were investigated. Both high-Re SMC models, applying the "equilibrium" conditions - the so-called "wall functions" - for bridging the viscous sublayer and buffer zone, and low-Re SMC models, accounting separately for the viscous and wall proximity effects, were applied. The results obtained show good agreement of the mean flow and turbulent quantities with available experimental and DNS/LES database cases in all characteristic flow subregions considered: separating, reattaching and recovery zones.

2 Introduction

Separation from a smooth surface and the accompanied recirculation, even when small, have a large influence on the overall flow behaviour, especially beyond and downstream of the separation location. The separation bubble is usually a result of an adverse pressure gradient and it may appear at the leading and/or trailing edges of the blades in gas turbines, compressor blades, aerofoils, etc. Due to separation and reattachment, the flow becomes complex, highly curved and usually highly turbulent, with anisotropy of the Reynolds-stress tensor affecting significantly the momentum equation. The success in computing such flows relies on the correct representation of turbulence anisotropy and the turbulence model's ability to predict large departures from equilibrium. Therefore, the modelling of a separation bubble on a smooth surface is a challenging task and important for the prediction of the component performance.

The following flows were considered, Fig. 1: (a) turbulent separation on a plane wall (DNS, Spalart and Coleman, 1997), (b) flow in an asymmetric plane diffuser (Exp., Buice and Eaton, 1997; LES, Kaltenbach et al., 1999) accounting for flow separation on an inclined wall, and (c) flow around a NACA 4412 airfoil (Exp., Coles and Wadcock, 1979), representing flow separation on a curved surface with a free reattachment point.

3 Turbulence Models

The computations were performed with two types of Reynolds-stress turbulence models: the basic high-Re version (Gibson and Launder - GL, [3]) and with the low-Re number version of the second-moment (Reynolds stress) closure model, denoted as HJ low-Re RSM (Jakirlić, [7]). The latter model satisfies all important requirements, with a specific emphasis on asymptotic and limiting states of turbulence (vanishing and very high turbulence Re numbers, two-component limit,...), reproducing the laminar-to-turbulent and reverse transition (by-pass and shear-generated transition with minimum background turbulence), appropriate reproduction of effects of extra strain rates (transverse shear, skew-induced three-dimensionality), high acceleration (including laminarization), high deceleration (approaching separation), swirl effects, mean compression, flow separation, recirculation and reattachment. The best illustration for the model's ability to account properly for a large deviation from the equilibrium conditions is given in Fig. 2a, where semi-log plots of the mean velocity are presented for several non-equilibrium flows featuring the specific phenomena mentioned above. Another interesting model feature accounts separately for the wall effect on anisotropy of stress bearing and dissipative scales by introducing both the turbulent stress and dissipation rate anisotropy (E and A, see e.g. [7] for definition) in addition to viscosity effects taken into account via Re_t. The strong difference in the anisotropy rates of both tensors ($a_{ij} = \overline{u_i u_j}/k - 2/3\delta_{ij}$ and $e_{ij} = \varepsilon_{ij}/\varepsilon - 2/3\delta_{ij}$) expressed in term of their two-componentality factors A and E is shown in Fig. 2b. The additional source term S_l introduced into the ε-equation plays a very important role for the flows considered. The term acts to suppress an excessive growth of the turbulence length scale. The term has a local character and was found particularly useful in reproducing the turbulence and streamline pattern around flow separation and reattachment, where the standard SCM closures (both high-Re and low-Re versions) produce anomalous behaviour, Figs. 3 (for further details see [6]). For the precise specification of the entire model see e.g. [7].

4 Numerical Method

All computations were performed with a computer code based on a finite volume numerical method for block-structured, body-fitted, non-orthogonal meshes (see e.g., Fig. 8a) for solving two-dimensional Reynolds-Averaged-Navier-Stokes equations. Block-wise local grid refinement was employed for the discretisation of the transport equations. Block interfaces are treated in a conservative manner, consistent with the treatment of inner cell-faces. Cell centered (collocated) variable arrangement and cartesian vector and tensor components are used as primary variables. The velocity-pressure coupling is ensured by the pressure-correction method based on the SIMPLE algorithm. Diffusion fluxes are approximated by central differences (CD). In order to minimise numerical error and retain/preserve the stability of the procedure when the SMC models are used, the higher order upstream-weighted approximation schemes (linear upwind scheme - LUDS and quadratic upwind scheme - QUICK) as well as total variation diminishing (TVD) - type schemes have been employed to approximate convection. The blending was implemented in the so-called deferred-correction manner. The equations are linearised and solved sequentially using an iterative ILU method. The work of Hadžić, [4], should be consulted for further details about the numerical scheme used. All specific details about the numerical procedure will be given for each of the flow cases computed.

5 Results and Discussion

Here only the turbulent flows separating from a solid wall will be presented. For the flows with a laminar separation bubble, followed by transition to turbulence, the work of Hadžić and Hanjalić, [5], can be consulted.

5.1 Turbulent Separation Bubble on a Plane Wall

In this flow, the separation bubble is created by an imposed large adverse pressure gradient (APG), induced by suction and reattaching rapidly under the equally large favourable pressure gradient (FPG) induced by blowing of the fluid into the free-stream at some downstream position, Fig. 1a. The bulk Reynolds-number based on the incoming free-stream velocity and the flow domain is $Re = UY/\nu = 22000$. The computational domain adopted for the RANS computation is a box of the length $L = 10$ and width $Y = 1$ (non-dimensional lengths), see Fig. 4, and is covered with an Cartesian numerical mesh of 100×100 CVs. The mesh is refined in the bubble region. The flow inlet is located at $X = 0$, where the DNS data for the mean velocity and the Reynolds-stresses are available and are used to specify the inlet conditions. Since the incoming boundary-layer at $Re_\theta \approx 300$ was not well developed (as stated by the authors of DNS), it posed uncertainties in reproducing the inflow conditions, particularly the dissipation rate profile (which was not provided by DNS), to which the downstream flow pattern is very sensitive. The dissipation ε at the inlet was computed by solving the ε model equation with imposed known DNS data and assuming zero gradients of all variables in stream-wise direction. In accordance with DNS, the top boundary suction and blowing velocity $V_{(x)}$ is anti-symmetric around $X = 6.6$. The zero vorticity $\partial U_{(x)}/\partial y = \partial V_{(x)}/\partial y$ boundary condition for the stream-wise mean velocity is imposed in an implicit manner in conjunction with Dirichlet boundary condition. For all turbulence variables, the Neumann boundary condition ($\partial\Phi/\partial y = 0$) was imposed. For computational details see [4].
The RANS computations are performed with the HJ low-Re-number, second-moment closure with and without the term S_l, yielding similar flow patterns in reasonable agreement with the DNS. However, without S_l, the dividing streamline shows an anomalous forward bending at the separation point and backward bending at reattachment (similar as in Fig. 3). The computed separation bubble, Fig. 4 , is somewhat thinner than in DNS, which may be a consequence of inadequate inflow conditions. The streamline anomaly without S_l leads to a shorter distance between separation and reattachment. The inclusion of S_l rectifies the anomaly and extends the bubble length at the wall, though both the separation and reattachment points are predicted somewhat more upstream than in DNS. This is seen in the plot of the friction factor (not shown here), which shows also some discrepancy in the recovery zone. The mean velocity profiles, however, agree very well with the DNS, Fig. 5. The computed turbulence stress components are similar to DNS (Fig. 5), exhibiting some discrepancies, especially in the profiles of the shear stress.

5.2 Turbulent Flow in an Asymmetric Plane Diffuser

The flow configurations with a fixed separation location are perhaps not the most appropriate for illustrating arguments which could be useful for studying separation phenomena pertinent to the aerodynamics of aircraft wings. Recently a 2-D, asymmetric diffuser flow with a smooth separation on the deflected bottom wall due to an adverse pressure gradient (Fig. 1b) came into focus. The flow possesses most of the major features of complex separating flows: separation of a boundary layer, generation of a curved free shear layer, recirculation zone, reattachment and recovery of the wall boundary layer. Because turbulence interactions are known to respond differently to changes in flow conditions, the ability to reproduce the turbulence anisotropy in response to strong perturbations at separation or reattachment is a serious test of models of turbulence dynamics. A

comprehensive experimental database for model validation ([1]) was recently enriched by the large eddy simulation (LES, [8]), which supplied properties that are inaccessible to experimental techniques. The flow served as a test case at the 8th ERCOFTAC/IAHR workshop on refined turbulence modelling.
The flow discharging into asymmetric diffuser with an expansion ratio of $4.7H$ represents a 2-D, fully-developed channel flow with Reynolds number 20000 based on the centerline velocity and the channel height H. The length of the diffuser itself is $21H$. The computational results (corresponding to the grid size 200x150 and 200x50 cv's for the low-Re and high-Re SMC models respectively) are compared with the Buice and Eaton's experiment and with the LES by Kaltenbach et al., Figs. 3 and 6-7. The conclusion about the form and size of the separation bubble could be drawn from the plots of the friction and pressure coefficients (Fig. 6) and streamwise velocity profiles, Fig. 7 left. The separation bubble obtained by both models is almost of the same length ($\approx 21H$) and compares well with experimental findings ($21.8H$). The LES results show a somewhat longer separation bubble. However the separation zone obtained by SMC models is shifted upwards. The application of the low-Re SMC model resulted in a thinner separation bubble, which is visible in a weaker back flow (Fig. 7 left). The same outcome was reported from the computation of the other two flows (section 5.1 and 5.3). A possible reason for that is a weaker stress anisotropy immediately above the bubble obtained by the low-Re model. It corresponds to a smaller flow blockage being connected to a weaker acceleration of the mean flow caused by a thinner separation zone. Both LES and the the low-Re model computations produced a small bubble on the first diffuser corner (sharp negative peak in the friction coefficient evolution), causing a boundary layer deceleration followed by higher velocity gradients (stronger velocity profile inflection) and finally by a peak in the shear stress component, Fig. 7 right.

5.3 Flow around a NACA 4412 Airfoil

We consider now the flow over the NACA 4412 airfoil at maximum lift, with the incidence angle of 13.87^o, and high Re-number (the chord-based $Re_c = U_o c/\nu = 1.52 \times 10^6$, where c is the chord length). Flows around airfoils involve several important physical phenomena such as: transition to turbulence, laminar separation bubble at the leading edge, separation bubble at the suction side, curvature of flow, turbulent wake. In the considered flow around a NACA 4412 airfoil the transition length is very short and the treatment of the transition proved to have little influence on the overall results.
Computations were performed with a four-block O-grid (Fig. 8a) with both the high-Re-number and the HJ low-Re-number second-moment closures (398×88 and 398×128 nodes, respectively), without imposing any artificial transition. The grid point nearest to the wall was at y^+ between 0.1 and 1.5 for the low-Re model and between 10 and 50 for the high-Re model with wall functions. The high-Re-number model gave almost identical results both for 1.5% and for 5% free stream turbulence. It should be noted that all computations were performed in a steady mode, which may not be fully suitable, particularly for a low free-stream turbulence. The computational domain adopted here was as shown in Fig. 8a. The inlet is located at $0.8c$ in front of the airfoil leading edge, the outlet at the distance of about $2c$ beyond the airfoil trailing edge and the top and bottom boundary are located at about $1c$ away from the airfoil. With respect to the experiments, the computational domain for which the results shown here are obtained is located in the middle of the experimental section. The inflow boundary was placed

sufficiently away upstream from the leading edge (two distances were tested $0.8c$ and $1.6c$, with no noticable difference in the results) i.e. in the region where the streamlines show no curvature. The uniform velocity profile was taken over the inflow cross-section with the magnitude corresponding to the free-stream velocity prescribed in experiment by Coles and Wadcock, [2]. The evolution of the turbulence was taken to correspond to the natural decay of turbulence: $\mathbf{D}k/\mathbf{D}t = -\varepsilon$.
Results obtained with 5% free stream turbulence are presented here. The pressure coefficient obtained by both models agree well with the experiments (Fig. 8d). A difference between the two sets of results is more visible in the mean velocity- and shear stress profiles, and, particularly, in the size of the separation bubble, Fig. 8b and c. Unlike some fully-turbulent flows (e.g. back-step), where both models result in a similar streamline pattern in the recirculation zone, here the low-Re-number models yield a substantially thinner bubble. Because of insufficient experimental data it is difficult to judge which streamline pattern is closer to reality. The mean velocity and shear stress profiles obtained by both models show also close agreement with experiments at most locations along the airfoil, and also in the near wake region, but the low-Re-number model seems to be slightly superior.

6 Conclusions

The performance of some popular single-point turbulence closures based on the solution of the transport equations for the Reynolds-stress tensor for predicting the boundary layer separation on the smooth plane, inclined and curved walls pertinent to flows over airfoils were investigated. Both Re-stress model versions applied show a potential for capturing complex flow phenomena, even the high-Re number closure. But if the flow is determined by the near wall properties (friction) and in particularly, if transitional phenomena occur, then the low-Re number model and integration up to the wall proved to be of decisive importance.

Acknowledgements

The authors thank P. Spalart and H.-J. Kaltenbach for making avaliable their DNS results of turbulent separated bubble i.e. LES results of the asymetric plane diffuser.

References

[1] C.U. Buice and J.K. Eaton. TSD-107, Stanford University, 1997.
[2] D. Coles and A.J. Wadcock. AIAA Journal, Vol. 17, 1979, pp. 321-329.
[3] M.M. Gibson and B.E. Launder. J. Fluid Mech., Vol. 86, 1978, pp. 491-511.
[4] I. Hadžić. Ph.D. Thesis, Delft University of Technology, 1999, p. 187.
[5] I. Hadžić and K. Hanjalić. Flow, Turbulence and Combustion, Vol. 63, 1999, pp. 153-173.
[6] K. Hanjalić and S. Jakirlić. Computers and Fluids, Vol. 22, No. 2, 1998, pp. 137-156
[7] S. Jakirlić. Ph.D. Thesis, FA Univ. Erlangen-Nuremberg, Germany, 1997, p. 398.
[8] H.-J. Kaltenbach, M. Fatica, R. Mittal, T.S. Lund and P. Moin. J. Fluid Mech., Vol. 390, 1999, pp. 151-185.
[9] S. Obi, K. Aoki and S. Masuda. 9th Symp. on Turbulent Shear Flows, Kyoto, Japan, August 16-19, 1993, pp. 305.1-305.5
[10] P.R. Spalart and G.N. Coleman. Eur. J. Mech., B./Fluids, Vol. 16, 1997, pp. 169-189.

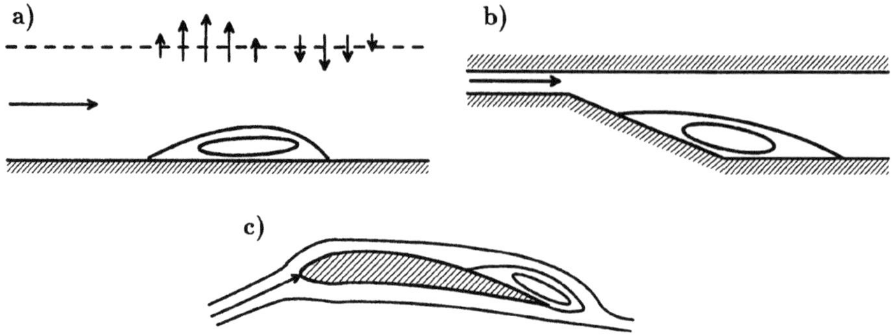

Figure 1 Schematic of the flow cases investigated

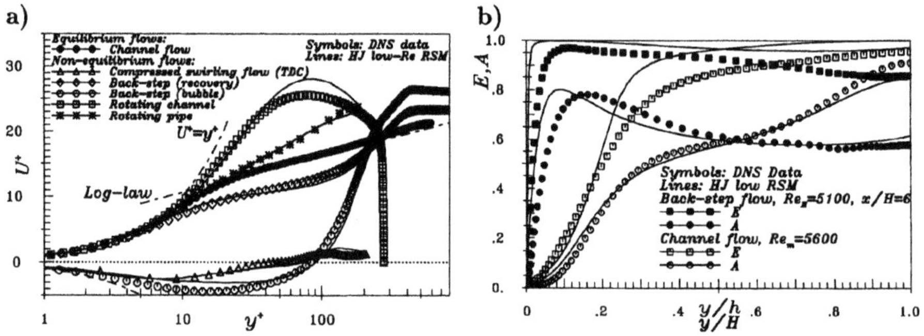

Figure 2 a) Semi-logarithmic plot of the mean axial velocities for different non-equilibrium flows and b) two-componentality parameters for the stress - A - and dissipation rate anisotropy tensors - E

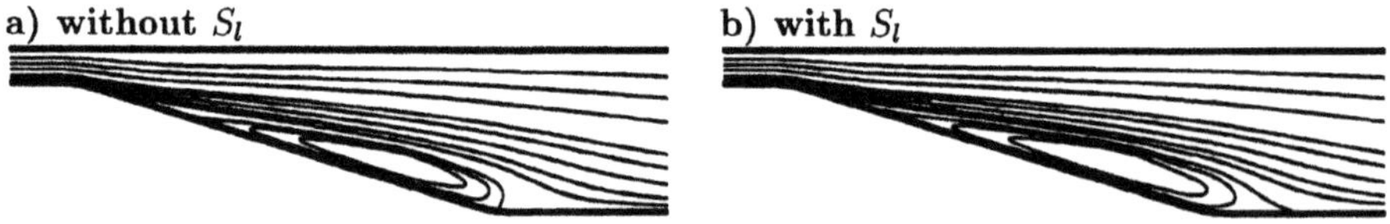

Figure 3 Plane asymmetric diffuser: effect of the length scale correction term S_l on the streamline pattern at reattachment (a high-Re-number SMC was employed)

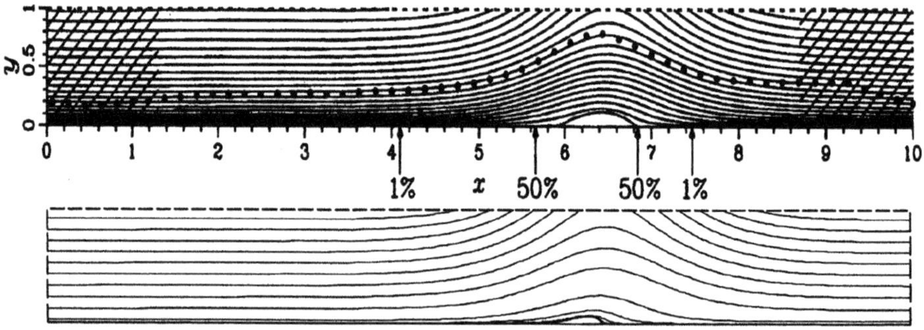

Figure 4 Turbulent separation bubble: streamline patterns obtained by DNS (upper) and by low-Re-number SMC (lower)

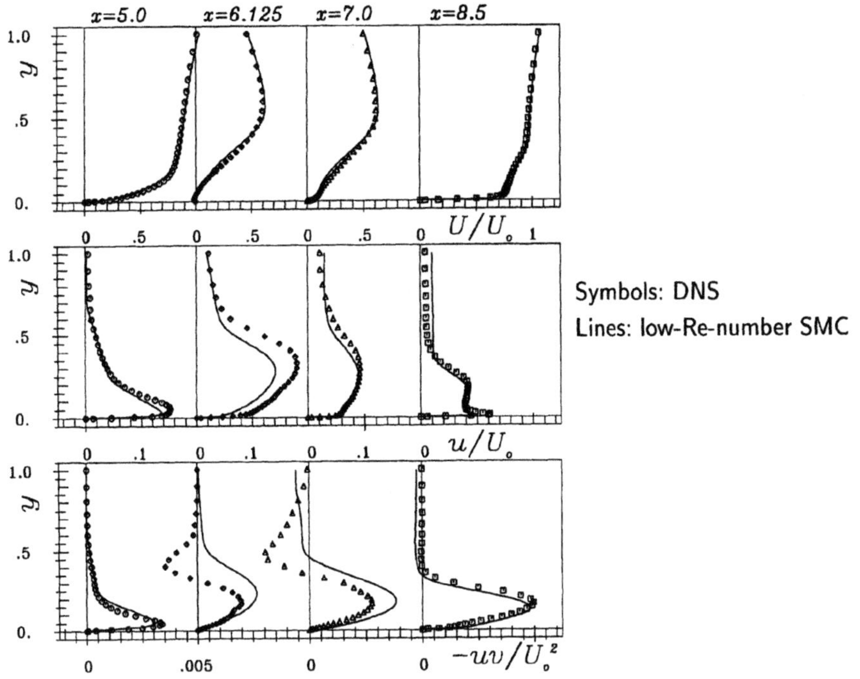

Figure 5 Turbulent separation bubble: mean-velocities and streamwise stress components at characteristic positions ($x = 5$ before separation, $x = 6.125$ in the middle of separation, $x = 7$ just beyond separation and $x = 8.5$ recovery of boundary layer).

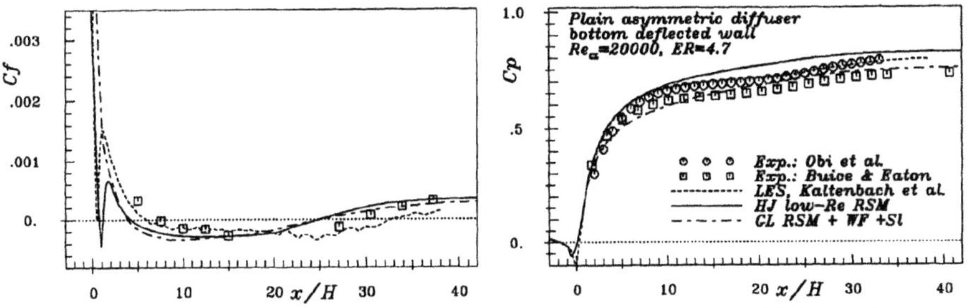

Figure 6 Plane asymmetric diffuser: friction (C_f) and b) pressure (C_p) coefficients

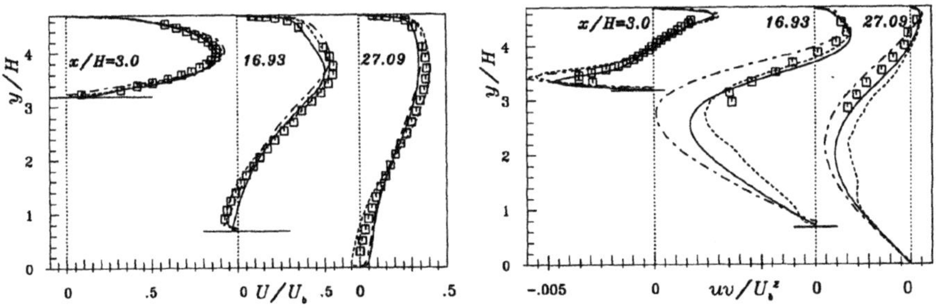

Figure 7 Plane asymmetric diffuser: streamwise velocities and shear stress components at selected locations: close to separation point $x/H = 3$, within the recirculation zone $x/H = 16.9$ and at reattachment $x/H = 27.09$. Key as Fig. 6

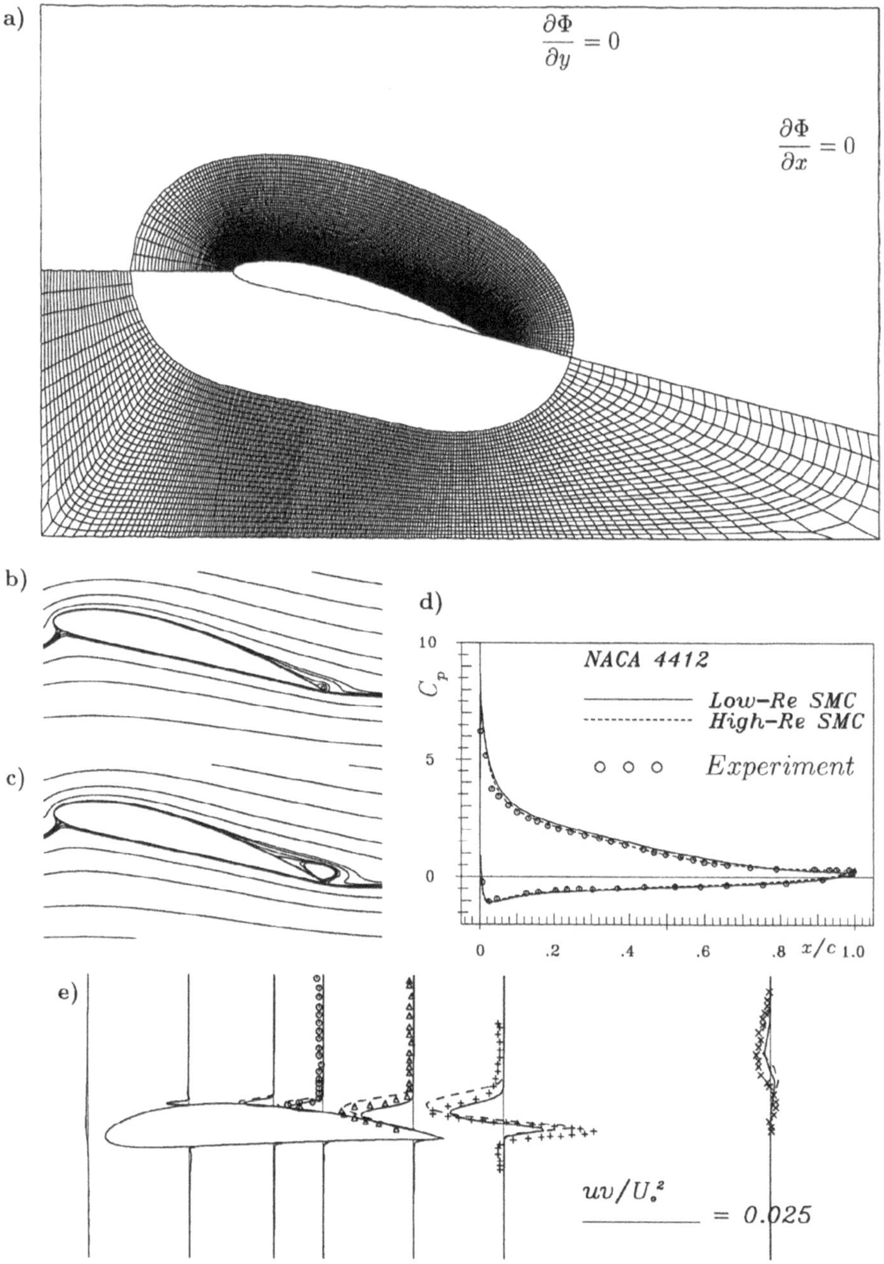

Figure 8 NACA 4412 airfoil: a) computational domain; numerical mesh of O type consists of four blocks, b) streamline patterns obtained with low-Re SMC and c) with high-Re SMC + WF, d) pressure coefficient C_p, and e) the shear-stress $\overline{uv}/U_o^2$ in the chord-wise oriented coordinate system at x/c = *-0.05, 0.25, 0.5, 0.642, 0.908, 1.1747, 1.95146*

Experimental Investigation on the Behaviour of Disturbances Emitted from a Point Source in a Laminar Boundary Layer

J. FREY, R. BUCKISCH, L. MÜLLER, R. GRUNDMANN
Technische Universität Dresden, Institut für Luft- und Raumfahrttechnik,
01062 Dresden, Germany

Summary

Within the framework of the DFG transition programme, the Institute for Aerospace Engineering (ILR) of the Dresden University of Technology (TUD) investigates the spreading of disturbances downstream of a point-shaped source in a laminar boundary layer on a flat plate.

The measurements were performed with hot wire probes on the transition test configuration in the low speed wind tunnel of the TUD. Two different configurations of a point source have been examined so far. A circular arrangement of six cannulas (0.2 mm wide) is not appropriate in the current set up because the disturbances were damped out immediately. A short slot (0.2x5 mm) however, has provided useful results.

The downstream development of amplitude and phase of the artificial disturbances is described in detail.

1 State of the Art

About 50% of the drag of a modern commercial aircraft is caused by skin friction. Due to the fact that turbulent boundary layers produce higher drag than laminar, the delay of transition is of particular interest.

The transition is preceded by destabilisation of the boundary layer due to the sensitivity against small disturbances. When these disturbances – called Tollmien-Schlichting waves (TSW) in the 2D-case – exceed a certain critical amplitude the flow state changes from laminar to turbulent. A passive way to postpone the transition is stabilising the boundary layer for instance by a negative pressure gradient (laminar airfoil). Active damping of the disturbances causes the critical amplitudes to be reached later.

This way of active damping has been investigated since the beginning of the 1980s, when Liepman, Brown and Nosenchuck [2, 3] made their experiments on the control of instability waves along a flat plate in water. They used periodically heated strips to induce and to damp the artificial disturbances as well as to damp the natural appearing. For their measurements they used hot film wall-shear probes.

Wiegand [4] described the transition process of a *wave train* emitted by a *harmonic point source*. These experiments were carried out in a laminar water tunnel at very low velocities (0.145 m/s) and frequencies (0.22..0.46 Hz).

Numerical calculations based on Direct Numerical Simulation (DNS) on point source disturbances were performed at the Institute for Aerodynamics and Gasdynamics (IAG), Stuttgart [6]. In cooperation with the ILR, the suitability of the chosen numerical or experimental methods and facilities of research can permanently be screened and verified.

Stability diagrams have set up available by the IAG for the Blasius boundary layer used by the ILR (fig. 1).

Further, similar experiments were carried out by Baumann [7] in the laminar wind tunnel of the TU Berlin. They focus on an adaptive sensor-actuator system on an airfoil. A significant delay of transition on the pressure side has been achieved. Disturbance induction is performed by loudspeakers as well as by flush-mounted surface actuators.

The transition test setup of the ILR has originally been designed for studies on the behaviour of oblique TSW. In the years 1995 to 1999 extensive measurements on this topic were done. Obliqueness angles of the emitted waves between 0° and 40° were examined. For angles up to 35° the artificial disturbances could be damped to a large degree.

Another focal point of this investigation was the visualisation of the development of disturbances. Therefore, one of the slots could be equipped with a particle generator. The method of laser light sheet provided data on the structure of the disturbances (spreading, wavelength, production of λ-vortices).

2 Experimental Setup

The Institute for Aerospace Engineering of the TUD runs a low speed wind tunnel of the Göttingen type with an open test section. It is powered by a two stage compressor with a total power of 300kW.

For transition experiments a 3-m-nozzle is installed. It has a contraction ratio of 7:1. In this configuration the wind tunnel has a low turbulence level of about 0.15 % mainly in a frequency range around 1.5 Hz which makes it appropriate for boundary layer research.

The test configuration (fig. 2) is based on a flat plate (1 m x 1 m x 40 mm) with an elliptical leading edge made of aluminium alloy of very low tension. It includes a movable flap (again 1 m x 1 m) for the adjustment of the downstream pressure gradient. Furthermore, the complete construction can change the overall incidence. With this configuration, a zero pressure gradient, laminar, two-dimensional boundary layer according to Blasius is achieved. Disturbance generators (slots and point source) are placed in a flush mounted rotatable insert to emit TSW at several oblique angles. They are powered by external loudspeakers. Measurements are carried out with hot wire anemometry. One or two velocity components can be detected at a time. The probes are positioned using a three axis linear traverse. There are special software programs for data aquisition and processing, in particular FFT-filters.

The experiments are carried out at a mean velocity of u=10 m/s and an excitation frequency of 140 Hz – due to the stability characteristics of the boundary layer (fig. 1). The velocity measurements were made with a sample rate of 1.4 kHz and averaged over a time of 25 sec for each position. A band pass filter in IIR-regime (Infinite Impulse Response) with 1 Hz band width (139.1..140.1 Hz) is used to calculate the portion of the artificial disturbance within the velocity signal.

3 Point Source by Cannulas (Point Source 1)

The characteristic of Point source 1 without wind tunnel flow showed a fountain-like jet of the air out of the cannulas - due to the small cross-section area of the source.

The velocity distribution versus the z-coordinate perpendicular to the wall has a TSW-typical profile in amplitude and phase. The very flat second maximum near the edge of the boundary layer indicates the three-dimensionality of the disturbance. Even some of the frequencies appearing naturally have been absorbed. However, the amplitudes are rather low and decrease in downstream direction.

A decrease of amplitude is observed independent of the excitation level. This was due to a too low energy transfer (up to an SPL of about 80 dB, U=2.5V) and to an immediate transition caused by roughness effects of the "fountain".

Additionally, a three-dimensional instability occurs because of the point shaped emission of the disturbance (fig. 6) causing a pair of longitudinal vortices (low speed streak) shedding from the source [4]. The phase distribution versus the spanwise coordinate (y) shows a lateral wavelength of 27 mm which is about the same in the longitudinal direction. So the emitted disturbances form a wave front with an angle of approximately 45°.

4 Point Source by a Short Slot (Point Source 2)

As an alternative the first slot of the disturbance configuration (x/l=0.2, fig. 2) has been shortened to a length of 5 mm. The remaining aperture has a cross-section area of 1 mm^2, which is six times that of the cannula. However, it can still be called small compared to the wave lengths and its dimension can be based on a virtual start point which is placed upstream.

The effects of the variation of the cross section area already show up without wind tunnel flow. The voltage necessary for the same SPL is clearly lower.

4.1 General Parameter Study

The properties of this source were examined in detail since extensive measurements along the centre line and spanwise direction were carried out. A distribution of energy in spanwise direction occurs before an amplification takes place. Linearity could be proved for very small SPLs (fig. 5).

The lateral distribution confirms a "running broad" of the disturbance with an appropriate energy transport. Because of this spanwise energy transport amplification only occurs over a smaller downstream range. The lateral wave length is again about 27 mm, but it increases in downstream direction such that the oblique angle decreases (fig. 8).

4.2 Spanwise Disturbance Spreading in Phase and Amplitude

Large-scale hot wire measurements in x-y-planes (parallel to the plate) allow to produce a clear picture of the spreading of the disturbances behind a point source. Phase plots are of particular interest as they show the course of the wave fronts (fig. 9, 10). Attention should be paid to the antimetric behaviour of the cross flow component v. The disturbance shows a wedge-shaped spreading under an opening angle of about 24°.

Besides that a downstream tendency to smaller oblique angles is noticed. This decrease of the three-dimensional share was observed in [4] as well when the disturbance conditions were similar especially the non-dimensional frequencies. This tendency is also verified by the measurements of the z-profiles at y/l=0.02 (fig. 11). As the downstream distance increases a second local maximum appears near the edge of the boundary layer, which is typical for two-dimensional TSW. In the DNS-calculations of a harmonic point source (IAG Stuttgart) a flattening of the wave fronts was observed too. This also can be interpreted as a result of more complex processes but the dominance of two-dimensional amplification (fig. 14 in [6]).

5 Conclusions and Outlook

Point source 1 as shown in the current set up is not suitable to influence the boundary layer transition. It is not possible to induce disturbances which can be observed over a sufficiently extended area. The small orifice of less than 0.2 mm^2 is assumed to be the main reason for that. Source 2, however, transports more energy into the boundary layer without causing the problem of roughness effects. This smoother induction of disturbances allows to observe the waves over longer distances. Clear differences from the measurements with the complete slot length and the wedge-shaped spreading prove the point-shaped emission.

In the proceeding research a high-resolution CCD-camera will be tested to verify the results gained with the hot wire measurement. For these experiments however, the problem of particle emission into the boundary layer has to be solved.
Additionally, it is planned to carry out damping experiments based on point source 2. Therefore, modifications are necessary to cope the oblique wave fronts with uneven amplitude distribution.
This research project is supported by the German Research Association (DFG) under GR 1388/3.

References

[1] H. Schlichting; K. Gersten. "Grenzschichttheorie". 9th edition, Springer-Verlag Berlin Heidelberg 1997

[2] H. W. Liepmann; G. L. Brown; D. M. Nosenchuck. "Control of Laminar-Instability Waves Using a New Technique". In: Journal of Fluid Mechanics, vol. 118, 1982, pp. 187-200

[3] H. W. Liepmann; G. L. Brown. "Active Control of Laminar-Turbulent Transition". In: Journal of Fluid Mechanics, vol. 118, 1982, pp. 201-204

[4] Th. Wiegand; H. Bestek, S. Wagner. "Transition Process of a Wave Train in a Laminar Boundary Layer". New Results in Numerical and Experimental Fluid Mechanics, 1997, pp.397-404

[5] F. N. Shaikh; M. Gaster. "The Non-Linear Evolution of Modulated Waves in a Boundary Layer". In: Journal of Engeneering Mathematics, Kluwer Academic Publishers, vol. 28, 1994, pp. 55-71

[6] Ch. Stemmer; M. Kloker; S. Wagner. "DNS of Harmonic Point Source Disturbances in an Airfoil Boundary Layer Flow". In: AIAA-Journal, AIAA 98-2436, 1998

[7] M. Baumann; D. Sturzebecher; W. Nitsche. "Experiments on Active Control of Boundary Layer Insataibilities on a Wing". New Results in Numerical and Experimental Fluid Mechanics II, Notes on Numerical Fluid Mechanics, vol. 72, Vieweg Verlag Braunschweig, 1999, pp. 22-29

Figures

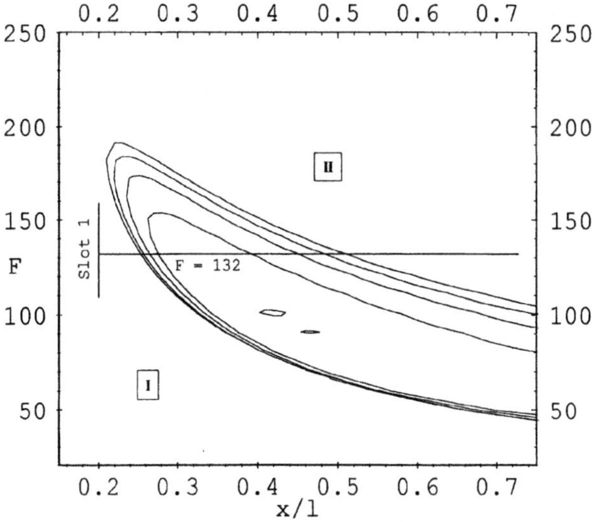

Figure 1 Stability diagram (curves of zero amplification) for the applied Blasius boundary layer for several oblique angles (0°,10°,20°,30°,40°)

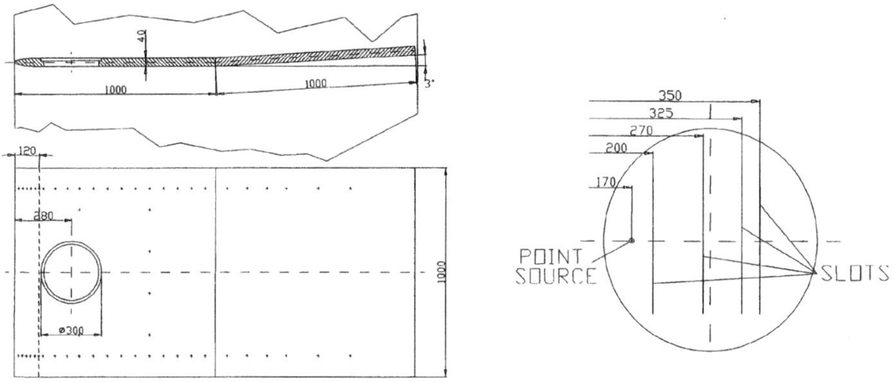

Figure 2 Transition test configuration– flat plate; disturbance insert

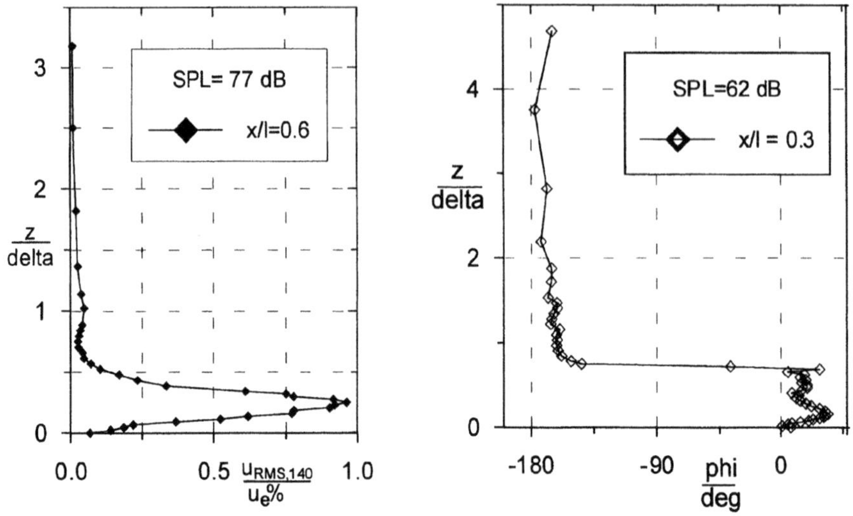

Figure 3, 4 Distribution of amplitude and phase difference of the $u_{RMS,140}$-values versus z

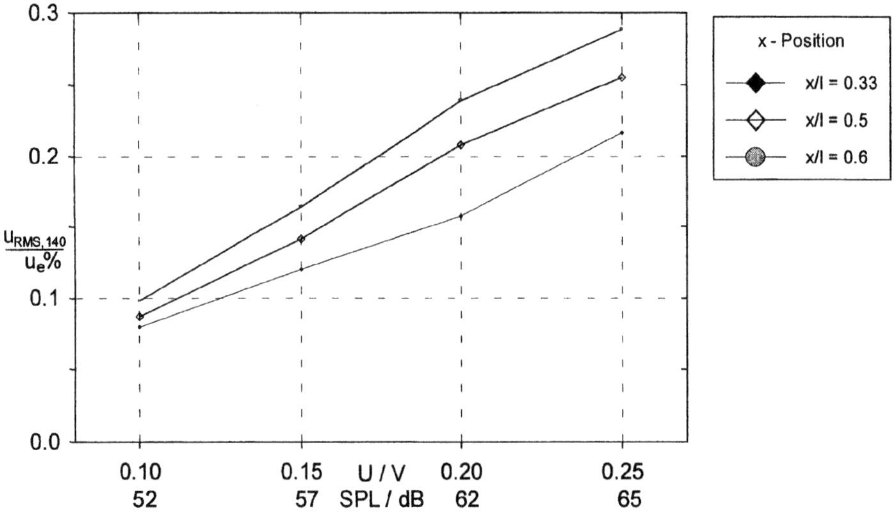

Figure 5 Point Source II: Linearity of $u_{RMS,140}$ versus excitation (voltage, SPL)

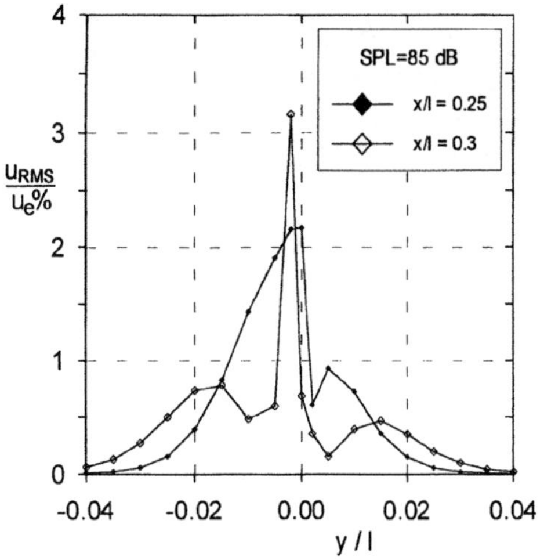

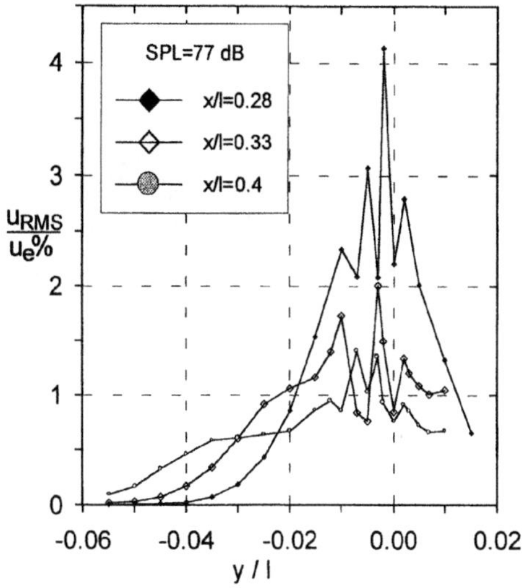

Figure 6,7 Amplitude of the $u_{RMS,140}$-signal downstream of point source 1 and 2 for comparison

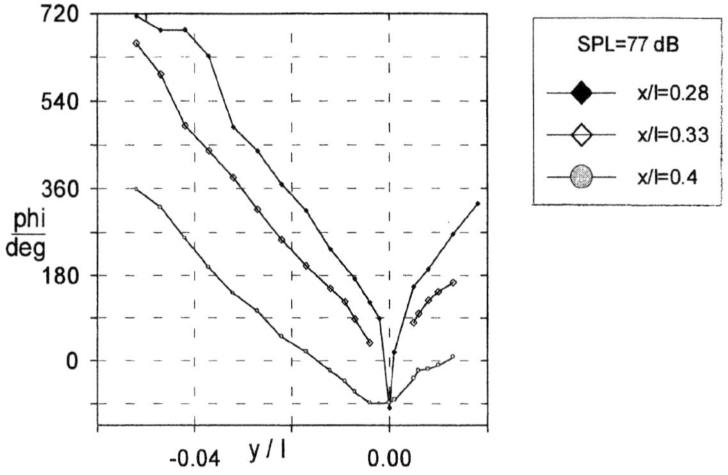

Figure 8 Spanwise distribution of the phase difference at several downstream positions (point source 2)

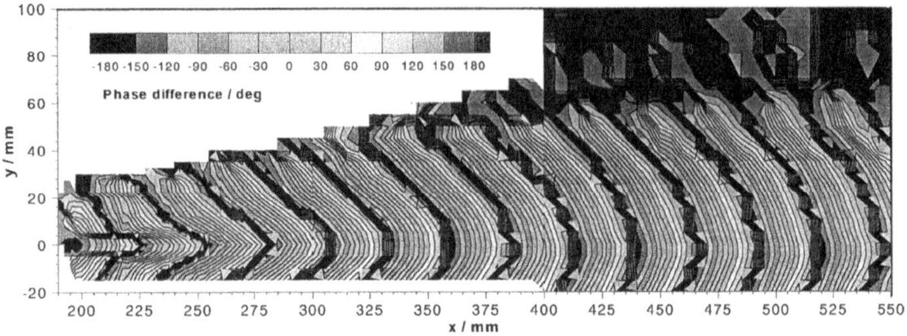

Figure 9 Level phase plot of the $u_{RMS,140}$-value downsteam of point source 2

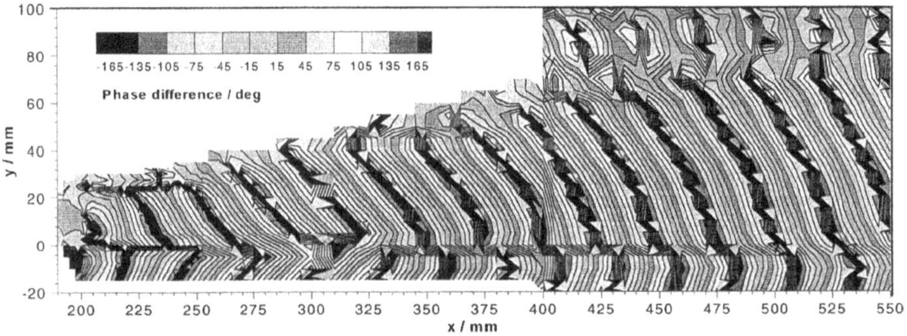

Figure 10 Level phase plot of the $v_{RMS,140}$-value downsteam of point source 2 notice antimetric behaviour

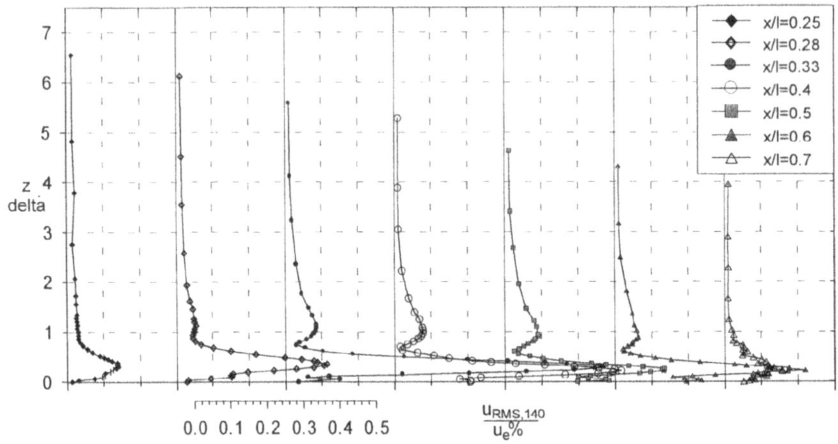

Figure 11 Downstream development of the disturbance outside the centre line (y/l=0.02, point source 2, SPL=73 dB)

Experimental/Numerical Investigation of the Influence of a Pressure Gradient on Acoustic Roughness Receptivity in the Boundary Layer of an Airfoil

S. Herr, A. Wörner, W. Würz, U. Rist, S. Wagner
Universität Stuttgart, Institut für Aerodynamik und Gasdynamik
Pfaffenwaldring 21, D-70550 Stuttgart, Germany

Summary

In the present paper, the influence of a pressure gradient on the acoustic roughness receptivity is investigated by means of wind tunnel experiments and Direct Numerical Simulations (DNS). The receptivity function for three different locations of a 3-D surface non-uniformity on an airfoil is calculated from the experimental and numerical data and compared to each other. The combined effects of the roughness acoustic receptivity and the amplification of the generated disturbances are taken into account in order to evaluate the position on the airfoil at which a surface non-uniformity is most dangerous in provoking an early transition.

Introduction

The problem of the boundary layer transition from laminar to turbulent flow still attracts much attention because of its fundamental and practical importance. In case of 2-D boundary layers, three main stages can be distinguished. First, there is the laminar flow receptivity to external (acoustic or vortical) perturbations. The second part is the region where the boundary layer instabilities develop according to linear stability theory. Finally, there is the nonlinear flow breakdown to turbulence. This paper is devoted to experimental and numerical investigations of the linear 3-D acoustic receptivity of a 2-D laminar boundary layer in the presence of a localized (3D) quasi steady surface non-uniformity. Pressure gradients strongly influence the amplification of disturbances in a transitional boundary layer. The location of transition is not only determined by the amplification of the disturbances, but also by their initial amplitude. As the initial amplitude itself is determined by the process of receptivity, it should be of interest whether a large scale pressure gradient may have an influence on the receptivity of a boundary layer to freestream perturbations.

Only little effort was spent on this problem so far. Most studies consider a Blasius flow as base flow [8], [6]. The 3-D vibrational receptivity of a 2-D self-similar boundary layer with an adverse pressure gradient was investigated experimentally in [5], and recently the influence of a favorable pressure gradient was studied in [4]. A numerical investigation on the influence of the pressure gradient on boundary layer receptivity can be found in [1]. In the present study, we would like to determine the role a large scale pressure gradient plays with regard to the acoustic receptivity of a boundary layer in the presence of a small scale surface non-uniformity on an airfoil.

In an earlier experiment and accompanying DNS, the acoustic receptivity function of a 3-D roughness element on an airfoil was determined quantitatively [13]. The receptivity function was defined as the ratio of the initial (i.e. at the surface non-uniformity) amplitude of the instability wave, generated in the boundary layer, to the amplitude of the external acoustic perturbation and the amplitude of the corresponding (localized) surface non-uniformity:

$$\tilde{G}_{av}(\alpha,\beta) = \frac{\tilde{B}_{inTS}(\alpha,\beta)}{\tilde{A}_{ac} \cdot \tilde{A}_{v}(\alpha,\beta)} \quad (1)$$

for every spanwise and streamwise wave number of the developing TS-wave train as proposed in [3]. The frequency for the external perturbation (f_{ac}) was chosen to be equivalent to the most amplified TS-frequency ($f_{ac} = 1088Hz$) and the surface roughness was located at the point of neutral stability of branch I for the investigated frequency. It is located at $\frac{s}{s_{max}} = 20\%$ on the symmetrical airfoil (XIS40MOD) [11] (figure 1). The boundary layer profile at this position ($H_{12} = 2.58$) is close to that of the Blasius boundary layer (figure 2), and therefore the accompanying DNS could be performed for the Blasius flow [9].
To be able to study the influence of a pressure gradient on the receptivity, two additional positions of the roughness element on the airfoil were chosen in a preliminary investigation. The position $\frac{s}{s_{max}} = 15\%$ corresponds to a favorable pressure gradient (figure 1). The accompanying DNS were performed for a Falkner-Skan self-similar boundary layer with Hartree parameter $\beta_H = 0.05$ (figure 2). The position $\frac{s}{s_{max}} = 30\%$ lies in the region of the instability ramp of the airfoil (figure 1) and the DNS were performed for a Falkner-Skan profile with Hartree parameter $\beta_H = -0.05$ (figure 2).
In the following section the numerical approach is outlined followed by a section concerning the experimental approach. Then the results for the three positions and considerations about practical applications are presented.

Numerical Approach

The DNS are based on the velocity-vorticity formulation of the complete Navier-Stokes equations for incompressible fluids. All spatial scales are non-dimensionalized using a reference length $\tilde{L}$ and all velocities using the freestream velocity $\tilde{U}_\infty$ at the inflow boundary, where $\tilde{}$ denotes dimensional variables. A sketch of the integration domain is shown in figure 3. The flowfield is discretized using fourth-order accurate finite differences in streamwise (x-) and wall-normal (y-) direction, and a spectral method in spanwise (z-) direction. The time integration is done using a fourth order, four step Runge-Kutta scheme. Details concerning the numerical method can be found in [7].

A pressure gradient is imposed on the boundary layer by prescribing the streamwise velocity at the upper boundary of the integration domain. For the study shown here, this is done using the Falkner-Skan ansatz for self similar boundary layers. This means that the streamwise velocity at the upper boundary u_e is set to:

$$u_e(x) = u_0 \cdot x^m, \quad \text{with} \quad m = \frac{\beta_H}{(2-\beta_H)}. \quad (2)$$

If $\beta_H < 0$ the flow is decelerated, which means that an adverse pressure gradient is imposed, whereas for $\beta_H > 0$ the flow is accelerated or a favorable pressure gradient is imposed.

The 'acoustic' wave is modelled using the analytical solution of the second Stokes' problem. This solution is added to the Falkner-Skan or, at zero pressure gradient, Blasius velocity profile, which is prescribed at inflow. The small roughness element within the equidistant cartesian grid is modelled by extrapolating the velocity at the lowest row of grid points in such a way as to fulfil the no-slip and the non-permeability condition at the surface of the roughness. This is done using fifth order polynomials in $y-$direction which are consistent with the finite difference representation of the flowfield.

In contrast to the experiments, the interaction of a planar 'acoustic' wave with a roughness element of one discrete spanwise wave number is calculated, and the complex receptivity function is evaluated for every spanwise wave number separately. For the 2-D case, the evaluation of the receptivity function requires four numerical simulations. In the first simulation the steady flow over a flat plate with a roughness element located at a certain distance from the leading edge is calculated using the unsteady Navier-Stokes equations starting with the flow over a flat plate without roughness. In the second simulation, the interaction of the 'acoustic' wave with the roughness is calculated using the previously calculated steady flow as initial condition. The difficulty now is to extract the Tollmien-Schlichting(TS) wave, which is created by the interaction of the 'acoustic' wave with the roughness, from the total solution. There are two problems. The first one is that the 'acoustic' wave itself has the same frequency as the created TS-wave, and the second is that there is a numerically created TS-wave resulting from an approximate inflow condition. These problems are solved using a method suggested by Crouch & Spalart [2]. Therefore, a third simulation is needed including the 'acoustic' wave but no roughness. In this simulation the TS-wave created at the inflow boundary is also present. So the TS-wave created at the roughness can be extracted from the total solution by subtracting the results of the third simulation from the results of the second one. Next, a Fourier analysis is used to determine the amplitude and phase part of the TS-wave at every streamwise position. The 'theoretical' amplitude of the TS-wave at the center of the roughness element is found by matching the amplitudes of the TS-wave created by the interaction with the TS-wave created in a fourth simulation (the so called reference simulation) by blowing and suction at the wall upstream of the roughness element. This can be seen in figure 5, where the amplification curve of a TS-wave which is created by the interaction of an 'acoustic' wave with a surface non-uniformity located at $Re_{\delta_1} = 1205$ within an adverse pressure gradient region is shown together with the results from the reference simulation. For the 3-D case, the third simulation is not necessary because the TS-wave can be distinguished from the 'acoustic' wave and the numerical TS-wave because of its different spanwise wave number.

Experimental Approach

The hot-wire measurements were carried out in the Laminar Wind Tunnel of the *Institut für Aerodynamik und Gasdynamik* [10]. The high contraction ratio of 100:1, as well as five screens and filters, result in a very low turbulence level, $\frac{u'}{u_\infty} < 2 \cdot 10^{-4}$. Downstream of the airfoil model a loudspeaker was mounted in a streamline shaped housing to insert the sound wave with a frequency $f_{ac} = 1088Hz$ as external perturbation. This sound wave had a sound pressure level of approximately $100dB$ in the test section, that is $16dB$ higher than the natural sound pressure level for the frequency range $10Hz - 5000Hz$ of the wind tunnel itself.

The roughness was modelled by the deflection of the membrane of a vibrating source [12]. The active diameter ($6mm$) was chosen to be approximately one half of the TS-wavelength (at f_{ac}) to allow a significant amount of amplitude at higher spanwise wave numbers. The membrane is vibrating with a very low frequency compared to the acoustic frequency $f_v : f_{ac} = 1 : 64$. This leads to an amplitude modulation of the resulting TS-wave, which is visible in the spectrum as discrete sidepeaks at frequencies $f_{TS1,2} = f_{ac} \mp f_v$. The TS-waves at these combination frequencies result from the scattering of the sound on the surface vibrator only. They are separated from the TS-amplitude at f_{ac} which is contaminated by signals from different sources (acoustic, probe vibration etc.). On the other hand, the ratio between the acoustic frequency and the frequency of the surface vibrations is large enough that the roughness can be assumed quasi stationary and the resulting receptivity function can be compared to DNS for stationary roughness. The shape of the membrane was measured and a double Fourier transformation was applied in streamwise and spanwise direction in order to determine the complex wave number spectrum of the surface vibration at the vibrational frequency:

$$\widetilde{A}_v(\alpha,\beta) = \int_{-\infty}^{\infty}\int_{-\infty}^{\infty} \widetilde{A}_m(s,z)\cdot e^{-i(\alpha s+\beta z)}ds\,dz\,. \tag{3}$$

In conjunction with equation 1, the resulting receptivity function is seen to be independent of the actual shape of the roughness. The ratios between the frequencies were chosen as integer powers of two ($f_v : f_{ac} : f_{samplerate} = 1:64:8$) and the frequencies were strictly phase locked because they were subdivided from a single quartz based clock. Therefore it was possible to use both the amplitude and the phase part of the Fast Fourier Transform of the hot-wire signal for further data processing. The measurement procedure was controlled by a PC and the results were monitored online. The complete setup is sketched in figure 4. Details on the setup, the measurement procedure and data evaluation can be found in [12].

Two main sets of data were collected, consisting of 13 respectively 5 scans in spanwise direction which covered the whole width of the TS-wave train developing downstream of the roughness. The time signals at every measurement point were analyzed with a FFT and the complex amplitude

$$\widetilde{A}_{TS,1/2,raw}(s_i,z) = A_{TS,1/2,raw}(s_i,z)\cdot e^{i\phi_{TS,1/2}(s_i,z)} \tag{4}$$

formed by the amplitude part ($A_{TS,1/2,raw}$) and the phase part ($\phi_{TS,1/2}$) for the combination frequencies was used for further processing. The amplitudes were normalized by the local freestream velocity U_δ. The phases were corrected in streamwise and spanwise direction by adding multiples of 360°. The wave train at the right combination frequency ($f_{TS} = f_{ac} - fv$) developing downstream of the source at $\frac{s}{s_{max}} = 15\%$ is plotted in figures 7 and 8. The complex Fourier transform

$$\widetilde{B}_{TS}(s_i,\beta) = \int_{-\infty}^{\infty} \widetilde{A}_{TS}(s_i,z)\cdot e^{-i\beta z}dz\,. \tag{5}$$

is applied to the spanwise scans in order to find the wave train in the wave number space. After this decomposition, the downstream development of waves with different spanwise wave numbers can be followed separately.

Similar to the procedure used in the DNS the initial amplitude at the position of the roughness is found by comparison to a reference amplification curve. The reference in the experimental case is the amplification as it is predicted by the linear stability theory. This is sketched in figure 6 for the wave number $\beta = \pm 0.269\frac{1}{mm}$ and the roughness located at $\frac{s}{s_{max}} = 15\%$. In the same figure, the phase part for the same wave number is plotted. The phases for all wave numbers can be interpolated by a straight line. The gradient of this line corresponds to the streamwise wave number α_r of the TS-wave. As a result, a dependence of α_r on β can be obtained.

The propagation angle can be found as $\theta = \arctan\frac{\beta}{\alpha_r}$. From the double Fourier decomposition of the shape of the membrane, the vibrational amplitudes that generate eigenmodes of the boundary layer can be found along the function $\alpha_r(\beta)$.

Results

The receptivity function as a function of the propagation angle θ is shown in figure 9 for the three investigated streamwise positions of the roughness element. The big symbols mark the results of the DNS which are in excellent agreement with the experimental results. The velocity fluctuations (B_{inTS}, A_{ac}) are normalized with the freestream velocity U_δ at the position of the roughness, and all linear dimensions (A_v, dz, ds) are normalized with the displacement thickness δ_1 of the boundary layer at the position of the roughness element which is indicated in figure 9. As a result of this normalization, the receptivity function depends on a local value of the boundary layer, the displacement thickness δ_1. In order to be able to study exclusively the dependence of the acoustic receptivity on the external pressure gradient, DNS were performed where all other parameters, particularly the Reynolds number Re_{δ_1}, based on the boundary layer displacement thickness, were kept constant at the location of the surface non-uniformity.

In figure 10 the receptivity function G_{av} for a 2-D roughness element located at $Re_{\delta_1} = 1205$ is plotted versus the wall shear stress, which is non-dimensionalized by the freestream velocity and the displacement thickness at the location of the roughness element. It can be seen that for the cases studied the acoustic receptivity depends only weakly on the external pressure gradient. The wall shear stress at the roughness lies between $0.54 - 0.62$ which leads to a change in the receptivity coefficient of about 15%. Similar to the results for vibrational receptivity, a favorable pressure gradient leads to an increase in the acoustic receptivity whereas an adverse pressure gradient reduces it. This means that the acoustic receptivity behaves in opposite to the instability which is decreased by a favorable pressure gradient and increased by an adverse one. This can be explained as follows: The wall shear stress $\frac{dU}{dy}$ is a function of the Hartree parameter β_H and decreases if the flow is decelerated. The roughness enters the receptivity mainly by the fact that the steady part of the streamwise velocity U at the lowest row of grid points scales with $h \cdot \frac{dU}{dy}$ (see also [1]). Hence, the receptivity in a decelerated flow ($\beta_H < 0$) is lower than in an accelerated flow ($\beta_H > 0$).

For practical application, it is of great interest to find the position on the airfoil where a roughness with constant height results in the highest TS-amplitudes far downstream. If we assume a disturbance with $A_{ac} \cdot A_v = \frac{1}{\delta_1}$ constant for every propagation angle θ, we find the initial TS-amplitudes as $B_{inTS} = G_{av} \cdot A_{ac} A_v$. We consider a control position far downstream of the roughness at $\frac{s}{s_{max}} = 45.2\%$. The N-factor from the roughness position to this control position, N_c, can be found in figure 1, and the amplitudes at the control position $B_{TS,45.2\%} = B_{inTS} \cdot e^{N_c}$ are plotted in figure 11. The relation between these amplitudes reflects the relation of the amplitudes resulting from a roughness of constant dimensional height placed at different streamwise positions close to the instability point of the airfoil. Figure 11 shows that a roughness element placed at branch I of the neutral stability curve ($\frac{s}{s_{max}} = 20\%$) leads to highest TS-amplitudes.

Conclusion

We have shown experimental as well as numerical results concerning the influence of a pressure gradient on the acoustic roughness receptivity in the boundary layer of an airfoil. The receptivity functions evaluated experimentally for three different positions of the surface non-uniformity on the airfoil and numerically for *Falkner-Skan* flows with three different Hartree parameters agreed very well. The values of the receptivity function did not change much when the position of the surface non-uniformity on the airfoil was varied. This is also due to the fact that the receptivity function as it is defined here depends on the length scale that is used for non-dimensionalization. Since the local displacement thickness at the location of the roughness was chosen here, the influence of the pressure gradient on the receptivity could not be completely separated from the influence of the growth of the boundary layer.

Therefore, DNS were performed where Re_{δ_1} at the location of the roughness as well as the dimensionless frequency F were kept constant so that we were able to study only the influence of the pressure gradient. This study showed that a favorable pressure gradient increases the acoustic roughness receptivity whereas an adverse pressure gradient slightly reduces it, which is in accordance with increased and decreased wall shear stress, respectively. This influence is contrary to the influence of the pressure gradient on the instability of the boundary layer. If we take into account that a combination of both effects determines the amplitude of the boundary layer disturbances far downstream, it turns out that the influence of the pressure gradient on the acoustic roughness receptivity is much weaker than its influence on the boundary layer stability. Therefore the most dangerous position for a surface non-uniformity on an airfoil with respect to transition is the location of branch I of the neutral stability curve.

Acknowledgement

This work was performed under grant of the German research council (DFG) and is a contribution to the research program 'Transition'.

References

[1] Choudhari M., Lian N., Streett C.L.: *Acoustic receptivity due to weak surface inhomogeneities in adverse pressure gradient boundary layers*; NASA Technical Memorandum 4577, 1995

[2] Crouch J.D., Spalart P.R.: *A study of non-parallel and nonlinear effects on the localized receptivity of boundary layers*; J. Fluid Mech., pp. 29-37, Vol. 290, 1995

[3] Gaster M., Grant I.: *An experimental investigation of the formation and development of a wave packet in a laminar boundary layer*; Proc. R. Soc. of London A; 347, pp. 253-269, 1975

[4] Ivanov A.V., Kachanov Y.S., Bake S., Neemann K.: *Influence of favorable pressure gradient on 3-D vibrational receptivity of boundary layer*, ICMAR Proc. Part 2, ITAM, Novosibirsk, pp. 78-83, 2000

[5] Kachanov Y.S., Koptsev D.B., Smorodskiy B.V.:*3-D stability and receptivity of two-dimensional self-similar boundary layer with adverse pressure gradient*; Laminar-Turbulent Transition, Proceedings IUTAM Symposium, Eds. Fasel H., Saric W., Berlin Springer-Verlag 1999

[6] Kosorygin V.S., Radetztsky R.H., Saric W.S.: *Laminar boundary layer sound receptivity and control*; Laminar-Turbulent Transition Vol. IV, Ed. Kobayashi R., Springer 1995

[7] Rist U., Fasel H.: *Direct numerical simulation of controlled transition in a flat-plate boundary layer* J. Fluid Mech., pp. 211-248, Vol. 298, 1995

[8] Saric W.S., Radetzky J.A., Hoos R.H. jr: *Boundary layer receptivity of sound with roughness*; Boundary Layer Stability and Transition, FED-Vol. 114, Eds. Reda, Reed, Kobayashi, ASME, 1991

[9] Wörner A., Rist U., Herr S., Würz W., Wagner S., Kachanov Y.S.: *Study of the acoustic receptivity of a Blasius boundary layer in the presence of a surface non-uniformity*; ECCOMAS 2000, Barcelona 11.-14.9.2000

[10] Wortmann F.X., Althaus D.: *Der Laminarwindkanal des Instituts für Aerodynamik und Gasdynamik der Technischen Hochschule Stuttgart*;Zeitschrift für Flugwissenschaften Nr.12 Heft 4, 1964

[11] Würz W.: *Hitzdrahtmessungen zum laminar-turbulenten Strömungsumschlag in anliegenden Grenzschichten und Ablöseblasen sowie Vergleich mit der linearen Stabilitätstheorie und empirischen Umschlagskriterien*; Dissertation, Universität Stuttgart, 1995

[12] Würz W., Herr S., Wagner S., Kachanov Y.S.: *Experimental investigation on 3-D acoustic receptivity of a laminar boundary layer in the presence of surface non-uniformities*; New Results in Numerical and Experimental FLuid Mechanics 2, Eds. Nitsche W., Heinemann H.-J., Hilbig R., Vieweg 1999

[13] Würz W., Herr S., Wörner A., Rist U., Wagner S., Kachanov Y.S.: *Study of 3-D wall roughness acoustic receptivity on an airfoil*; Laminar-Turbulent Transition, Proceedings IUTAM Symposium, Eds. Fasel H., Saric W., Berlin Springer-Verlag 1999

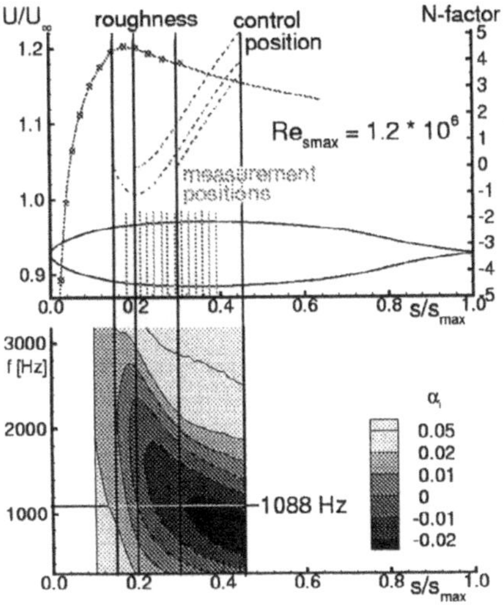

Figure: 1 Velocity distribution and position of the roughness in stability diagram

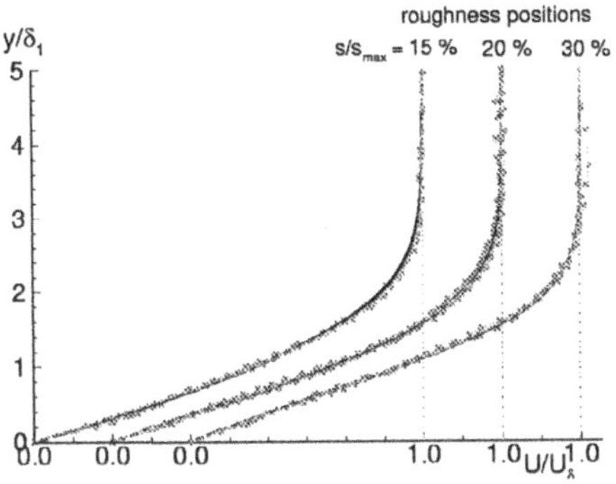

Figure: 2 Boundary layer profiles at the chosen streamwise positions of the roughness

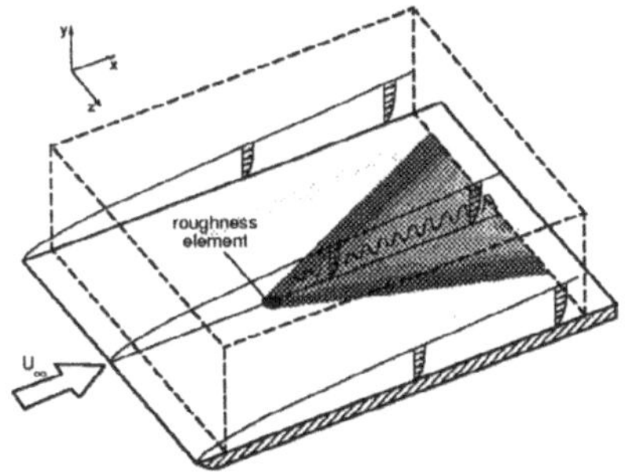

Figure: 3 Integration domain for the DNS

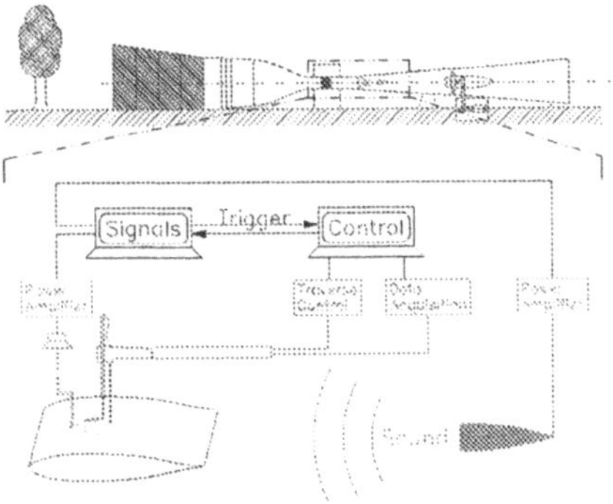

Figure: 4 Experimental setup at the Laminar Wind Tunnel

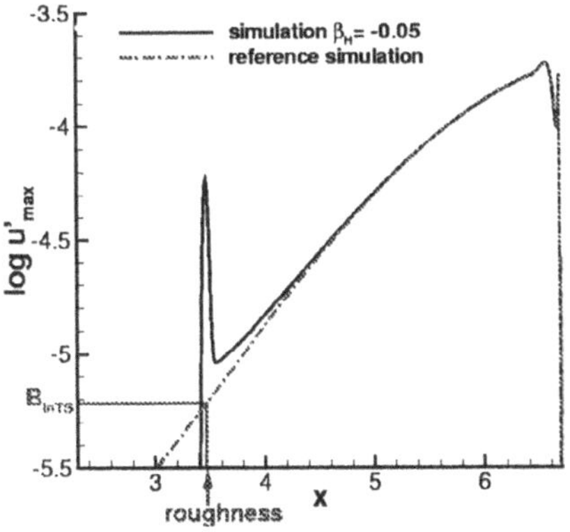

Figure: 5 Amplification as calculated by DNS

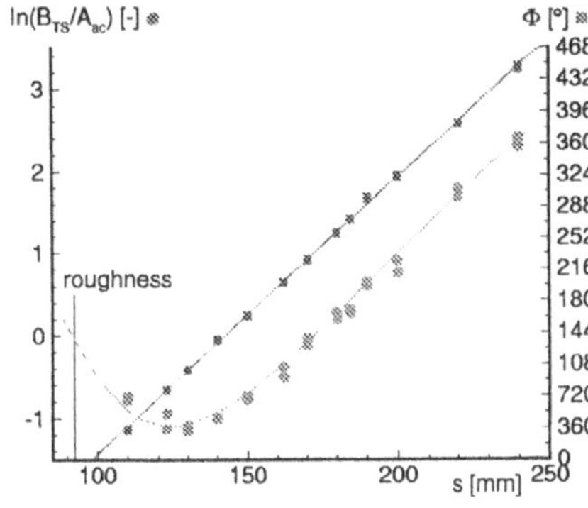

Figure: 6 Downstream development of amplitude and phase for $\beta = 0.269\frac{1}{mm}, \theta = 25^o$

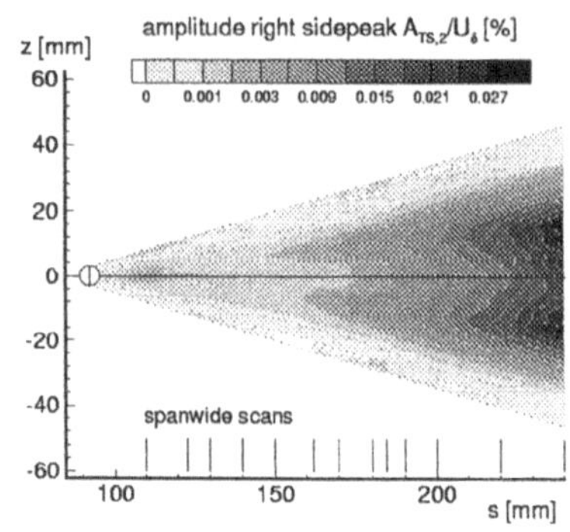

Figure: 7 Amplitude part of the wave train (roughness at $\frac{s}{s_{max}} = 15\%$)

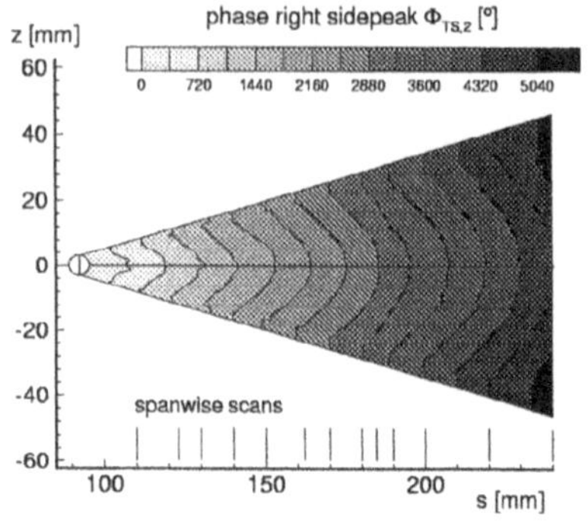

Figure: 8 Phase part of the wave train (roughness at $\frac{s}{s_{max}} = 15\%$)

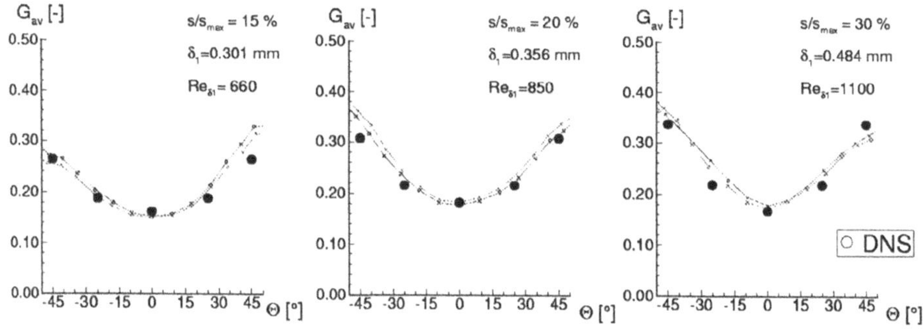

Figure: 9 Receptivity function for roughness at different streamwise positions

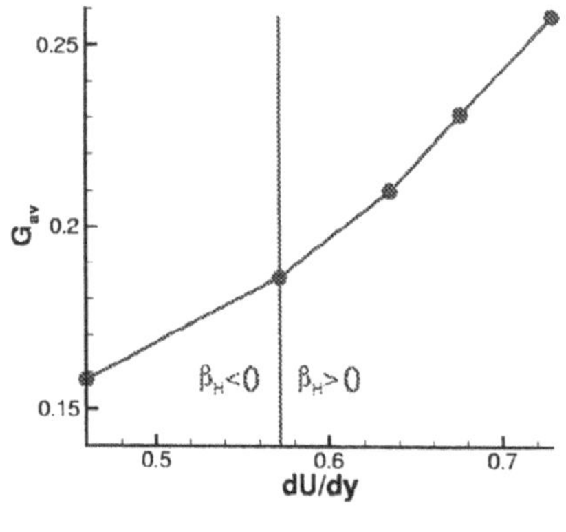

Figure: 10 Receptivity function for different Hartree parameters β_H

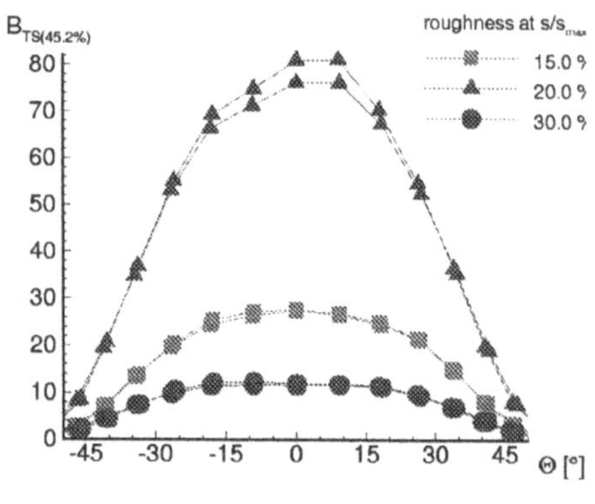

Figure: 11 TS-amplitudes at control position for roughness at different streamwise positions

Interference of Wave Trains with Varying Phase Relations in a Decelerated Two-Dimensional Boundary Layer

Christian Stemmer, Markus Kloker
Institut für Aerodynamik und Gasdynamik, Universität Stuttgart
Pfaffenwaldring 21, 70550 Stuttgart

Summary

The investigation of laminar-turbulent boundary-layer transition for adverse pressure gradient flows is of particular interest for the evaluation of transition scenarios with respect to natural transition on an airfoil and the prediction of the transition location. An examination of transition scenarios initiated by a spanwise array of harmonic point sources shall be given in this article. The differences caused by the harmonic excitation of the adjacent point sources with a phase shift forms the major part of the work presented here. Direct numerical simulations are used to investigate the spatial downstream development of the excited wave trains in a two-dimensional boundary layer adapted to the flow present on the wing glove of an experimental sailplane of the TU Berlin.

1 Introduction

This work is the continuation of a cooperative DFG (German Research Council) special research program on transition (compare [1, 2, 3]) where the downstream development of a single harmonic point source disturbance was the focus of attention. As an extension of the problem, the overlapping of spanwise periodic wave trains with timewise phase shift was investigated. The underlying idea for this study is to check step by step whether the laminar breakdown pattern induced by the considered disturbance generators can be correlated to the structures appearing with natural transition on an airfoil. A somewhat related study for a Blasius boundary layer can be found in [4].

2 Numerical Scheme

The numerical scheme is based on the complete unsteady three-dimensional Navier-Stokes equations in vorticity-velocity formulation for flat plate flow. The discretization is realized with fourth-order finite differences in the downstream (x) and wall-normal (y) directions. The spanwise direction (z) is discretized in Fourier space with a spectral ansatz. The non-linear terms are calculated in physical space using fully de-aliased transformations. For the discretization in time, a classical fourth-order Runge-Kutta scheme is employed.

The present boundary-layer flow is split in a steady two-dimensional base flow as present on the wing glove of a sailplane prepared for the experimental investigation of multiple wave trains with multiple frequencies. The measured pressure distribution from in-flight experiments is used for the velocity boundary condition at the free-stream upper boundary (see fig. 1) to determine the steady base flow with a two-dimensional pseudo-temporal Navier-Stokes code. The resulting flow is the base for the three-dimensional unsteady calculation implemented in a disturbance formulation. For a further detailed description of the numerical method employed, please refer to [5, 6, 7]. All variables are non-dimensionalized using the reference values of $\tilde{L} = 61.0$ mm and $\tilde{U}_\infty = 23.6$ m/s. The grid spacings in the equidistant grid are $\Delta x = 0.00385$, $\Delta y = 0.0005$ and $\Delta z \leq 0.02$ (max. number of Fourier modes is 20).

3 Numerical Results

3.1 Base Flow Properties and Disturbance Generation

The wall pressure values obtained from the TU Berlin flight experiments were translated to free-stream velocities using Bernoulli's equation under Prandtl's boundary layer assumption that the pressure is constant across the boundary layer. The corresponding velocity distribution is shown in figure 2. The base flow exhibits an adverse pressure gradient over the entire integration domain. This is also reflected in the shape factor development $H_{12}(x)$ presented in the same figure. Up to $x = 8.0$, the shape factor remains below $H_{12} = 2.7$ indicating moderate deceleration. Downstream of $x = 8.0$, H_{12} increases up to 3.2 at $x = 10.8$, still well below the critical value for separation.

The disturbances are introduced in the experiments and in the simulations at a downstream position of $x = 4.0$ ($x/c = 20\%$). A spanwise row of point sources was placed in the simulations with a hole spacing of $\Delta z = 24$mm translating to hole positions at $z = n \cdot 1/2\lambda_z$, with $\lambda_z = 0.787$ in the simulations. The phase shift between the adjacent point sources was set to $\varphi = 0$, $\pi/4$, $\pi/2$ and π (see table 1) with the disturbance frequency $\tilde{f} = 600$Hz translating to a dimensionless frequency of $\beta = 2\pi\tilde{f}\tilde{L}/\tilde{U}_\infty = 9.7$ in the calculations. The point-source disturbances at $z = 0$ and $z = \pm 0.787$ (even n) are locked in phase as well as the ones at $z = \pm 0.393$ (odd n) which is due to the symmetric calculation with respect to the spanwise direction.

In the numerical scheme, a disturbance strip is present, imposing modelled blowing and suction through a finite hole as a boundary condition at the wall for the wall-normal velocity. The maximum wall-normal velocity for the point-source disturbance was $v'_{max} = 1.0 \times 10^{-4}$ for all presented cases. Blowing and suction is applied simultaneously to assure that no net mass flow is introduced into or subtracted from the flow at any time instance. The emerging wave trains overlap each other during their downstream spatial evolution. The influence of the phase relation between the adjacent point sources on the laminar-turbulent breakdown is then investigated in detail.

3.2 Linear Stability Properties

Linear-Stability-Theory calculations deliver stability properties for the base flow as shown in figures 3 and 4. The stability diagram for two-dimensional TS-disturbances (Figure 3) shows that the frequency range with the largest integral amplification over the considered downstream extent is around $\beta = 9.7$. This frequency will equally be investigated in the before mentioned flight experiments. For this frequency, the amplification rates for three-dimensional waves with the spanwise wave number γ are shown in figure 4. It is obvious that weakly oblique waves are as amplified as 2D modes up to $x = 8.0$. At the beginning ($x < 5.0$), they are amplified even stronger than two-dimensional waves. In general, the neutral curve is located between $35 < \gamma < 50$ (obliqueness angle θ ranges approximately from 60° to 68°).

3.3 DNS results

Vorticity at the wall

Considering the spanwise disturbance vorticity ω'_z at the wall, the influence of the varying phase relations on the transition process can be studied and compared to upcoming measurements involving surface sensors. In the following figures 5 and 6, instantaneous representations at the beginning of an excitation cycle are shown.

First, the initial disturbance development downstream of the excitation is shown in figure 5 depicting how the overlapping of the wave trains forms. For results for a single wave train in a decelerated boundary-layer flow, please refer to [7]. The comparison with case 0 (see table 1) is given where only half the number of disturbance sources, at $z = n\lambda_z$, are excited. The small circles at $x = 4.05$ in figure 5 show the locations of the point sources. The upper part (case 0) shows the overlapping of the wave trains that are further apart than for all other cases 1-4. At first, the wave trains develop into boomerang shaped "islands" of ω'_z before they merge downstream of $x = 5.5$. The centre parts of the wave trains develop into wave fronts aligned perpendicular to the streamwise direction. On the other hand, the closer, likewise in-phase point sources (lower part of figure 5) interact much earlier and initially develop higher amplitudes of the spanwise disturbance vorticity ω'_z halfway between the wave trains (best to be seen for $4.7 \leq x \leq 6.0$).

For all cases (cases 1 to 4 in figure 6), where closer spacing of the point sources is present, it is notable, that the highest spanwise-vorticity values at the wall arise between adjacent point sources. This continues in the transitional region, where wedge-shaped areas of increased wall vorticity (dark areas) are parted by areas of relative quietness (downstream of $x = 8.4$). For case 1, with all point sources in phase, eight wedges appear for a spanwise distance of $z = 1.57 \mathrel{\widehat{=}} 2 \times \lambda_z$ (two for every overlapping of adjacent wave trains). With increasing phase shifts, the number of wedges decreases to just one per overlapping region. It is also worth mentioning that for case 3 with the phase shift of $\varphi = \pi/2$, the disturbance development downstream of the point sources lagging behind ($z = \pm 0.393$) is very calm whereas the development downstream of the point sources leading in time is stronger than in the other cases. For the present choice of parameters, the phase shift of $\varphi = \pi/2$ is not a mere side-by-side of point sources but a distinguished case. For case 4, where the point sources are actuated out-of-phase ($\varphi = \pi$), the pattern at the wall resembles the one obtained by an oblique breakdown case initiated by a wave pair (1,$\pm$1), where (h,k) denotes a disturbance with the frequency $h\beta$ and the spanwise wave number $k\gamma$. Through wave superposition, the disturbance modes with even multiples of the basic spanwise wave number get cancelled out.

RMS results

For the investigation of the fluctuation intensity, the root-mean-square values of the downstream disturbance velocity at its wall-normal maximum, $u'_{rms}(y)|_{max}$, is considered (figure 7).

At the centre position right behind one of the point sources at $z = 0.0$, the u'_{rms}–values are largest for case 4 with the phase shift of $\varphi = \pi$ (figure 7 left). However, the overall maxima of the u'_{rms}-values can be found at different positions for the 4 cases (figure 7 right). Especially, case 3 shows the first steep ascent of the streamwise velocity fluctuations. We note, that for the scaling with the local edge velocity, the rms-values have to be divided by 1.6. It is pointed out that the development of the overlapping wave trains obviously is not a mere sum of a multitude of single wave trains. If that were the case, the maximum amplitudes would have to be expected approximately downstream of the point sources (i.e. $z = 0.0$) for the case 1 where all sources are operated in phase. Quite contrary, the largest disturbance amplitudes are found *between* the wave trains for a phase shift of $\varphi = \pi/2$.

Time averaged flow

The steady disturbance share $\sum_k u'(0,k)$ shows where the flow is accelerated and where it is decelerated in the time average through the point source disturbances in nonlinear stages. Figure 8 demonstrates the steady disturbance share at three distinct spanwise positions ($z = 0$ upper left, $z = 0.25\lambda_z$ lower left and $z = 0.5\lambda_z$ lower right) and at the spanwise position where the largest amplitudes could be observed (upper right). For $z = 0.0$ (upper left), the flow is decelerated for all cases. The maximal deceleration is at about $1/3\delta$ (where δ is the boundary layer thickness). Only very close to the wall, acceleration can be observed for case 2 and 4.

At the spanwise position with the largest observable amplitudes for the steady disturbance share (Figure 8 upper right), a clear acceleration close to the wall is present. Above the accelerated region, a deceleration takes place ($4.0 < y\sqrt{Re} < 8.0$) and hence a strong time constant shear layer ($\partial u'/\partial y$) is generated near $y\sqrt{Re} = 3$, which is the largest for cases 3 and 4. Thus, the largest u'_{rms}-amplitudes occur where time mean of ω'_z is strongest inside the boundary layer. Examplary for all cases, a spanwise–wall-normal distribution of the three steady disturbance velocities is presented in figure 9. Four vortices are detectable with the strongest one holding the core just above the location of the maximal u'-amplitude at $y = 1.3 \times 10^{-2}$ and $z = -0.11$.

4 Conclusions

The downstream interaction of disturbances emanating from spanwise aligned harmonic point sources operated with varying phase shifts in a 2-D APG boundary layer has been investigated by direct numerical simulations.

It could be shown that the shape of the irregular "transition line" is a function of the phase shift. A phase shift of $\varphi = \pi/2$ leads to largest local disturbance amplitudes u_{rms} with the most pronounced sawtooth pattern of the wall shear.

Further investigations have to show how these structured at the wall alter with varying disturbance frequencies and how they compare to the picture of natural transition on the airfoil.

Acknowledgements

The financial support of this research work by the Deutsche Forschungsgemeinschaft (DFG) under contract KL 890/4-1 is gratefully acknowledged.

References

[1] Suttan, J., Baumann, M., Fühling, S., Erb, P., Becker, S., Lienhart, H., Müller, W. and Stemmer, C., In-flight research on laminar-turbulent transition – a joint universitary research project. In: 2nd *European Forum on Laminar Flow Technology, Bordeaux, France*, pp. 10.18–10.41 (1996).

[2] Suttan, J., Baumann, M., Fühling, S., Erb, P., Becker, S. and Stemmer, C., In-Flight Research on Boundary Layer Transition – Works of the DFG-University Research Group. In: H. Körner and R. Hilbig (eds.), *Notes on Numerical Fluid Mechanics*, vol. 60, pp. 343–350. Vieweg Verlag, Wiesbaden (1997), 10th Stab Symposium 96, Braunschweig.

[3] Stemmer, C., Suttan, J., Kloker, M. and Nitsche, W., Point-source induced transition in free flight. In: W. Nitsche, H.-J. Heinemann and R. Hilbig (eds.), *Notes on Numerical Fluid Mechanics*, vol. 72, pp. 458–465. Vieweg Verlag, Wiesbaden (1999), 11th Stab Symposium 98, Berlin.

[4] Seifert, A. and Wygnanski, I., On the interaction of wave trains emanating from two point sources in a Blasius boundary layer. In: *Boundary Layer Transition and Control*, pp. 7.1–7.13 (1991), Conference organised by The Royal Aeronautical Society, Peterhouse College, Cambridge, UK.

[5] Rist, U. and Fasel, H., Direct numerical simulation of controlled transition in a flat-plate boundary layer. *J. Fluid Mech.*, 298, 211–248 (1995).

[6] Kloker, M., Konzelmann, U. and Fasel, H., Outflow boundary conditions for Spatial Navier-Stokes Simulations of Transition Boundary Layers. *AIAA J.*, 31, 620–628 (1993).

[7] Stemmer, C., Kloker, M. and Wagner, S., Navier-Stokes Simulation of Harmonic Point Disturbances in an Airfoil Boundary Layer. *AIAA J.*, **38(8)**, 1369–1376 (2000).

Table 1 phase relation between the point source disturbances for the four cases simulated at the spanwise positions indicated in the first line

case	$z = 0$	$z = \pm 0.5\lambda_z$	$z = \pm\lambda_z$
0	$\varphi = 0$	—	$\varphi = 0$
1	$\varphi = 0$	$\varphi = 0$	$\varphi = 0$
2	$\varphi = 0$	$\varphi = \pi/4$	$\varphi = 0$
3	$\varphi = 0$	$\varphi = \pi/2$	$\varphi = 0$
4	$\varphi = 0$	$\varphi = \pi$	$\varphi = 0$

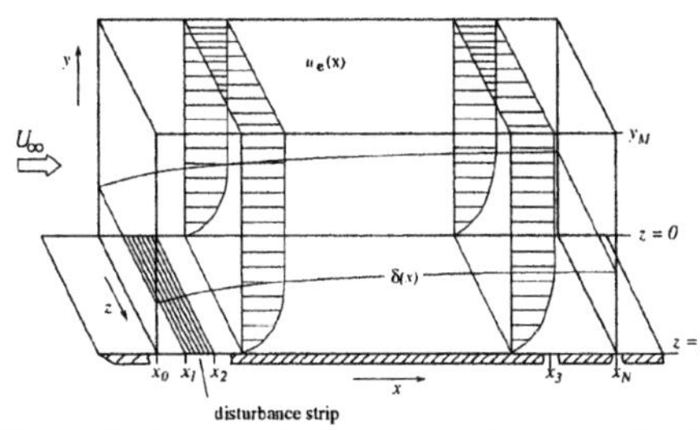

Figure 1 Rectangular integration domain with disturbance strip

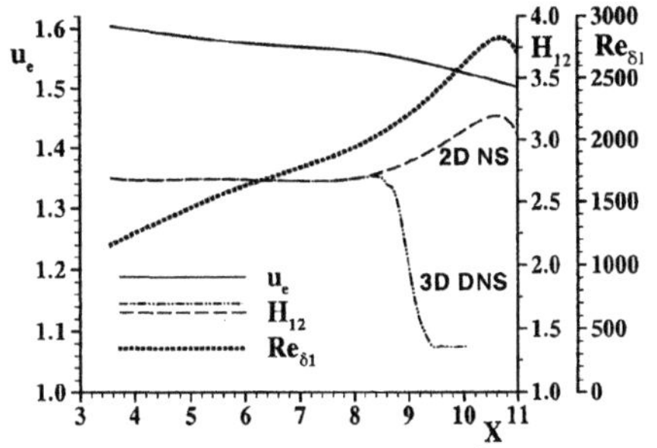

Figure 2 Free-stream velocity $u_e(x)$, shape factor H_{12} and Reynolds number based on the displacement thickness Re_{δ_1} for the entire streamwise extent of the integration domain

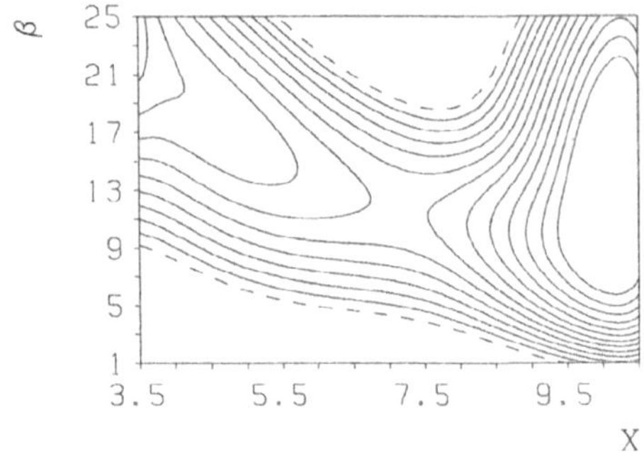

Figure 3 Stability diagram (spatial LST) for two-dimensional waves for the sailplane boundary-layer flow. Isolines of $\alpha_i = \frac{d}{dx}\ln(A/A_0)$, $\Delta\alpha_i = 0.25$; dashed curve is the neutral curve $\alpha_i = 0$.

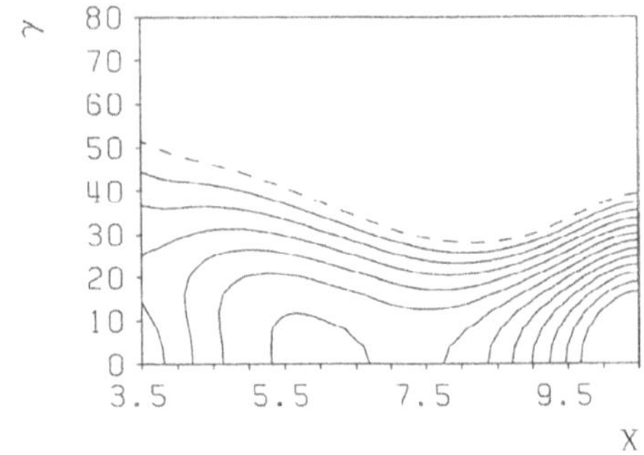

Figure 4 Stability diagram for three-dimensional waves for the sailplane boundary-layer flow for the frequency $\beta = 9.74$; γ – spanwise wave number. $\gamma = 20$ roughly corresponds to a wave angle of 45°.

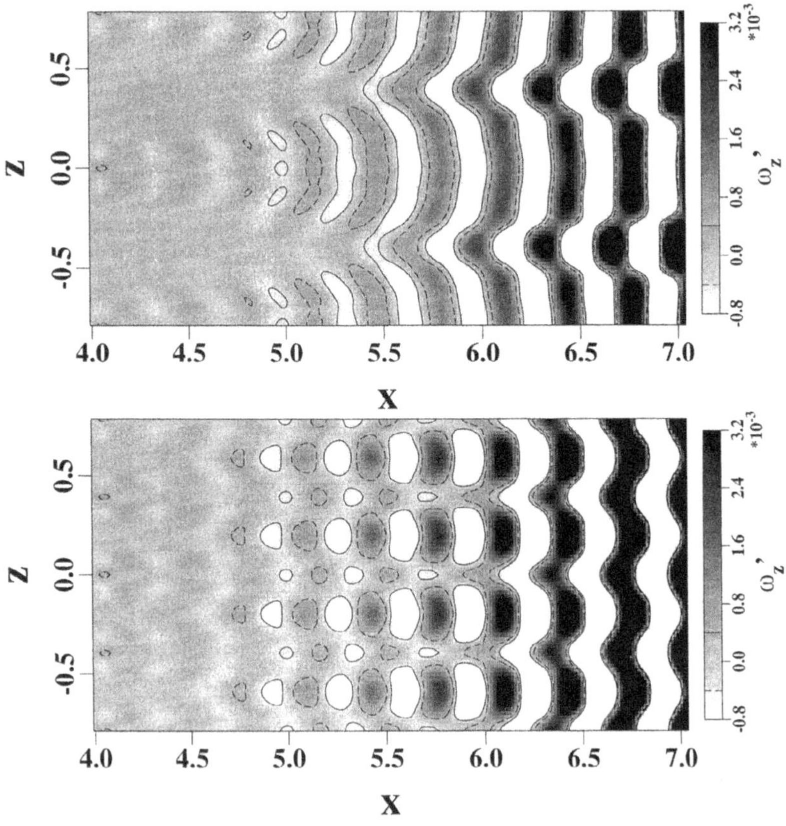

Figure 5 Instantaneous spanwise vorticity ω'_z at the wall for case 0 and case 1. Upper graph represents case 0 with half the number of point sources excited whereas the lower graph represents case 1. The disturbance of adjacent point sources is realized in-phase; contours at $\omega'_z = \pm 0.0004$, negative value contours are dashed.

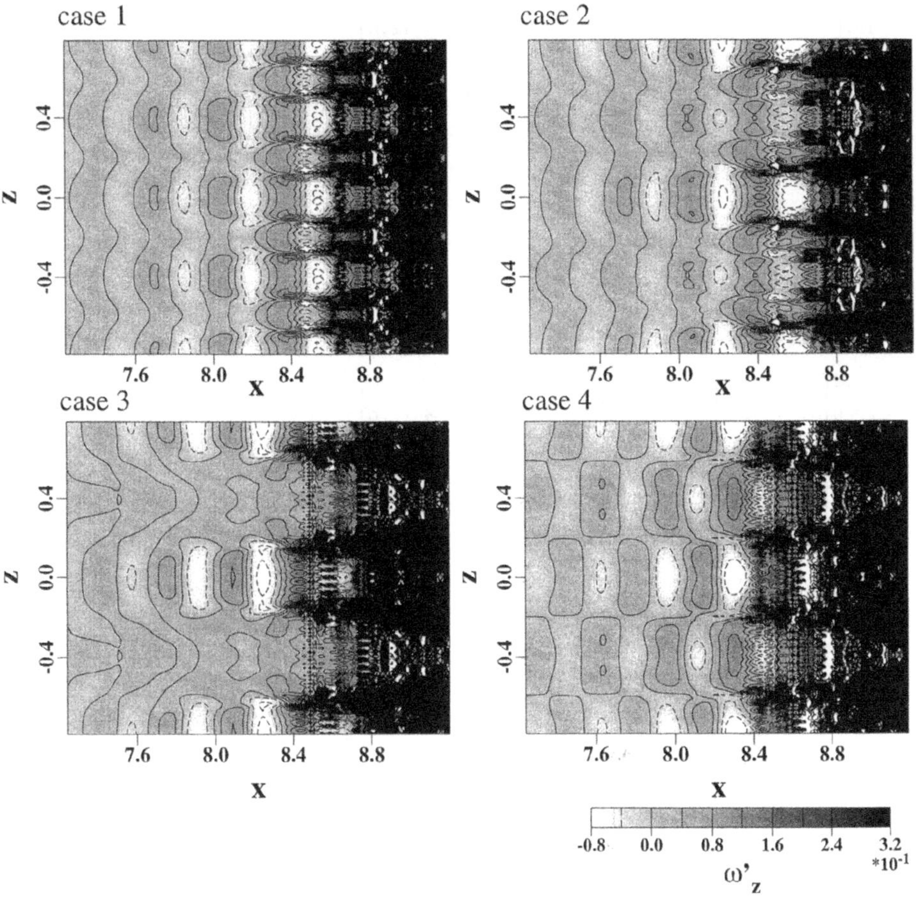

Figure 6 Instantaneous spanwise vorticity ω_z' at the wall for cases 1 to 4 (see table 1). top left: case 1; top right: case 2; lower left: case 3; lower right: case 4

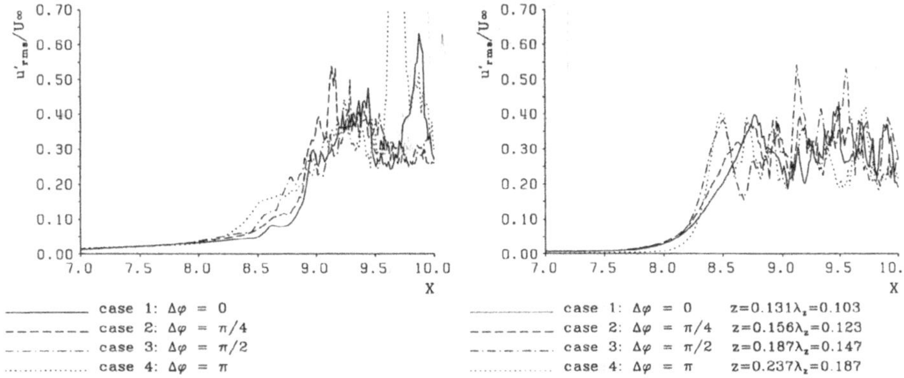

Figure 7 Wall-normal maximum of the rms-value of the downstream velocity $u'_{\rm rms}(y)|_{max}$ at various spanwise positions: left: $z/\lambda_z = 0$; right: $z/\lambda_z \widehat{=}$ streak location (as indicated in the respective legend).

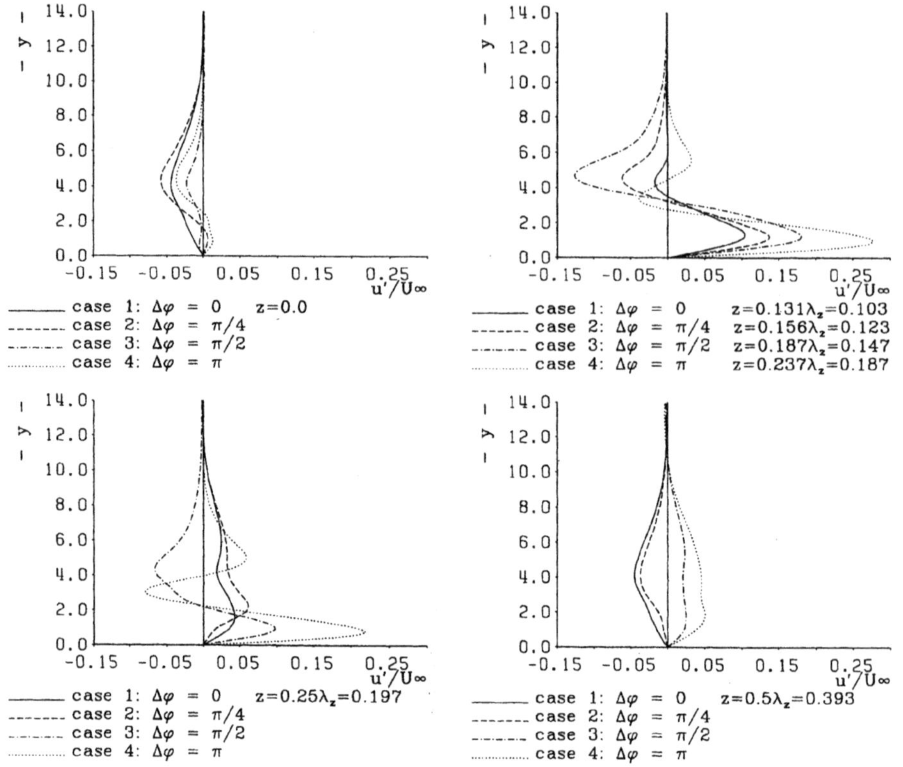

Figure 8 Amplitude of the steady share of the downstream disturbance velocity, $\overline{u}'$, at x=8.5 at different spanwise positions: upper left: $z/\lambda_z = 0$; upper right: $z/\lambda_z \widehat{=}$ streak location (as indicated in the respective legend); lower left: $z/\lambda_z = 0.25$; lower right: $z/\lambda_z = 0.5$. The boundary layer thickness ranges from $y = 14.5$ to $y = 15.0$ for the considered cases (y is stretched by $\sqrt{Re_L} = 309.8$).

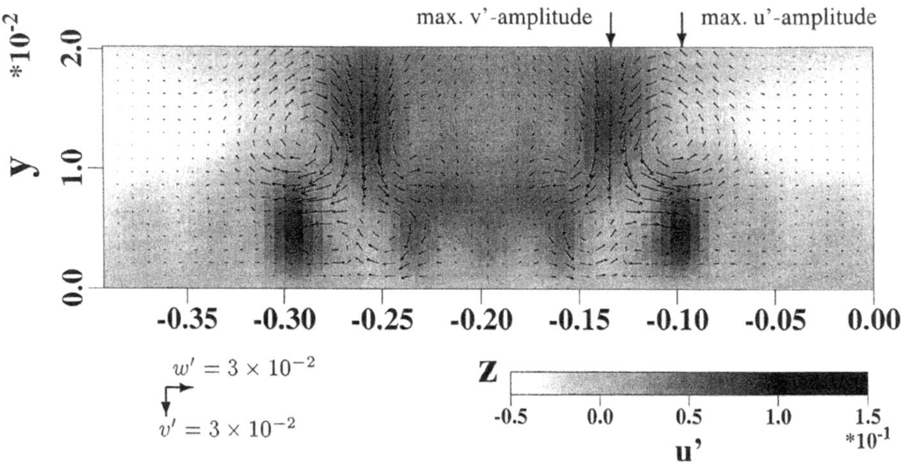

Figure 9 Steady disturbance velocities: u' (colorcoded), v' and w' (as vectors) at $x = 8.5$ for case 1.

Secondary Instability of Steady and Unsteady Crossflow Waves

P. Wassermann, M. Kloker, S. Wagner

Institut für Aerodynamik und Gasdynamik
Universität Stuttgart
Pfaffenwaldring 21, D-70550 Stuttgart, Germany

Summary

The crossflow induced laminar-turbulent transition of a 3-D flat-plate boundary-layer flow is investigated by means of spatial direct numerical simulations based on the complete incompressible 3-D Navier-Stokes equations and a combined 6th-order compact Finite-Difference / spanwise Fourier-spectral scheme. The considered spanwise-invariant laminar base flow is a model of the boundary-layer flow in the front region of an infinite swept wing, with favorable and ensuing adverse chordwise pressure gradient. The secondary instability of *(i)* a saturated steady crossflow-vortex packet and *(ii)* a saturated unsteady crossflow-wave packet is investigated in detail. For *(i)*, transition delay by means of a 3-D upstream flow deformation is presented.

Introduction

On a swept-back airfoil the chordwise acceleration of the potential flow induces an inboard-oriented crossflow component inside the boundary layer perpendicular to the mean flow direction. The crossflow velocity profile $W_s(y)$, y being the wall-normal coordinate, is inflectional and causes a strong primary instability of the flow with respect to so-called crossflow (CF) modes. At low freestream turbulence conditions, steady crossflow waves, also denoted as vortices (CFVs) excited by surface nonuniformity are found to be dominant (see [1], [6]). Upon downstream vortex saturation, local inflectional mean velocity profiles $U_s(z)$, z being the spanwise coordinate, are formed and subsequent "explosive" secondary instability mechanisms typically lead to transition. The question arises whether this phenomenon is caused by a strong convective or a global instability with respect to unsteady modes. "Global" means that a disturbance is amplified in time while remaining within a fixed finite location in space, causing inevitably a nonlinear disturbance state (example: bluff-body wake). On the other hand, a moderate turbulence environment, resulting in dominant crossflow waves, can also be present. In this case the transition mechanisms are unclear up to date and need further investigation.

Numerical Results

The results are gained with the carefully tested 6th-order compact Finite Difference / spanwise Fourier-spectral scheme for spatial DNS in vorticity-velocity formulation of unsteady 3-D boundary-layer flows in a rectangular integration domain. All variables are nondimensionalized by the reference length $\bar{L} = 0.05m$ and the freestream velocity $\bar{U}_\infty = 30m/s$; the Reynolds number is $Re = \bar{U}_\infty \bar{L}/\bar{\nu} = 100000$. The wall-normal coordinate y and velocity v are stretched by $\sqrt{Re}$. First, the spanwise invariant steady base flow is calculated; next, defined disturbances are introduced in a disturbance strip at the wall and the three-dimensional, asymmetric, unsteady disturbance flow is calculated (for further details see [2], [4], [8] and [9]). For the discussed simulations the spatial resolution $\Delta x = 0.0018$, $\Delta y = 0.088$ and $\Delta z = 2\pi/2K\gamma = 0.0043$ ($K = 16$) has been used.

Base flow

The chordwise edge-velocity distribution of the base flow resembles the potential flow in the front region of a swept wing with favorable and ensuing adverse chordwise pressure gradient (see Fig. 1, cf. [7], [8]). It is defined analytically to facilitate the calculation of the boundary condition $u_e(x)$ at the upper boundary. The sweep angle is $\varphi_\infty = 45^o$ and the crossflow component attains a maximum value of 13.3%.

1 Steady Primary Disturbance

1.1 Nonlinear saturation of a CFV packet

For the calculation of a nonlinearly saturated CFV scenario (case *VPB*) a packet of steady modes $(0,k), k = 1\text{-}4$ is introduced at $x_2 = 0.68$ with an amplitude of $A_{v\ (0,k)} = 5.0 \cdot 10^{-3} \cdot \sqrt{Re}$ and a fundamental spanwise wave number of $\gamma_0 = 45$ ($\gamma_0 \delta_1(x_2) \approx 0.1$, (h,k) denotes a Fourier mode with frequency $h\beta_0$ and spanwise wavenumber $k\gamma_0$). The spanwise phase of each mode is chosen in order to model local "blowing" with momentum, but no net-mass-flow input. The Fourier analysis results in Fig. 2 (left) show that the disturbance components are strongly amplified first by primary instability and saturate with different amplitudes, with the mode (0,2) clearly dominating. Accordingly two streamwise vortices per spanwise wavelength establish (clockwise-rotating when looking downstream); due to the superposition of the other CFV modes, especially the odd k modes, one of the vortices is stronger and further away from the wall (see Fig. 3, $\zeta \approx 0.11$). Thus, low momentum fluid from the near-wall region is brought up into faster regions resulting in a strong steady deformation of the meanflow profiles $u_s(y)$, $u_s(z)$ with relatively large y- and z-gradients.

1.2 Secondary Instability initiated by the CFV packet

So far, solely a steady disturbance input has been considered and the flow field, calculated by a high-order time-stepping scheme, finally settled to a steady nonlinear disturbance state. Naturally, transition to turbulence can only be caused by unsteady disturbances. In low turbulence free-flight conditions a sudden breakdown of the dominating CFVs is observed and the question arises whether this phenomenon is caused by a strong convective or a global instability with respect to unsteady modes. For clarification the saturated CFV packet has been combined with a low-amplitude background pulse disturbance to simulate the natural background (case *VPB+B*). The periodic background pulse, consisting of harmonic waves with frequencies from $\beta = 10$ up to $\beta = 300$ ($\Delta\beta = 10$, $\beta = 2\pi f = 2\pi \bar{f} \bar{L}/\bar{U}_\infty, \bar{f}$ dimensional frequency in Hz) for $k = 2$ and an excitation amplitude of $A_{v\ (h,2)} = 5.0 \cdot 10^{-5} \cdot \sqrt{Re}$ for each component, has been excited at $x_3 = 2.0$, shortly upstream of the CFV packet's saturation point. As for primary instability, the most amplified CF mode has $\beta \approx 14$. After a short transient region the background disturbances undergo strong amplification, especially the component $\beta = 160$, which attains the largest amplitude in this region (Fig. 4). This mode grows over the smaller frequency modes, although its receptivity is smaller. The maximum spatial amplification rate is three times the maximum primary amplification of the most unstable CF mode. The visible growth of the low frequency as well as the higher frequency ($\beta \geq 240$) modes is clearly nonlinearly generated. Isocontours of the $\tilde{u}_s$-amplitude for $\beta = 20$ and $\beta = 160$ in a crosscut in the vortex-oriented coordinate system reveal the origin of the secondary instability (Fig. 5). For the high-frequency (HF) instability the location of the largest amplitude perfectly coincides with the minimum of the spanwise gradient of the time averaged meanflow velocity component (Fig. 3, bottom, $\zeta \approx 0.09$) rather than with the largest wall-normal gradient at $\zeta \approx 0.11$. Hence, this secondary instability mode, which has been found in secondary instability analysis as well (see [5]), is called a 'z'-mode. A different picture arises for the low-frequency (LF) secondary instability. Exactly at the ξ-position shown here ($\xi = 3.2, \zeta = 0.09 \rightarrow x = 2.43$) a kink can be observed in the amplitude-curve in Fig. 4 indicating a change of the dominating local amplitude maxima. In Fig. 5 two distinct maxima appear at $\zeta = 0.05$ and $\zeta = 0.09$, the first coupled to a local maximum of the spanwise gradient, and the second one, nonlinearly generated from the HF instability modes, which becomes dominant downstream. Thus, also in the disturbance shape the feedback-effect from higher to lower frequencies can be identified. Several supplementing simulations with single primary vortex modes of different spanwise wave numbers complement the findings about this transition mechanism and can be summarized as follows. The localized Secondary Instability (SI) is fixed to the primary vortical structure and appears in the low-momentum upwelling region, where especially the spanwise gradients in the meanflow shows extrema. However the *criterion for the onset* of the SI is the local mean flow deceleration. Whenever it exceeds a threshold value of about $0.4 \cdot \tilde{u}_{s,e}$, SI sets in. Two distinct types, both being of convective nature (cf. [9]), can be observed: the stronger amplified high-frequency SI situated at the position of maximum negative spanwise gradient, and the low-frequency SI, which is connected to the maximum positive spanwise gradient. Due to the full disturbance spectrum (in span and time) manifold nonlinear interactions take place and a feedback-effect from HF-modes to LF-modes can be observed. The characteristics of the SI strongly depends on the development and nonlinear saturation of the primary CFV modes with respect to the most amplified frequencies, amplification rates, initiation point and the dominance of one type of the SI. Summarizing it can be pointed out that a "harmonic" spanwise distribution of CFV induced vortical strength is less prone to trigger SI than a more modulated distribution with the same average value. Thus, a packet of CFV-modes is always more "dangerous" than a single CFV mode with respect to SI.

1.3 Transition Delay by Upstream Flow Deformation (UFD)

There exist several strategies to possibly delay transition in order to reduce the skin friction of aerodynamic vehicles. One of the most elegant is trying to deform the mean flow by "easy means" in such a way that the disturbances triggering transition would not grow. Motivated by the above conclusive findings the strategy of UFD is considered here. The idea is based on the introduction of an only weakly unstable nonlinear CF vortex with comparably small spanwise wavelength. The investigations above have shown that small wavelength vortices are less unstable with respect to secondary instability, which is the key element in the breakdown of CFV-dominated scenarios. The CFV modes with wavenumbers greater than the most amplified mode generate typically only primary stable, higher spanwise harmonics and are damped further downstream. Furthermore, there is a strong influence of CFV modes on each other in nonlinear stages, i.e. the mode attaining about 10% amplitude first, dominates the scenario by suppressing the other modes. Considering all these findings the idea arises to enforce a large amplitude CFV mode (0,3) ($\gamma = 135$, 1.5 times the wavenumber of the most amplified CFV mode) upon the flow to suppress the most amplified and most destabilizing mode (0,2) (in the sense of secondary instability) and therefore to delay the secondary instability mechanisms. In practice this method is relatively easy to apply by attaching a regular roughness row near the leading edge of the wing, or, to be more flexible, to use a flexible, spanwise strip controllable from inside the wing. Basic experimental evidence for the working of such a method has been reported by Saric et al. [6] without conclusive explanation of the working principle.

In the simulations presented here this strategy is firstly applied to the CFV packet case *VPB* to investigate the effect on the CFV packet components. Secondly, the UFD is applied to case *VPB+B* to quantify the change in the secondary stability properties and thus the delay of transition.

In case *UFD-5* (Fig. 2, right) an excitation amplitude A_v for the UFD mode (0,3) of 5% at $x_1 = 0.57$ has been chosen. The UFD mode is amplified first, generating also a large 2-D mean flow distortion, and attains a maximum physical amplitude of 29% at $x = 1.8$; thereafter it decays somewhat. Compared to case *VPB*, the development of the other steady modes is clearly suppressed and the UFD mode remains the largest mode throughout the whole integration domain.

Next, this steady scenario has been combined with the low amplitude, periodic background puls disturbance, excited at $x_3 = 2.0$. The downstream amplitude development is shown in Fig. 6. All unsteady modes are neutrally stable or even damped, and transition is shifted downstream far beyond $x = 3.5$ (transition at $x = 2.7$ before). Complementing simulations with varying excitation positions x_3 for the background disturbances substantiate these results and exclude that the secondary instability sets in further upstream.

An explanation for the excellent working of the UFD can be found looking at the mean flow deceleration shown in Fig. 7. The maximal deceleration, the criterion for the SI onset, significantly decreases with the imposition of the UFD. This effect is much stronger than could be expected from the different saturation levels of the dominating spectral modes, because in physical space all modes, also the ones being nonlinearly generated, superimpose and possibly intensify the distortion of the mean flow. The deceleration in the UFD case attains its maximum value far upstream and does not exceed the crucial value of about $0.4 \cdot \tilde{u}_{s,B,e}$.

2 Unsteady Primary Disturbances

2.1 Nonlinear saturation of a low-frequency CFW packet

Experimental investigations in wind tunnel facilities with different turbulence levels have shown that under medium- to high-turbulence flow conditions unsteady CF waves may be dominant (see [1]). Hence, for the following simulations the disturbance velocity distribution of case *VPB* (Fig. 2, left, Insert) has been combined with timewise oscillation at $\beta_0 = 5$ ($\approx 1/3 \cdot \beta(\alpha_{i,max})$), resulting in the excitation of the spectral modes $(1, \pm k)$, $k = 1 - 4$, with an amplitude of $A_{v\ (1,k)} = 5.0 \cdot 10^{-3} \cdot \sqrt{Re}$ each. The downstream amplitude development in Fig. 8 shows that the growth of the outward travelling waves (wrt. the curvature of the potential streamlines, $k > 0$) is greater than of the inward travelling waves ($k < 0$). The modes (1,2) and (1,3) are dominating, (1,2) attaining a maximal amplitude of 20% at $x = 2.4$, whereas the inward travelling waves do not surpass an amplitude of 3%. The disturbance state finally settles to an unsteady equilibrium solution and laminar breakdown is not at hand.

2.2 Secondary Instability initiated by the CFW-Packet

Following the results of section 1.2 the question arises, whether the mechanism found for steady primary disturbances are the same for slowly moving primary disturbances as considered here. Therefore the saturated scenario has been combined with the same low-amplitude, periodic background pulse disturbance as before to investigate the secondary instability properties of the unsteady equilibrium solution. The results in Fig. 9 are quite similar to case *VPB+B*. After a short transient region most of the background disturbances undergo a strong growth, and the component with $\beta = 160$ is strongest amplified again. This secondary instability triggers transition, which sets in at $x \approx 2.3$.

Conclusions

Crossflow-instability induced laminar-turbulent transition of a swept-wing-like boundary-layer flow with decreasingly favorable chordwise pressure gradient has been investigated by spatial DNS.
First, a packet of four steady crossflow vortex modes has been considered as a primary disturbance, modelling the effect of localized, spanwise periodic disturbances on the wing surface. The packet undergoes strong growth downstream up to amplitude saturation, and the induced strong three-dimensional deformation of the meanflow gives rise to a convective secondary instability with respect to high-frequency unsteady modes. The triggering secondary disturbance mode is a 'z'-type mode, i.e. it is spatially connected to the location of the spanwise gradient of the deformed mean flow. The criterion for the onset of the secondary instability however is the maximal local mean flow deceleration, which has to exceed 40% of the boundary-layer edge velocity.
Moreover it could be shown for the first time in a simulation that a significant transition delay is possible in a 3-D boundary layer by taking advantage of the underlying nonlinear mechanisms. A 3-D nonlinear upstream flow deformation with a steady crossflow vortex mode of subcritical spanwise wavelength is superimposed upon the flow suppressing the "natural" transition mechanisms.
Second, a packet of harmonic low-frequency CF waves has been considered as primary disturbance to model medium-turbulence flow conditions. Without additional multi-frequency background pulse, only a saturated equilibrium solution establishes with no breakdown within the considered integration domain. With background pulse the observed secondary instability mechanisms triggering transition are the same as observed for steady primary disturbances. Thus the secondary instability in 3-d boundary layers is connected to (steady or unsteady) conditions in physical space, rather than to priming modes in Fourier space as known from 2-d boundary layers.

Acknowledgements

The financial support by the German Research Council (DFG) under contract number KL 890/2-3 is gratefully acknowledged.

References

[1] Bippes, H.: Environmental Conditions and Transition Prediction in 3-D Boundary Layers. AIAA-97-1906.

[2] Bonfigli, G.; Kloker, M.: Spatial Navier-Stokes simulation of crossflow-induced transition in a 3-d boundary layer. In Nitsche, W.G.; Heinemann, H.-J.; Hilbig, R., (eds.): *New Results in Numerical and Experimental Fluid Dynamics II.* Proc. 11. AG STAB/DGLR Symposium (1998), NNFM 72, Vieweg Verlag, Braunschweig, 1999.

[3] Jeong, J.; Hussain, F.: On the Identification of a Vortex. *J. Fluid Mech.*, Vol. 285, pp. 69-94, Cambridge University Press (UK), 1995.

[4] Kloker, M.: A robust high-resolution split-type compact FD scheme for spatial direct numerical simulation of boundary-layer transition. *Applied Scientific Research*, 59 (4), pp. 353-377. Kluwer Acad. Publishers, NL, 1998.

[5] Koch, W.; Bertolotti, F.P.; Stolte, A.; Hein, S.: Nonlinear equilibrium solutions in a three-dimensional boundary layer and their secondary instability. *J. Fluid Mech.*, Vol. 406, pp. 131-174, Cambridge University Press (UK), 2000.

[6] Saric, W.S.; Carrillo, Jr.; Reibert, M.S.: Leading-Edge Roughness as a Transition Control Mechanism. AIAA-98-0781.

[7] Spalart, P.R.; Crouch, J.D.; Ng, L.L.: Numerical study of realistic perturbations in 3-d boundary layers. In *Proc. AGARD Conf.: Application of Direct and Large Eddy Simulation to Transition and Turbulence*, AGARD-CP-551, pp. 30.1 – 30.10., Chania, Crete, Greece, 1994.

[8] Wassermann, P.; Kloker, M.: Direct Numerical Simulation of the Development and Control of Boundary-Layer Crossflow Vorticies. In Nitsche, W.G.; Heinemann, H.-J.; Hilbig, R., (eds.): *New Results in Numerical and Experimental Fluid Dynamics II.* Proc. 11. AG STAB/DGLR Symposium (1998), NNFM 72, Vieweg Verlag, Braunschweig, 1999.

[9] Wassermann, P.; Kloker, M.: DNS of Laminar-Turbulent Transition in a 3-D aerodynamic Boundary-Layer Flow. In Krause, E.; Jäger, W., (eds.): *High Performance Computing in Science and Engineering '00.* Springer-Verlag, Berlin, 2000.

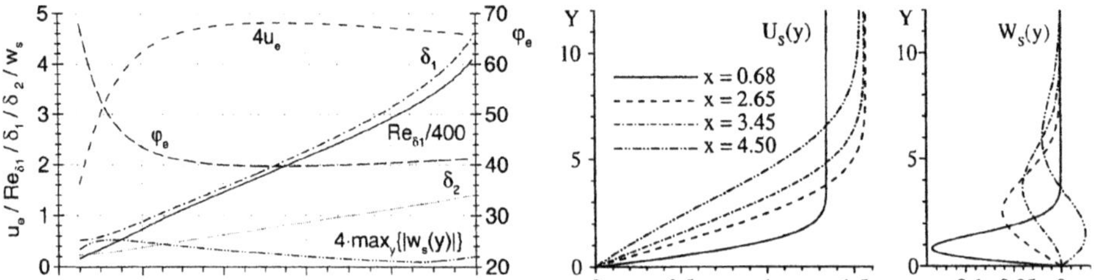

Fig. 1: Left: Base flow parameters in the chordwise coordinate system (δ_1 denotes the displacement thickness, δ_2 the momentum thickness). Right: Velocity profiles in the streamwise coordinate system at various downstream positions.

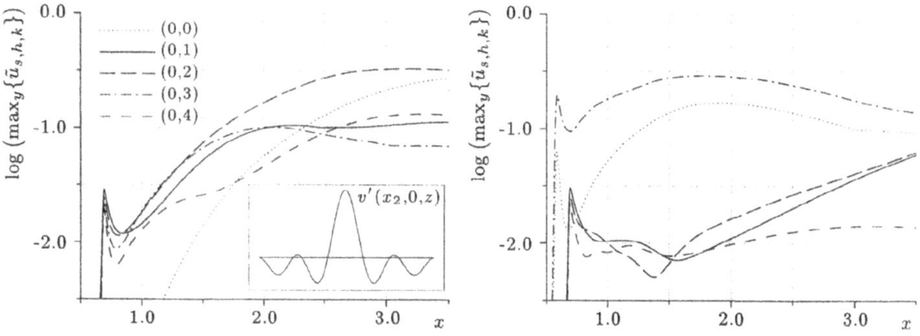

Fig. 2: Downstream (t-z)-modal amplitude development ($\tilde{u}_{s,h,k}$-maximum over y, $\tilde{u}_s = u_s/u_{s,B,e}$) of the CFV packet. Left: Vortex packet of case *VPB*. Insert: spanwise distribution of the forcing v'-velocity in the x-center ($x = x_2$) of the disturbance strip. Right: Vortex packet of case *VPB* plus upstream flow deformation (vortex mode (0,3), excited at $x = 0.57$).

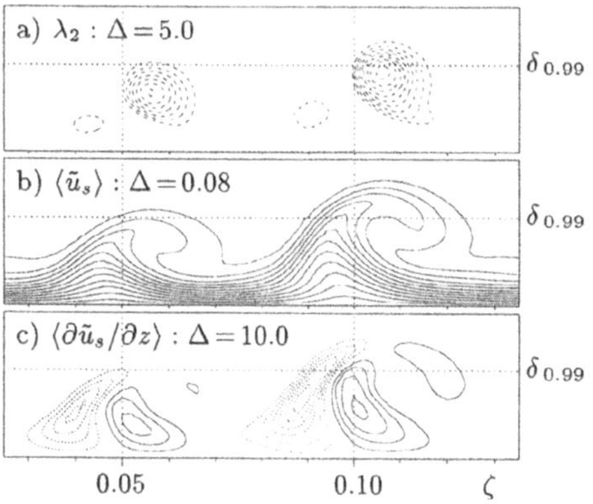

Fig. 3: Flow field distorted by the nonlinearly saturated CFV packet in a crosscut in the vortex-oriented system at $\xi = 3.2$ ($\zeta = 0.09 \rightarrow x = 2.43$). From top to bottom: a) λ_2 isolines (indicating local pressure minima, see [3]); b) deformed mean flow; c) spanwise gradient of the deformed meanflow (negative isolines are dashed, approx. one fundamental, spanwise wave length is shown). To scale.

Fig. 4: Downstream t-modal amplitude development ($\tilde{u}_{s,h}$-maximum over y and z) for Case *VPB+B*. Frequencies from $\beta = 0$ to $\beta = 320$ ($\Delta\beta = 20$) are shown. The curves left of the vertical lines are gained from second time-derivative analysis. Insert: time-dependence of the forcing v'-velocity in the center of the disturbance strip.

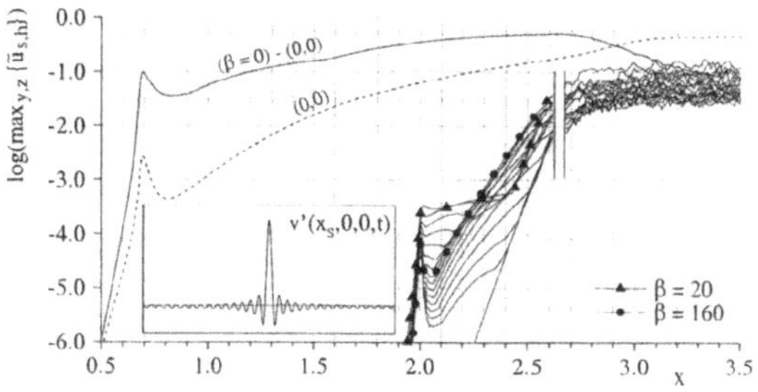

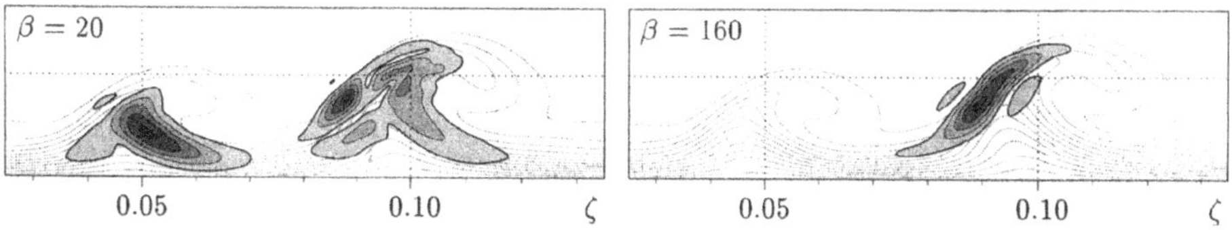

Fig. 5: Isocontours of normalized $\tilde{u}_{s,h}$-amplitude for $\beta = 20$ (left) and $\beta = 160$ (right) for case *VPB+B* in a crosscut at $\xi = 3.2$.

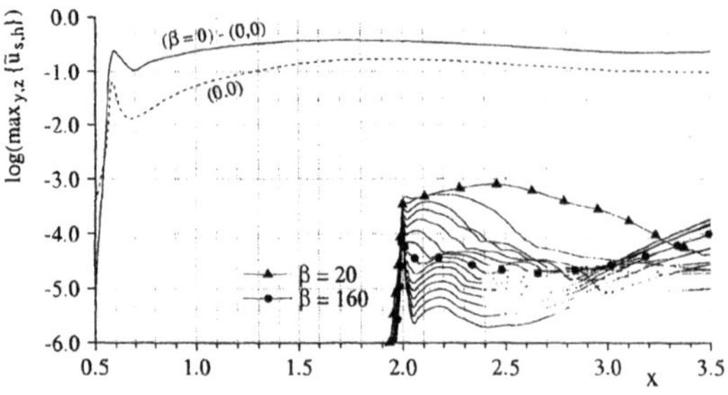

Fig. 6: Downstream t-modal amplitude development ($\tilde{u}_{s,h}$-maximum over y and z) for case *UFD-5+B*: vortex packet plus upstream flow deformation plus background pulse. Frequencies from β=0 to β=320 ($\Delta\beta$=20) are shown. The curves for $\beta \geq 20$ are gained from second time-derivative analysis.

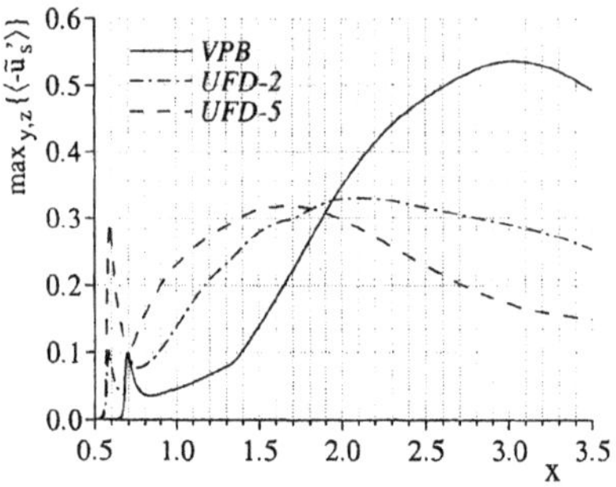

Fig. 7: Mean-flow distortion by the nonlinearly saturated CFV modes. Downstream development of the maximum mean flow deceleration ($\langle -\tilde{u}_s' \rangle$-maximum over y and z) with and without upstream flow deformation:
VPB: CFV packet, without UFD
UFD-2: CFV packet plus UFD, A_{UFD}=0.02
UFD-5: CFV packet plus UFD, A_{UFD}=0.05

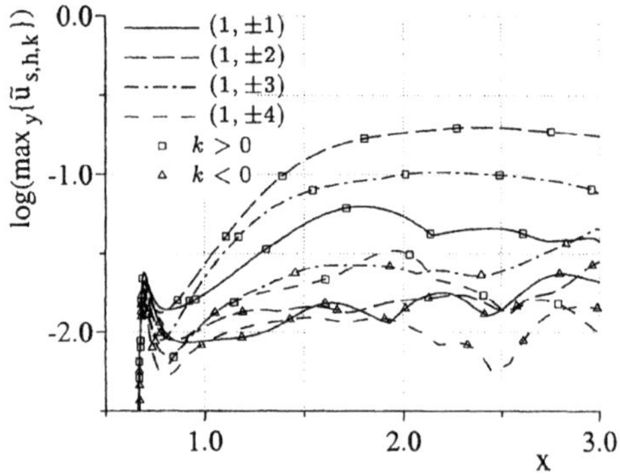

Fig. 8: Downstream t-z-modal amplitude development ($\tilde{u}_{s,h,k}$-maximum over y) for Case *WP5* (β = 5.0).

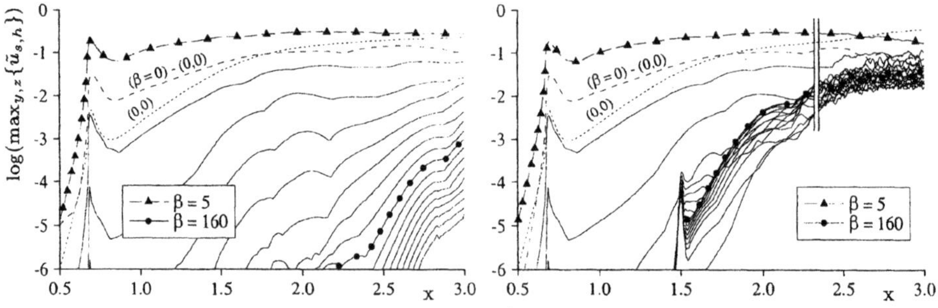

Fig. 9: Downstream t-modal amplitude development ($\tilde{u}_{s,h}$-maximum over y and z) for Case *WP5* (left) and *WP5+B* (right). Frequencies from β=0 to β=320 ($\Delta\beta$=20) and the frequency β=5 are shown. The curves for $\beta \geq 20$ are gained from second time-derivative analysis.

Examination of the Eddy-Viscosity Concept Regarding its Physical Justification

Holger Frahnert[1], Uwe Ch. Dallmann[2]
German Aerospace Center - DLR, Institute of Fluid Mechanics,
Bunsenstraße 10, D-37073 Göttingen, Germany
[1]Holger.Frahnert@dlr.de, [2]Uwe.Dallmann@dlr.de

Summary

In industrial application, the eddy-viscosity concept is the root of nearly every turbulence model in use since the whole class of zero-equation up to two-equation models relies on this hypothesis. Even in large eddy simulation, the eddy-viscosity hypothesis is the base of many subgrid-scale models. In view of that a clear understanding of the theoretical basis and practical perspective of the concept is desirable. Valuable information on this issue is accessible by a comparative study of the principal axes and the eigenvalues of both the real and the modeled Reynolds-stress tensor. Conclusions drawn from this analysis are independent of any particular turbulence model but result from the structure of the underlying eddy-viscosity ansatz. The procedure is exemplified with plane channel flow. Important in practice is the modeling of complex flow problems with separation. Despite of the apparent simplicity of the case regarded, the insights gained from the present analysis are influential to problems of practical interest. In summary, the eddy-viscosity ansatz is judged unsustainable from both a physical and an engineering predictive point of view.

1 Introduction

A fundamental link of turbulent-flow calculations of industrial relevance, as there are simulations of wing-body configurations, airfoil high-lift conditions or flow separation in turbine cascades, to geometrically simple cases, like plane channel flow, is the common approach to model the Reynolds-stress tensor. The need to handle this closure task arises from the averaging procedure of the Navier-Stokes equations. Not evaluating on the assumptions incorporated in this first step towards a statistical modeling of turbulence, the acceptance of averaging on non-linear differential equations shall be presumed initially. Accompanied by the formal decomposition of a physical quantity into its mean value and the fluctuation about it, this procedure results in the familiar Reynolds-averaged Navier-Stokes equations

$$\begin{gathered}\partial_t \bar{\mathbf{u}} + \nabla \cdot (\bar{\mathbf{u}} \otimes \bar{\mathbf{u}}) + \nabla \cdot \left(\overline{\mathbf{u}' \otimes \mathbf{u}'}\right) = -\nabla \bar{p} + \nu \Delta \bar{\mathbf{u}} \\ \nabla \cdot \bar{\mathbf{u}} = 0\end{gathered} \tag{1}$$

which are stated for incompressible flows. They are intended to constitute evolutionary equations for the mean values of the physical quantities but exhibit a further term accounting for the convection of velocity fluctuations by the fluctuations themself. This term is conventionally formulated as the divergence of the Reynolds-stress tensor

$$\tau \equiv \overline{(\mathbf{u}' \otimes \mathbf{u}')} \tag{2}$$

which is not known at all and subject to modeling. A way of trying to evaluate the unknowns is to hypothesize a relation between them and the mean-flow quantities, primarily to make the Reynolds-averaged Navier-Stokes equations solvable in a self-consistent manner. This is a second step, after averaging, towards statistical turbulence modeling and it is the very subject of the present consideration.

2 The eddy-viscosity ansatz

If the Reynolds-stress tensor is to be related to features of the mean velocity field, preliminary analysis determines the scope that such a relation needs to be cast in. Firstly, it should not depend on the velocity field itself. That is because, the Reynolds-stresses represent correlations of velocity fluctuations which are relative to the mean. Therewith, a Gallilei transformation to the frame of reference should keep the modeled Reynolds-stresses unchanged. Secondly, it may depend on the mean velocity gradients but not on higher derivatives. This is because, the Navier-Stokes equations are of second order. Being concerned with the definition of appropriate boundary conditions, the divergence of the Reynolds-stress tensor, modeled in terms of mean flow quantities, ought to be limited in order by this upper bound. Hence, a hypothesis on a relation between the modeled Reynolds-stress tensor and the mean velocity field does evolve out of the general ansatz:

$$\tilde{\tau} = \mathcal{F}(\nabla \otimes \bar{\mathbf{u}}) \tag{3}$$

The simplest possible realization of this scope is obtained by means of a linear constitutive equation. Due to the symmetry of the Reynolds-stress tensor, the function $\mathcal{F}$ of the mean velocity gradient tensor should likewise be formulated. Following the original approach of Boussinesq [1], this is achieved by assuming proportionality between the Reynolds-stress tensor and the mean rate-of-strain tensor. For dimensional reasons, the proportionality factor is postulated to constitute an eddy viscosity. While this could always be defined in tensorial form, a scalar representation shall be adopted for the present analysis as it reflects the common approach to statistical turbulence modeling in industrial application.

A further term needs to be incorporated to account for an energetic inconcistency of the relation so far. This is because, the trace of the mean rate-of-strain tensor is governed by the continuity equation which amounts to be identically zero. On the other hand, the trace of the Reynolds-stress tensor specifies twice the turbulence kinetic energy which is finite in general. To balance the relation, any tensor with an appropriate trace might be included, the simplest choice of which is a diagonal one having three equal eigenvalues. This results in the formulation of the energy correction by means of the tensorial identity **I** times a scalar. With the turbulence kinetic energy, the eddy viscosity and the mean velocity gradient tensor being symbolized by k, ν_t and **A** respectively, the familiar eddy-viscosity ansatz for the Reynolds-stress tensor is established:

$$\tilde{\tau} = \frac{2}{3} k\, \mathbf{I} - \nu_t \left[\mathbf{A} + \mathbf{A}^T\right] \quad \text{with} \quad \mathbf{A} = \nabla \otimes \bar{\mathbf{u}} \tag{4}$$

That relation is the backbone of a broad range of eddy-viscosity models which are evoked to determine ν_t. These models, relying on the eddy-viscosity ansatz, are classified into zero-equation models, like Baldwin-Lomax, up to two-equation models, like k-ϵ or k-ω, a survey of which is given in [2].

The present analysis is qualified for accomplishment without reference to a particular model. That is to say, the conclusions drawn are independent and therefore apply to any eddy-viscosity model which is used in conjunction with ansatz (4).

3 Eigensystem analysis

A tensor is uniquely determined by its principal axes and the corresponding eigenvalues. With symmetric tensors, the principal axes are orthogonal in pairs and all eigenvalues are real. This applies to the Reynolds-stress tensor. For statistically two-dimensional flows in general, it degenerates like

$$\tau = \begin{bmatrix} \overline{u'^2} & \overline{u'v'} & \overline{u'w'} \\ \overline{u'v'} & \overline{v'^2} & \overline{v'w'} \\ \overline{u'w'} & \overline{v'w'} & \overline{w'^2} \end{bmatrix} \quad \Rightarrow \quad \tau_{\overline{2D}} = \begin{bmatrix} \overline{u'^2} & \overline{u'v'} & 0 \\ \overline{u'v'} & \overline{v'^2} & 0 \\ 0 & 0 & \overline{w'^2} \end{bmatrix} \tag{5}$$

The principal axes of any statistically two-dimensional Reynolds-stress tensor $\tau_{\overline{2D}}$ are determined by the eigenvectors

$$\mathbf{e}_{1,2} = \frac{1}{\text{norm}} \begin{bmatrix} \overline{u'^2} - \overline{v'^2} \pm \sqrt{(\overline{u'^2} - \overline{v'^2})^2 + 4\overline{u'v'}^2} \\ 2\overline{u'v'} \\ 0 \end{bmatrix} \quad \text{and} \quad \mathbf{e}_3 = \frac{1}{\text{norm}} \begin{bmatrix} 0 \\ 0 \\ \overline{w'^2} \end{bmatrix} \tag{6}$$

while the corresponding eigenvalues are given through

$$\lambda_{1,2} = \frac{1}{2} \left[\overline{u'^2} + \overline{v'^2} \pm \sqrt{(\overline{u'^2} - \overline{v'^2})^2 + 4\overline{u'v'}^2} \right] \quad \text{and} \quad \lambda_3 = \overline{w'^2} \tag{7}$$

The strategy to be followed consists in the inspection of both the modeled Reynolds-stress tensor, from ansatz (4), and the real one by means of their respective eigensystems. This complies with a comparison of the tensors as a whole. From direct numerical simulation the Reynolds-stress tensor is immediately known. Further application of DNS data to the right hand side of relation (4) allows for the evaluation of the eigensystem that results from the adoption of the eddy-viscosity ansatz. An accompanying consideration of a k-ω model computation is done for cross-check.

4 Test case : Channel flow

Utilization of an eigensystem analysis for the investigation of the eddy-viscosity ansatz shall be demonstrated with plane channel flow. This is because, it is a minimal test case that a turbulence model may cope with and it widely allows for analytic considerations to be accomplished on both the evolutionary equations and the eddy-viscosity ansatz itself. The latter is due to the stationarity and partial homogeneity of plane channel flow which results in the mean velocity gradients to behave like

$$\begin{array}{lll} \partial_x \overline{u} = 0 & \partial_x \overline{v} = 0 & \partial_x \overline{w} = 0 \\ \partial_y \overline{u} \neq 0 & \partial_y \overline{v} = 0 & \partial_y \overline{w} = 0 \\ \partial_z \overline{u} = 0 & \partial_z \overline{v} = 0 & \partial_z \overline{w} = 0 \end{array} \quad \text{with} \quad \begin{array}{l} \partial_x \overline{p} = \text{constant} \\ \partial_y \overline{p} = \text{unconstrained} \\ \partial_z \overline{p} = 0 \end{array} \tag{8}$$

if the y-direction is defined wall-normal. Evaluation of the eddy-viscosity ansatz (4) with these conditions on the mean velocity gradient tensor yields a modeling of the Reynolds-stress tensor in accordance with

$$\tilde{\tau} = \begin{bmatrix} 2/3\,k & -\nu_t\,\partial_y\overline{u} & 0 \\ -\nu_t\,\partial_y\overline{u} & 2/3\,k & 0 \\ 0 & 0 & 2/3\,k \end{bmatrix} \tag{9}$$

In a statistical flow simulation, the elements of (9) are determined by use of some eddy-viscosity model. However, it turns out that the particular values of these elements are not significant for the fixing of the principal axes. For the latter, formal substitution of the Reynolds-stresses in equation (6) with the elements of (9) results in

$$\tilde{\mathbf{e}}_{1,2} = \frac{1}{\text{norm}} \begin{bmatrix} \pm 2\,|\,\nu_t\,\partial_y\overline{u}\,| \\ -2\,\nu_t\,\partial_y\overline{u} \end{bmatrix} \quad \Rightarrow \quad \frac{1}{\sqrt{2}} \begin{bmatrix} \pm 1 \\ 1 \end{bmatrix} \tag{10}$$

The far right hand side is given in simplified notation due to a possible sign exchange in the y-component of the eigenvectors not being significant to the system of the principal axes. Also, the z-component as well as the third eigenvector are not treated further at all because they are trivial in statistically two-dimensional flows. Equation (10) reveals that the principal axes of the Reynolds-stress tensor in plane channel flow, determined by use of the eddy-viscosity ansatz, are fixed, irrespective of the Reynolds number and the selection of the eddy-viscosity model. For comparison, the Reynolds-stresses obtained from a direct numerical simulation are inserted into equation (6) to determine the principal axes of a real Reynolds-stress tensor. The computation utilized for that was performed at a Reynolds number $Re_\tau = 180$, based on channel half width and wall friction velocity. While the results obtained either way are displayed in figure 1, their interpretation is deferred to the end of this section.

A consideration of the eigenvalues is suitable to rate on the significance that the orientation of the principal axes does have in a statistical flow simulation. If any eigenvalues are equal then the Reynolds-stress tensor is isotropic in the sub-space spanned by the corresponding principal axes. For that to happen, their orientation in that sub-space might be of no concern. However, effective is the divergence of the Reynolds-stress tensor. That implies that the orientation of the principal axes at a given location can be disregarded only if the corresponding eigenvalues are equal in a surrounding of that location. In anticipation of the analysis result, this condition is satisfied nowhere in the channel. For the following it appears useful to analyse pairs of eigenvalues and for statistically two-dimensional problems only one shall be considered. For that, an absolute measure of isotropy be defined by

$$\Lambda = \frac{2\,\lambda_1\lambda_2}{\lambda_1^2 + \lambda_2^2} \tag{11}$$

It is derived from the area $\pi\lambda_1\lambda_2$ of the ellipse that is spaned by two principal axes, scaled by their eigenvalues, which is normalized by the area πr^2 of a circle with $r^2 = \frac{1}{2}(\lambda_1^2 + \lambda_2^2)$. An isotropic situation is identified by $\Lambda = 1$. Respectively, the ratios $\lambda_1/\lambda_2 = 2$ and $\lambda_1/\lambda_2 = 10$ are indicated by $\Lambda = 0.8$ and $\Lambda \approx 0.2$, for example. With the Reynolds-stress tensor becoming completely anisotropic in the sub-space, the absolute measure approaches zero.

For the eigenvalues of the Reynolds-stress tensor resulting from adoption of the eddy-viscosity ansatz, substitution of ansatz (9) into equation (7) yields

$$\tilde{\lambda}_{1,2} = \frac{2}{3}\,k \pm |\overline{u'v'}| \quad \text{with} \quad \overline{u'v'} \approx -\nu_t\,\partial_y\overline{u} \tag{12}$$

Contrary to the result for the principal axes, the influence of the eddy-viscosity model does not cancel out immediately for the eigenvalues. Therefore, data for k, ν_t and $\partial_y\overline{u}$ from a k-ω model computation at $Re_\tau = 180$ are inserted into (12) to evaluate the eigenvalues in wall-normal direction at a fixed x-location and to determine the absolute measure of isotropy. The corresponding graph is given in figure 2 by a dotted line. For comparison, the left part of equation (12) shall be evaluated directly with data for k and $\overline{u'v'}$ from the previous direct numerical simulation. Therewith, the eigenvalues, resulting from the eddy-viscosity ansatz alone, are determined without reference to a particular model. The corresponding graph of Λ is given in figure 2 by a dashed line. The quantitative difference in both graphs of the figure is of minor concern at present. It is due to a mismatch in the turbulence kinetic energy while the Reynolds-shear-stresses match very well. The respective data is shown in figure 3. Important to note is a qualitative feature in figure 2, which is: both graphs approach $\Lambda = 1$ at the wall. That means, the modeled Reynolds-stress tensor is forced to become isotropic towards the wall. This situation changes dramatically if the eigenvalues of the real Reynolds-stress tensor are calculated straightforwardly by means of equation (7), again evaluating with data from the direct numerical simulation. In that case the absolute measure of isotropy is zero at the wall which corresponds to a completely anisotropic situation. The respective graph is given in figure 2 by a solid line. That opposite trend can be explained on a physical basis as follows.

A limiting wall behaviour of the Reynolds-stresses is given in [3] according to

$$\begin{aligned} &\overline{u'^2} = u_1\,\hat{y}^2 + \ldots \qquad &&\overline{v'^2} = v_1\,\hat{y}^4 + \ldots \qquad &&\overline{w'^2} = w_1\,\hat{y}^2 + \ldots \\ &\overline{u'v'} = s_1\,\hat{y}^3 + \ldots &&\hookrightarrow &&k = k_1\,\hat{y}^2 + \ldots \end{aligned} \tag{13}$$

whereby only leading-order terms are kept here and $\hat{y}$ denotes the distance from the wall. Primarily, these relations express the physically reasonable approach of the respective quantities towards the wall. With them, equation (12) evaluates as

$$\tilde{\lambda}_{1,2} = \frac{2}{3}(\,k_1\,\hat{y}^2 + \ldots) \pm |\,s_1\,\hat{y}^3 + \ldots\,| \tag{14}$$

It should be noted that equation (12) solely is the result of assuming the eddy-viscosity ansatz (4) as the basic relation for expressing the Reynolds-stress tensor and evaluating this by means of equation (7) for the eigenvalues. Hence, regarding to the leading-order term of equation (14), the eigenvalues of the modeled Reynolds-stress tensor close to the wall behave like:

$$\tilde{\lambda}_1 \propto \hat{y}^2 \quad \text{and} \quad \tilde{\lambda}_2 \propto \hat{y}^2 \tag{15}$$

To determine the eigenvalues of the real Reynolds-stress tensor, the limiting relations (13) are to be inserted into equation (7) directly to yield:

$$\lambda_{1,2} = u_1\,\hat{y}^2 + v_1\,\hat{y}^4 + \ldots \pm \sqrt{(\,u_1\,\hat{y}^2 + v_1\,\hat{y}^4 + \ldots)^2 + 4\,(\,s_1\,\hat{y}^3 + \ldots)^2} \tag{16}$$

With neglection of the higher-order terms again, the near-wall behaviour of the real Reynolds-stress tensor follows:

$$\lambda_1 \propto \hat{y}^2 \quad \text{and} \quad \lambda_2 \propto 0 \tag{17}$$

The trend of the graphs in figure 2 becomes obvious when the near-wall relations (15) and (17) are inserted into equation (11). The actual scaling of the principal axes by the eigenvalues is certainly case dependent. Contrary, what has been focused on here is a qualitative feature of the modeled and the real eigensystem. If the limiting relations (13) are assumed to hold for all Reynolds numbers then the outcome of the analysis does persist accordingly. From that it may be generally concluded that the eddy-viscosity ansatz forces a prediction of the near-wall behaviour of the Reynolds-stress tensor that contradicts physical facts. However, immediate consequences from the adoption of the concept for statistical turbulence modeling in practice are subsequently to be derived from an inspection of the principal axes.

The eigenvalue analysis and especially the solid line in figure 2 does reveal that the Reynolds-stress tensor is anisotropic everywhere in the channel. Therefore, the change of both the eigenvalues and the orientation of the principal axes make an impact on the force-density balance of the Reynolds-averaged Navier-Stokes equations. From figure 1 and with equation (10) it is evident that the principal axes of the modeled Reynolds-stress tensor do not change at all and that they are misaligned almost everywhere, irrespective of the Reynolds number and the turbulence model used with ansatz (4). Of course, the question arises why this fundamentally wrong performance is not obviously reflected in the results of statistical flow simulations. Due to the stationarity and partial, that is streamwise and spanwise, homogeneity of plane channel flow, the Reynolds-averaged Navier-Stokes equations decouple and can be integrated like ordinary differential equations in the wall-normal direction:

$$x\text{-momentum} \quad \Rightarrow \quad \bar{u} = \frac{\partial_x \bar{p}}{2\nu}(y^2 - 1) + \frac{1}{\nu}\int_{-1}^{y} \overline{u'v'}\, dy \tag{18}$$

$$y\text{-momentum} \quad \Rightarrow \quad \bar{p} = -\,\overline{v'^2} + f(x) \tag{19}$$

This implicitly takes advantage of the case specific form of the force-density vector which results from the divergence of the Reynolds-stress tensor. With the partial homogeneity of plane channel flow, only changes of the Reynolds-stresses in wall-normal direction do occur. Therefore, the contribution of the Reynolds-stress tensor to the balance of the equations of motion is, in the present case, determined by single elements of that tensor. Explicitly, with equations (5) and (9), the force-density vector $\nabla \cdot \tau \approx \nabla \cdot \tilde{\tau}$ evaluates to

$$[\, \partial_y(\overline{u'v'}) \,,\, \partial_y(\overline{v'^2}) \,,\, 0 \,] \;\approx\; [\, \partial_y(-\nu_t\, \partial_y \overline{u}) \,,\, \partial_y(2/3\, k) \,,\, 0 \,] \tag{20}$$

Therewith, an eddy viscosity can be determined for the x-component to match the Reynolds-shear-stress and to reproduce the mean velocity profile. That is possible in general since the zeros of $\overline{u'v'}$ and $\partial_y \overline{u}$ happen to coincide. If only this latter condition were to be fulfilled for both the Reynolds-shear-stress and any arbitrary quantity, which currently is the derivative of the mean velocity profile, then a proportionality factor could be determined to yield exact match. The coincidence of the zeros is necessary but not particularly supportive for the eddy-viscosity hypothesis. With the various failures in recovering substantial properties of the Reynolds-stress tensor itself, the eddy-viscosity ansatz is obviously invalid for plane channel flow. An example of results for the mean velocity and the mean static pressure is shown in figure 4, comparing a direct numerical simulation with a k-ω model computation. The prediction of the mean static pressure is wrong as much as the normal Reynolds-stress $\overline{v'^2}$ does not equal $2/3\, k$. Qualitatively similar results are to be expected for nearly parallel shear flows, like flat plate boundary layer.

The special situation of partial homogeneity, in the sense above, allows for one component of the force-density vector to be matched by means of the free parameter ν_t. Therewith, the misprediction of the principal axes as a whole is hidden if attention is addressed to the mean velocity profile only. In general, the divergence of the Reynolds-stress tensor represents the projection of its change onto each spatial direction. That is to say, the orientation of the principal axes connected with a proper scaling by the corresponding eigenvalues is necessary for the modeled Reynolds-stress tensor to contribute correctly to the force-density balance of the Reynolds-averaged Navier-Stokes equations. If similarly severe mispredictions of the eigensystem, as seen for plane channel flow, do occur in more complex flow situations, then neither physical quantity is likely to be predicted correctly. Although, it might be adjusted for a specific problem. That is because, the elements of the force-density vector are generally determined by more than one, that is to say three, elements of the Reynolds-stress tensor. This makes it difficult to transfer the case specific calibration of a turbulence model to changing flow conditions. The predictive unreliability of turbulence models in engineering application is suspected mainly to be due to the inappropriate performance of the eddy-viscosity ansatz with respect to the eigensystem.

5 Conclusions

The eddy-viscosity ansatz (4) has been subject to investigation with the elementary case of plane channel flow. It is regarded far from being a physically reasonable assumption. This is disturbing as much as in industrial application it is the vital concept of a large class of statistical turbulence models, ranging from Baldwin-Lomax up to k-ϵ and k-ω. Independent of any such turbulence model, basically the inadequacy of the eddy-viscosity ansatz to render the eigensystem of the Reynolds-stress tensor degrades the predictive reliability of respective turbulent-flow simulations. The phenomenological success of the eddy-viscosity ansatz with regard to the mean velocity profile does hide its extensive failure in the modeling of the Reynolds-stress tensor itself. The fundamentally incorrect performance of the concept is suspected to develop its full impact all the more the problem under investigation becomes of practical relevance. Eddy-viscosity models may be appreciated pragmatically as case dependent expert tools. But above that, they are not justified by physical reasoning. Already in [4] it has been pointed out what still are current problems in turbulent shear flows and they are likely to persist as long as the eddy-viscosity ansatz (4) is taken for the foundation stone of turbulence modeling. These conclusions shall be understood in a positive sense as they highlight a vantage point for physically reasonable turbulent-flow simulations. However, the necessary point of restart might lie even deeper at the Reynolds-averaging itself. Parallels of the conclusions drawn to subgrid-scale modeling in large-eddy simulation are obvious. An investigation of the eddy-viscosity ansatz in this regard has been published in [5].

Acknowledgements

H. F. is indebted to Prof. Dr. D. Ronneberger, University of Göttingen, for giving a hint on an initially erroneous conjecture on the near-wall eigenvalue analysis. This contribution is greatfully acknowledged. Thanks are due to Dr. C. Wagner, DLR, for the surrender of his DNS data and to Dr. R. Kessler, DLR, for providing us with the code for the k-ω model computations.

References

[1] J. Boussinesq: *Essai sur la Theorie des Eaux Courantes.* Mem. Pres. Acad. Sci. Prais **23**, 46 (1887)

[2] W. Haase, E. Chaput, E. Elsholz, M. A. Leschziner, U. R. Müller (Eds.): *ECARP - European Computational Aerodynamics Research Project: Validation of CFD Codes and Assessment of Turbulence Models.* Notes on Numerical Fluid Mechanics vol. 58, Vieweg (1997)

[3] J. Kim, P. Moin, R. Moser: *Turbulence statistics in fully developed channel flow at low Reynolds number.* J. Fluid Mech. **177** 133-166, (1987)

[4] S. Corrsin: *Some current problems in turbulent shear flows.* Proc. 1st Naval Hydro-Symp, Natn. Acad. Sci. / Natn. Res. Counc., Pub. 515, 373-400 (1957)

[5] T. S. Lund, E. A. Novikov: *Parametrization of subgrid-scale stress by the velocity gradient tensor.* Annual Research Briefs, CTR, 27-43 (1992)

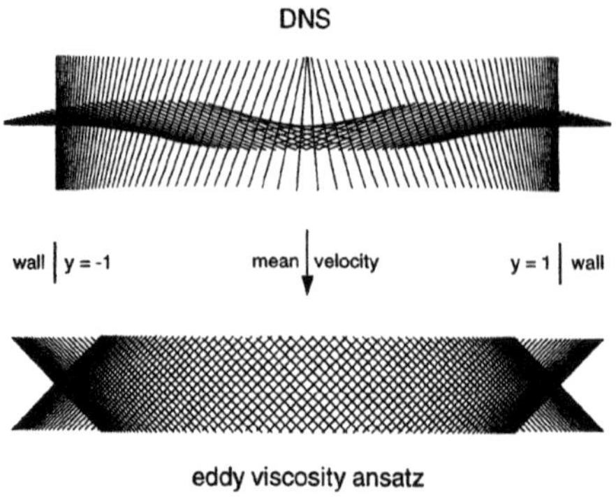

Figure 1 Reynolds-stress tensor: wall-normal distribution of the principal axes at a fixed x-location.

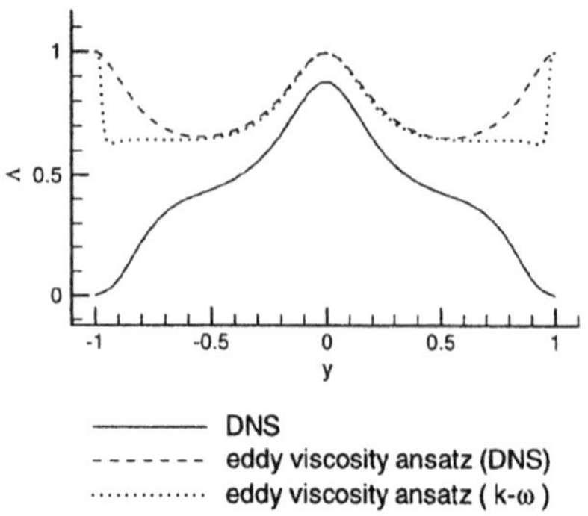

Figure 2 Isotropy of either Reynolds-stress tensor defined by: $\Lambda = 1 \Rightarrow$ isotropic; $\Lambda \neq 1 \Rightarrow$ anisotropic

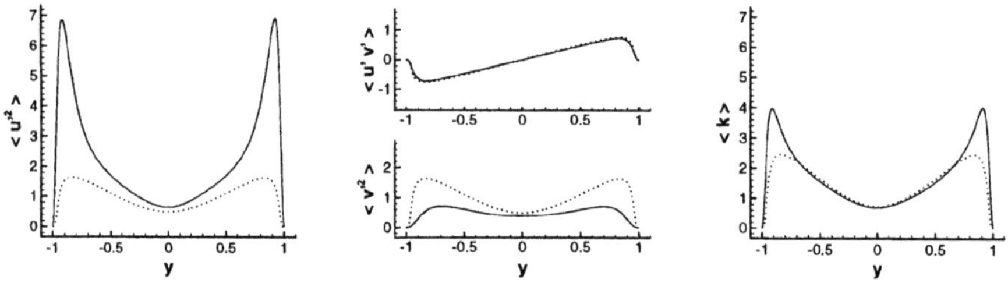

Figure 3 Elements of the Reynolds-stress tensor in plane channel flow, from direct numerical simulation (solid line) and k-ω model computation (dotted line) at $Re_\tau = 180$.

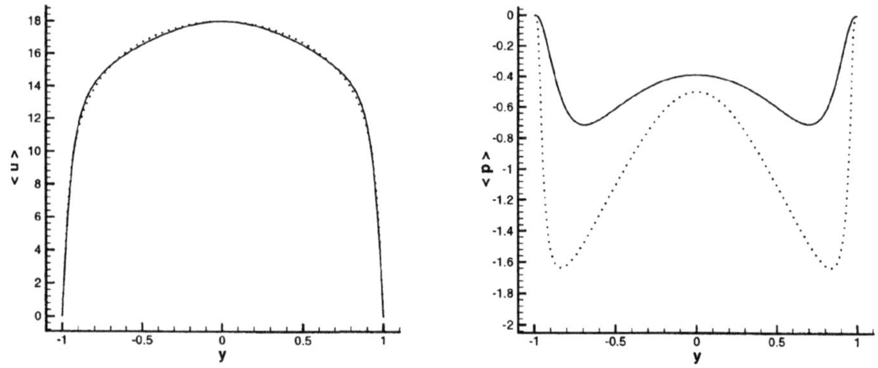

Figure 4 Mean velocity and mean static pressure in plane channel flow, from direct numerical simulation (solid line) and k-ω model computation (dotted line) at $Re_\tau = 180$.

Mathematical Fundamentals / Numerical Simulation

A Numerical Method for Compressible and Low Mach Number Fluid Flow

Andreas Meister
Fachbereich Mathematik
Schwerpunkt Differentialgleichungen und Dynamische Systeme
Universität Hamburg
Bundesstrasse 55, 20146 Hamburg, Germany

Summary

The development of stable and highly accurate numerical methods which can be employed to simulate incompressible as well as compressible flow fields represents an exciting and important task of the future in computational fluid dynamics. The paper is devoted to the investigation and extension of a compressible finite volume method to the low Mach number regime. The scheme represents a box-type method on unstructured grids and the numerical flux function is given by a special combination of the AUSMDV and the Lax-Friedrichs scheme. Nevertheless, the proceeding can be straightforwardly employed within different existing high resolution schemes.

1 Introduction

Current flow solvers are usually based on the equations of either compressible or incompressible fluid flow whereby the transition between both regimes is neglected. Briley et. al. [1] and Merkle et. al. [2] reported of performance and accuracy problems in the case that standard compressible methods are used for the simulation of low Mach number flows. Sesterhenn [3] investigated this disadvantage with the view on rounding errors and Volpe [4] described the influence of mesh adaption.

All these research activities prove that the extension of a compressible solver to the incompressible regime requires an appropriate mathematical analysis of the governing equations in order to extract the main features of the physical quantities in the limit of interest. Asymptotic expansions represent one of the most powerful and efficient tools in the backbone of such assignments and are widely used in almost all areas of science to get a deeper insight concerning the behaviour of solutions of partial differential equations. During the past five years a comprehensive study of the Euler equations in the low Mach number limit has been performed [5, 6]. The results of this asymptotic analysis can be used in a numerical method for the solution of the Euler and Navier-Stokes equations where the global Mach number is strongly varying with respect to time.

Recently, Guillard and Viozat [7] preconditioned the numerical dissipation within the Riemann-solver of Roe. Thereby, a discrete asymptotic single scale analysis is employed to prove the validity of this slightly modified Roe method in the low Mach number regime. This procedure overcomes the disadvantage of standard preconditioning techniques [8, 9, 10] which usually restricts the underlying numerical method to the simulation of steady flow fields except the scheme is involved in a dual time stepping environment [11, 12, 13].

The paper presents the extension of a compressible finite volume method to the low Mach number regime. First, we give a brief survey on the results of a multiple scale analysis of the Euler equations [5, 6]. Thereafter, the finite volume approximation of the governing equations is presented. Using a discrete asymptotic analysis one can prove that variations within the leading order velocity field implies variations inside the first order pressure distribution. In contradiction to the continuous asymptotic analysis these variations are associated with the mesh size instead of the mach number. The results are confirmed by means of a flow about the RAE2822 airfoil. Independent of the numerical flux function used, the only possibility for an unphysical merging of the physical quantities with respect to the asymptotic expansion can be incorporated within the numerical dissipation. By means of a simple preconditioning of the numerical dissipation the unphysical variation inside the pressure field vanish and reliable results in the sense of the continuous asymptotic analysis are obtained. We observe this analytic behaviour also in the numerical test case.

2 Governing Equations

Concerning the study of physical quantities in the low Mach number limit we assume that the nondimensional formulation of the governing equations consists exclusively of physical variables ϕ satisfying $\phi = O(1)$ if the Mach number tends to zero. Note that we employ an independent pressure reference $\hat{p}_{ref}$ instead of $\hat{p}_{ref} = \hat{\rho}_{ref}\hat{v}^2_{ref}$ usually employed in the case of compressible fluid flow. Finally, the two dimensional Euler equations become (see [16])

$$\partial_t \mathbf{u} + \sum_{\ell=1}^{2} \partial_{x_\ell} \mathbf{f}_\ell(\mathbf{u}) = \mathbf{0} \text{ in } \mathbf{G} \times I\!R^+, \tag{1}$$

where $\mathbf{G} \subset I\!R^2$ represents the spatial domain, $\mathbf{u} = (\rho,\rho v_1,\rho v_2,\rho E)^T$ denotes the vector of conserved variables, and the convective fluxes are given by $\mathbf{f}_\ell(\mathbf{u}) = (\rho v_\ell,\ \rho v_\ell v_1 + \mathrm{M}^{-2}\delta_{1\ell}p,\ \rho v_\ell v_2 + \mathrm{M}^{-2}\delta_{2\ell}p,\ (\rho E + p)v_\ell)^T$ with $\mathrm{M}^2 = (\hat{\rho}_{ref}\hat{v}^2_{ref})/(\hat{p}_{ref})$. Furthermore, the equation of state for an ideal gas can be written in the form $p = (\gamma - 1)\rho\left(E - \frac{\mathrm{M}^2|\mathbf{v}|^2}{2}\right)$, where γ denotes the ratio of specific heats.

A comprehensive survey on the results of a single as well as multiple scale asymptotic analysis is presented in [5, 6]. We briefly summerize the statements which are important with respect to the investigation of the finite volume scheme presented in the following. Taking into account the characteristic wave speeds in direction $\mathbf{n}$

$$v_{\mathbf{n}} = O(1) \text{ and } v_{\mathbf{n}} \pm \frac{c}{\mathrm{M}} = O\left(\frac{1}{\mathrm{M}}\right) \quad \mathrm{M} \to 0$$

it is obvious that the propagation of information is associated with two different length scales, namely

$$\eta = \mathbf{x} \text{ and } \xi = \mathrm{M}\mathbf{x}.$$

This observation motivates the consideration of multiple length scales. Assuming that each physical quantity ϕ can be expressed in the form of a multiple length scale asymptotic expansion

$$\phi(\mathbf{x},t;\mathrm{M}) = \phi^{(0)}(\eta,\xi,t) + \mathrm{M}\phi^{(1)}(\eta,\xi,t) + \mathrm{M}^2\phi^{(2)}(\eta,\xi,t) + o(\mathrm{M}^2),\ \mathrm{M} \to 0$$

it is proven that under convenient additional requirements (see [6]) the pressure becomes

$$p(\mathbf{x},t;\mathrm{M}) = p^{(0)}(t) + \mathrm{M}p^{(1)}(\xi,t) + \mathrm{M}^2p^{(2)}(\eta,\xi,t) + o(\mathrm{M}^2),\ \mathrm{M} \to 0. \tag{2}$$

3 Numerical Method and Results

In order to rewrite the Euler equations in a convenient manner for standard compressible flow solvers we define the auxiliary variables

$$\widetilde{p} = \frac{p}{\mathrm{M}^2}, \quad \widetilde{c} = \frac{c}{\mathrm{M}} \text{ and } \widetilde{E} = \frac{E}{\mathrm{M}^2}.$$

Thus the system (1) becomes

$$\partial_t \widetilde{\mathbf{u}} + \sum_{\ell=1}^{2} \partial_{x_\ell} \widetilde{\mathbf{f}}_\ell(\widetilde{\mathbf{u}}) = \mathbf{0} \text{ in } \mathbf{G} \times \mathbb{R}^+, \tag{3}$$

where $\widetilde{\mathbf{u}} = (\rho,\rho v_1,\rho v_2,\rho\widetilde{E})^T$, $\widetilde{\mathbf{f}}_\ell(\widetilde{\mathbf{u}}) = (\rho v_\ell,\ \rho v_\ell v_1 + \delta_{1\ell}\widetilde{p},\ \rho v_\ell v_2 + \delta_{2\ell}\widetilde{p},\ (\rho\widetilde{E} + \widetilde{p})v_\ell)^T$ and $\widetilde{p} = (\gamma - 1)\rho\left(\widetilde{E} - \frac{|\mathbf{v}|^2}{2}\right)$.

Smooth solutions of the system of Euler's equations in general exist only for short time and well-known phenomena like shock waves and contact discontinuities develop naturally. Hence we introduce the concept of weak solutions. A bounded set $\sigma \subset \mathbb{R}^2$ is said to be a control volume if Gauss' integral theorem is applicable to functions defined on σ. The mapping $\widetilde{\mathbf{u}}$ is called a weak solution of the system (3) if

$$\frac{d}{dt}\int_\sigma \widetilde{\mathbf{u}}\, d\mathbf{x} + \sum_{\ell=1}^{2} \int_{\partial\sigma} \widetilde{\mathbf{f}}_\ell(\widetilde{\mathbf{u}}) n_\ell \, ds = \mathbf{0} \tag{4}$$

is valid on every control volume σ with outer unit normal vector $\mathbf{n} = (n_1,n_2)^T$. With respect to the numerical solution of equation (4) the space part $\mathbf{G}$ and the time part $\mathbb{R}^+$ have to be discretised. We are starting from an arbitrary conforming triangulation $\mathcal{T}_h$ which is called the primary mesh and consists of triangles $\mathcal{T}$. Although a triangle itself can be used as a control volume we consider boxes σ_i which are defined as the open set including the node $\mathbf{x}_i = (x_{i1},x_{i2})^T$ (see figure 1 (left)) and bounded by the straight lines l_{ij}^k, $k = 1,2$, connecting the midpoint of the edge with the point $\mathbf{x}_s = (x_{s1},x_{s2})^T$ (see figure 1 (right)). Here $\mathbf{x}_s$ is defined by

$$\mathbf{x}_s = \sum_{m\in\{i,j,k\}} \alpha_m^s \mathbf{x}_m \text{ with } \alpha_m^s = \frac{1}{2\left(|l_i| + |l_j| + |l_k|\right)} \sum_{\substack{\overline{m}\in\{i,j,k\} \\ \overline{m}\neq m}} |l_{\overline{m}}|.$$

The union of these boxes σ_i, $i = 1,\ldots,I$ is called the secondary mesh.

Introducing the cell average

$$\widetilde{\mathbf{u}}_i(t) = \frac{1}{|\sigma_i|}\int_{\sigma_i} \widetilde{\mathbf{u}}(\mathbf{x},t)\, d\mathbf{x}$$

the integral form with respect to an inner box σ_i can be written as

$$\frac{d}{dt}\widetilde{\mathbf{u}}_i(t) = -\frac{1}{|\sigma_i|} \sum_{j\in N(i)} \sum_{k=1}^{2} \int_{l_{ij}^k} \sum_{\ell=1}^{2} \widetilde{\mathbf{f}}_\ell(\widetilde{\mathbf{u}}(\mathbf{x},t)) n_\ell \, ds,$$

where $N(i)$ denotes the index set of all neighbouring boxes of the control volume σ_i. Note that in the case of a boundary box σ_i (i.e. $\partial\sigma_i \cap \partial\mathbf{G} \neq \emptyset$) the boundary conditions have to be considered additionally. In order to evaluate the integral we employ the AUSMDV scheme H^{AUSMDV} proposed by Liou and Wada [14] which achieves desirable properties like no numerical dissipation on stationary as well as moving contact discontinuities, preserving enthalpy in steady flows and high resolution for shock waves. Interpretating the cell average $\widetilde{\mathbf{u}}_i(t)$ as a piecewise constant function on σ_i and using a simple explicit time marching procedure leads to the first order scheme

$$\widetilde{\mathbf{u}}_i^{n+1} = \widetilde{\mathbf{u}}_i^n - \frac{\Delta t}{|\sigma_i|} \sum_{j \in N(i)} \sum_{k=1}^{2} |l_{ij}^k| H^{AUSMDV}(\widetilde{\mathbf{u}}_i^n, \widetilde{\mathbf{u}}_j^n, \mathbf{n}_{ij}^k). \tag{5}$$

with $\widetilde{\mathbf{u}}_i^n = \widetilde{\mathbf{u}}_i(t^n)$ and $t^{n+1} = t^n + \Delta t$. A detailed derivation of primary and secondary grid methods can be found in [15].

Let us focus the attention on the properties of the scheme in the low Mach number regime. Therfore, we introduce a discrete asymptotic expansion

$$\phi_i = \phi_i^{(0)} + \mathrm{M}\phi_i^{(1)} + \mathrm{M}^2\phi_i^{(2)} + o(\mathrm{M}^2),\ \mathrm{M} \to 0,$$

for each physical quantity $\phi \in \{\rho, v_1, v_2, p\}$ into (5). Identifying terms multiplied by the same power of M we obtain the leading order continuity equation in the form

$$0 = \sum_{j \in \mathcal{N}(i)} \sum_{k=1}^{2} \omega_{ij} \left(p_i^{(0)} - p_j^{(0)}\right) |l_{ij}^k| \,. \tag{6}$$

Since $\omega_{ij} > 0$ it is obvious that the maximum as well as the minimum of the discrete leading order pressure field $p_i^{(0)}$ is always located at the boundary of the computational domain. Consequently, if $p^{(0)}$ is assumed to be constant at the boundary it follows that $p_i^{(0)} = p_j^{(0)}$ for all $i,j \in \{1,\ldots,I\}$ in agreement with the continuous case. Especially for a structured mesh and a constant leading order density distribution the first order continuity equation can be written as

$$p_k^{(1)} - p_j^{(1)} = \beta\left(v_{1k}^{(0)} - v_{1j}^{(0)}\right)$$

where $\beta > 0$ and $k,j \in N(i)$. This statement shows that variations within the leading order velocity field $\mathbf{v}^{(0)}$ directly cause variations in the first order pressure distribution which are associated with the grid size instead of the Mach number, see (2).

In order to investigate the above mentioned properties numerically we consider a flow past the RAE2822 airfoil at varying free stream Mach numbers between 10^{-6} and 0.1. We define $\Delta\mathbf{x}_{max}$ to be the minimal spatial distance between p_{max} and p_{min} denoting the maximum and minimum pressure value, respectively. Without loss of generality we consider the case of a spatially varying pressure distribution. In the absense of long wave effects, which are associated with $p^{(1)}$, we obtain $\Delta\mathbf{x}_{max}^{-1} = O(1)$ which leads to

$$\frac{p_{max} - p_{min}}{p_{max}\Delta\mathbf{x}_{max}} = O(\mathrm{M}^2). \tag{7}$$

In the presense of acoustic effects, which are associated with $p^{(1)}$, equation (2) yields $\Delta\mathbf{x}_{max}^{-1} = O(\mathrm{M})$ and consequently (7) is likewise valid. Unfortunately, the numerical results presented in figures 2 and 3 prove the unphysical behaviour

$$\frac{p_{max} - p_{min}}{p_{max}\Delta \mathbf{x}_{max}} = O(\mathrm{M}).$$

To overcome this failure we replace the standard AUSMDV method in (5) by

$$H(\widetilde{\mathbf{u}}_i^n,\widetilde{\mathbf{u}}_j^n,\mathbf{n}_{ij}^k) = (1-\delta)H_P^{LF}(\widetilde{\mathbf{u}}_i^n,\widetilde{\mathbf{u}}_j^n,\mathbf{n}_{ij}^k) + \delta H^{AUSMDV}(\widetilde{\mathbf{u}}_i^n,\widetilde{\mathbf{u}}_j^n,\mathbf{n}_{ij}^k). \tag{8}$$

The weight $\delta \in \{0,1\}$ is chosen to be 1 for all $\mathbf{x} \in \mathbf{G}$ where the local Mach number of the flow field is greater than 0.1. Furthermore, the weight is set to be 1 everywhere in $\mathbf{G}$ if the freestream Mach number is greater than 0.1. The involved modified Lax-Friedrichs scheme writes as

$$H_P^{LF}(\widetilde{\mathbf{u}}_i^n,\widetilde{\mathbf{u}}_j^n,\mathbf{n}_{ij}^k) = \frac{1}{2}\sum_{\ell=1}^{2}\left(\widetilde{\mathbf{f}}_\ell(\widetilde{\mathbf{u}}_i) + \widetilde{\mathbf{f}}_\ell(\widetilde{\mathbf{u}}_j)\right) - \mathbf{D}_P(\widetilde{\mathbf{u}}_i^n - \widetilde{\mathbf{u}}_j^n)$$

whereby the dissipation matrix $\mathbf{D}_P$ is obtain by preconditioning the original matrix by means of a preconditioning matrix which can be expressed with respect to the entropy form as $\mathbf{P} = \mathrm{diag}\{\mu,1,1,1\}$. Thereby, $\mu = \mu(\mathbf{x},t)$ has to satisfy $\mu = O(\mathrm{M})$ for $\mathrm{M} \to 0$. Performing a discrete asymptotic analysis for the method (5) involving the new numerical flux function (8) yields the leading and first order continuity equation in the form

$$0 = \sum_{j\in\mathcal{N}(i)}\sum_{k=1}^{2}\underbrace{\overline{\omega}_{ij}}_{>0}\left(p_i^{(m)} - p_j^{(m)}\right)\left|l_{ij}^k\right|, \quad m = 0,1.$$

Equivalent to the preceding argumentation we obtain $p_i^{(m)} = p_j^{(m)}$ for all $i,j \in \{1,\ldots,I\}$ and $m = 0,1$. This statement coincides with the continuous asymptotic analysis if appropriate boundary and initial conditions are considered. The asymptotic statements are confirmed by the numerical results presented in figures 2 and 3. Note that the pressure fluctuations shown in figure 3 meet the condition (7). For a detailed derivation of the above described proceeding we refere to [16].

Finally, we present two test cases in order to ensure the validity of our new method for transonic flows. Thereby, we increase the spatial accuracy of the scheme by means of a second order weighted essential non-oscillatory (WENO) method recently constructed on unstructured grids by Friedrich [17].

The first test case shows a transonic steady-state flow around the RAE2822 airfoil at a free stream Mach number $\mathrm{Ma} = 0.75$ and an angle of attack $\alpha = 3°$. The density distribution obtained is shown in figure 4. From figure 5 one can learn about the good correspondence between the computed C_p-distribution and the results presented during the GAMM workshop [18].

The next case is concerned with the transonic flow past the SKF1.1 airfoil [19] at $\mathrm{Ma} = 0.65$ under $\alpha = 3°$ (see fig. 6). In order to validate our numerical method by the experimental data, it must be corrected for wall interferences [20]. To overcome the high value of the lift coefficient, we use $\alpha = -0.25°$ for our scheme. From the comparison (fig. 7) between the computed and the experimental C_p-values, it is obvious that our numerical data is a good approximation of the described physical phenomenon.

4 Conclusion

A mathematical and numerical investigation of the AUSMDV scheme embedded in a finite volume method is presented, which proves the failure of this algorithm in the low Mach number regime. Thereafter, a combination of the AUSMDV method and a preconditioned form of the Lax-Friedrichs scheme is developed. The validity of the new approach is proven by means of an asymptotic analysis as well as numerical experiments.

References

[1] Briley, W. R., McDonald, H. and Shamroth, S. J., A Low Mach Number Euler Formulation and Application to Time-Iterative LBI Schemes. *AIAA J.*, **21(10)**, 1467–1469 (1983).

[2] Merkle, C. L. and Choi, Y.-H., Computation of Low-Speed Compressible Flows with Heat Addition. *AIAA J.*, **25(6)**, 831–838 (1987).

[3] Sesterhenn, J., *Zur numerischen Berechnung kompressibler Strömungen bei kleinen Mach-Zahlen*. Ph.D. thesis, Swiss Federal Institute of Technology, Diss. ETH No. 11334 (1995).

[4] Volpe, G., Performance of Compressible Flow Codes at Low Mach Numbers. *AIAA J.*, **31(1)**, 49–56 (1993).

[5] Klein, R., Semi-Implicit Extension of a Godunov-Type Scheme Based on Low Mach Number Asymptotics I: One-Dimensional Flow. *J. Comput. Phys.*, 121, 213–237 (1995).

[6] Meister, A., Asymptotic Single and Multiple Scale Expansions in the Low Mach Number Limit. *SIAM J. Appl. Math.*, **Vol. 60(1)**, 256–271 (1999).

[7] Guillard, H. and Viozat, C., On the Behaviour of Upwind Schemes in the Low Mach Number Limit. *Computers and Fluids*, Vol. 28, 63–86 (1999).

[8] Merkle, C. L. and Choi, Y.-H., Computation of Low-Speed Compressible Flows with Time-Marching Procedure. *Int. J. Num. Meth. Eng.*, Vol. 25, 293–311 (1988).

[9] Turkel, E., Preconditioned Methods for Solving the Incompressible and Low Speed Compressible Equations. *J. Comput. Phys.*, 72, 277–298 (1987).

[10] Choi, D. and Merkle, C. L., The Application of Preconditioning in Viscous Flows. *J. Comput. Phys.*, 105, 207–223 (1993).

[11] Shuen, J.-S., Chen, K.-H. and Choi, Y., A Coupled Implicit Method for Chemical Non-equilibrium Flows at All Speeds. *J. Comput. Phys.*, 106, 306–318 (1993).

[12] Weiss, J. M. and Smith, W. A., Preconditioning Applied to Variable and Constant Density Flows. *AIAA J.*, **33(11)**, 2050–2057 (1995).

[13] Vierendeels, J., Riemslagh, K. and Dick, E., A Multigrid Semi-implicit Line-Method for Viscous Incompressible and Low-Mach-Number Flows on High Aspect Ratio Grids. *J. Comput. Phys.*, 154, 310–341 (1999).

[14] Wada, Y. and Liou, M.-S., A Flux Splitting Scheme with High-Resolution and Robustness for Discontinuities. *AIAA Paper*, 94–0083 (1994).

[15] Meister, A. and Sonar, T., Finite-volume schemes for compressible fluid flow. *Surv. Math. Ind.*, 8, 1–36 (1998).

[16] Meister, A., Analyse und Anwendung Asymptotik-basierter numerischer Verfahren zur Simulation reibungsbehafteter Strömungen in allen Mach-Zahlbereichen (2000), Eingereichte Habilitationsschrift, Universität Hamburg.

[17] Friedrich, O., Weighted Essential Non-Oscillatory Schemes for the Interpolation of Mean Values on Unstructured Grids. *J. Comput. Phys.*, 144, 194–212 (1998).

[18] Rizzi, A. and Viviand, H. (eds.), *Numerical Methods for the Computation of Inviscid Transonic Flows with Shock Waves*, no. 3 in Notes on Numerical Fluid Mechanics, Braunschweig, Wiesbaden (1981). Vieweg.

[19] Stanewsky, E. and Thibert, J. J., Airfoil SKF 1.1 with maneuver flap. AGARD AR 138, A5-1 - A5-29 (1979).

[20] Stanewsky, E., private communication (1999).

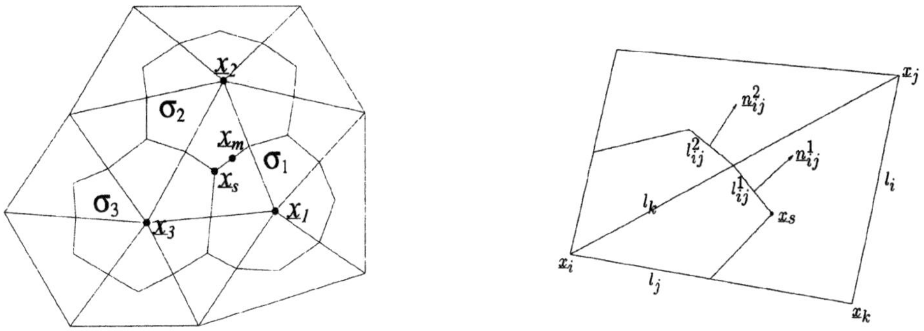

Figure 1 General form of a control volume of the triangulation $\mathcal{T}_h$ (left) and its boundary (right)

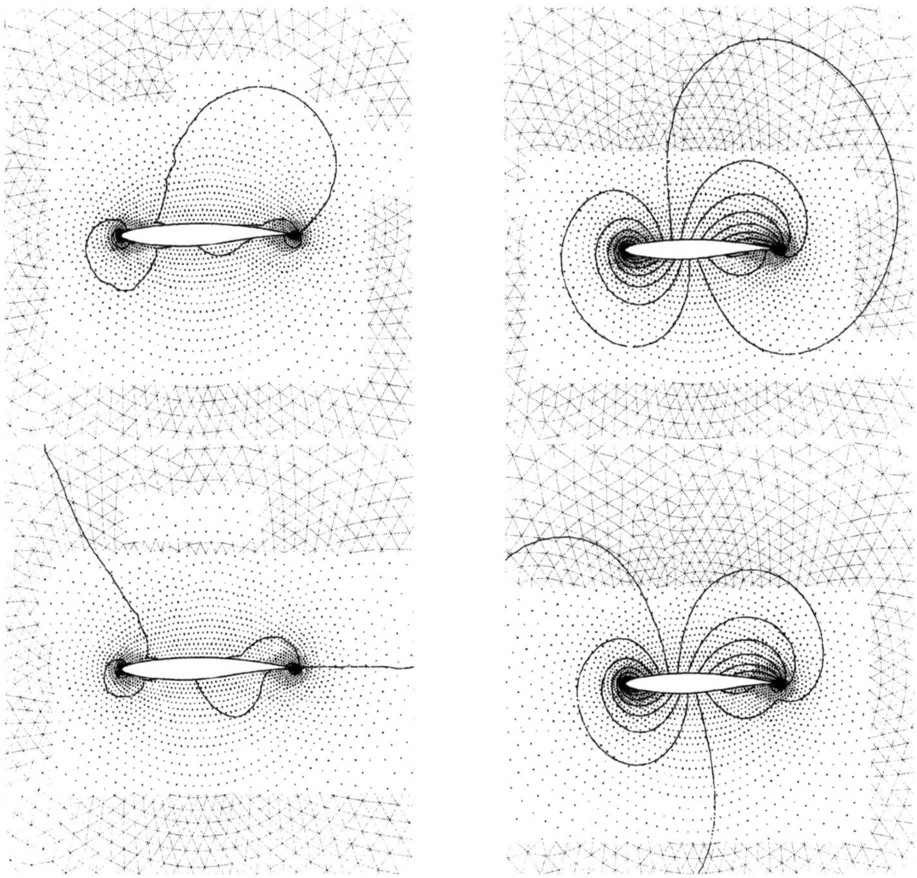

Figure 2 Isolines of the pressure distribution computed by the AUSMDV (left) and the new approach (right) for different Mach numbers (10^{-2} (top), 10^{-5} (bottom)).

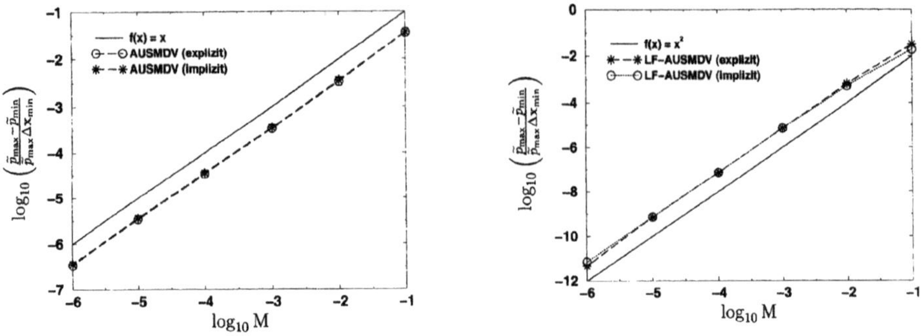

Figure 3 Pressure fluctuation $\frac{p_{max}-p_{min}}{p_{max}\Delta x_{max}}$ versus the Mach number for the AUSMDV scheme (left) and the modified method (right).

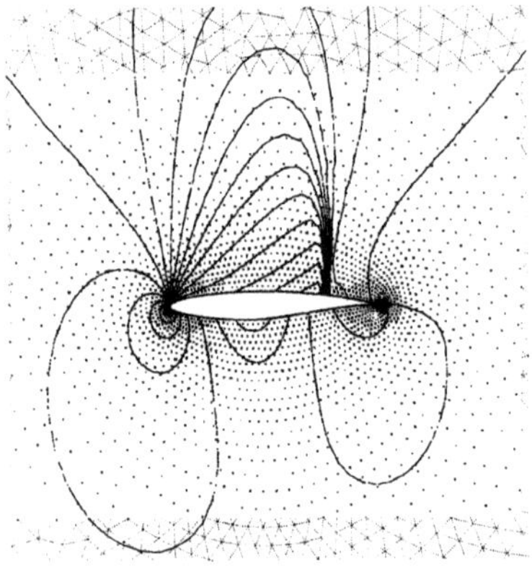

Figure 4 Isolines of the density distribution

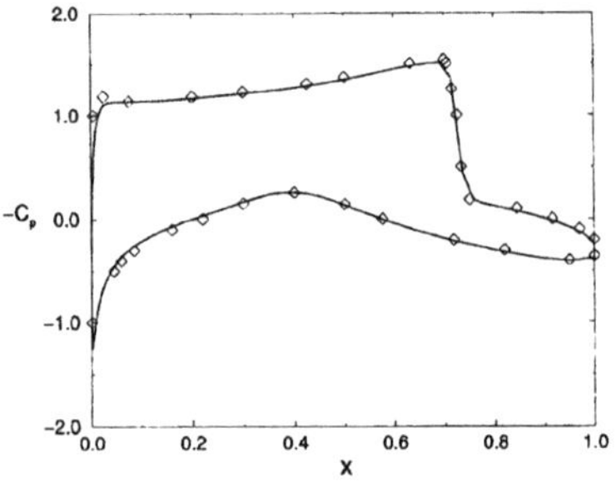

Figure 5 Comparison of the C_p-distribution, – Finite volume scheme, ◇ Reference data [18].

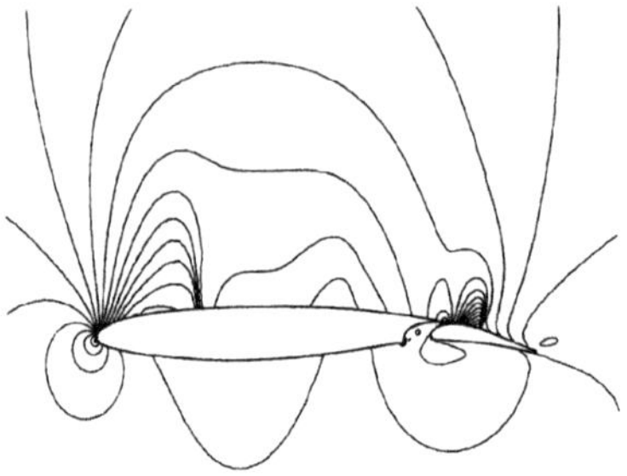

Figure 6 Isolines of the density distribution

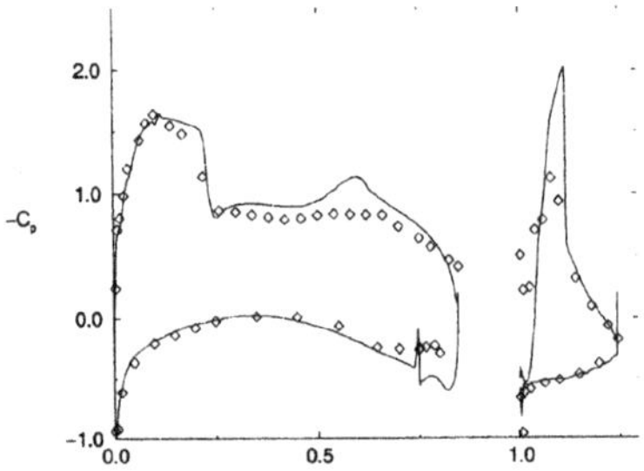

Figure 7 Comparison of the C_p-distribution, – Finite-volume scheme, ◇ Reference data [19].

Turbulent Two-Dimensional Flow Around a Flexible Membrane Airfoil

U. BUNGE, T. RUNG, F. THIELE
Hermann–Föttinger–Institut für Strömungsmechanik, Technische Universität Berlin,
Müller–Breslau–Straße 8, D–10623 Berlin, Germany

Summary

The two-dimensional, incompressible flow at Reynolds number $Re = \frac{U_\infty \cdot c}{\nu} = 1.3 \cdot 10^6$ around an inextensible, flexible membrane airfoil (*sail*) with varying excess length ε is examined solving the Reynolds-Averaged Navier-Stokes (RANS) equations on grids deforming according to the sail movement within fixed outer boundaries. Results are presented for fully turbulent conditions employing closure models of different degree of complexity in comparison to experimental and analytical results. Good agreement can be found for low angle of attack. However, for higher angle of attack, approaching onset of separation and beyond, the predictive accuracy varies significantly with the representation of turbulence in the presence of strong unsteady phenomena.

1 Introduction

Only recently the turbulent viscous flow equations have been introduced as the fluid dynamic model in membrane airfoil theory. An advanced study of flexible, massless membrane wing aerodynamics incorporating the RANS equations has been presented in [1]. Unlike an airfoil which is usually regarded as a rigid 2D structure, the sail *shape* is a function of both incidence and Reynolds number in which the shape itself interacts with the pressure distribution in a rather complicated manner.

There are several references for experimental results, e.g. [2], which will be taken for comparison, and a variety of theoretical investigations exist, such as [3, 4, 5, 6].

The computation of turbulent flow around a flexible 2D sail poses a challenge for the flow solver as well as the structural-mechanics solver, since the simulation has to be time-accurate on a domain with moving boundaries. The prediction of the aerodynamic forces has to be as exact as possible in order to allow for a realistic representation of the sail shape that deforms according to these loads. As classical inviscid theory only holds for small ranges of angles of incidence with small excess lengths, the application of RANS methods becomes necessary for this task.

A chain model [4] is considered for the sail in conjunction with a RANS method instead of the sail equation that is usually used.

2 Method and Governing Equations

2.1 Flow Solver

For the computation of the 2D flow around the sail an in-house finite-volume (FV) method is used, which is based on an implicit procedure of second order accuracy in time and space. Due to the fully implicit scheme, true time steps are possible and steady flow simulations are unconditionally stable. Complex two-dimensional geometries and local mesh refinements can be resolved by semi-structured grids, i.e. using hanging nodes, on the basis of general curvilinear coordinates. Moving meshes are implemented based upon the space-conservation law (SCL), such that forced unsteadiness caused by moving boundaries can be captured.

The cartesian tensor components as well as all scalar variables are stored in the control-volume centers. Diffusive terms are approximated applying central difference schemes whereas convective terms are approximated by upwind biased, bounded, monotonic functions of higher order [7]. The governing equations of the incompressible mean flow field, the continuity equation and the transport equation(s) for turbulent quantities are solved sequentially. The solution is iterated to convergence using a compressible (all-speed) pressure correction scheme of the SIMPLE type [8]. Introducing apparent pressures and viscosities, a generalised Rhie & Chow interpolation avoids an odd-even decoupling of pressure, velocity, and Reynolds stress $\overline{u_i u_j}$ [9].

The computational modelling of turbulent effects is based on a variety of models featuring different degrees of complexity to compute the Reynolds stresses arising in the averaging process. Amongst the many implemented modelling practices are standard isotropic Boussinesq-viscosity approaches employing either one or two additional transport equations ($k-\epsilon$, $k-\omega$). Moreover, their extension to non-linear constitutive stress-strain relations, as well as (non-linear) full Reynolds-stress transport models are available. In the present study, however, attention is restricted to the assessment of one- and two- equation models: the standard (SA) and a modified (SALSA) Spalart-Allmaras one-equation model [10], and two different $k-\omega$ two-equation models [11].

Due to the implementation of grid fluxes arising from the formulation of transport equations on moving grids, the step from grids moving in a rigid-body manner to deforming grids does not require any changes in the part of the code solving the equations for the fluid.

2.2 Sail Shape

The model presented in this paper describes the behaviour of an inextensible, massless sail using the structural model of a chain given in [4]. This description is preferred since an analytical function for the membrane shape as in the sail equation does not easily allow for a computation of the exact movement of the sail's material points. Equilibrium of forces is evaluated at each grid node of the sail, see Fig.1 (right), leading to the following equation:

$$\underline{F}(i-1) = \underline{a}(i)l(i) + \underline{F}(i)\,, \quad \text{with } l(i) = \sqrt{(\underline{x}(i) - \underline{x}(i-1))^2}\,, \tag{1}$$

where $\underline{F}(i)$ stands for the force vector acting on the hinge connecting neighbouring sail-elements. Assuming that no friction occurs in the hinge and that the aerodynamic load $\underline{a}(i)$ acts in the center between $\underline{x}(i-1)$ and $\underline{x}(i)$, the moment equilibrium is given as:

$$0 = \left[\frac{1}{2}l(i)\underline{a}(i) + \underline{F}(i)\right] \times \left[\underline{x}(i) - \underline{x}(i-1)\right]\,. \tag{2}$$

Note that there are no time dependent terms included, thus inertia and damping are completely neglected. This simplification has become necessary due to a lack of experimental data concerning mass distribution and structural damping. However, the iterative procedure to converge the solution of these equations has to be started with a relaxation factor of 0.001 that can slowly be raised each step up to a maximum of 0.5.

2.3 Grid-Deformation Method

The starting point for the grid-deformation method is the momentum conservation equation valid for a solid body in steady state without any volume loads:

$$0 = \frac{\partial T_{ij}}{\partial x_j}, \tag{3}$$

where $\underline{\underline{T}}$ represents the stress tensor. Displacements are imposed on the continuum (grid) via given displacement values on the boundaries.

Assuming ideally linearly elastic material, the stress-strain relation becomes

$$T_{ij} = E_{ijkl} \frac{1}{2} \left(\frac{\partial v_k}{\partial x_l} + \frac{\partial v_l}{\partial x_k} \right) \tag{4}$$

with only linear terms in the strain-displacement relation where v_i are displacements in the direction of the coordinates and E_{ijkl} is the generalised Hooke-material tensor enabling an arbitrary link of displacement derivatives and stresses (anisotropy). However, the grid will be idealised as isotropic material with zero Poisson's ratio reducing the material tensor to an isotropic parameter E aligning stress and strain tensor. This is done in order to avoid transverse contraction of grid cells that could result in a loss of grid quality.

It has to be remarked, that although the material is considered as isotropic, it is not homogeneous, i.e., the scalar E, representing Young's modulus or the material stiffness, can vary in space. Thus, the grid stiffness depends on grid-cell volume or deformation undergone (strain energy).

It is desirable to keep the grid orthogonal in near-wall regions, where it is usually very fine, e.g. Fig.2 (right), but in order to balance deformation it is necessary to avoid strain weakening, so one possible formulation for E could be:

$$E = \left(\frac{minVOL}{VOL} \right)^{\alpha} \left(\frac{D_{ii}^2}{minD_{ii}^2} \right)^{\beta} \quad \text{with} \quad D_{ij} = \frac{1}{2} \left(\frac{\partial v_i}{\partial x_j} + \frac{\partial v_j}{\partial x_i} \right), \tag{5}$$

where min stands for the minimum value in the whole computational domain and the exponents α and β can be adjusted to requirements: if it is desirable to keep a section of the grid with very fine grid cells nearly undeformed, α has to be raised, whereas a higher β tends to distribute the deformation of the grid equally. The best formulation for the cell stiffness will depend on the application, but there is no general set of parameters α and β that constitutes an optimum.

In order to implement this algorithm in the previously described code, the same FV method is applied for the calculation of grid deformation where the grid is treated as a continuum [12]. In the absence of convective and transient terms, diffusive terms are approximated applying central difference schemes and a harmonic interpolation of cell-center values on cell faces.

2.4 Coupled Solution

Loose coupling is achieved by a sequential solution of the governing equations for the fluid domain for every time step based upon the sail shape from the previous step. Applying the loads of the converged flow field, a new sail shape is computed and the iteration is proceeded by one time step, Fig.3.

Though there are more elaborate methods available for coupling, this simple one was chosen since there is no time-dependent term in the sail equation implemented as yet and time-step refinement is more straightforward than any coupling over partitioned time steps. Moreover, the method imposes no constraint on the time-step size consistent with the implicit method for the flow. An iterative procedure is set up with the following steps:

1. With an initial shape obtained from [5] and a grid generated manually around the sail or with a grid and sail shape from a preceding time step, the unsteady flow computation is proceeded by one time step.

2. The force $\underline{a}(i)$ due to the flow solution determined in step 1 is evaluated on each element.

3. With the aerodynamic force, equations (1) and (2) are considered. Equation (1) gives an explicit expression for the nodal forces that can be calculated depending on the force on the first node which can be arbitrarily initialised. In conjunction with equation (2), the length of each sail element has to be kept constant. Along with the boundary condition of fixed sail ends, a corrected force can be determined as a constant to start with equation (1) again. This iterative process is continued until a certain level of convergence is reached.

4. The displacements resulting from the change in sail shape are imposed on the grid and a new grid is computed.

5. Having a new sail shape and a corresponding grid, the procedure can be proceeded by one time step (return to step 1).

3 Results

The computations are started at 3° angle of attack. After convergence, the angle is raised step by step up to a chosen maximum of 20° far beyond separation, and on another "computational branch" decreased step by step to an angle, where the algorithm fails to give a solution. In this case the incidence angle has become negative below a critical value, where tension in the sail is close to zero and luffing occurs. Lowering the angle further, a hysteresis would occur, schematically shown in Fig.4. Thus, the sail shape becomes unstable and the deformations exceed the validity of the method, which cannot yet handle luffing that occurs earlier at higher excess lengths [13]. However, the algorithm gives stable solutions for negative incidence. The onset of numerical divergence roughly matches the angle where luffing occurs in the experiments.

Computational results are presented as diagrams of lift coefficient vs. angle of attack for three different excess lengths, Fig.1 (left), $\varepsilon = 0.21\%$, $\varepsilon = 0.77\%$, and $\varepsilon = 1.14\%$ obtained on grids with 16.000 control volumes covering a physical domain of $\frac{X}{c}$ from -10 to 11 and $\frac{Y}{c}$ from -9 to 9 at $\frac{X}{c} = -10$ and -12 to 12 at $\frac{X}{c} = 11$, Fig.2. This number was found sufficient when evaluating the effects of grid refinement on the prediction of lift for these excess lengths. Computations are compared to experimental results [2] and a linear approximation

$$C_l \cong 2\pi\alpha \pm 0.7\ (100\,\varepsilon)^{\frac{1}{2}}\ . \qquad (6)$$

The $\pm$ sign stands for a change in curvature when the sail snaps through, where the negative branch represents the sail bent downwards. The computations were performed with five different turbulence models marked by different symbols in Fig.5-7.

At low angles of attack good agreement can be found with the linear approximation as well as with the experimental results for all excess lengths. For $\varepsilon = 0.21\%$, the shortest sail, no distinctive maximum in computed lift can be seen, Fig.5, whereas for longer sails the loss in lift after separation increases and a maximum lift coefficient originates roughly at $7^o - 8^o$, its height dependent on the sail length, Fig.6 and 7. The experiments show a different behaviour and neither model gives the decreasing lift-coefficient for angles of attack greater than 13° in case of $\varepsilon = 0.21\%$. Approaching the onset of separation, the computational results depend strongly on subtle details pertaining to the representation of turbulence and the computed separation is overestimated for all turbulence models. The one-equation models capture most of the experimental results but fail to render the effects of massive separation, whereas the other models tend to produce more separation resulting in a lower lift coefficient at lower angle of attack.

A problem occurred at an angle of attack between 6° and 8° for the two-equation models, as the computations did not converge for ε exceeding 0.21%. Grid-refinement posed even more problems with stability in these cases, revealing that the effects of inertia and damping on the sail shape computation will have to be studied. The results for lift with separation are averaged values as the flow becomes unsteady before full separation has developed. Fig.3 shows the extremely unsteady behaviour of the system at these incidence angles for $\varepsilon = 1.14\%$. Unfortunately, this behaviour is not represented in any experimental result, so nothing can be said about the physical reliability of the frequency computed, which is roughly 0.3Hz.

With higher excess length ε not only the maximum lift but also the sail deformation grows, Fig.9, and the grid-deformation algorithm has to meet these requirements. Especially difficult are the sail ends where it is desirable that the grids on the other side of the fixed ends (without sail) rotates around the fixed points reverse to the movement of the sail, rather than being translated in the direction of the movement. Unfortunately, the latter is the natural behaviour of a continuous material, so that an enhanced method of prescribing the grid-motion is under development.

The pressure distribution depicted in Fig.10 illustrates the changes in the distribution of loads with increasing incidence angle. At 0°, the pressure distribution indicates a small separation bubble on the lower side, for all other angles the distribution does not change significantly on the lower side. The maximum of lift is reached roughly around 7°, where C_P reaches its highest negative value on the upper suction side of the sail. From 10° on, full separation can be observed on the upper side and the unsteady behaviour can also be observed in the instantaneous distributions.

4 Conclusions

The chain model works in conjunction with the RANS-method and all turbulence models investigated are able to capture the experimental results at low angle of attack and the weakly non-linear area is predicted with success. In case of massive separation, however, severe deviations and differences between the turbulence models show up and the general tendency of the present approach is towards a conservative result underestimating the lift. Hence, more research is considered necessary and additional effects will have to be taken into account, such as elasticity of the sail, transitional effects and three-dimensionality. The introduction of elasticity will lead to sail elements with non constant length and enable the computation of a tension in the sail. Moreover, work in progress is directed towards higher-order closure models and obtaining and evaluating more experimental data. Another step will be the introduction of (if not physical, at least artificial) damping and inertia into the sail algorithm to stabilise the procedure which will be necessary for the computation of a full hysteresis including luffing. An examination of the effects of grid refinement on the frequency of oscillations and sail shape is also considered necessary, as well as developing an enhanced method of prescribing the grid motion instead of only computing it using the described structural analogy, which can nonetheless still be used for additional smoothing.

Acknowledgement

This research was partly funded by the EU (UNSI, 'Unsteady Viscous Flow in the Context of Fluid-Structure Interaction', Project Ref: BRPR-CT97-0583) and the Deutsche Forschungsgemeinschaft (DFG-SFB 557 'Beeinflussung komplexer turbulenter Scherströmungen').

References

[1] Smith, R; Shyy, W. Computation of aerodynamic coefficients for a flexible membrane airfoil in turbulent flow: A comparison with classical theory. *The Physics of Fluids*, 8(12):3346 – 3353, December 1996. University of Florida.

[2] Greenhalgh, S.; Curtiss, H. C.; Smith, B. Aerodynamic Properties of a Two-Dimensional Inextensible Flexible Airfoil. *AIAA-journal*, 22(7):865–870, July 1984.

[3] Sugimoto, T.; Sato, J. Aerodynamic Characteristics of Two-dimensional Membrane Airfoils. *Trans. Japan Soc. Aero. Space Sci.*, 34(104):88–100, August 1991.

[4] de Matteis, G; de Socio, L. Nonlinear Aerodynamics of a Two-Dimensional Membrane Airfoil with Seperation. *J. Aircraft*, 23(11):831–836, November 1986.

[5] Jackson, P. S. A Simple Model for Elastic Two-Dimensional Sails. *AIAA journal*, 21(1):153–155, January 1983. Technical Notes.

[6] Vanden-Broeck, J.-M.; Keller, J. B. Shape of a sail in a flow. *The Physics of Fluids*, 24(3):552–553, March 1981.

[7] Lien, F.S.; Leschziner, M.A. Upstream Monotonic Interpolation for Scalar Transport with Application to Complex Turbulent Flows. *Int. Journal for Numerical Methods in Fluids*, 19:527–548, 1994.

[8] Patankar, S.V. *Numerical Heat Transfer and Fluid Flow*. McGraw Hill, New York, 1980.

[9] Obi, S.; Perić, M.; Scheurer. Second Moment Calculation Procedure for Turbulent Flows with Collocated Variable Arrangement. *AIAA Journal*, 29(4):585–590, 1991.

[10] Rung, T. Erweiterung von Eingleichungs-Turbulenzmodellen für lokales Nichtgleichgewicht. Internal Report 03/98, TU Berlin, Hermann-Föttinger-Institut für Strömungsmechanik, 1998.

[11] T. Rung and F. Thiele. Computational modelling of complex boundary–layer flows. In *Proc. 9th Intern. Symposium on Transport Phenomena in Thermal–Fluid Engineering*, pages 321–326, Singapore, 1996.

[12] Bunge, U. Die Behandlung der Strukturmechanik mittels der FV-Methode. Internal Report 01/99, TU Berlin, Hermann-Föttinger-Institut für Strömungsmechanik, April 1999. download: http://hodgson.pi.tu-berlin.de /SUB_PAGES /research /papers /POSTSCRIPTFILES /Struktur.ps.gz.

[13] B.G Newman. Aerodynamic Theory for Membranes and Sails. *Prog. Aerospace Sci.*, 24(1):1–27, 1987.

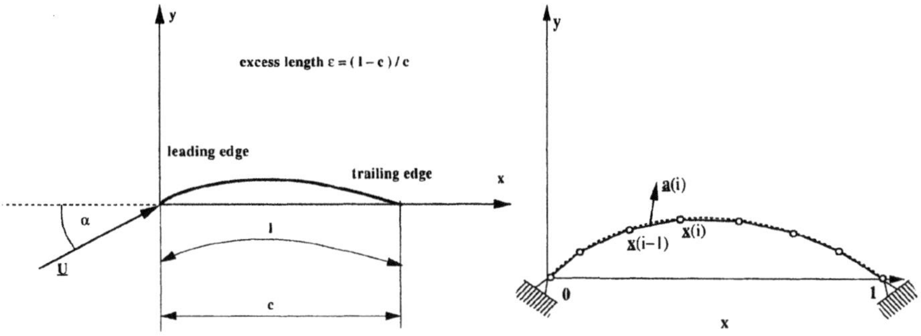

Figure 1: Two-dimensional sail (left) and structural model (right)

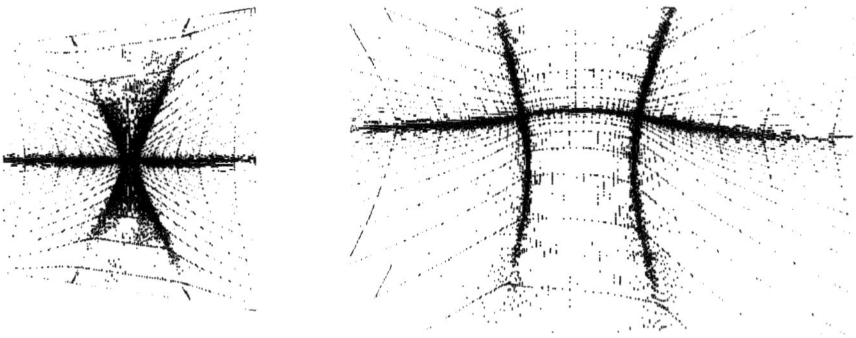

Figure 2: A (coarse) computational grid (left) and detail of grid around sail (right)

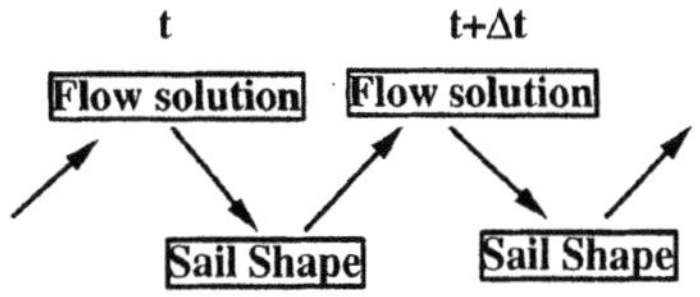

Figure 3: Coupling of flow and sail computations

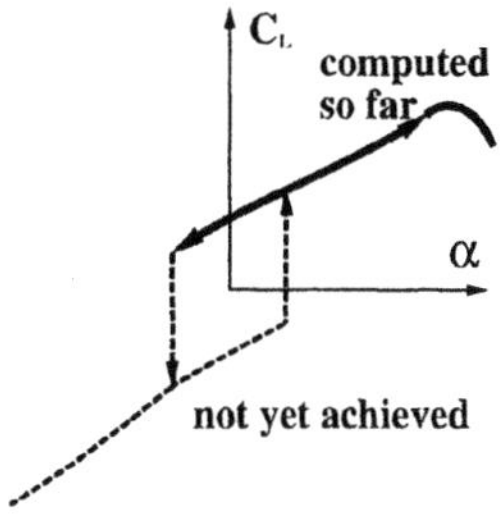

Figure 4: Principle of hysteresis loop of lift coefficient vs. angle of attack

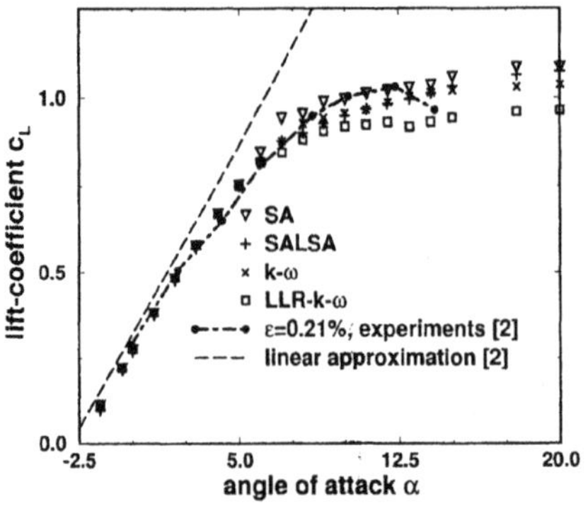

Figure 5: Lift coefficient vs. angle of attack for excess length $\varepsilon = 0.21\%$

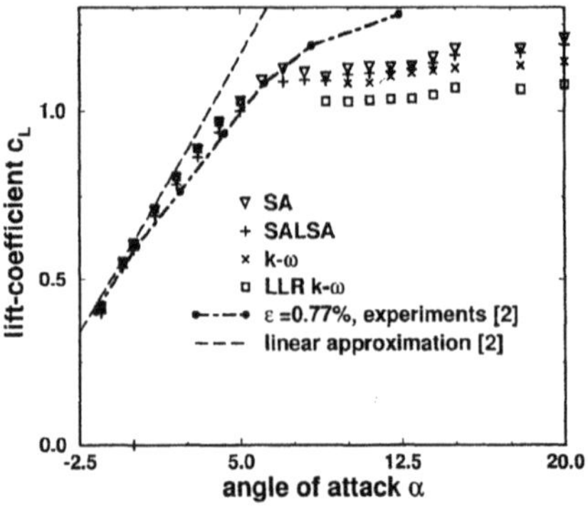

Figure 6: Lift coefficient vs. angle of attack for excess length $\varepsilon = 0.77\%$

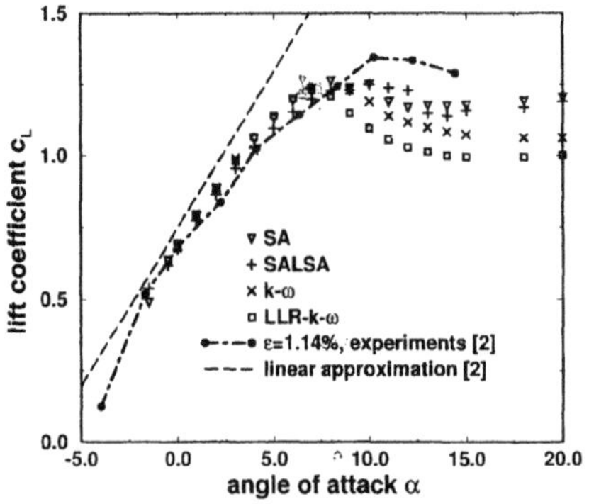

Figure 7: Lift coefficient vs. angle of attack for excess length $\varepsilon = 1.14\%$

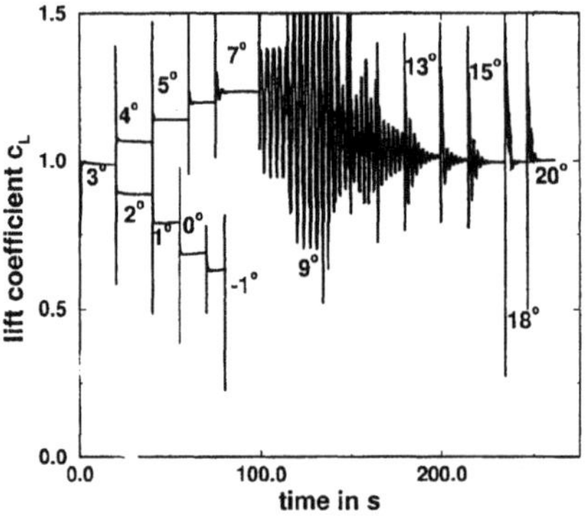

Figure 8: Evolution of lift coefficient vs. time for varying angle of attack during computation

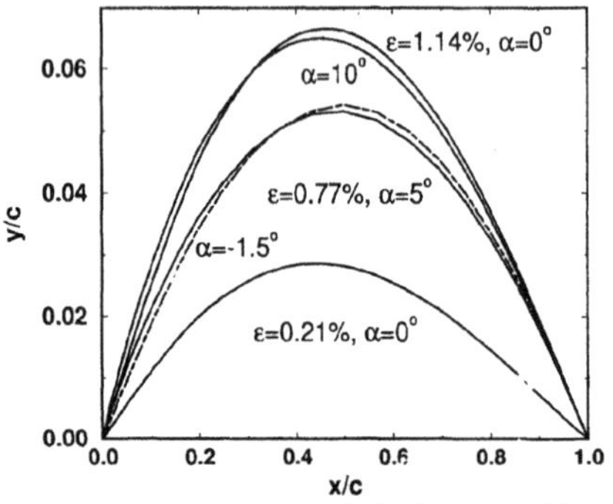

Figure 9: Comparison of sail shape for different excess lengths and angles of attack

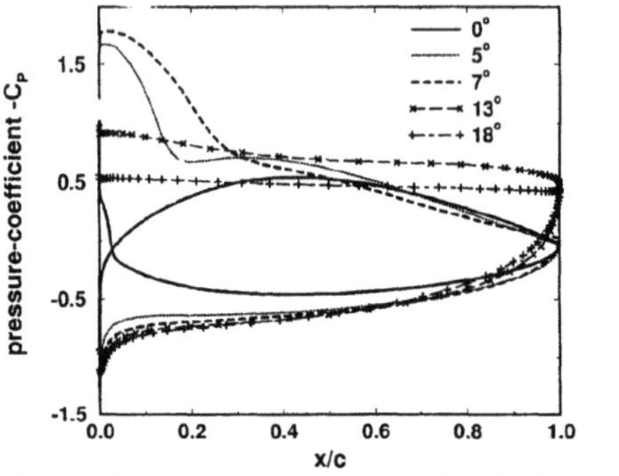

Figure 10: Instantaneous pressure distribution for $\varepsilon = 1.14\%$ for different angles of attack

Solution of the Euler Equations on Locally Adaptive B-Spline Grids

F. Bramkamp, J. Ballmann
Lehr- und Forschungsgebiet für Mechanik, RWTH Aachen
52056 Aachen, Germany

Summary

A numerical scheme for solving the Euler equations on locally h-adaptive grids with hanging nodes in two space dimensions is presented. The method consists of an integrated framework, including a novel grid generation technique using B-Splines, advanced adaptation criteria based on multiscale analysis and a flow solver which is capable to operate on arbitrary unstructured meshes with polygonal elements. An implicit finite volume method is developed to solve the conservation laws, employing a higher-order upwind scheme for the conservative fluxes. A variety of examples for stationary, inviscid flows have been studied to validate the new method. Firstly, the subsonic and transonic flow in a channel with a circular arc bump is investigated to assess the perfomance of the implicit scheme and to demonstrate the quality of the solution obtained with the adaptive algorithm. Secondly, the flow around a wing section has been considered in subsonic Mach number regime to study the advantage of the present method for smooth flows.

1 Introduction

The outlined investigations are a project part of the collaborative research center SFB401: *Modulation of Flow and Fluid/Structure Interaction at Airplane Wings*, RWTH Aachen, Germany [1], whose aim it is to efficiently simulate the viscous flow around flexible wings in cruise and high-lift configuration. The aerodynamics of high-lift configurations is extremely complex. Amongst others, confluent boundary layers and wakes, turbulent separated flow regions and compressibility effects are encountered. In order to resolve these effects by a numerical scheme, a strongly heterogeneous discretization is required to account for the different scales. A uniform discretization that is determined by the finest scale is not appropriate, since it requires a prohibitive amount of memory resources and computational time. Therefore, algorithms have to be developed with computational effort and memory requirements that are proportional to the complexity of the underlying problem, i.e., only in regions where small-scales physical effects are present the discretization has to be refined, and elsewhere a much coarser discretization will be sufficient. This task is being addressed adequately by locally adaptive schemes. Adaptive methods efficiently utilize available resources and may provide more accurate and reliable solutions. Further, the grid generation process is greatly facilitated, since the initial -coarse- mesh has solely to be adjusted to the geometry but not to the flow solution. Efficient and reliable adaptive strategies require an integrated framework consisting of the components: grid generation, adaptation and flow solver. These items will be addressed within this paper.

2 Grid Generation

The computational domain is initially tessellated by a coarse multiblock structured grid. In two space dimensions, quadrilaterals are employed as basic element type. The mesh is locally adapted to the solution according to the principle of h-adaptation. Hanging nodes, which occur due to local refinement are maintained, i.e. no transition elements are introduced. Hence, the adaptation remains local and the basic element type is being preserved. The grid generation is based on a B-Spline technique [2], which is particularly suited for adaptive algorithms. Each block of the mesh is represented by a B-Spline function of the form

$$\mathbf{x}(\mathbf{s}) = \sum_{\mathbf{i}} \mathbf{P}_{\mathbf{i}} \prod_{l=1}^{d} N_{i_l}^{m}(s_l) \tag{2.1}$$

with control points $\mathbf{P}_{\mathbf{i}}$. d denotes the dimension of the space. The order m of the B-Spline function N_l^m is chosen as 4. Since the grid is stored as a function, only a few control points are necessary to represent each block. Geometric quantities required by the flow solver may consistently be evaluated at any location within the domain. In particular, a truly nested sequence of locally adapted cells can be provided, which is required by the applied adaptation criteria, based on multiscale analysis.

3 Adaptation Criteria

The mesh adaptation is performed by means of multiscale techniques similar to those used in data compression. Here, we follow a strategy originally developed by A. Harten [3]. Starting point is a sequence of *nested* grids $\mathcal{G}_j := \{\Omega_{j,i}\}_i$, not necessarily structured, and corresponding averages $\overline{\mathbf{u}}_j := \{\mathbf{u}_{j,i}\}_{i\in\mathcal{G}_j}$, $j = 0,\ldots,L$, see Fig. 1. By means of this sequence, we successively decompose the averages $\overline{\mathbf{u}}_L$ on the finest level L into a sequence of averages $\overline{\mathbf{u}}_0$ on the coarsest level and details $\overline{\mathbf{d}}_j$, $j = 0,\ldots,L-1$, see Fig. 2. The averages and the details are interpreted as L_2–scalar products of the conservative quantities $\mathbf{u}$ and the L_1–weighted box function $\varphi_{j,i} := \frac{1}{|\Omega_{j,i}|}\chi_{\Omega_{j,i}}$ and appropriate wavelet functions $\psi_{j,i,e}$, $e \in E$, i.e.,

$$\mathbf{u}_{j,i} := (\mathbf{u},\varphi_{j,i})_{L_2(\Omega)},\ \mathbf{d}_{j,i,e} := (\mathbf{u},\psi_{j,i,e})_{L_2(\Omega)}$$

where E indicates the set of subcells corresponding to $\Omega_{j,i}$. For reasons of simplicity we assume a regular refinement, i.e. the cardinality of E always corresponds to the number of subcells minus one. The construction of the wavelet functions $\psi_{j,i,e}$ is subject to certain constraints, namely, reversibility of multiscale decomposition, local support of wavelet functions, vanishing moments and stability.

The detail coefficients can be interpreted as differences, which become negligibly small in regions where the solution is sufficiently smooth. We introduce the set of significant details $\mathcal{D}_\varepsilon := \{(j,i,e) : |d_{j,i,e}| > \varepsilon_j\}$ where $\varepsilon_j = 2^{j-L}\varepsilon$ is a level–dependent threshold value. For the presented applications, the parameter ε, which determines the sensitivity of the adaptation, is typically chosen within the range $\varepsilon \in \{10^{-3}, 10^{-2}\}$.

Since the flow field at hand is supposed to be heterogeneous, the number of significant detail coefficients, i.e., $N_D := \#\mathcal{D}_\varepsilon$, is much smaller than the number of averages N_L corresponding to the finest scale. The ultimate goal is to perform the multiscale transformation such that the complexity with respect to computational time as well as memory requirements corresponds to N_D rather than N_L. For details see [4].

By means of $\mathcal{D}_\varepsilon$ we determine a locally refined grid with hanging nodes denoted by $\mathcal{G}_\varepsilon$, i.e., $\bigcup_{(j,i)\in\mathcal{G}_\varepsilon} \Omega_{j,i}$ and the corresponding sequence of averages $\overline{\mathbf{u}}_{L,\varepsilon} := \{\overline{\mathbf{u}}_{j,i}\}_{(j,i)\in\mathcal{G}_\varepsilon}$. The adapted grid $\mathcal{G}_\varepsilon$ is determined by proceeding from coarse to fine. If there exists a significant detail $\mathbf{d}_{j,i,e}$, $e \in E$ corresponding to the cell $\Omega_{j,i}$ the grid is locally refined.

4 Governing Equations

In the present study, the fluid motion is governed by the time dependent Euler equations for a compressible gas, which can be expressed in conservation form for a control volume Ω with boundary $\partial\Omega$ and outward unit normal $\mathbf{n}$ as:

$$\int_\Omega \frac{\partial \mathbf{u}}{\partial t} dV + \oint_{\partial\Omega} \mathbf{F}^c(\mathbf{u})\mathbf{n}\, dS = \mathbf{0} \tag{4.2}$$

with initial values $\mathbf{u}(\mathbf{x},t_0) = \mathbf{u}_0(\mathbf{x})$, $\mathbf{x} \in \Omega$ and boundary conditions $\mathbf{u}(\mathbf{x},t)|_{\partial\Omega} = \mathcal{B}(\mathbf{x},t)$, $\mathbf{x} \in \partial\Omega$. The state vector of conserved flow variables $\mathbf{u}$ and the physically conservative flux $\mathbf{F}^c$ are defined by

$$\mathbf{u} = \begin{pmatrix} \rho \\ \rho\mathbf{v} \\ \rho e_{tot} \end{pmatrix}, \mathbf{F}^c(\mathbf{u}) = \begin{pmatrix} \rho\mathbf{v} \\ \rho\mathbf{v}\circ\mathbf{v} + p\mathcal{I} \\ \rho h_{tot}\mathbf{v} \end{pmatrix} \tag{4.3}$$

where ρ denotes the density, p the static pressure, $\mathbf{v}$ the velocity vector, e_{tot} the total energy and h_{tot} the total enthalpy. The system is closed by the equation of state for a perfect gas $p = \rho(\gamma - 1)(e_{tot} - 1/2\,|\mathbf{v}|^2)$, where γ is the ratio of specific heats, which is taken as 1.4.

5 Finite Volume Method

5.1 Data Structure

The discretization of the governing equations (4.2) is based on a cell centered finite volume scheme. Caused by the occurance of hanging nodes, the grid is treated as a fully unstructured mesh with arbitrary polygonal control volumes. The data structure of the flow solver is primarily based on the faces of the grid. A face based data structure has the important benefit that there are no limitations on the number of faces, which can be connected to a cell. In particular, the evaluation of fluxes and their distribution to cells can efficiently be implemented by sweeps over the faces. Further, the face based infrastructure provides a sufficient amount of information to establish a *compact* stencil for a linear reconstruction of the flow variables. Hence, the solution algorithm is fully grid-transparent, which makes it a very fexible tool that can be combined with any type of grid. Presently, we focus on grids in two space dimensions.

5.2 Conservative Fluxes

The conservative fluxes are determined by solving quasi-one-dimensional Riemann problems at the cell interfaces. Within the present paper, the HLLC scheme due to Batten and Leschziner [5] has been applied, which is capable of exactly preserving isolated shocks, contact, and shear waves. Further, the entropy condition is inherently enforced by the scheme.

5.3 Higher Order Method in Space

In order to obtain second-order accuracy in space, for each control volume Ω_i a linear reconstruction of the flow variable $w \in \{\rho,\rho u,\rho v,\rho e_{tot}\}$ is determined

$$w(\mathbf{x})|_{\Omega_i} = w_i + \phi_i \nabla w_i \cdot (\mathbf{x} - \mathbf{c}_i) \; , \tag{5.4}$$

where w_i represents its mean value at the centroid $\mathbf{c}_i$ of Ω_i and ϕ_i denotes a limiter function. The gradient ∇w_i is calculated by a multi-dimensional Taylor series expansion around the centroid of Ω_i

$$w_i + \nabla w_i \cdot (\mathbf{x}_j - \mathbf{c}_i) = w_j \Rightarrow \mathcal{A} \nabla \mathbf{w}_i = \mathbf{\Delta w}_i \tag{5.5}$$

where

$$\mathcal{A} = \begin{bmatrix} \Delta x_1 & \Delta y_1 \\ \vdots & \vdots \\ \Delta x_{N_i} & \Delta y_{N_i} \end{bmatrix}, \mathbf{\Delta w}_i = \begin{pmatrix} \Delta w_1 \\ \vdots \\ \Delta w_{N_i} \end{pmatrix} \tag{5.6}$$

with $\Delta(\bullet)_j = (\bullet)_j - (\bullet)_i \, , j \in N_i$. N_i is the set of neighbouring cells of Ω_i, which support the reconstruction of w. In practice, equation (5.6) represents an overdetermined system of equations for the gradient of w_i. It is solved in a least squares sense.

At local extrema and discontinuities, the recovery polynom may generate new extrema and therefore yield oscillations in the numerical solution. In order to circumvent this problem, limiter functions with TVD (total variation diminishing) property are used. The slope limiter from Venkatakrishnan [6] is employed.

5.4 Time Integration

For steady fluid flow, the solution is advanced in time by an implicit backward Euler method. Local time steps are applied. For the flux balance of the cell i we introduce the so called residual vector

$$\mathbf{R}_i := \oint_{\partial\Omega_i} \mathbf{F}^c(\mathbf{u}) \, \mathbf{n} \, dS \, . \tag{5.7}$$

Herewith, the conservation laws (4.2) read in semi-discrete form as

$$\widehat{\mathbf{R}}^{n+1} = \mathbf{0} \quad \text{with} \tag{5.8}$$

$$\widehat{\mathbf{R}}^{n+1} = \mathbf{R}^{n+1} + \frac{\Omega}{\Delta t}\left(\mathbf{u}^{n+1} - \mathbf{u}^n\right) \quad \text{and} \tag{5.9}$$

$$\widehat{\mathbf{R}}^{n+1} := \widehat{\mathbf{R}}\left(\mathbf{u}^{n+1}\right) . \tag{5.10}$$

Equation (5.8) is solved by a Newton scheme. For stationary flows we take one Newton iteration per physical time step. In this case the Newton scheme is:

$$\widehat{\mathcal{J}}(\mathbf{u}^n)\, \Delta\mathbf{u}^n = -\mathbf{R}(\mathbf{u}^n) \tag{5.11}$$

with

$$\widehat{\mathcal{J}} = \frac{\Omega}{\Delta t}\mathcal{I} + \left.\frac{\partial \mathbf{R}}{\partial \mathbf{u}}\right|^n , \quad \Delta\mathbf{u}^n := \mathbf{u}^{n+1} - \mathbf{u}^n . \tag{5.12}$$

The linear system of equations (5.11) is solved by an iterative Krylov subspace method. In this study, we employ the GMRES algorithm [7], preconditioned by an incomplete LU-factorization. Typically, a Krylov subspace method requires only the inner product of a Jacobian matrix $\mathcal{J}$ and a vector ν, and not $\mathcal{J}$ explicitely [8]. This product can also be approximated by a difference quotient of the form

$$\mathcal{J}(\mathbf{u})\,\nu \approx \frac{\mathbf{R}(\mathbf{u}+\epsilon\nu)-\mathbf{R}(\mathbf{u})}{\epsilon}, \quad \text{with} \quad \epsilon \in \mathbb{R}. \tag{5.13}$$

The difference parameter ϵ is chosen as

$$\epsilon = \hat{\epsilon}\frac{\sqrt{1.0+||\mathbf{u}||_2}}{||\nu||_2} \tag{5.14}$$

with $\hat{\epsilon} = 10^{-8}$ in our application. The advantage of using a matrix free version of a Krylov subspace method is, that the matrix vector product may consistently be discretized according to the evaluation of the residual vector, that represents the right hand side of equation (5.11). Nevertheless, a Jacobian is still required for the applied $ILU(n)$ preconditioner. The preconditioning matrix is based on a first order accurate scheme in space. The Jacobian is derived numerically [9], using one sided difference operators of the form:

$$\frac{\partial \mathbf{F}_i(\mathbf{u})}{\partial \mathbf{u}_j} = \lim_{\tilde{\epsilon}\to 0}\frac{\mathbf{F}_i(\mathbf{u}+\tilde{\epsilon}\mathbf{e}_j)-\mathbf{F}_i(\mathbf{u})}{\tilde{\epsilon}}, \quad \text{with} \quad \tilde{\epsilon} \in \mathbb{R}, \tag{5.15}$$

where $\mathbf{F}_i$ is the ith component of the numerical flux vector and $\mathbf{e}_j$ is the jth unit vector, associated with the dependent variable $\mathbf{u}_j$. The infrastructure related to the Newton-Krylov method is based on the PETSc [10] library of Argonne National Laboratory. For further details of the numerical method and validation of the spatial accuracy of the scheme see [11].

6 Numerical Results

6.1 Subsonic and Transonic Channel Flow

The first example considers the GAMM test case [12] of inviscid flow at Mach 0.5 and Mach 0.85 over a 4.2% circular arc bump in a channel with height 2.073 times the chord length. The inflow and outflow boundaries were placed two chord lengths upstream and downstream of the bump, respectively. Firstly, computations for Mach 0.5 were carried out on a non-adaptive mesh with 320x30 cells to assess the convergence of the implicit scheme for smooth flows. Fig. 3 shows the isobars for this case. Fig. 4 illustrates the convergence history for four different *CFL* numbers, namely *CFL*= $10, 10^2, 10^3$ and 10^4. The residual is monitored in the L_1 norm of the density. The rate of convergence is greately accelerated with increased *CFL* number. For *CFL*> 10^4, no further improvement could be observed. *Remark:* The scheme remained unconditionally stable even for *CFL*= 10^{20}.

Secondly, the channel configuration is investigated in transonic Mach number regime. Inflow Mach number is 0.85. The domain is tesselated by three blocks, each of them intitially consisting of 5x5 cells. The computation is initialized on a grid, corresponding to one level of uniform adaptation, since the adaptive algorithm must be capable of coarsening the mesh. The computation was carried out with *CFL*= 70. Fig. 5 shows the isobars for the present flow condition. The solution exhibits very crisp shock resolution and highly resolved stagnation areas. The isobars smoothly cross interfaces between different grid levels. Fig. 6 presents the locally adapted grid. The mesh consists of 14910 cells, including 8 levels of refinement. Such a high resolution is not feasible using non-adaptive grids. A uniform discretization of the domain, corresponding to a grid at level 8, would consist of 1228800 cells. Of course, meshes do not use to have the same fine resolution in the complete domain, nevertheless it is interesting to recognize that the adaptive scheme requires 1.21% of this amount.

6.2 SFB401 Cruise Configuration

The next test case investigates the flow around the airfoil of the cruise configuration of the collaborative research center SFB401. Results for $M = 0.176, \alpha = 0^0$ are presented only to demonstrate the utilization of the adaptive algorithm for subsonic flow. The computation is started on a grid consisting of four blocks, each with a resolution of 10x10 cells. Adaptation is executed each time the residual decreased 5 orders of magnitude, based on the initial residual. Convergence is measured in the L_1 norm of the density. Fig. 7 shows, that during the initial phase the number of cells increases with each adaptation step, until a maximum number of 10000 cells is reached after 6 cycles. Subsequently, the mesh is coarsened again - particularly in the wake region -. After 8 adaptation steps, the number of cells remains almost constant at ≈ 7000, containing 7 grid levels. The solution quality just marginally improves with further adaptation. We were not able to obtain a fully stationary grid and repeatedly refinement/derefinement cycles were observed. Fig. 8 shows the locally adapted grid after 15 cycles of adaptation. The stagnation area as well as the trailing edge are highly resolved. Fig. 9 illustrates the corresponding pressure distribution around the airfoil. *Note:* Non-smooth variation of isolines is attributed to visualization rather than to the flow solution itself.

7 Conclusion

A new method for solving the Euler equations on locally adaptive grids with hanging nodes in two space dimensions has been presented. The spatial discretization is carried out using a higher-order upwind method. Time integration is based on a fully implicit Newton-Krylov type scheme. The convergence of the implicit method has been studied for subsonic flow in a channel with a circular arc bump. For the same geometric configuration, the adaptive algorithm provided excellent solution quality within transonic Mach number regime. In particular, a crisp resolution of the shock and highly resolved stagnation areas were obtained. It was found, that in this case the adaptive method only requires about 1.21% of the number of cells, compared with a uniform mesh, based on the finest level of the adapted grid. Further, the application of the adaptive algorithm for subsonic flow around a wing section was demonstrated. It was found, that the applied adaptation criteria provided a reliable tool to detect the significant flow features, in subsonic as well as in transonic flow.

Acknowledgement

The financial support from the Deutsche Forschungsgemeinschaft (DFG) is gratefully acknowledged. Thanks to Mr. P. Lamby and Mr. M. Hesse for providing the computational grids.

References

[1] Ballmann, J., *Strömungsbeeinflussung und Strömungs-Struktur-Wechselwirkung an Tragflügeln, Erste Ergebnisse des SFB401 der RWTH Aachen*, DGLR-JT99-189, DGLR Jahrbuch 1999

[2] Brakhage, K.H., Müller, S., *Algebraic-Hyperbolic Grid Generation with Precise Control of Intersection of Angles*, Int. J. Numer. Meth. Fluids. To appear

[3] Harten, A., *Multiresolution algorithms for the numerical solution of hyperbolic conservation laws*, Comm. Pure Appl. Math. 12, vol. 48 (1995), 1305–1342

[4] Gottschlich–Müller, B., Müller, S., *Adaptive Finite Volume Schemes for Conservation Laws based on Local Multiresolution Techniques*, in: Hyperbolic Problems: Theory, Numerics, Applications, eds. M. Fey and R. Jeltsch, Birkhäuser, 1999, pp. 385–394

[5] Batten, P., Leschziner, M.A., Goldberg, U.C., *Average-State Jacobians and Implicit Methods for Compressible Viscous and Turbulent Flows*, J. Comp. Phys., 137, 1997, pp. 38–78

[6] Venkatakrishnan, V., *Convergence to Steady State Solutions of the Euler Equations on Unstructured Grids with Limiters*, J. Comp. Phys., 118, 1995, pp. 120–130

[7] Saad, Y., *Iterative Methods for Sparse Linear Systems*, PWS, Boston, 1996

[8] Grotowsky, I.M.G., Ballmann, J., *Efficient Time Integration of Navier-Stokes Equations*, Computers & Fluids, vol 28, no 2, 1999, pp. 243–263

[9] Vanden, K.J., Orkwis, P.D., *Comparison of Numerical and Analytical Jacobians*, AIAA Journal, vol 34, no 6, 1996

[10] Balay, S., Gropp, W., McInnes, L.C., Smith, B.F., *PETSc 2.0 Users Manual*, Tech. Rep. ANL-95/11 - Revision 2.0.28, Argonne National Laboratory, 2000, `http://www-fp.mcs.anl.gov/petsc/`

[11] Bramkamp, F., Ballmann, J., Müller, S., *Development of a Flow Solver Employing Local Adaptation Based on Multiscale Analysis on B-Spline Grids*, in: Proceedings of 8th Annual Conf. of the CFD Society of Canada, 2000, pp. 113-118

[12] Rizzi, A., Viviand, H. (Eds.), *Numerical Methods for the Computation of Inviscid Transonic Flow with Shock Waves*, Notes on Numerical Fluid Mechanics, vol 3, Vieweg, 1981

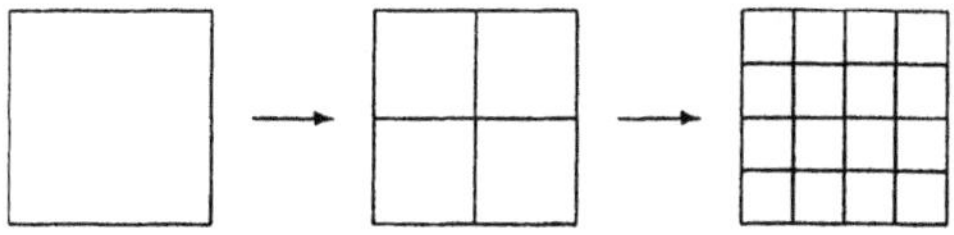

Figure 1 Sequence of nested grids

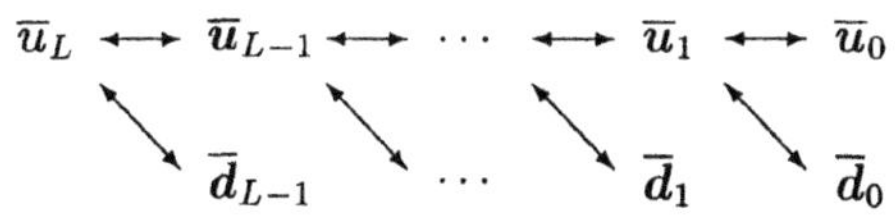

Figure 2 Multiscale Transformation

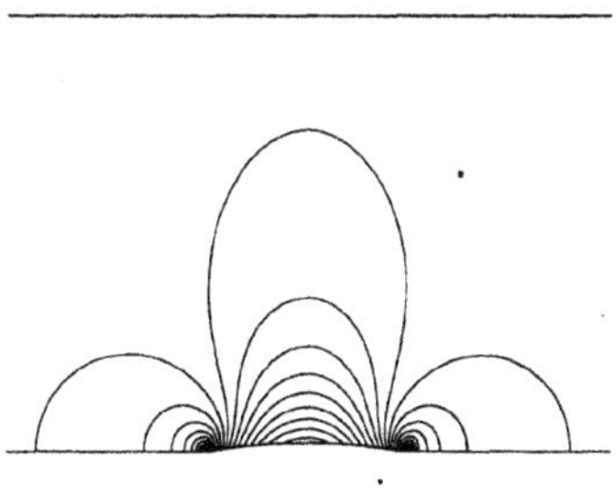

Figure 3 Isobars of channel with circular arc bump (detailed view), $M_\infty = 0.5$, $c_{p,min} = -0.26$, $c_{p,max} = 0.34$, $\Delta c_p = 0.03$

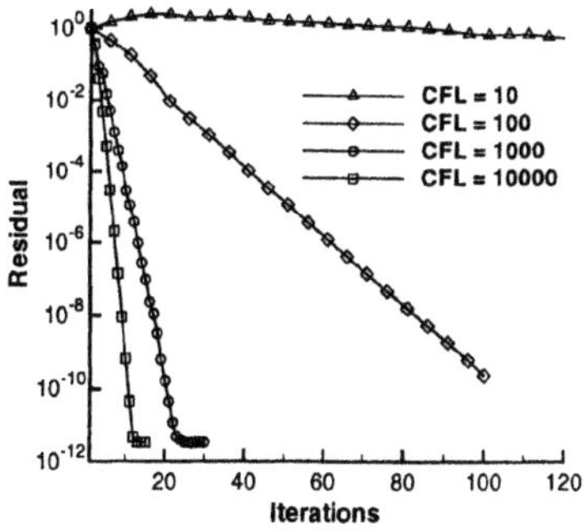

Figure 4 Convergence histories for channel with circular arc bump, $M_\infty = 0.5$

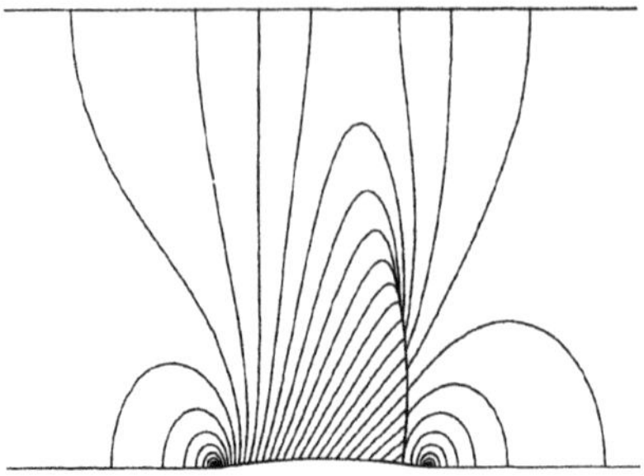

Figure 5 Isobars of channel with circular arc bump (detailed view), $M_\infty = 0.85$, $c_{p,min} = -0.94$, $c_{p,max} = 0.67$, $\Delta c_p = 0.05$

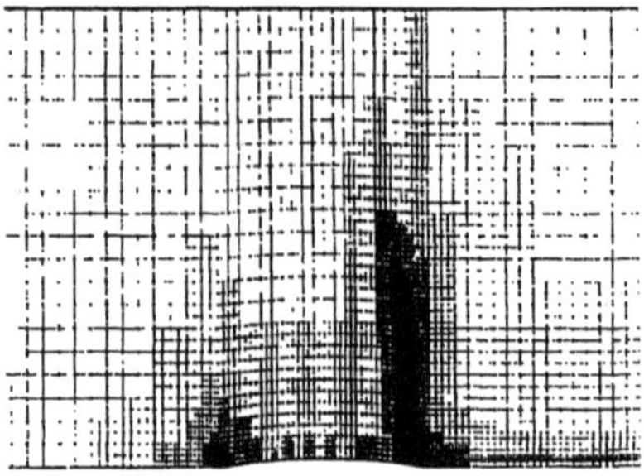

Figure 6 Locally adapted grid for channel with circular arc bump, $M_\infty = 0.85$, number of cells: 14910, number of grid levels: 8

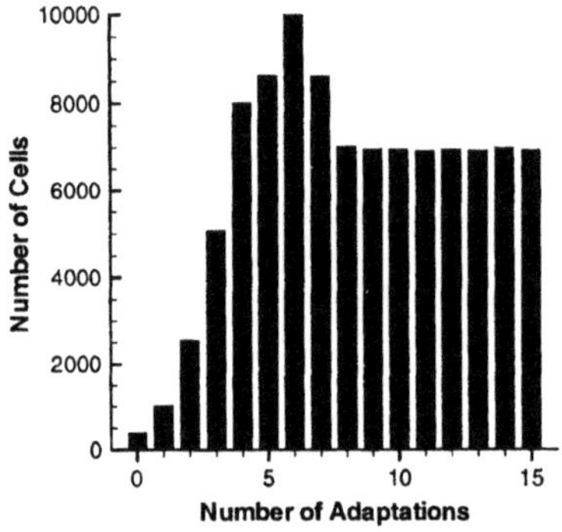

Figure 7 Number of grid cells during adaptation process for the SFB401 cruise configuration

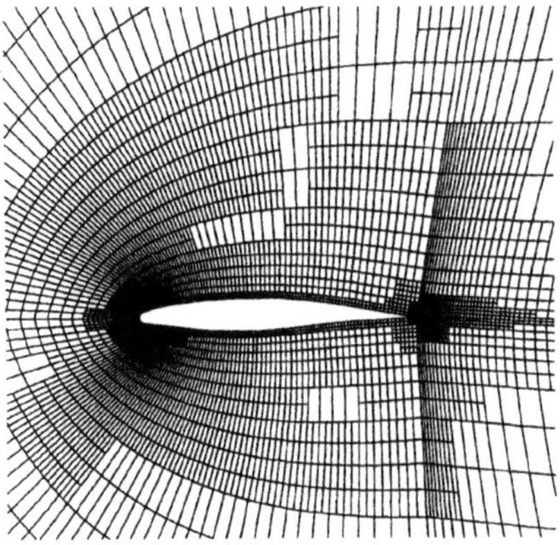

Figure 8 Locally adapted grid of the SFB401 cruise configuration, $M_\infty = 0.176$, $\alpha = 0°$, number of cells: 6991, number of grid levels: 7

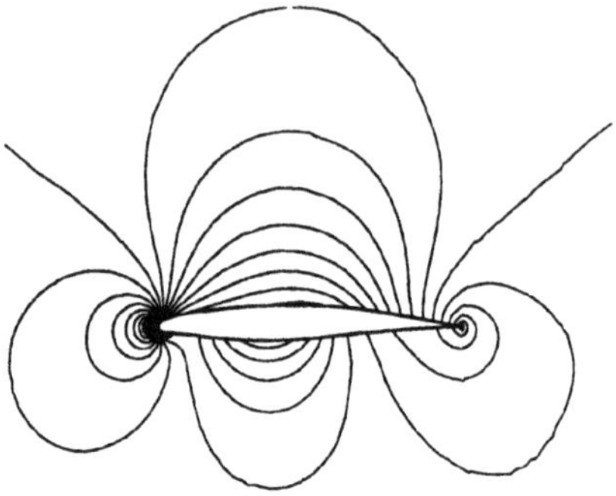

Figure 9 Isobars of the SFB401 cruise configuration, $M_\infty = 0.176$, $\alpha = 0°$, $c_{p,min} = -0.46$, $c_{p,max} = 0.99$, $\Delta c_p = 0.05$

An Algorithm to Treat Flows Bounded by Arbitrarily Shaped Surfaces with Cartesian Meshes

F. Tremblay and R. Friedrich
Fachgebiet Strömungsmechanik, Technische Universität München
Boltzmannstr. 15, D-85748 Garching, Germany

Summary

A method to compute flows around arbitrarily shaped surfaces with cartesian grids is presented. It is implemented in the framework of a second-order finite volume method based on staggered variable arrangement. The method uses Dirichlet velocity boundary conditions for cells located closest to the body surface in order to satisfy the wall no-slip and impermeability conditions. The velocity values are obtained by spatial interpolation/extrapolation of field values via third order Lagrangian polynomials. The method has been validated for steady and unsteady laminar flows and also for a turbulent pipe flow. Validation tests show that the method preserves the second order accuracy of the code MGLET and that reliable solutions can be obtained.

1 Introduction

The feasibility to compute flows over arbitrarily shaped bodies with cartesian grids is very attractive and of great practical importance since typically a cartesian code is anywhere between 10 and 50 times more economical in terms of both CPU time and memory requirements than a code solving the Navier-Stokes equations in curvilinear coordinates [1]. One can thus afford to do a computation with more grid points and still achieve appreciable savings in computational resources. Another important aspect is the complete elimination of the need to produce a body-fitted grid, a task that is not trivial and can consume an important amount of time.

The code MGLET, used here, is based on a finite volume formulation of the Navier-Stokes equations on staggered Cartesian non-equidistant grids. The spatial discretization is central and of second order accuracy for the convective and diffusive terms. For the time advancement of the momentum equations, an explicit second-order time step is used which is central with respect to the convective terms. The pressure solver uses a multigrid method based on a point-wise velocity-pressure iteration as described by Hirt [2].

The most straightforward method to represent the no-slip condition on a body surface within a cartesian grid is to apply zero velocity at the cell which is the closest to the surface of the body. The difference between the actual and exact body geometry is at most half a cell. This was the method employed by Manhart and Wengle [3] among others. The drawback of this method is that it is only first order accurate. This drawback can be removed by modifying the discretization near the body in order to take into account the cells which are cut. Although more accurate schemes are obtained, like in [4] and [5], their computations were two-dimensional and the extension of their methods to 3D geometries is not easy because a cell can be cut in many different ways. Other researchers [6],[7] tried to preserve the same discretization in all the domain, even including the cells inside the body. Forcing is applied at the location of the body in order to represent its effect on the fluid. Much simpler and general formulations are obtained. The major problem encountered is the impossibility to enforce strictly the impermeability condition which could possibly lead to erroneous solutions at a stagnation point for example. In the present approach we apply Dirichlet velocity boundary conditions on each cell located in the immediate vicinity of the body surface. The boundary conditions are applied in such a way that the physical location of the surface and its velocity are best represented. The cells beyond the body surface are excluded from the computation by using a masking array. The discretization remains the same for all cells. A similar approach was derived independantly by Gullbrand et al. [8] but in the context of finite differencing on regular grids.

2 Handling of arbitrarily shaped bodies

The first step is to obtain a representation of the surface of interest. A very versatile way of doing it is by using an unstructured mesh made of triangles. All geometries can be represented this way. Once the body is available, a preprocessing step is done. During this phase a masking array is initialized, which has values of one for cells within the flow and zero for cells belonging to the body. This array is used to block the pressure cells inside the body considered so that they do not contribute to the computation. Boundary conditions have to be applied to the velocity components at the interface between masked and unmasked pressure cells in order to have a well defined problem. In the second and last phase of the preprocessing step, these boundary conditions are defined as a linear function of the velocities computed within the domain. The coefficients weighting the computed velocities are determined and need no longer to be updated during the run.

2.1 Preprocessing STEP 1: Masking the pressure cells

Blocking pressure cells depends on whether the intersection between computational cells and triangles, representing the body surface, satisfies certain conditions which are described below. These conditions have to be simple and unique. We start specifying a triangle by its 3 vertex points (see figure 1) :

$$\vec{P}_1 = (x_1,y_1,z_1)\,,\vec{P}_2 = (x_2,y_2,z_2)\,,\vec{P}_3 = (x_3,y_3,z_3)$$

They define a plane in which a point P lies, that is given by its position vector $\vec{P} = (x,y,z)$ and satisfies:

$$\left(\vec{P}-\vec{P}_1\right)\cdot\left(\left(\vec{P}_2-\vec{P}_1\right)\times\left(\vec{P}_3-\vec{P}_1\right)\right)=0 \tag{2.1}$$

From eq.(2.1), we get the equation of the plane

$$a\,(x-x_1)+b\,(y-y_1)+c\,(z-z_1)=0 \tag{2.2}$$

where

$$
\begin{aligned}
a &= (y_2 - y_1)(z_3 - z_1) - (z_2 - z_1)(y_3 - y_1) \\
b &= (z_2 - z_1)(x_3 - x_1) - (x_2 - x_1)(z_3 - z_1) \\
c &= (x_2 - x_1)(y_3 - y_1) - (y_2 - y_1)(x_3 - x_1)
\end{aligned}
$$

This plane may have up to three intersection points with coordinate lines passing through the cell center C (the pressure point) and these points may lie inside or outside the cell volume. We call D_1, D_2 and D_3 the intersection points with the (x,y,z)-directions, respectively. Their coordinates are :

$$
\begin{aligned}
D_1 &= \left(\frac{-b\,(y_c - y_1) - c\,(z_c - z_1)}{a} + x_1, y_c, z_c \right) \\
D_2 &= \left(x_c, \frac{-a\,(x_c - x_1) - c\,(z_c - z_1)}{b} + y_1, z_c \right) \\
D_3 &= \left(x_c, y_c, \frac{-a\,(x_c - x_1) - b\,(y_c - y_1)}{c} + z_1 \right)
\end{aligned}
$$

where $C\ :\ (x_c, y_c, z_c)$ is the cell center. If one of these 3 points lies within the pressure cell and within the triangle, then this pressure cell is blocked out of the computational domain and does not contribute to the numerical solution. Verifying if a point lies in a triangle is done in the following way: Let us consider a point D that has been found to lie on the plane defined by a triangle (see figure 3).

$$
\begin{aligned}
\vec{n}_1 &= \left(\vec{P_1P_2} \times \vec{P_1P_3} \right) \times \vec{P_1P_2} \\
\vec{n}_2 &= \left(\vec{P_1P_3} \times \vec{P_1P_2} \right) \times \vec{P_1P_3} \\
\vec{n}_3 &= \left(\vec{P_2P_3} \times \vec{P_2P_1} \right) \times \vec{P_2P_3}
\end{aligned}
$$

The point D is inside the triangle if $\left(\vec{n}_1 \cdot \vec{d}_1\right)$, $\left(\vec{n}_2 \cdot \vec{d}_2\right)$ and $\left(\vec{n}_3 \cdot \vec{d}_3\right) \geq 0$.
Once all the cells have been checked, we obtain a thin layer of blocked cells along the surface of the body. The interior of the body must also be blocked. This is easily done by a "painting" algorithm, i.e. the user specifies one cell which is in the flow field (open cell), and by iteratively marking the neighbours of the open cells unless they are blocked, we can paint the exterior of the body and thus the interior is easily identified.

2.2 Preprocessing STEP 2: Determining Dirichlet velocity boundary conditions

Once we have obtained the blocking array of the pressure cells, the blocking arrays of the velocities can be deduced in a straightforward manner. Each velocity component belonging to a blocked pressure cell is also blocked. In order to represent the effect of the body on the surrounding flow, a few layers (enough to have a complete stencil of the discretization at the first open cell) of blocked velocities are interpolated/extrapolated using the no-slip condition that prevails at the surface of the body. We use third order Lagrangian polynomials to achieve this goal.

In figure 4,x_p is the location of the point to obtain an extrapolated variable, x_1 is the location of the surface of the body and x_2, x_3, x_4 are the positions of the neighbouring points. ϕ_p is the variable to be extrapolated (say a velocity component), ϕ_1 is the value at the wall which should be zero in case of a no-slip condition, ϕ_2, ϕ_3, ϕ_4 are the values taken from the neighbouring points.
The value of the extrapolated variable will be :

$$\phi_p = \sum_{i=1}^{4} \alpha_i \phi_i$$

where

$$\alpha_i = \prod_{j=1, j \neq i}^{4} \frac{(x_p - x_j)}{(x_i - x_j)}$$

We can define the distance Δ_k from the extrapolated point to the wall as $\Delta_1 = |x_p - x_1|$, e.g.. If the variable at a point can be extrapolated from more than one direction, each direction is weighted by a multiplication factor γ_{dir}:

$$\gamma_{dir} = \frac{\beta_{dir}}{\sum_{k=1}^{Ndir} \beta_k}$$

where

$$\beta_{dir} = \frac{\prod_{k=1, k \neq dir}^{Ndir} \Delta_k}{\Delta_{dir}}$$

This formulation ensures that if a point is exactly on a wall, it will have the wall value. Also, the sum of the multiplication factors is always 1. From these simple formulas, all the coefficients used to set the velocities at the surface of the body are determined and used during the actual computation.

3 Validation tests

3.1 Cylindrical Couette flow

This test case was performed in order to evaluate if the method preserves the second order accuracy of MGLET, a finite volume code which uses second order central schemes [1]. This test case also demonstrates the possibility of handling surfaces with non-zero velocity boundary conditions. In this particular configuration, the inner cylinder has a diameter of 1 meter and is rotating while the outer one has a diameter of 6 meters and is kept at rest. The Reynolds number based on the rotation velocity of the inner cylinder is equal to 120. Four equidistant grids were used to compute the cylindrical Couette flow (which in principle is a one-dimensional problem), the grid spacing was varied from 0.1m to 0.025m.

In figures 5 and 6, it is nearly impossible to distinguish our solution from the analytical one, only the coarser grid produces noticeable differences. This shows that sufficiently accurate solutions can be obtained, and that the method is suited for surfaces having non-zero velocity. Figure 8 shows that the method is second order accurate.

3.2 Circular cylinder between parallel plates

A steady and an unsteady 2D test case have been computed. The configuration chosen is a cylinder with a circular cross-section mounted between 2 parallel plates with no-slip conditions. A sketch can be seen in figure 7. The cylinder is not exactly centered between the plates, so that lift is produced. This configuration was studied by Schäfer et al. [9]. For the steady case, the Reynolds number is 20, and for the unsteady case, it is 100.

In table 1, the drag and lift coefficients C_D, C_L, the recirculation length L_a and the pressure difference ΔP between the front and the back of the cylinder are presented for the steady case.

Table 2 presents the results for the unsteady case. The quantities computed were: the maximum lift and drag coefficient, the Strouhal number and the maximum pressure difference between the front and the back of the cylinder. It should be noted that the last grid used was stretched so that the grid spacing is smaller near the cylinder.

From this validation test, we conclude that our results are in good agreement with computations done with other codes and methods, the results of which were scattered between the lower and upper bounds and often beyond the limits.

3.3 Turbulent pipe flow

In order to demonstrate the applicability of the method to turbulent flows, we computed the fully developed flow in a circular pipe at $Re_\tau = 180$ directly, resolving all turbulent scales. We compare our results to the results obtained using a code [10] for helical as well as cylindrical coordinates. The grid was chosen so that the closest point is at around $y^+ = 0.5$ away from the wall and that there are approximatively 10 points below $y^+ = 10$ at a position with the coarsest resolution, located at $\Theta = 45$ degrees. In the middle of the channel, the grid spacing is $\Delta y^+ = 5.0$. The grid can be seen in figure 12, only one quarter of the pipe is shown. Our conservative decisions gave rise to a grid which had 256x256 points in a cross section of the pipe. In the streamwise direction, the grid spacing is $\Delta x^+ = 7.0$, so that 256 points were used. Figures 9,10 and 11 show the mean velocity profile, the RMS of velocity fluctuations and the RMS of velocity fluctuations with a logarithmic x-axis. No artefacts of our method can be seen in the solution near the wall. The overall agreement is very good. Using a cartesian grid removes the singularity at the centerline associated with a cylindrical coordinate system. We must confess that the method works more efficiently in flow situations around bodies since there is less overhead of blocked cells (e.g. flow around a circular cylinder or an airfoil), in our computation of the pipe flow one can notice in figure 12 that the regions where the grid is most refined lie outside of the computational domain. In any case, the DNS results obtained with the present method are in good agreement with previous DNS data, which confirms the suitability of the present technique for DNS of turbulent flows.

4 Conclusions

In the test cases computed, we have demonstrated the capability of the cartesian method to provide reliable solutions in an efficient way for flows around bodies of arbitrary shape. Although the test cases presented here are geometrically simple, the present method can handle much more complicated surfaces which are of industrial relevance.

5 Acknowledgement

The first author gratefully acknowledges the support of this work in the framework of the EC project Alessia.

References

[1] Manhart, M., Deng, G., Hüttl, T., Tremblay, F., Segal, A., Friedrich, R., Piquet, J. and Wesseling, P., The minimal turbulent flow unit as a test case for three different computer codes. In: E. Hirschel (ed.), *Vol. 66, Notes on numerical fluid mechanics*, pp. 365–381. Vieweg-Verlag, Braunschweig (1998).

[2] Hirt, C., Nichols, B. and Romero, N., SOLA – A numerical solution algorithm for transient fluid flows. In: *Los Alamos Sci. Lab.*, Los Alamos (1975).

[3] Manhart, M. and Wengle, H., Large-eddy simulation of turbulent boundary layer flow over a hemisphere. In: V. P.R., L. Kleiser and J.-P. Chollet (eds.), *Direct and Large-Eddy Simulation I*, pp. 299–310, Dordrecht (1994). ERCOFTAC, Kluwer Academic Publishers.

[4] Yang, G., Causon, D. and Ingram, D., Cartesian Cut-Cell Method for Axisymmetric Separating Body Flows. *AIAA Journal*, **37(8)**, 905–911 (1999).

[5] Forrer, H., Second Order Accurate Boundary Treatment for Cartesian Grid Methods. Research Report No. 96-13, Seminar für Angewandte Mathematik, Eidgenössische Technische Hochschule, Zürich (1996).

[6] Mohd-Yusof, J., Combined immersed-boundary/B-spline methods for simulations of flow in complex geometries. In: *Annual Research Briefs - 1997*, pp. 317–327. Center for turbulence research, Stanford (1997).

[7] Glowinski, R., Pan, T. and Periaux, J., Fictitious Domain Methods for Incompressible Viscous Flow around Moving Rigid Bodies. In: *Proceedings of MAFELAP, Brunnel University, UK, 25-28th June 1996* (1996).

[8] Gullbrand, J., Bai, X. and Fuchs, L., High Order Cartesian Grid Method for Calculation of Incompressible Turbulent Flows. *Int. J. Numer. Meth. Fluids*, p. to appear (2000).

[9] Schäfer, M. and Turek, S., Benchmark Computations of Laminar Flow Around a Cylinder. In: E. Hirschel (ed.), *Notes on Numerical Fluid Mechanics*, pp. 547–566. Vieweg-Verlag (1996).

[10] Hüttl, T., *Direkte Numerische Simulation turbulenter Strömungen in gekrümmten und tordierten Rohren.* Ph.D. thesis, TU München (1999).

Table 1 Integral parameters for the steady flow over a cylinder between parallel plates

Dimensions	Cells	C_D	C_L	L_a	ΔP
270x82	22 140	5.587	0.0115	0.0831	0.1162
330x123	40 590	5.572	0.0109	0.0839	0.1170
420x164	68 880	5.570	0.0109	0.0841	0.1172
	lower bound	5.57	0.0104	0.0842	0.1172
	upper bound	5.59	0.0110	0.0853	0.1176

Table 2 Integral parameters for the unsteady flow over a cylinder between parallel plates

Dimensions	Cells	C_{Dmax}	C_{Lmax}	St	ΔP_{max}
440x82	36 080	3.27	0.8820	0.299	2.426
880x164	144 320	3.23	0.987	0.302	2.486
700x200	140 000	3.22	1.000	0.302	2.496
	lower bound	3.220	0.9900	0.2950	2.460
	upper bound	3.240	1.0100	0.3050	2.500

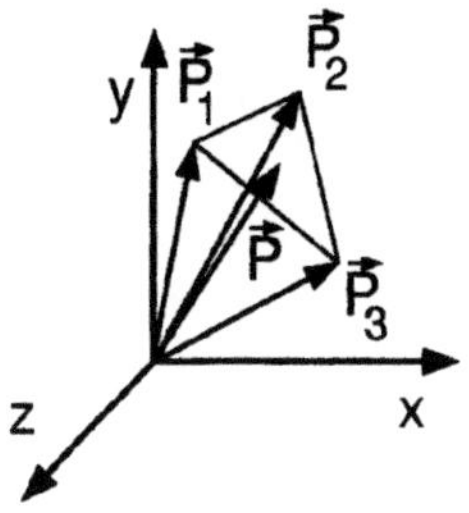

Figure 1 Definition of a triangle

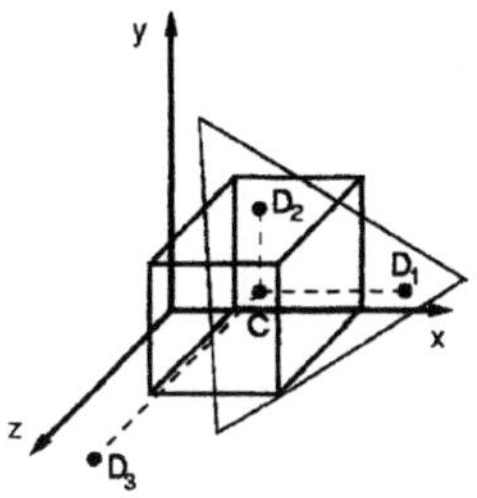

Figure 2 Intersection of a triangle and a cell

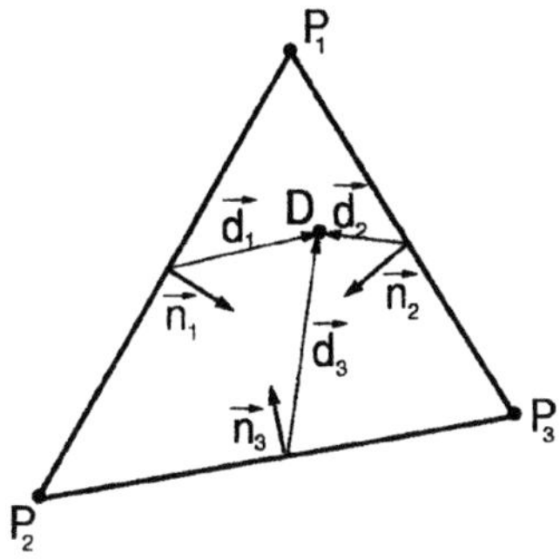

Figure 3 A point lying in a triangle

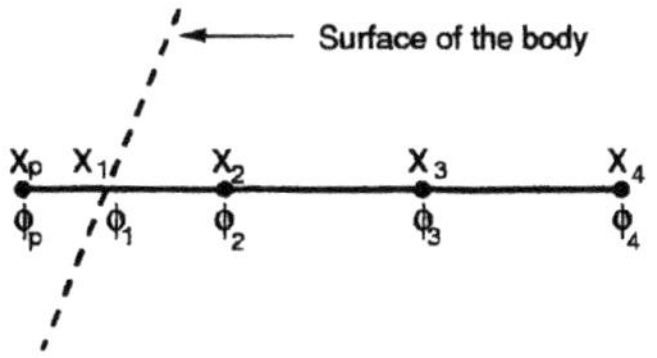

Figure 4 Using computed velocities to determine the boundary conditions

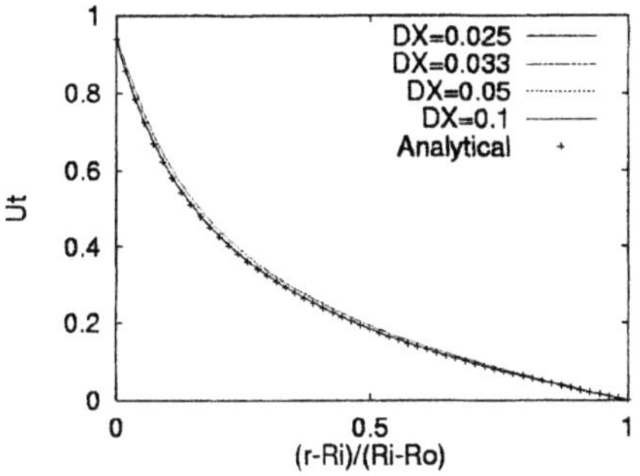

Figure 5 Radial distribution of the tangential velocity in a laminar cylindrical Couette flow for different grid spacing

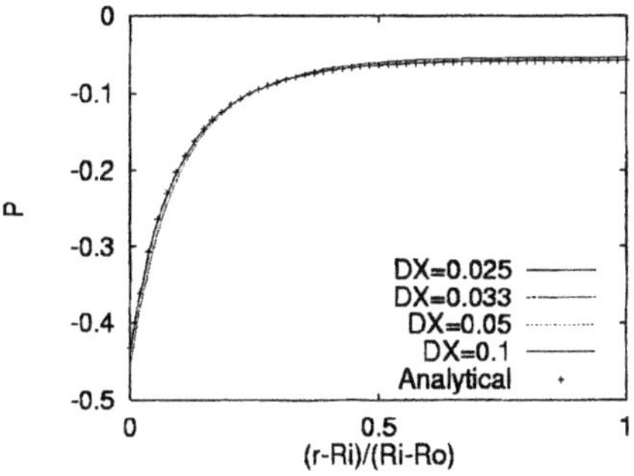

Figure 6 Radial distribution of the pressure in a laminar cylindrical Couette flow for different grid spacing

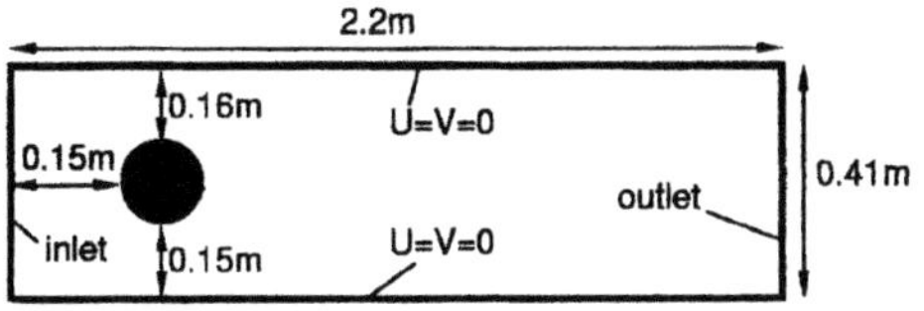

Figure 7 Geometry for the 2D laminar test case

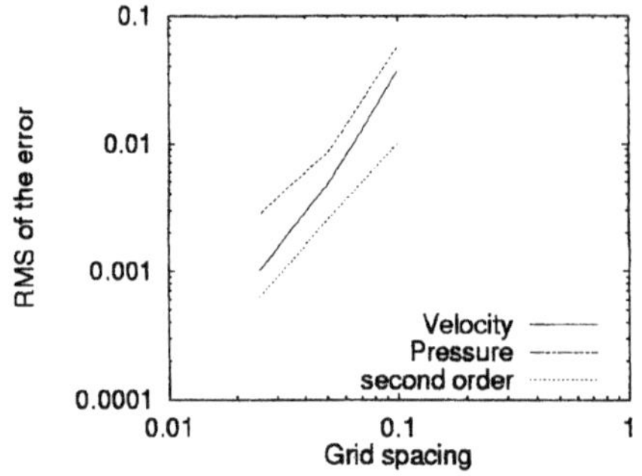

Figure 8 RMS of the error as a function of grid spacing for the laminar cylindrical Couette flow

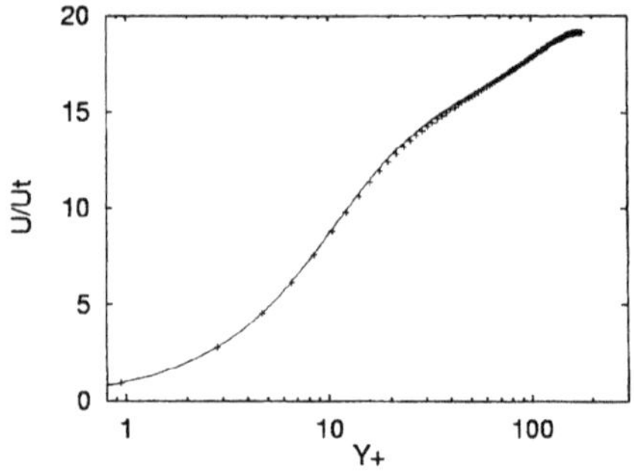

Figure 9 Turbulent pipe flow: Mean axial velocity profile. Lines: cartesian method, Symbols: cylindrical code.

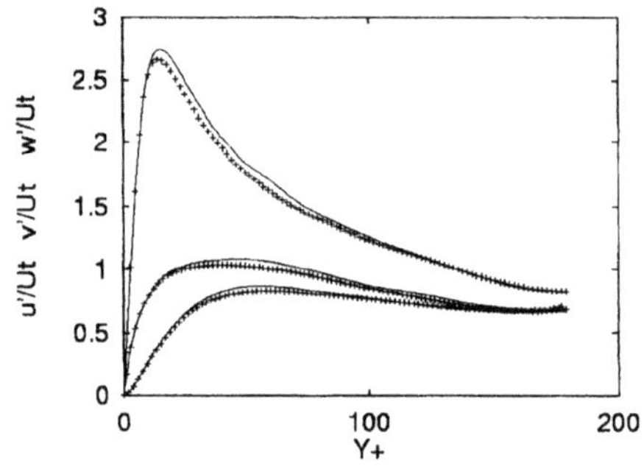

Figure 10 RMS of the velocity fluctuations Lines: cartesian method, Symbols: cylindrical code.

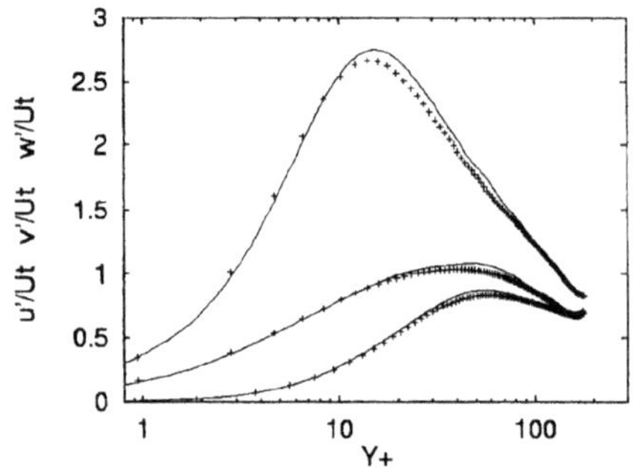

Figure 11 RMS of the velocity fluctuations Lines: cartesian method, Symbols: cylindrical code.

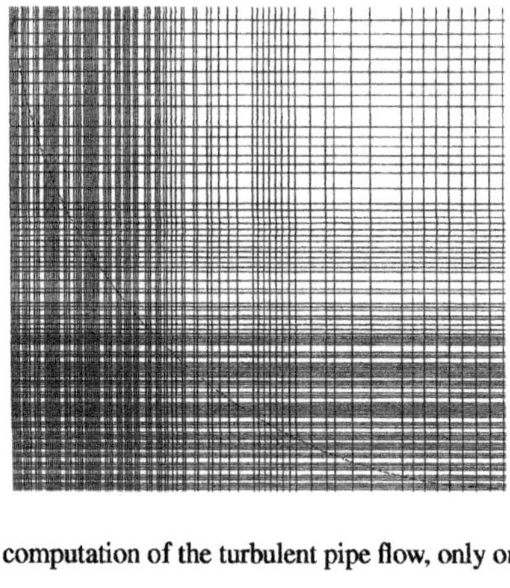

Figure 12 Grid used for the computation of the turbulent pipe flow, only one quarter of the pipe shown

Grid Refinement Study with a UHCA Wing-Body Configuration Using Richardson Extrapolation and Grid Convergence Index GCI

M. RAKOWITZ
German Aerospace Center (DLR), Institute of Design Aerodynamics,
Lilienthalplatz 7, 38108 Braunschweig, Germany

Summary

The prediction of aerodynamic drag for complex configurations using CFD is a nontrivial matter. The computed drag is affected by numerical discretization and physical modelling error. The discretization error due to the finite computational grid can be estimated by grid extrapolation techniques. Aerodynamic coefficients can be extrapolated to within 1% of the grid independent value using a grid refinement approach. The GCI can be between 5% - 15% for lift and drag. Reliable evaluations of grid-independent solutions and error bands require high-quality grids on all levels of grid convergence studies.

1 Introduction

The prediction of aerodynamic drag is of major importance in the development of new aircraft. In order to use Computational Fluid Dynamics (CFD) in an industrial environment, the reliability of the numerical results has to be established. One of the major error sources in numerical solutions is the discretization error, which to a large extent is influenced by grid quality and grid resolution. The aim is to reduce this error ideally to zero, i.e. to achieve grid independent solutions. The direct computation of grid independent solutions requires extremely fine grids in the case of non-adapted meshes (appr. 1.5 thousand million cells) and about 30 million cells in case of high quality, solution-adapted cells [1]. This is still at the limit of current computing resources, and is of no practical use in an industrial environment. To circumvent these immense requirements with current computing resources, one can use sequences of successively refined structured grids for extrapolations of grid independent aerodynamic coefficients and an estimation of the numerical error. In [2] two mesh families are generated with the coarsest grids having $0.26e6$ and $1e6$ cells which appeared to be not sufficiently fine enough. The present report is a continuation of [2] and follows the approach of Roache [3] and [4]. The aim is to estimate grid-converged aerodynamic coefficients and to provide error bands around them, as it is impossible to estimate the numerical error from single-grid studies. It will be shown in this study, that the cancellation of numerical and physical modelling errors can lead to false conclusions about the quality of a single grid computation in cases of good correlation with experimental results. As it is important to achieve asymptotic convergence on all levels of a grid refinement study, the high quality on all three grid levels is achieved using non-integer grid refinement, as grid doubling (or halving) leads to excessively large grid sizes for complex 3D-applications. The aim is to reduce the computational effort while maintaining numerical accuracy. The wing-body configuration used in this investigation is chosen to be a typical representative of geometries encountered in the civil aerospace industry.

2 Megaliner Geometry

The geometry under investigation is an UHCA wing-body configuration at high Machnumber wind tunnel conditions in cruise (Ma: 0.85, Re: 2.68e6, C_L= 0.5 at $\alpha = 2.84deg$). The CFD-representation of the wind tunnel model includes aeroelastic wing-twist. The outer wing has a relatively thin section with a small nose radius. The blunt trailing edge of the wing is closed from 90% of the chord on using Bezier-splines to meet grid generation and flow solver requirements. This is shown in [5] to yield good results for transonic cases.

3 CFD Software

The utilized structured CFD-software package is MEGAFLOW [6] which comprises the multi-block-stuctured grid generator MegaCads [7] and the RANS flow solver FLOWer [8]. MEGA-FLOW is developed by DLR in collaboration with DASA-Airbus, several German universities and other research facilities for the simulation of viscous turbulent flows around transport aircraft at cruise and in take-off/landing configuration.

3.1 Flow Solver

The FLOWer code solves the three-dimensional compressible Reynolds-averaged Navier-Stokes equations in integral form. Turbulence is modelled by either algebraic or transport equation models. The spatial discretization uses a central cell-vertex finite volume formulation. Dissipative terms are explicitly added in order to damp high frequency oscillations and to achieve sufficiently sharp resolution of shock waves. The dissipative operator is comprised of second and fourth differences scaled by the largest eigenvalue following Jameson, Schmidt and Turkel [9] and Martinelli [10]. On smooth meshes, the scheme is second order accurate in space. Time integration is carried out by an explicit hybrid multistage Runge-Kutta scheme. For steady state calculations the integration is accelerated by local time stepping, enthalpy damping for inviscid flows, and implicit residual smoothing. These techniques are embedded in a multigrid algorithm. The code allows two dummy layers around each block in order to maintain second order accuracy in space at block intersections.

3.2 Grid Generation

The structured multiblock mesh has a C-O-H topology for high resolution. The grid generation takes advantage of the replay capability of MegaCads, important parameters like cell numbers and spacings are controlled by variables which allow fast modifications to grids. This was used for the generation of six computational meshes in [2]. From these six grids, the computations on grid A1 (16.85e6 cells), A2 (2.1e6 cells) and B1 (7.85e6 cells) are used for comparison in tables 1 and 2. The same grid design is used to generate the mesh sequence of the present work with non-integer grid refinement with grids C1-C3 (6.59e6, 4.8e6 and 3.37e6 cells, see table 1). The grid refinement ratios r_{12} : 1.111 and r_{23} : 1.125 are obtained in cell numbers in all three dimensions and relevant spacings, like first spacing and spacings at block intersections. The first spacing on the coarse grid C3 was chosen to achieve a y^+ of appr. 1, resulting in smaller values of y^+ on the finer grids C2 and C3. Cell numbers and placement are chosen build on the experience made in [2] with the aim to assure a minimum resolution on grid C3 which is better then the resolution of grid A2. The grid construction includes a manual adaption of the boundary layer thickness and grid clustering at shocks and other regions with high flow gradients. The far field distance is four times the fuselage length in every direction. Figure 1 shows a cutaway of the multiblock structured grid. More details on grid generation can be found in [2].

4 Richardson Extrapolation and Grid Convergence Index GCI

Richardson extrapolation is used by Zingg [11], [12] to achieve extrapolated values for lift and drag of 2D airfoil sections. The general Richardson extrapolation for the grid-converged value $f_{h=0}$ is:

$$f_{h=0} \cong f_1 + \frac{f_1 - f_2}{r^p - 1} \tag{4.1}$$

with p being the theoretical or observed order of accuracy, r the grid refinement ratio, f_1 a fine grid solution and f_2 being the solution on the next coarser grid level. The Grid Convergence Index GCI provides an empirical error band around the asymptotic numerical value, i.e. the zero mesh size value. The GCI on the fine grid is defined as

$$GCI = \frac{F_s|\epsilon|}{r^p - 1} \tag{4.2}$$

F_s is a factor of safety which is 3 for a two-grid study and 1.25 for three-grid studies with an observed order of accuracy p. ϵ is $(f_2 - f_1)/f_1$. For three-grid studies

$$GCI_{coarse} \cong r^p GCI_{fine} \tag{4.3}$$

is an indication that all three grids have a sufficient resolution. Richardson Extrapolation and Grid Convergence Index are described in detail in [3] and [4].

5 Computations

The Navier-Stokes computations are carried out with algebraic Baldwin-Lomax turbulence modelling (with Degani-Schiff modification [13]) for the grid refinement studies. A farfield vortex correction is applied and transition is fixed at 0.15 of the chord on the upper and lower wing surface. All computations in this study use $k^2 = 1/4$ and $k^4 = 1/64$ in the scalar artificial dissipation model [9]. All grid refinement computations are carried out at cruise conditions (see section 2 and [2]). The high Mach number and geometric originalities like a relatively sharp leading edge and low thickness with aft maximum thickness location of the outer wing lead to a strong shock at aft position on the upper wing side.

To use a small safety factor F_S of 1.25 in three-grid studies, the observed order of accuracy p should be known with some confidence [3], as the observed order can be lower than the formal order of a CFD-method. Theoretically, the observed order can be extracted from three-grid studies. The experience of [2] and the present work shows, that this seems to be problematic for complex 3D flows. Therefore the results in tables 2 - 5 for Richardson Extrapolation and GCI are presented for orders p = 1.75 and 2, and F_S = 1.25 and 3. As the first order influence of shocks and stagnation lines should be local due to high grid resolution, the assumption of an observed order of accuracy between 1.75 and 2 should be reasonable.

6 Results

Table 1 shows total lift and drag normalized with their respective experimental values for grids A1, A2 and B1 of [2] compared to the results obtained with meshes C1-C3 of the present work. It can be observed that the numerical drag is decreasing with increasing mesh size. This is a feature of the algebraic turbulence model of Baldwin-Lomax which moves the shock downstream of the experimental location with increasing mesh size (see also figs. 2(a) - 2(d)). This interaction of mesh size and physical modelling ability can lead to erroneous conclusions about solution quality with single-grid studies with meshes like C2, which is only 0.2% below the experimental value due to the finite mesh. Looking at table 2 one notices immediately that the extrapolated, grid converged drag for grid C2 is about 4% below the experimental drag, i.e. the computed drag contains about 4% numerical and therefore nonphysical dissipation. This observation would be impossible with a computation on one grid only. It is known from experience, that one can achieve computed drag values for wing-body configurations for grids with about $3e6$ - $5e6$ cells using the Baldwin-Lomax turbulence model which are very close to experimental results.
In table 2 the Richardson extrapolated (eqn. 4.1) lift and drag coefficients are normalized with their respective experimental values using an order of convergence p = 2. The results of the present report are compared to the extrapolations of the two finest mesh sequences of [2]. For lift the three extrapolations are 0.6% and for drag 0.5% apart. Considering the typical levels of lift and drag for UHCA wing-body configurations, this is a very good result.
Table 3 shows the GCI's for grids C1 and C2 using the assumed order p = 2 and two safety factors F_S of 1.25 and 3 (see section 4 and 5). Assuming that the observed order of accuracy is close to 2, this yields a 4% error band for drag on the fine grid which is quite low. Using the very conservative factor of safety of F_S = 3 gives a GCI of 10% for drag on the fine grid. There is a difference of about 10% between $r^p GCI_{fine}$ and GCI_{coarse} which is considered to indicate sufficient asymptotic convergence for a multiblock-structured three-grid study in [3].
In Table 4 the extrapolations to zero mesh size are repeated with p = 1.75. For lift the three extrapolations are 0.9% and for drag 0.7% apart, similar to table 2.
The results of table 5 for the GCI with p = 1.75 are also close to the values computed with p = 2. Again the difference of $r^p GCI_{fine}$ and GCI_{coarse} is about 10%, indicating near asymptotic convergence. Figs. 2(a) - 2(d) show the C_P-distributions at four spanwise stations of the grids C1-C3 compared to experiment and the fine grid solution A1 with 16.85$e6$ cells. There are minor differences in shock position and shock steepness inboard which grow slightly in the outboard direction. The resolution of aerodynamic flow features indicated in these figures is high.

7 Conclusions

The results discussed in the preceding section show consistent extrapolations of grid independent aerodynamic coefficients using Richardson Extrapolation. The target to use non-integer grid refinement with reduced mesh size compared to integer grid refinement [2] for the extrapolation of zero mesh size coefficients has been achieved. The Richardson extrapolations with three different mesh families differ by less than 1%. The GCI's have a range of 4-6% using a safety factor F_S = 1.25 and 9-15% for the very conservative F_S = 3 for the mesh family of the present study. The asymptotic convergence for all three grids has been achieved with engineering accuracy. Experience so far shows difficulties in evaluating the precise observed order of accuracy for complex 3D flows. It was shown that a careful grid refinement study can provide useful insights about numerical errors of CFD computations for 3D applications which are not possible with one-grid calculations. Future work will evaluate the possibility of extracting observed orders of accuracy by investigating nearby flow problems, e.g. 2D flows and wing flows. Also the transference of Richardson Extrapolation and GCI to unstructured methods will be investigated.

References

[1] B. Oskam and J.W. Sloof: "Recent Advances in Computational Aerodynamics at NLR", AIAA 98-0138, (1998).

[2] M. Rakowitz: "Grid Refinement Study with a Megaliner Wing-Body Configuration", ECCOMAS 2000 Barcelona, (2000).

[3] P.J. Roache: *Verification and Validation in Computational Science and Engineering*, Hermosa Publishers, P.O. Box 9229, Albuquerque, New Mexico 87119, USA, (1998).

[4] NPARC Alliance CFD Verification and Validation Web Site: Performing a Grid Convergence Study, http://www.grc.nasa.gov/www/wind/valid/.

[5] E. Monsen and R. Rudnik: "Investigation of the Blunt Trailing Edge Problem for Supercritical Airfoils", AIAA 95-0089, (1995).

[6] N. Kroll , C.-C. Rossow, K. Becker and F. Thiele: "MEGAFLOW - A Numerical Flow Simulation System", 21st ICAS Congress, Melbourne, Australia, Sept. (1998).

[7] O. Brodersen, M. Hepperle, A. Ronzheimer, C.-C. Rossow and B. Schöning: The Parametric Grid Generation System MegaCads. Proc. of 5th Intern. Conf. on Numerical Grid Generation in Computational Field Simulation. Ed. Soni B.K. et. al., Mississippi, pp. 353-362 (1996).

[8] J. Raddatz et. al.: "FLOWer115 - Installation and User Handbook", DLR Braunschweig, (1998).

[9] A. Jameson, W. Schmidt and E. Turkel: "Numerical Solutions of the Euler Equations by Finite Volume methods Using Runge-Kutta Time-Stepping Schemes", AIAA Paper 81-1259, (1981).

[10] L. Martinelli: "Calculations of Viscous Flows with a Multigrid Method", PhD Dissertation, MAE Department, Princeton University, (1987).

[11] D.W. Zingg: "Grid Studies for Thin-Layer Navier-Stokes Computations for Airfoil Flowfields", AIAA 92-0184, (1992).

[12] D.W. Zingg: "Viscous Airfoil Computations Using Richardson Extrapolation", AIAA 91-1559-CP, (1991).

[13] D. Degani and L.B. Schiff: "Computation of Turbulent Supersonic Flows Around Pointed Bodies Having Crossflow Separation", *Journal of Computational Physics 66, 173-196*, (1986).

Table 1 Computed Lift and Drag Coefficients divided by Experimental Values

$Grid$	$No.ofCells$	C_{Lcomp}/C_{Lexp}	C_{Dcomp}/C_{Dexp}
$A1$	16,850,944	1.048	0.976
$A2$	2,106,368	1.016	1.013
$B1$	7,848,576	1.039	0.985
$C1$	6,586,000	1.034	0.990
$C2$	4,801,194	1.027	0.998
$C3$	3,372,032	1.018	1.009

Table 2 Richardson Extrapolated Zero Mesh Size Lift and Drag Coefficients

$RE(p=2)$	$C_{Lextrapol}/C_{Lexp}$	$C_{Dextrapol}/C_{Dexp}$
$A1-A2$	1.059	0.963
$A1-B1$	1.062	0.961
$C1-C2$	1.065	0.958

Table 3 GCI_{fine} for Grid C1 and GCI_{coarse} for Grid C2 with Order p = 2

$C_L(p=2)$	$F_S=1.25$	$F_S=3$	$C_D(p=2)$	$F_S=1.25$	$F_S=3$
GCI_{fine}	3.74%	8.98%	GCI_{fine}	4.06%	9.74%
GCI_{coarse}	4.11%	9.87%	GCI_{coarse}	5.45%	13.07%
r^pGCI_f/GCI_c	1.12		r^pGCI_f/GCI_c	0.92	

Table 4 Richardson Extrapolated Zero Mesh Size Lift and Drag Coefficients

$RE(p=1.75)$	$C_{Lextrapol}/C_{Lexp}$	$C_{Dextrapol}/C_{Dexp}$
$A1-A2$	1.061	0.960
$A1-B1$	1.063	0.959
$C1-C2$	1.070	0.953

Table 5 GCI_{fine} for Grid C1 and GCI_{coarse} for Grid C2 with Order p = 1.75

$C_L(p=1.75)$	$F_S=1.25$	$F_S=3$	$C_D(p=1.75)$	$F_S=1.25$	$F_S=3$
GCI_{fine}	4.33%	10.40%	GCI_{fine}	4.70%	11.28%
GCI_{coarse}	4.77%	11.45%	GCI_{coarse}	6.32%	15.17%
r^pGCI_f/GCI_c	1.09		r^pGCI_f/GCI_c	0.89	

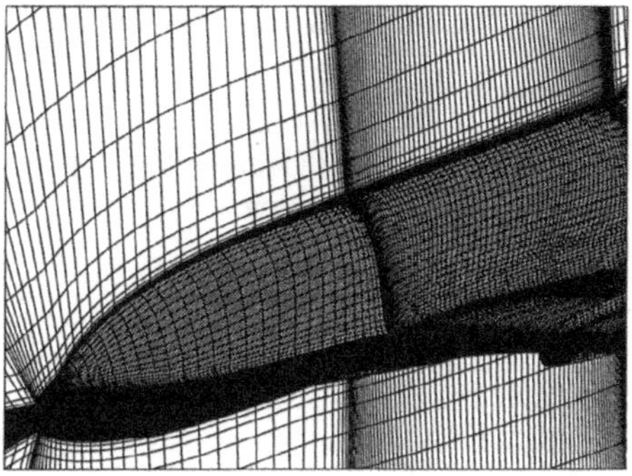

Figure 1 Multiblock-Structured Grid

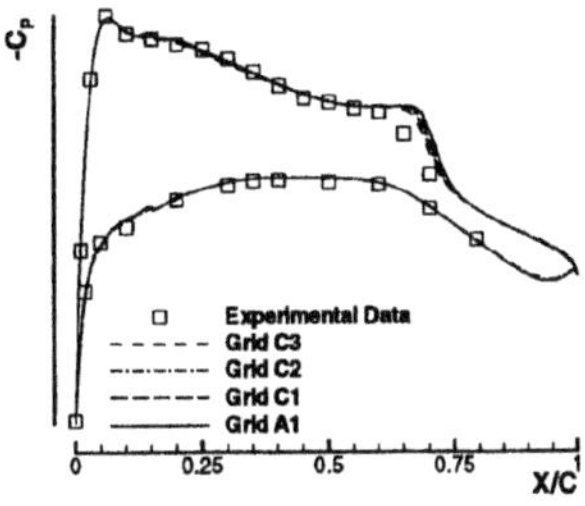

(a) η : 0.12

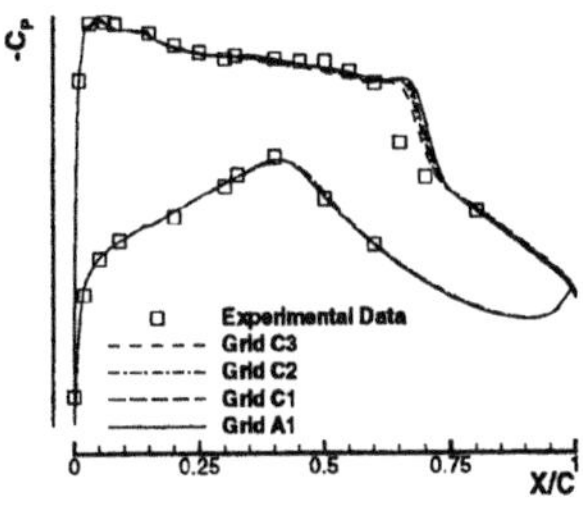

(b) η : 0.36

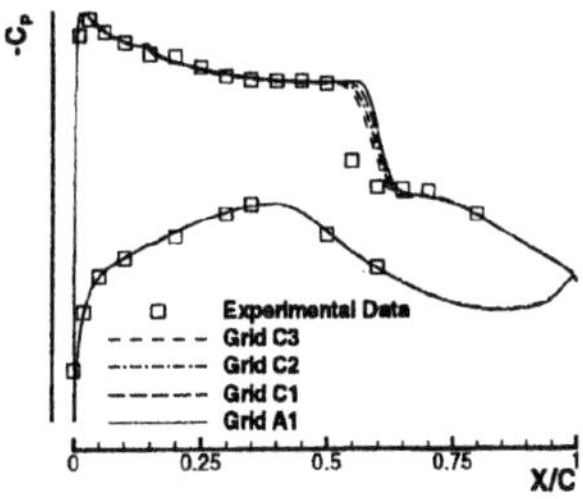

(c) η : 0.57

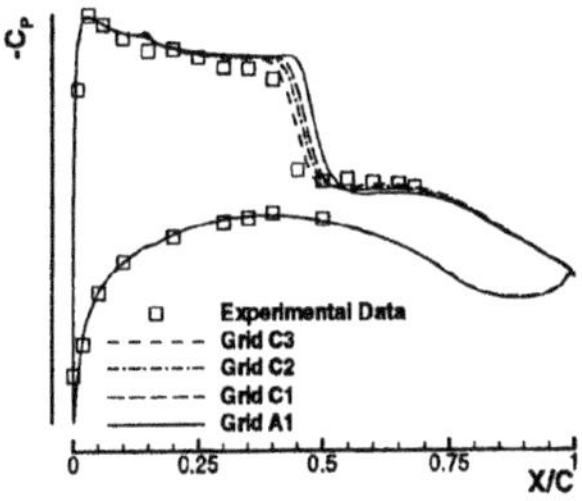

(d) η : 0.85

Figure 2 $C_P(x/c)$ of *mesh sequence C* compared to Experiment and Fine Grid Solution A1

Convergence Acceleration on Unstructured Meshes

C.-C. Rossow
DLR, Institut für Entwurfsaerodynamik, Lilienthalplatz 7, D-38022 Braunschweig, Germany

Summary

An implicit residual smoothing technique to enhance convergence properties of Runge-Kutta time stepping schemes on unstructured meshes is developed. In contrast to standard implicit residual smoothing schemes applied on structured meshes, no direct information about the cell aspect ratio is required. The development of the acceleration technique is based on the Mach number based Advection Pressure Splitting (MAPS) spatial discretization. With the new implicit smoothing, maximum CFL numbers comparable to those on structured meshes are obtainable for unstructured meshes, and in combination with multigrid acceleration, convergence rates similar to structured codes are achieved. In order to allow the solution for incompressible flows, the implicit residual smoothing is adapted to low Mach number preconditioning.

1. Introduction

One of the main drivers towards more efficient flow solvers is the use of multigrid schemes in combination with Runge-Kutta timestepping [1]. The efficiency of such codes is further enhanced by augmenting the explicit time stepping scheme through implicit residual smoothing techniques: on the one hand, implicit residual smoothing extends the stability region of the basic time stepping scheme; on the other hand, a properly formulated implicit residual smoothing reduces the stiffness resulting from high aspect ratio cells [2]. For highly stretched cells, the cell aspect ratio has to be evaluated for the smoothing coefficients, and efficient methods for the solution of viscous flows were developed for structured meshes [2]. For unstructured meshes however, the situation is different. In the case of inviscid flows, implicit smoothing with constant coefficients was successfully implemented [3]. For viscous flows, the implementation is not as straightforward, because the local information on the aspect ratio is not directly available for determining the locally varying smoothing coefficients. Therefore, most viscous unstructured methods with Runge-Kutta time integration are employed without any means to increase the CFL number, and typical CFL numbers are in the range of 1.5 - 2.0 [4]; in contrast to structured meshes, where CFL numbers of 6.0 - 7.5 are frequently used [1,2].

In the present contribution, an alternative formulation for the smoothing coefficients will be developed. Based on the MAPS scheme of [5], a simplified implicit version will be derived as smoothing technique. This technique will first be implemented into a structured code for direct comparison to the standard implicit smoothing of [2].

Since the solution for incompressible flows with compressible codes gains more and more attention, the new formulation for the smoothing coefficients will be adapted to the use of preconditioning techniques [6], to also allow the computation of low Mach number flows. Following the assessment in the framework of the structured code, the new implicit residual smoothing will be implemented into an unstructured code to demonstrate the applicability to unstructured meshes.

2. MAPS Spatial Discretization

The equations to be solved are the compressible Navier-Stokes equations. In the numerical solution procedure, for the viscous terms central discretization leads to stable numerical approximations and therefore no conceptual difficulties arise. For the convective terms however, the characteristics of information propagation have to be taken into account. In [5], the MAPS scheme was derived as a simple but robust and accurate means for discretization. Since MAPS follows the common dimensional-splitting approach where only the flux normal to a cell interface is regarded, for brevity only the one-dimensional scheme will be outlined. The convective flux $\vec{F}_{i-1/2}$ through a cell interface at point $i-1/2$ reads:

$$\begin{aligned}
\vec{F}_{i-1/2} &= \vec{F}^{\phi}_{i-1/2} + \vec{F}^{P}_{i-1/2};\\
F^{\phi}_{i-1/2} &= \frac{1}{2}\bar{c}\left[(\overline{M} - M_{Dif} - M_{up})\cdot\vec{\phi}_i + (\overline{M} - M_{Dif} + M_{up})\cdot\vec{\phi}_{i-1}\right]\cdot S,\\
\vec{F}^{P}_{i-1/2} &= \frac{1}{2}\left[\left(1-\beta^P_i\right)\cdot\vec{P}_i + \left(1+\beta^P_{i-1}\right)\cdot\vec{P}_{i-1}\right]\cdot S.
\end{aligned} \tag{1}$$

where

$$\begin{aligned}
&\vec{\phi} = [\rho,\, \rho u,\, \rho H]^T \;;\quad \vec{P} = [0,\, p,\, 0]^T,\quad \vec{W} = [\rho,\, \rho u,\, \rho E]^T \quad,\\
&\bar{c} = \frac{1}{2}(c_i + c_{i-1}),\qquad \overline{M} = \frac{1}{2}(M_i + M_{i-1}) \quad,\\
&M_{Dif} = \frac{1}{2}\beta_M\left(|M|_i - |M|_{i-1}\right),\qquad M_{up} = \max(M_i, M_{i-1}) \quad,\\
&\beta^P_i = \beta_p\cdot sign(M_i),\qquad \beta^P_{i-1} = \beta_p\cdot sign(M_{i-1}) \quad.
\end{aligned} \tag{2}$$

Here ρ, u, p, E, H, and c denote density, velocity in x-direction, pressure, total energy and total enthalpy per unit mass, and speed of sound, respectively. The coefficients β_M and β_P are given in [5], and S represents the cell face area at $i-1/2$.

The terms comprising the convective flux $\vec{F}_{i-1/2}$ may be rearranged to yield:

$$\vec{F}_{i-1/2} = q^{(-)}_{i-1/2}\cdot\vec{\phi}_i + q^{(+)}_{i-1/2}\cdot\vec{\phi}_{i-1} + p^{(-)}_{i-1/2}\cdot\vec{P}_i + p^{(+)}_{i-1/2}\cdot\vec{P}_{i-1}\ , \tag{3}$$

where

$$\begin{aligned}
q^{(-)}_{i-1/2} &= 0.5\cdot\bar{c}\cdot S\cdot(\overline{M} - M_{Dif} - M_{up}),\quad q^{(+)}_{i-1/2} = 0.5\cdot\bar{c}\cdot S\cdot(\overline{M} - M_{Dif} + M_{up}),\\
p^{(+)}_{i-1/2} &= 0.5\cdot S\cdot(1+\beta^P_{i-1}),\quad p^{(-)}_{i-1/2} = 0.5\cdot S\cdot(1-\beta^P_i).
\end{aligned} \tag{4}$$

3. Basic Implicit Residual Smoothing Technique

3.1 Quasi-Implicit formulation of 1D Euler equations

A time-implicit discretization of the one-dimensional Euler equations is given by

$$\frac{\Delta\vec{W}}{\Delta t} + \frac{\vec{F}^{n+1}_{i+1/2} - \vec{F}^{n+1}_{i-1/2}}{Vol} = 0, \tag{5}$$

where $\vec{F}^{n+1}$ denotes the flux vector of the unknown state at $t^{n+1} = t^n + \Delta t$. A simplified quasi-implicit formulation of eq. (5) may be obtained when expressing the flux vector $\vec{F}^{n+1}$ by using eq. (1) as:

$$\vec{F}^{n+1}_{i-1/2} = q^{(-)}_{i-1/2} \cdot \vec{\phi}^{n+1}_i + q^{(+)}_{i-1/2} \cdot \vec{\phi}^{n+1}_{i-1} + p^{(-)}_{i-1/2} \cdot \vec{P}^{n+1}_i + p^{(+)}_{i-1/2} \cdot \vec{P}^{n+1}_{i-1}, \tag{6}$$

where the coefficients $q^{(-)}$ and $q^{(+)}$ are evaluated using the known flow variables at time level t^n, and by the introduction of:

$$\begin{aligned} \vec{\phi}^{n+1} &= \vec{\phi}^n + \Delta\vec{\phi} \approx \vec{\phi}^n + \Delta\vec{W} , \\ \vec{P}^{n+1} &= \vec{P}^n + \Delta\vec{P} . \end{aligned} \tag{7}$$

Substituting eq. (6) and eq. (7) into eq. (5), and rearranging terms leads to

$$\begin{aligned} &-\frac{\Delta t}{Vol} q^{(+)}_{i-1/2} \cdot \Delta\vec{W}_{i-1} + \left[1 + \frac{\Delta t}{Vol}(q^{(+)}_{i+1/2} - q^{(-)}_{i-1/2})\right] \cdot \Delta\vec{W}_i + \frac{\Delta t}{Vol} q^{(-)}_{i+1/2} \cdot \Delta\vec{W}_{i+1} \\ &= \vec{R}^n_i - \frac{\Delta t}{Vol}\left[- p^{(+)}_{i-1/2} \cdot \Delta\vec{P}_{i-1} + (p^{(+)}_{i+1/2} - p^{(-)}_{i-1/2}) \cdot \Delta\vec{P}_i + p^{(-)}_{i+1/2} \cdot \Delta\vec{P}_{i+1}\right], \end{aligned} \tag{8}$$

where the explicit residual $\vec{R}^n$ is given by

$$\begin{aligned} \vec{R}^n = -\frac{\Delta t}{Vol} \cdot \quad &\left[q^{(-)}_{i+1/2} \cdot \vec{\phi}^n_{i+1} + \left(q^{(+)}_{i+1/2} - q^{(-)}_{i-1/2}\right) \cdot \vec{\phi}^n_i - q^{(+)}_{i-1/2} \cdot \vec{\phi}^n_{i-1}\right. \\ &\left. + p^{(-)}_{i+1/2} \cdot \vec{P}^n_{i+1} + \left(p^{(+)}_{i+1/2} - p^{(-)}_{i-1/2}\right)\vec{P}^n_i - p^{(+)}_{i-1/2} \cdot \vec{P}_{i-1}\right] . \end{aligned} \tag{9}$$

Note that in the derivation of eq. (8) two major simplifications where made: First, in eq. (6) it was assumed that the coefficients $q^{(+)}$ and $q^{(-)}$ were constant in time. Second, in eq. (7) the unknown state $\vec{\phi}^{n+1}$ is approximated by using the changes $\Delta\vec{W}$ instead of $\Delta\vec{\phi}$. A correct implicit treatment would lead to

$$\vec{F}^{n+1} = \vec{F}^n + \frac{\partial\vec{F}}{\partial\vec{W}}\Delta\vec{W} , \tag{10}$$

instead of eqs. (6) and (7). However, the use of a fully implicit scheme with eq. (10) requires the evaluation and inversion of the jacobian matrix $\partial\vec{F}/\partial\vec{W}$, and then leads to a system of coupled, implicit equations for the updates $\Delta\vec{W}$. Instead, the system of equations (8) provides scalar implicit equations for the updates $\Delta\vec{W}$, where the only coupling between the equations is caused by the changes of pressure $\Delta\vec{P}$.

3.2 Derivation of basic implicit residual smoothing

Implicit residual smoothing is employed to enhance the performance of Runge-Kutta time stepping schemes [1, 2]. The basic idea of implicit residual smoothing is as follows: consider an m-stage Runge-Kutta scheme:

$$\vec{W}^{(m)}_i = \vec{W}^{(0)}_i + \vec{R}^{(m-1)}_i . \tag{11}$$

The explicit residual $\vec{R}^{(m-1)}_i$ is then replaced by a smoothed residual $\bar{\vec{R}}^{(m-1)}_i$

$$\vec{W}^{(m)}_i = \vec{W}^{(0)}_i + \bar{\vec{R}}^{(m-1)} , \tag{12}$$

where $\tilde{\vec{R}}_i^{(m-1)}$ is obtained from an implicit equation of the type:

$$-\varepsilon_{i-1}\cdot\tilde{\vec{R}}_{i-1}^{(m-1)}+(1+\varepsilon_i)\cdot\tilde{\vec{R}}_i^{(m-1)}-\varepsilon_{i+1}\cdot\tilde{\vec{R}}_{i+1}^{(m-1)}=\vec{R}_i^{(m-1)}\ . \tag{13}$$

Note that only scalar equations need to be solved, and a simple implementation results. Inspecting eq. (8), a similarity to eq. (13) may be observed. Interpreting the changes $\Delta\vec{W}$ as smoothed residuals $\tilde{\vec{R}}$ to be used in a Runge-Kutta scheme, eq. (8) may be rearranged to:

$$\varepsilon_{i-1}\cdot\tilde{\vec{R}}_{i-1}+(1+\varepsilon_i)\cdot\tilde{\vec{R}}_i+\varepsilon_{i-1}\cdot\tilde{\vec{R}}_{i+1}=\vec{R}_i-\vec{S}_P\ , \tag{14}$$

where the smoothing coefficients ε are now given by:

$$\varepsilon_{i-1}=-\frac{\Delta t}{Vol}q_{i-1/2}^{(+)}\ ,\quad \varepsilon_i=\frac{\Delta t}{Vol}\left(q_{i+1/2}^{(+)}-q_{i-1/2}^{(-)}\right),\quad \varepsilon_{i+1}=\frac{\Delta t}{Vol}q_{i+1/2}^{(-)}\ ,$$
$$S_P=\frac{\Delta t}{Vol}\left[-p_{i-1/2}^{(+)}\cdot\Delta\vec{P}_{i-1}+\left(p_{i+1/2}^{(+)}-p_{i-1/2}^{(-)}\right)\cdot\Delta\vec{P}_i+p_{i+1/2}^{(-)}\cdot\Delta\vec{P}_{i+1}\right]. \tag{15}$$

Eq. (14) represents a set of scalar equations for the smoothed residuals $\tilde{\vec{R}}$ (or the changes of variables $\Delta\vec{W}$ according to eq. (8)). In order to avoid a coupling of the equations, the pressure changes $\Delta\vec{P}$ are included in the source term $\vec{S}_P$. This term is evaluated using the explicit residuals $\vec{R}$ (or explicit changes of variables $\Delta\vec{W}$) :

$$\Delta p=(\gamma-1)\left[\frac{q^2}{2}\Delta\rho-u\Delta(\rho u)+\Delta(\rho E)\right]\ , \tag{16}$$

where q^2 represents the square of the flow speed. Eq. (14) is solved by point-jacobi iteration, as was already proposed in [3] for constant coefficients on unstructured meshes. Usually, 3-5 iterations are sufficient. Note that the extension of the method to higher dimensions and arbitrary cell shapes is trivial: only cell face quantities $q^{(+)}$, $q^{(-)}$, $p^{(+)}$, $p^{(-)}$ are needed, which are evaluated using only flow components normal to the cell face, i.e. in a quasi-one dimensional manner.

Implementing this smoothing technique into a numerical code, the coefficients $q^{(+)}$, $q^{(-)}$, $p^{(+)}$, $p^{(-)}$ are evaluated only in the first stage of an m-stage Runge-Kutta scheme. Similarly, the pressure source term S_P is evaluated only in the first stage and then kept constant. Note that the cell aspect ratio is not explicitly required, but the metrics are incorporated into the coefficients $q^{(+)}$, $q^{(-)}$, $p^{(+)}$ and $p^{(-)}$.

3.3 Results for basic implicit smoothing

In order to assess the performance of the new implicit smoothing, the method was implemented in the framework of the structured code used in [5] and [7], where the MAPS discretization was developed. For time integration, the code uses a 5-stage Runge-Kutta scheme and multigrid convergence acceleration. Since the code uses structured meshes, the implicit smoothing technique of [2] was used in [5] and [7]. Results are obtained for viscous flow around the RAE 2822 airfoil at $M_\infty = 0.73$, $\alpha = 2.79°$, and Re = 6,500,000. The computational mesh consists of 320x64 mesh cells, and a 4 level W-cycle is used in the multigrid acceleration. Since the steady state solutions are identical to those presented in [5], only convergence histories will be presented here. Fig. 1 shows a comparison of the convergence using the standard implicit residual smoothing of [2] (curve (1), central smoothing) and the convergence employing the newly derived smoothing technique (curve (2), upwind smoothing). For the standard technique, CFL numbers of 7.5 were obtained, whereas for the new technique a CFL number of 6.0 could be achieved. Despite the lower CFL number, the new smoothing scheme required less multigrid iterations to achieve the same level of convergence as the standard scheme. The amount of CPU time required for convergence is roughly the same for both methods.

4. Improvement of Basic Implicit Smoothing

Additional numerical investigations revealed, that the new smoothing derived above only works in combination with the basic MAPS of [5], and an application in combination with low speed preconditioning [6] is not possible. In [7], the basic MAPS scheme was augmented by additional dissipative terms to improve the computation of grid aligned and low speed flows. The resulting MAPS+ discretization scheme bears a significant similarity to the Roe-scheme, however avoiding the necessity for an entropy fix and the Roe-averaging to evaluate cell face quantities. It was shown that the additional terms are vital for the application of low speed preconditioning. These additional dissipative terms consist of spatial differences of pressure and normal interface velocity. All additional dissipative terms are scaled by a factor $(1-|M_0)|$, where M_0 is given by $M_0=\min(|M|,1)\cdot sign(M)$, with M denoting the directional Mach number normal to the cell interface. Thus, the additional terms are activated at low normal interface velocities, providing additional damping in such situations. Note that these additional terms cause a strong coupling of the governing equations.

In order to retain a smoothing technique based on scalar equations and to incorporate low speed preconditioning, the influence of the additional terms on the smoothing coefficients $q^{(+)}$ and $q^{(-)}$ is boldly approximated by

$$q_{i-1/2}^{(-)}=0.5\cdot\bar{c}\cdot\bar{S}\cdot\left[\bar{M}-M_{Dif}-M_{up}-(1-|M_0|)\frac{c'}{\bar{c}}\right],$$
$$q_{i-1/2}^{(+)}=0.5\cdot\bar{c}\cdot\bar{S}\cdot\left[\bar{M}-M_{Dif}+M_{up}+(1-|M_0|)\frac{c'}{\bar{c}}\right] \tag{17}$$

where c' denotes the preconditioned speed of sound [6,7]. In Fig. 1, the curve (3), denoted upwind-c smoothing, shows the convergence behaviour with the modified coefficients. Still, convergence is faster than for the standard implicit smoothing. Using the coefficients given by eq. (17), the smoothing may now also be applied with the MAPS+ discretization, and works in combination with preconditioning. This will be demonstrated later for unstructured meshes.

The newly derived implicit smoothing technique requires roughly the same amount of CPU-time as the standard technique. But due to the coefficients $q^{(+)}$, $q^{(-)}$, $p^{(+)}$, $p^{(-)}$, and the source term S_P, additional storage is needed. In order to reduce the storage requirements, the smoothing coefficients are simplified to:

$$q_{i-1/2}^{(-)}=0.5\cdot\bar{c}\cdot\bar{S}\cdot\left[-M_{up}-(1-|M_0|)\frac{c'}{\bar{c}}\right]\ ,\quad q_{i-1/2}^{(+)}=0.5\cdot\bar{c}\cdot\bar{S}\cdot\left[M_{up}+(1-|M_0|)\frac{c'}{\bar{c}}\right]\ =\ -q_{i-1/2}^{(-)} \tag{18}$$

Thus, only one coefficient per cell face needs to be stored, and the pressure source term dropped out completely. In Fig. 1, the convergence with the simplified smoothing is denoted by curve (4), central-c smoothing. The convergence rate is still better than for the standard implicit smoothing (curve (1)), and the storage requirements are now similar. Note that the CFL number remains at 6.0.

5. Results on Unstructured Meshes

The implicit residual smoothing with the simplified coefficients of eq. (17) was implemented into a generic unstructured research code (NOUGAT: Node Oriented Unstructured Generic Algorithm Testing). This code uses a dual mesh metric, prepared by the preprocessing part of the DLR τ-code [4]. For the spatial discretization of the convective terms, the MAPS+ discretization is employed, and second order reconstruction is based on least squares gradients. Time integration is performed with a 5-stage Runge-Kutta scheme augmented by agglomeration multigrid. The influence of turbulence is modeled with the Spalart-Allmaras turbulence model following the implementation of [4]. Fig. 2 shows the convergence history for transonic viscous flow around the RAE 2822 airfoil using the same freestream conditions as in the investigations on structured meshes. The residual is decreased by 5 orders of magnitude in less than 400 multigrid cycles. Total forces are converged after 100 MG cycles. This convergence behaviour compares well with the efficiency observed on structured meshes. Fig. 3 gives a view of the field pressure distribution with the underlying hybrid mesh. The number of surface points corresponds closely to the number of surface points used in the structured case, and in total about 14,000 points were used. Fig. 4 shows a comparison of the convergence of the present method and the DLR-τ code [4] on the identical mesh: the speed-up, especially for the total forces, is clearly visible.

An important issue for numerical methods in flow simulation is the extension of the Mach number regime towards incompressible flows. Preconditioning techniques such as described in [6] offer a means to extend compressible codes to incompressible flows. Therefore, the new implicit smoothing was adapted to incorporate the preconditioned speed of sound, see eqs. (17) and (18). Fig.5 shows the convergence history for the flow around the RAE 2822 airfoil at M_∞=0.001 and $\alpha = 8.0°$. Note that no reduction of CFL number was necessary in comparison to the compressible case, and total forces converge also as rapidly.

6. Conclusions

A new implicit residual smoothing technique suitable for unstructured meshes was developed. The main distinction from commonly used residual smoothing techniques is that the cell aspect ratio is no longer explicitly required, thus abandoning the restriction to structured meshes. In order to derive the new method, a quasi-implicit version of the basic MAPS discretization was employed. Successive development of the new scheme led to a smoothing technique which is applicable for both the MAPS and the MAPS+ discretizations, and low Mach number preconditioning may be used as well. The storage requirements of the method are the same as for standard implicit residual smoothing, and on structured meshes similar convergence rates at comparable CFL numbers were achieved. Implementation into an unstructured code demonstrated that with the new smoothing technique, CFL numbers and convergence rates as commonly achieved on structured meshes can be established. This holds for compressible as well as for nearly incompressible flows.

7. References

[1] Jameson, A.: *Multigrid Algorithms for Compressible Flow Calculations*, MAE Report 1743, Princeton University, Text of Lecture given at 2nd European Conference on Multigrid Methods, Cologne, October 1985.

[2] Martinelli, L.: *Calculations of Viscous Flows with a Multigrid Method*, Ph.-D. Dissertation, MAE Department, Princeton University, October 1987.

[3] Jameson, A.; Baker, T.J.; Weatherhill, N.P.: *Calculations of Inviscid Transonic Flow over a Complete Aircraft*, AIAA Paper 86-0103, January 1986.

[4] Gerhold, T.; Friedrich, O.; Evans, J.; Galle, M.: *Calculation of Complex Three-Dimensional Configurations Employing the DLR-τ-Code*, AIAA-Paper 97-0167, January 1997.

[5] Rossow, C.-C.: *A Simple Flux Splitting Scheme for Compressible Flows*, Contributions to the 11th STAB/DGLR Symposium Berlin 1998, Notes on Numerical Fluid Mechanics, Vieweg Publishers, Braunschweig, Wiesbaden, Vol. 72, pp.355-362, 1999.

[6] Turkel, E.: *Preconditioning Techniques in Computational Fluid Dynamics*, Annual Reviews in Fluid Mechanics, Vol. 31, pp. 385-416, 1999.

[7] Rossow, C.-C.: *A Flux Splitting Scheme for Compressible and Incompressible Flows*, Journal of Computational Physics, Vol. 164, pp. 104-122, 2000.

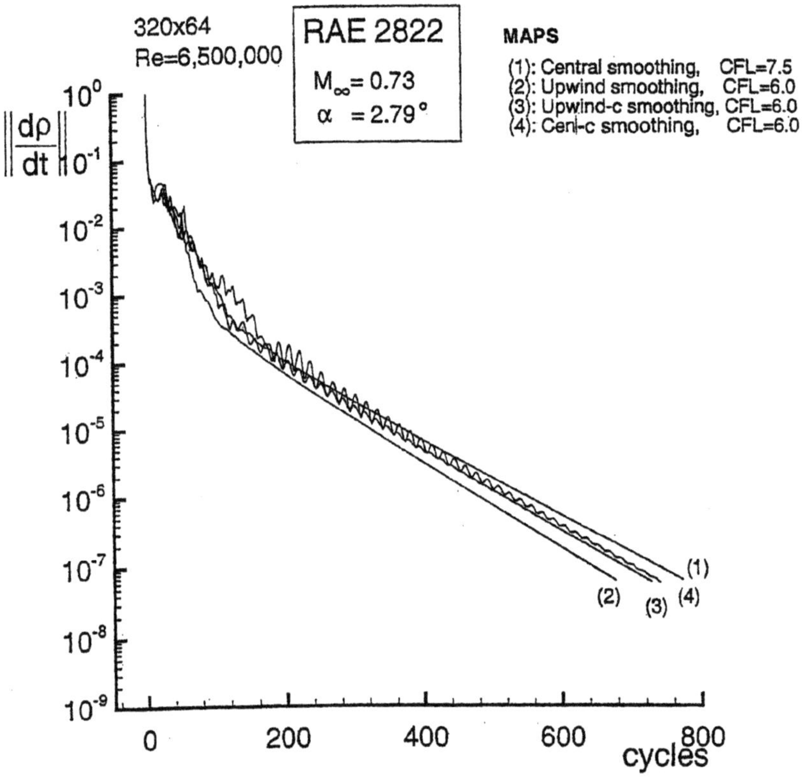

Fig. 1: Convergence histories for different smoothing schemes (structured)

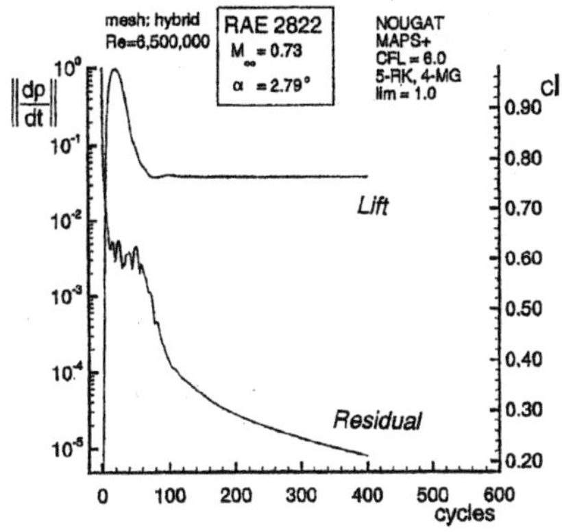

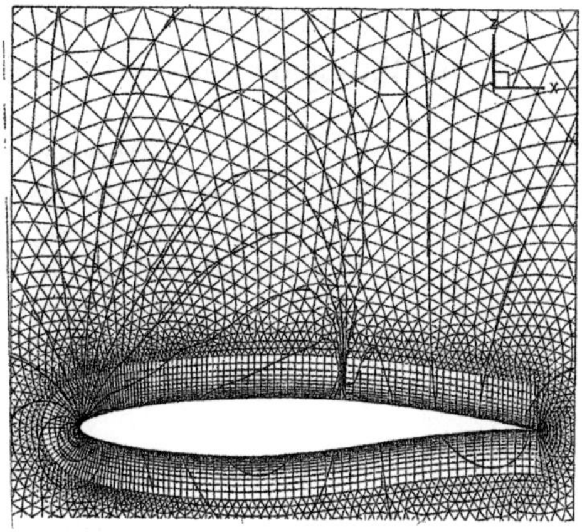

Fig. 2: Convergence (hybrid, transonic) Fig. 3: Pressure distribution on hybrid mesh

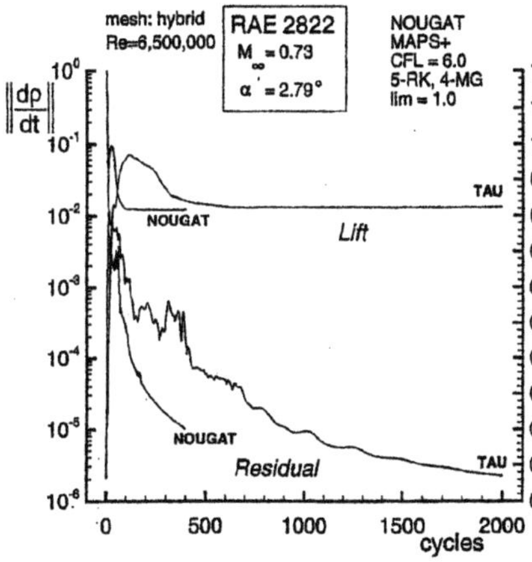

Fig. 4: Convergence (NOUGAT/TAU)

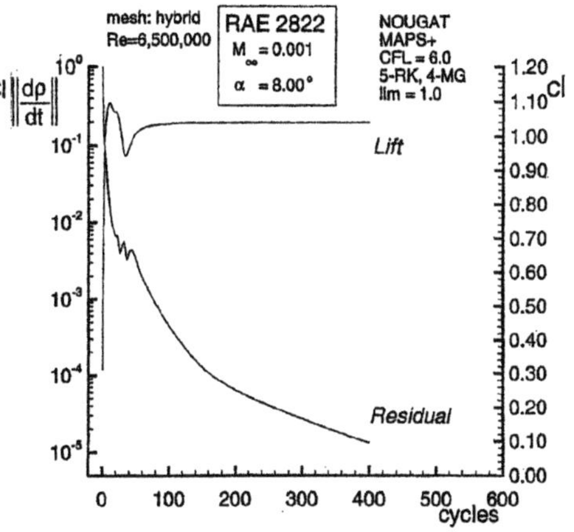

Fig. 5: Convergence (hybrid, incomp.)

Enhancement of a Navier-Stokes Flow Solver for Patched Grids with Non-Coincident Grid Nodes

Thorsten Schwarz
German Aerospace Center (DLR),
Institute of Design Aerodynamics,
Lilienthalplatz 7, 38108 Braunschweig, Germany
e-mail: thorsten.schwarz@dlr.de

Summary

This paper presents two methods for the treatment of block interfaces with non-coincident grid nodes. The first approach uses linear interpolation to transfer flow data between the adjacent blocks. This technique permits an arbitrary distribution of grid nodes at the interface. A new procedure enables an accurate interpolation near body surfaces. But this approach is not flux conservative. The second method requires a fixed refinement ratio but ensures flux conservation across the interface. The methods are designed to simplify the grid generation process and to be used for grid adaptation. Several examples demonstrate the gain in flexibility with either technique.

1 Introduction

The generation of a structured multi block grid around a complex configuration is still a time consuming process. The mesh must resolve all flow phenomena accurately while grid distortion must be avoided and the number of grid points should be minimized. The grid generation procedure could be simplified, if grid lines must not coincide at the interface between two neighbouring grid blocks. This would allow to generate the grid blocks independently from their neighbours. But this also requires special procedures for the flow solver in order to establish the intergrid communication across this kind of block interface. Beside the increasing flexibility in grid generation, interfaces with hanging grid nodes are also an important ingredient to extend existing multi-block flow solvers for an automatic grid adaptation.

Several approaches were developed in the past to extend flow solvers for interfaces with hanging grid nodes. Most studies were concentrated on a flux conservative treatment [1], [2], [3], [4], [5]. A non-conservative method is described in [6].

This paper presents two different approaches for grid interfaces with hanging nodes: The first part describes an interface condition which permits arbitrary point distributions but is not flux conservative. A special technique allows an accurate transfer of flow data near body surfaces. In the second part, some preliminary results for a flux conservative treatment for interfaces with fixed refinement ratio will be presented.

2 Numerical Method

All computations are carried out using the DLR flow solver FLOWer [7]. FLOWer solves the unsteady, compressible three-dimensional Reynolds averaged Navier-Stokes equations on block structured meshes. The approximation of the governing equations follows the method of lines which decouples the discretization of space and time. Two different spatial discretizations based on a finite volume method are available. For the first method, the flow variables are calculated at the vertices of the grid cells. The control volume is built by the eight cells surrounding the grid node. For the second approach, the conservative variables are calculated at the centre of the cell and the control volume is given by the respective cell. For both spatial discretizations, the fluxes through the cell faces are approximated using a central discretization operator. Dissipation is added by a blend of first and third differences. For the integration in time, an explicit five stage Runge-Kutta time stepping scheme is used. The convergence can be accelerated by implicit residual smoothing, local time steps and multigrid.

3 Treatment of Interfaces with Coincident Grid Nodes

The solution of the Navier-Stokes equations on one single block of a multi-block mesh requires the specification of boundary conditions at the block faces. In FLOWer the boundary conditions are prescribed for two layers of dummy cells which are added at each grid face. This technique permits to use the same spatial discretization operator for the interior of a grid block as well as at the grid boundaries. At the interface between two blocks with coincident grid nodes, the grid coordinates of the first/second dummy layer are taken from the first/second interior grid plane of the adjacent block. This ensures an identical cell geometry of the dummy layers and the cells of the neighbouring block.

In order to exchange flow variables across a cut interface first a grid block sends flow data of the cells near the cut to the adjacent block. Then the block receives data which were sent by the neighbouring block. The data are stored into the dummy layers. In the following a grid block which sends information will be called 'donor' block. A block which receives data will be called 'target' block. For a full exchange of flow data every block is both donor and target block.

4 Non-Conservative Interface Treatment

A simple way to establish intergrid communication across interfaces with hanging grid nodes is to transfer flow information by interpolation from the interior planes of the donor grid into the dummy layers of the target grid. Since grid-planes are considered, a interpolation formula for two dimensions is sufficient for three dimensional flow calculations. In order to maintain second order accuracy, the flow variables must be interpolated by bilinear interpolation:

$$\begin{aligned} f(c_1,c_2) &= (1-c_1)(1-c_2)\cdot f_{i,j} + c_1(1-c_2)\cdot f_{i+1,j} \\ &+ (1-c_1)c_2\cdot f_{i,j+1} + c_1c_2\cdot f_{i+1,j+1} \end{aligned} \tag{4.1}$$

where f denotes any grid or flow data which is interpolated, c_1 and c_2 denote the interpolation coefficients, see figure 1.

4.1 Calculation of Interpolation Coefficients

At the beginning of a flow calculation the interpolation coefficients c_1 and c_2 in eq. (4.1) are unknown. If the flow variables are assumed to be located at the grid nodes, the coefficients can be calculated from the coordinates of the grids at the interface. The basic procedure is as follows: For every grid point P at the interface of the target block the interface grid of the donor mesh is searched for the cell which encloses P. Next the coordinates of the enclosing cell $\vec{x}_{i,j}, \vec{x}_{i+1,j}, \vec{x}_{i,j+1}, \vec{x}_{i+1,j+1}$ with $\vec{x} = (x,y,z)$ are used to build the right hand side of eq. (4.1). The arising system of three equations describes a surface $\vec{x}(c_1,c_2)$ in three dimensional space with the coordinates c_1,c_2. Now the coordinates of P are projected onto this surface which permits to calculate the interpolation coefficients c_1 and c_2 analytically.

This approch is not sufficient if one or more edges of a grid interface are placed on a body surface. This is illustrated by figure 2 showing an interface with non-coincident grid nodes located at the nose region of a wing (see also figure 4). Due to the discretization with a finite number of grid nodes, the surface definitions of the grids at the interface differs. As a consequence a target node having a distance δ_1 from the surface definition of the target grid interpolates flow data from the donor grid at a location which has a distance $\delta_2 \neq \delta_1$ from the surface definition of the donor grid. The difference in the wall distances δ_1 and δ_2 may cause large errors in the transferred flow data especially if high flow gradients exist near the body surface, for example inside a boundary layer.

In order to enable an accurate calculation of interpolation coefficients a new grid is generated for the interface of the donor grid. It contains all grid points of the original donor mesh but also all grid points of the edges of the target mesh, see figure 3. The grid is built using the following procedure: First the grid at the interface of the donor mesh is copied into a temporary array. Then the grid is compared to the interface grid of the target block. New grid lines are introduced into the temporary grid in order to shift the grid to the grid points at the edge of the target grid. A smooth distribution of grid lines is achieved, if only those points of the new grid lines are shifted to the target points, which are located near the edge. The other new points are placed on the lines which connect the original grid nodes. At last, for every point of the new grid the index and the local coordinates related to the original grid are stored in a table.

The interpolation coefficients are now calculated based on the new grid. Afterwards the coefficients must be recomputed for the original donor grid. This is a look up operation in the table generated in the last step. If all interpolation coefficients are calculated the temporary grid is removed since the flow is computed using the original mesh.

The method described above can be applied for interfaces with an arbitrary distribution of grid points. The interface itself can have an arbitrary shape. For the cell centred discretization a similar procedure was developed. But the method is more complicated, since the creation of a new interface grid is necessary for the donor and the target mesh.

4.2 Test Case for Noncoincident Interface Treatment

The new interpolation method is now applied to an infinite wing with NACA 0012 airfoil. The original mono-block Navier-Stokes grid is subdivided into two blocks, see figure 4. For one block every even grid line in chordwise direction is removed from the original grid, whereas for the second block, every odd grid line is deleted. This introduces an interface with hanging grid nodes between the two blocks. The point distribution normal to the wall is still identical for both grids. The flow is calculated for a Mach number $M = 0.8$ and an incident angle $\alpha = 1.25°$. The Baldwin-Lomax turbulence model is used.

Figure 5,*left* presents the friction distribution for a computation without using the correction for interface edges at body surfaces. The flow is clearly disturbed near the interface. The disturbances are avoided if the correction is applied, see figure 5,*right*.

4.3 Application of the Method

The generation of a grid around a cooling fan is often a difficult task, because all the grid lines needed to resolve the boundary layer and the wake of a blade must be continued into the grid around the following blade. Furthermore, the fan blades are highly twisted, see figure 6. Both constraints often prevent the generation of a smooth grid. The new boundary condition for hanging grid nodes significantly simplifies the grid generation for such kind of applications, since the grid lines do not have to be continued at block interfaces. Figure 7 shows one plane of the 3D grid and demonstrates, that the flow is not disturbed by the non-coincident block interface, from [8].

5 Conservative Grid Interface Treatment

The interpolation of flow variables at block interfaces does not preserve the conservation of mass, momentum and energy. In literature it is discussed that the lack of conservation has a negative influence on the convergence behaviour and causes spurious reflections of shocks at the interface. Therefore, a conservative treatment for the non-coincident grid node condition is under development.

In order to reduce the coding effort, it was decided to develop a conservative treatment only for interfaces with a fixed refinement ratio. This means that the grid lines of one block match with every n-th grid line on the neighbouring block. The refinement might be anisotropic. Up to now the conservative interface treatment is only available for Euler computations with the cell centred discretization.

The following procedure is used in FLOWer to ensure the conservation of fluxes, see figure 8: The cell faces of the coarser grid are subdivided into segments which fit exactly with the faces of the cells of the finer grid. Now the convective fluxes and the fluxes due to the artificial dissipation across the segments are calculated. The fluxes for the coarse cells are obtained by summation of the individual fluxes through all respective segments.

5.1 Channel with 10% Bump

The first test case is a channel flow with a ten percent bump and an inflow Mach number of 0.675. Figure 9,*top* shows the four times adapted grid with 2352×8 cells where adaptation is controlled using a pressure sensor. The results are compared with the solution for a globally fine grid with 12288×8 cells, see figure 9,*bottom*. The grid density of the fine mesh is identical to the finest grids of the adapted mesh. Figure 10 compares the convergence behaviour for a calculation on the fine mesh and for the adapted grid with conservative and non-conservative interfaces respectively. It turns out that the convergence on the adapted grid is hardly influenced by the different treatments of the block interfaces. The gain in convergence compared to the globally refined grid is also obvious. To prove the conservation properties, the mass conservation error

$$error = \frac{|F_{in} + D_{in} - (F_{out} + D_{out})|}{|F_{in} + D_{in}|} \tag{5.2}$$

is calculated. In eq. (5.2) F and D denote the sum of the convective and dissipative fluxes at the inlet and the exit of the channel. Figure 11 demonstrates that the scheme is indeed mass conservative, since the mass flux error reduces to machine accuracy. In contrast, the mass flux error for the non-conservative method is about 0.1%. The comparison of the Mach number distribution shows hardly any difference between the three calculations, see figure 12. The deviation at the outflow boundary is caused by the shear layer behind the shock, which was not marked for adaptation.

5.2 Flow around Airfoil

In the future, the grid adaptation with conservative block interfaces will be applied for complex airplane configurations. As a first test case for such kind of applications the flow around a NACA 0012 airfoil is computed. Figure 13 shows a three times adapted grid with 15 blocks and 5696×8 cells in total. A globally refined grid would have 20480×8 cells. The pressure distribution shown in the figure was calculated for M=0.8 and α=1.25°. This test case also demonstrates the need for an automatic tool for mesh adaptation since the adapted mesh was obtained by coarsening the fine grid and establishing the block connectivity manually. This is not practicable for realistic applications.

6 Conclusion

Two boundary conditions for block interfaces with non-coincident grid nodes were presented. The first method enables an arbitrary distribution of grid lines. It is based on bilinear interpolation. A special technique allows an accurate interpolation near body surfaces. The method is implemented for the cell centred and the cell vertex discretization. The simplification for the grid generation process is demonstrated for a fan flow calculation.

The second approach establishes a flux conservative interface between two adjacent grid blocks. This method allows an anisotropic 1:n refinement and is currently available only for the Euler equations with cell centred discretization. Results for a channel flow test case are in good agreement with conventional mono-block calculations.

In the future, the conservative method will be extended for the Navier-Stokes equations and for the cell vertex scheme. It is planned to apply the method to complex 3D airplane configurations. The grid adaptation will also be automatized.

References

[1] G. H. Klopfer, G. H. Molvik: *Conservative Multi-Zonal Interface Algorithm for the 3-D Navier-Stokes Equations*, AIAA Paper 91-1601-CP

[2] J. Bell, M. Berger, J. Saltzmann, M. Welcome: *Three-Dimensional Adaptive Mesh Refinement for Hyperbolic Conservation Laws*, SIAM Journal for Scientific Computations, Vol 15, No 1, pp. 127-138, January 1994

[3] A. Lerat, Z. N. Wu: *Stable Conservative Multidomain Treatments for Implicit Euler Solvers*, Journal of Computational Physics, Vol. 123, pp. 45-64, 1996

[4] M. Delanaye, Y. Liu: *Quadratic Reconstruction Finite Volume Schemes on 3D Arbitrary Unstructured Polyhedral Grids*, AIAA Paper 99-3259

[5] J. Bohbot, G. Grondlin, D. Darracq : *A Parallel Multigrid Conservative Patched/Sliding Mesh Algorithm for Turbulent Flow Computation of 3D Complex Aircraft Configurations*, AIAA Paper 2001-1006

[6] R. T. Biederon, J. L. Thomas: *A Generalizied Patched-Grid Algorithm with Application to the F-18 Forebody with Actuated Control Strake*, Computing Systems in Engineering, Vol 1, Nos 2-4, pp. 563-576, 1990

[7] N. Kroll, C. C. Rossow, K. Becker, F. Thiele: *The MEGAFLOW project*, Aerospace, Science and Technology, Vol. 4, pp. 223-237, 2000

[8] M. Mauß, D. Lohmann, H. Saathoff, U. Stark: *Experimentelle und numerische Untersuchung an Kfz-Kühlerventilatoren mit Schaufeln unterschiedlicher Sichelung sowie mit und ohne Prallplatte*, VDI-Berichte 1591, VDI Verlag, pp. 459-475, 2001

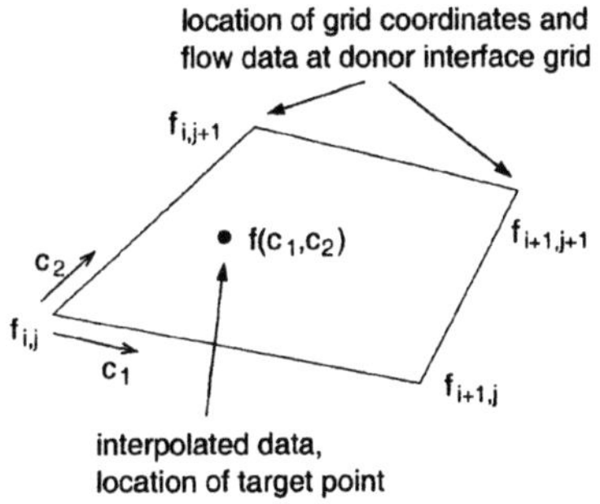

Figure 1 Interpolation of data, the flow variables are located at the nodes of the grids

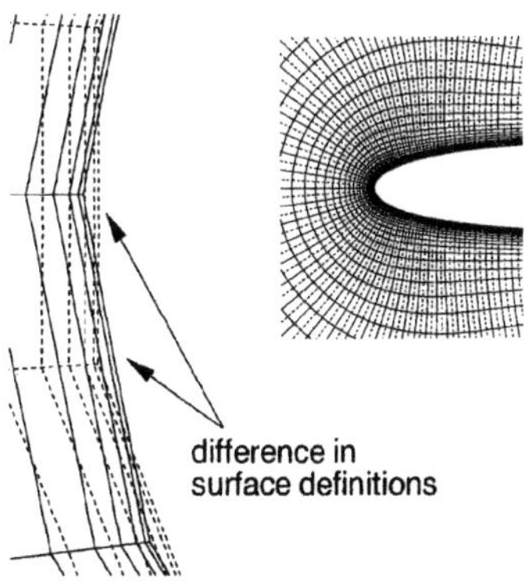

Figure 2 Non-coincident interface between two grids at the nose of a wing, (see also figure 4)

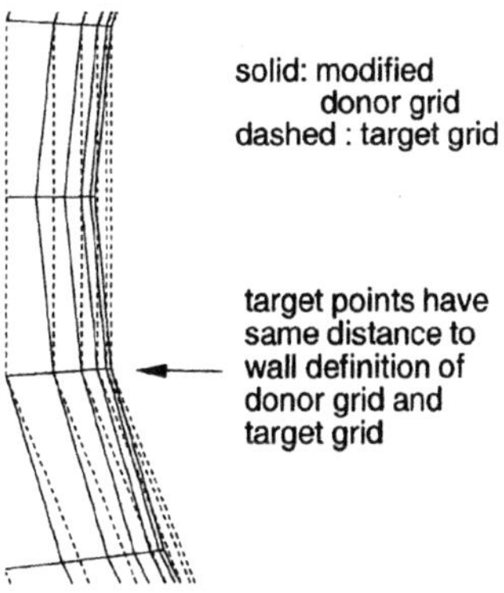

Figure 3 Accurate interpolation using the modified donor grid

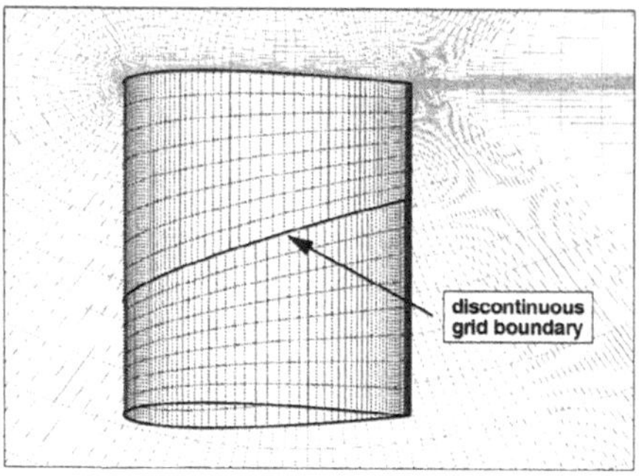

Figure 4 NACA0012-wing with discontinuous grid lines

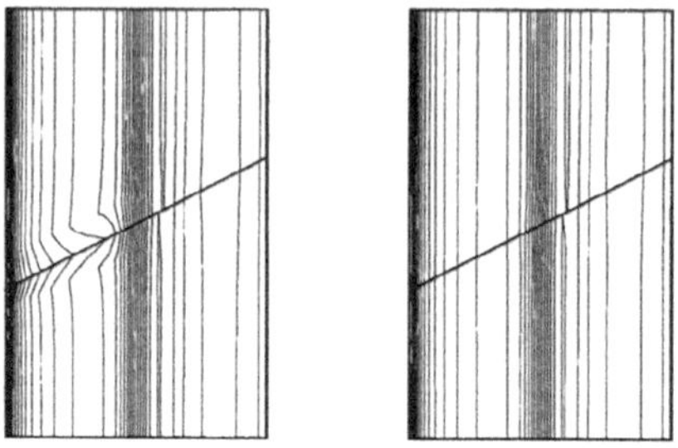

Figure 5 Friction distribution on upper side of NACA 0012-wing, left: without correction of interpolation coefficients near wall, right: correct interpolation

Figure 6 Overall view of fan

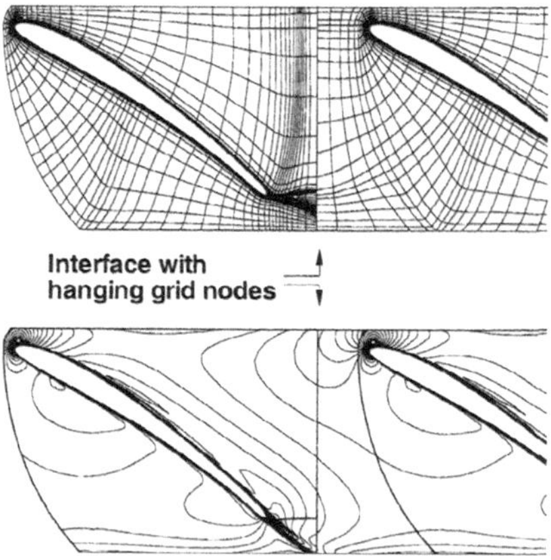

Figure 7 Application of non-coincident node interface to fan flow calculation, top: grid near hub, for clarity, every other grid line is omitted, bottom: Mach number distribution

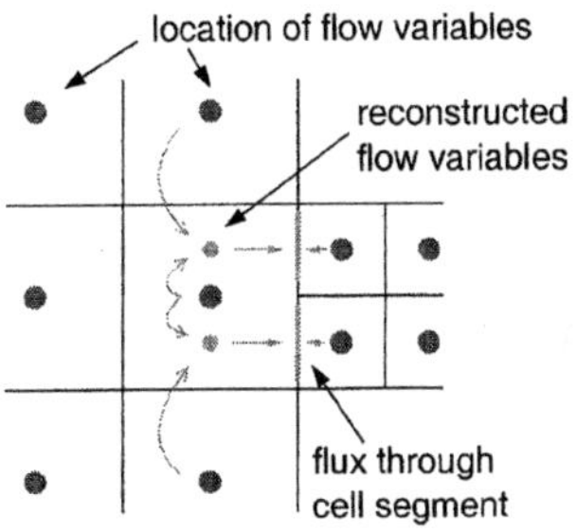

Figure 8 Conservative flux calculation at non-coincident interface

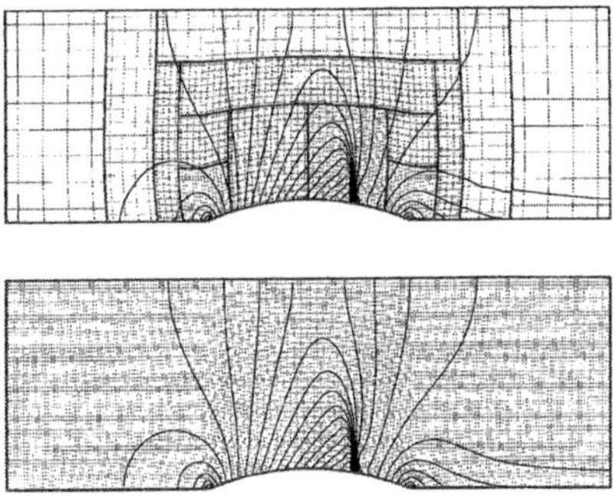

Figure 9 Channel flow with 10% bump, top: adapted grid, bottom: fine grid, plotted are iso-pressure lines

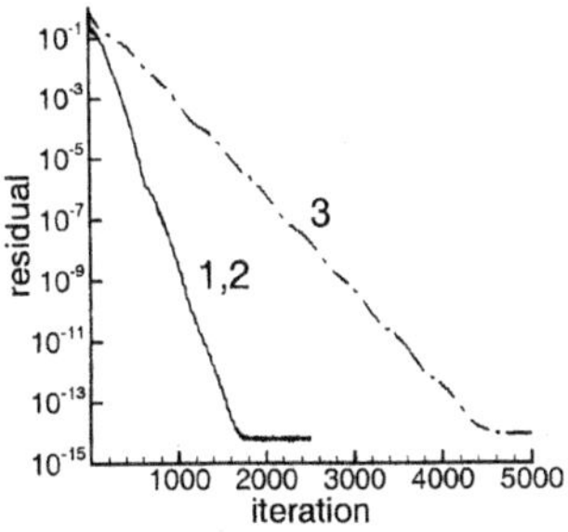

Figure 10 Convergence for channel flow calculation, the conservative method and the non-conservative method give nearly identical convergence behaviour, 1: conservative method, 2: bilinear interpolation, 3: fine grid

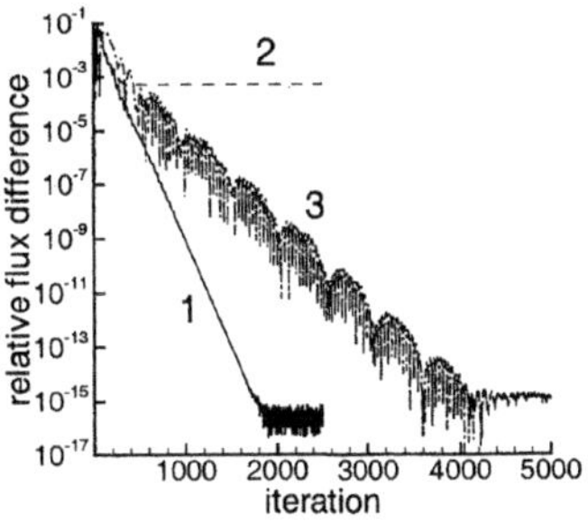

Figure 11 Conservation error for channel flow calculation, 1: conservative method, 2: bilinear interpolation, 3: fine grid

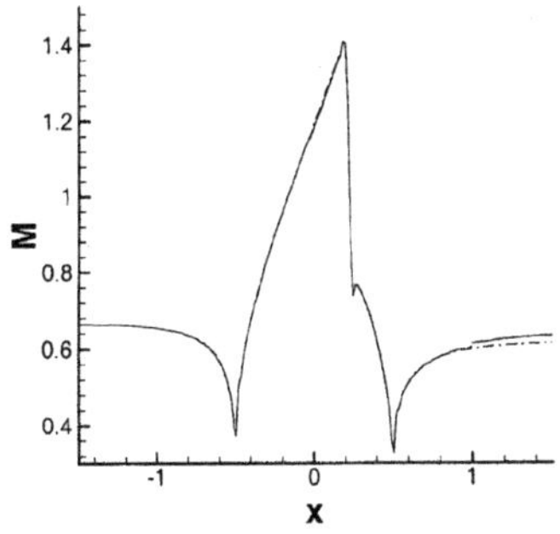

Figure 12 Mach number distribution at lower wall of channel solid: conservative method, dashed: bilinear interpolation, dash-dotted: fine grid

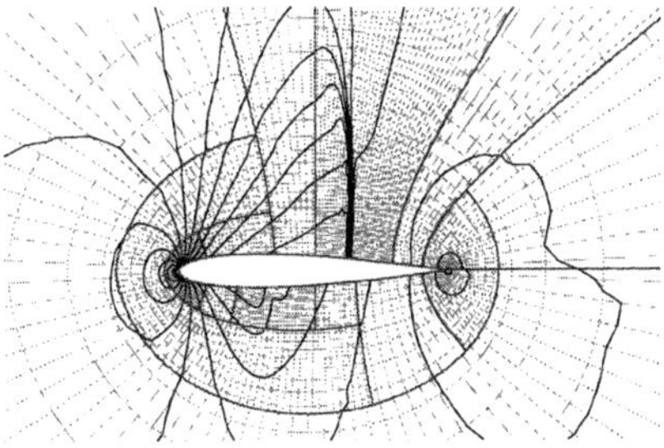

Figure 13 Adapted airfoil grid, plotted are iso-pressure contours

Bluff Bodies

Flow and Turbulence Structures in the Wake of a Simplified Car Model (Ahmed Modell)

H. Lienhart, C. Stoots, S. Becker

Lehrstuhl für Strömungsmechanik (LSTM)
Universität Erlangen-Nürnberg
Cauerstr. 4, 91058 Erlangen, Germany

Summary

The aim of the "Models for Vehicle Aerodynamics" (MOVA) Project is to develop, refine, and validate the latest generation of turbulence models for selected examples encountered in vehicle aerodynamics. The validation of turbulence models requires the availability of detailed experimental data. These quantitative data should cover the most critical flow regions around a bluff car-shaped body and they should give physical quantities that can directly be correlated to the results of numerical simulations. Such experimental data were measured in the LSTM low speed wind tunnel using a 2-component laser-Doppler anemometer (LDA) mounted on a traversing system and a simplified model of a car (Ahmed model). Measurements were made for two rear vehicle body slant angles (25° and 35°) at a bulk air velocity of 40 m/s. This paper serves as a synopsis of the major results of this experimental investigation.

Introduction

The flow region which presents the major contribution to a car's drag, and which poses severe problems to numerical predictions and experimental studies as well, is the wake flow behind the car. The location at which the flow separates determines the size of the separation zone, and consequently the drag force. Clearly, a more exact simulation of the wake flow and of the separation process are essential for the correctness of drag predictions. However, a real-life automobile is a very complex shape to model or to study experimentally. Therefore the MOVA consortium partners (TU Delft, University of Manchester, LSTM, Electricite de France, AVL List, and PSA Peugeot Citroen) agreed to study the vehicle shape employed by Ahmed [1], known as the Ahmed model. Figure 1 is a schematic of the Ahmed model, with actual dimensions in mm included. Two different rear body slant angles (25° and 35°) were considered, which happen to bracket the critical angle of 30° at which separated flow occurs within the wake of the slant (see Figure 2, c_W, c_R, c_K, c_B, c_S represent coefficients of total drag, friction drag, nose pressure drag, slant pressure drag, and base pressure drag, respectively).

The Ahmed model was mounted in the test section of the LSTM wind tunnel (Figure 3) and detailed measurements of velocity profiles were made around this body. The experiments were performed in the LSTM low speed wind tunnel, a closed return facility which can be configured with an open or a closed test section. The present studies were conducted in a ¾ open test section (i.e., floor, but no sides or ceiling) with a blockage ratio of 4%. The wind tunnel can generate flow velocities from 3 to 55 m/s with average turbulence intensities of less than 0.25%. All measurements concerning the Ahmed model were taken at bulk air velocities of 40 m/s. To ensure constancy of the test section bulk velocity and air temperature, a computer-based feedback control system was utilised.

Hot-wire measurements of the velocity profiles 400 mm upstream of the Ahmed model were obtained to serve as a inlet condition for the numerical simulations. The measurements were performed by a two-component hot-wire system, which was rotated to obtain the third component. The results showed that the test section inlet velocity profile was well defined and controlled, which guaranteed reliable measurements for the model validation database. Flow visualization using oil streaks was performed for both model shape angles. The visualization showed complex three-dimensional flow patterns and confirmed earlier findings that a small change in the slant angle around the critical 30° causes a dramatic change in the flow pattern. Comparison of Figures 4 and 5 illustrate these changes. The formation of clinging vortices and flow reversal for a slant angle of 25° is evident in the photograph of oil streaks in Figure 4. Attached flow is maintained in this case while at a slant angle of 35° (Figure 5) detached flow is obvious.

A two-component laser-Doppler anemometer (LDA) was installed on an existing three-dimensional computer controlled traversing system (rotation of the sending/receiving optics offer an additional axis of traversing). The LDA was composed of DANTEC fiberoptic-based optics and electronics. Two different laser wavelengths were used: 514.5 nm (green) and 488 nm (blue). The system was of backscatter orientation, the only optical configuration possible for these measurements due to the proximity of the measurement grid to the body surface. The laser was a water-cooled, 5 watt argon-ion Spectra Physics Model 2060. Beam splitting and frequency shifting were provided by DANTEC FiberFlow optics. Signal analysis and signal processing were accomplished via a DANTEC Model 57N20 Burst Spectrum Analyzers (BSA) and DANTEC BSA Flow software. A PC computer provided measurement control and data acquisition / storage. LDA measurements were made for all three components of velocity in the symmetry plane from upstream of the Ahmed model to some distance downstream behind the closure of the wake. LDA measurements were also made in several transverse planes in the wake.

LDA Measurement Results

Figure 6 attempts to show the different measurement positions employed in this effort. In summary, there were 7,500 discrete measurement positions located in 13 unique planes. Because higher order statistical moments (i.e., Reynolds stresses, $\overline{u'u'u'}, \overline{u'u'v'}$, etc.) were of interest, each measurement location had to be sampled twice -- first for the U and V components of velocity, then for the U and W components. Preliminary statistical analysis indicated that approximately 40,000 measurement realisations were necessary at each location for statistically significant results. This translates to a maximum measurement time of approximately 5 minutes at each location. Following the procedure outlined in [2], for a 95% confidence interval the statistical uncertainty in the outer flow (far from the Ahmed model) mean velocity was less than 0.005% of the local mean velocity. Within the wake of the Ahmed model, where mean velocities can approach zero and turbulence intensities are very high, an estimate of mean velocity measurement uncertainty is rather arbitrary. An order of magnitude estimate for the calculated 95% confidence interval was 1% for mean and 1.5% for rms quantities. The LDA system measured turbulence intensities as small as 0.8% in the outer flow region. This value represents a LDA lower threshold for measured turbulence intensity and includes inaccuracies due to measurement technique, wind tunnel fluctuations, traverse system vibrations, etc..

Figure 7 serves as a good overview of typical velocity distributions around the 25° slant Ahmed model. The mean velocity vectors along the line of symmetry of the rear slant indicate that no separation of the flow is occurring. The two counter-rotating trailing vortices are shown in the three transverse planes of turbulent kinetic energy (TKE) contour plots. Peaks in TKE occur in the centres of the vortices. These vortices are responsible for maintaining attached flow at the slant up to a slant angle of approximately 30°. Vortical structures extend more than 500 mm beyond the end of the Ahmed model.

Figure 8 is a direct comparison of the region of attachment / detachment and recirculation at the rear of the Ahmed model for the two slant angles of interest. These measurements were made along the plane of symmetry of the Ahmed model. The intention of this figure is to indicate the major differences in the near-wake flow fields for the two slant angles studied.

More detailed near-wake measurements were made but omitted from this figure to avoid obscuring the wake structures. The only additional information provided by the more detailed measurements concerns a very small recirculation region in the upper part of the 25° slant surface. Beyond this region, the flow then reattaches to the slanted surface, then develops a second, larger recirculation region along the lowermost part of the rear. Observations for the 35° case are quite different, however. The flow detaches along the slanted surface and develops a single, significantly larger recirculation region within the wake.

Mean velocity vectors and profiles of turbulent kinetic energy (TKE) for the symmetry plane of the Ahmed 25° degree slant angle model are displayed in Figure 9. Peak TKE values are within the small recirculation zone at the rear of the body. Note that the Ahmed model only disturbs the outer flow for a relatively small distance from its surface.

Finally, Figure 10 shows the downstream development of the counter-rotating trailing vortex system produced by the 25° slant Ahmed model. The contours represent regions of constant magnitude downstream velocity. At 80 mm downstream from the trailing edge of the Ahmed model, there is a large and strong region of recirculation back towards the surface. Although the recirculation has disappeared by x = 200 mm, there is still a large streamwise velocity deficit. At x = 500 mm the location of the cores of the trailing vortices can still be distinguished by deficits in streamwise velocity.

Figure 11 is an example of profiles of second - and third - order moments for the wake region of the Ahmed model. The contours of various moments display very good symmetry.

Conclusions

Flow visualization, hot wire, and LDA measurements have been performed around an Ahmed model for two different rear slant angles: 25° and 35°. These two slant angles bracket a critical instability when the flow detaches from the slanted surface. LDA measurements include all three components of mean and RMS velocity as well as most second (Reynolds stresses) and third order moments. The measurements clearly show the differences in flow attachment and recirculation for the two different slant angles considered. The quantitative information provided by these measurements should prove invaluable for developing, testing, and validating computer models of the aerodynamics of vehicular wake regions

Acknowledgements

The support of the presented work by the European Community under the Industrial & Materials Technologies Program (Brite-EuRam III), contract number BRPR-CT98-0624, is gratefully acknowledged. Special thanks to Dr. Ahmed and the DLR for providing us with the wind tunnel model.

References

[1] Ahmed, S.R., Ramm G.: Some Salient Features of the Time-Averaged Ground Vehicle Wake. SAE Technical Paper 840300, 1984.

[2] Bendat, J.S., Piersol, A.G.: Random Data Analysis And Measurement Procedures. John Wiley & Sons, New York, 1986.

Figures

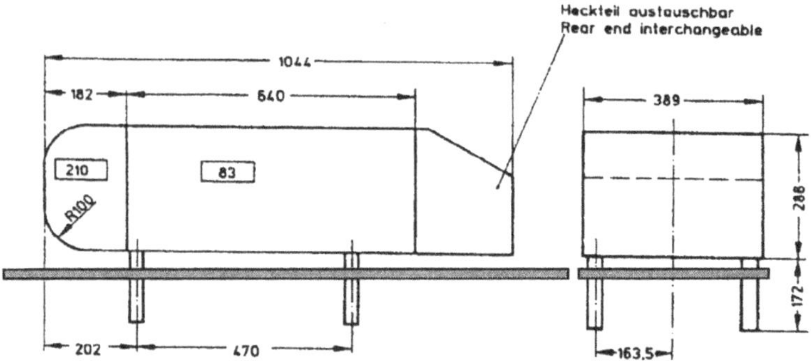

Figure 1. Schematic of the bluff body (Ahmed, 1984) shape.

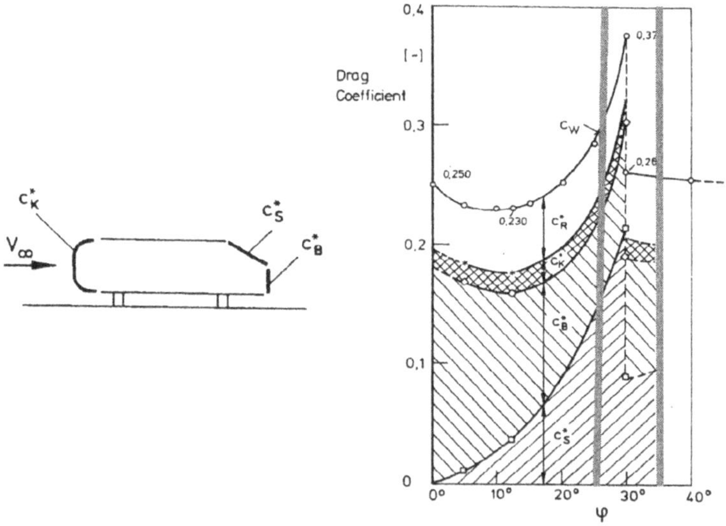

Figure 2. Drag coefficients for the Ahmed model for various slant angles (Ahmed, 1984).

Figure 3. Photograph of the Ahmed model mounted in the LSTM low-speed wind tunnel.

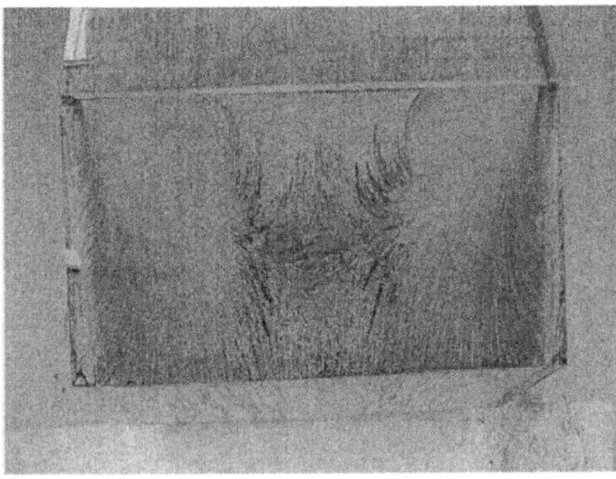

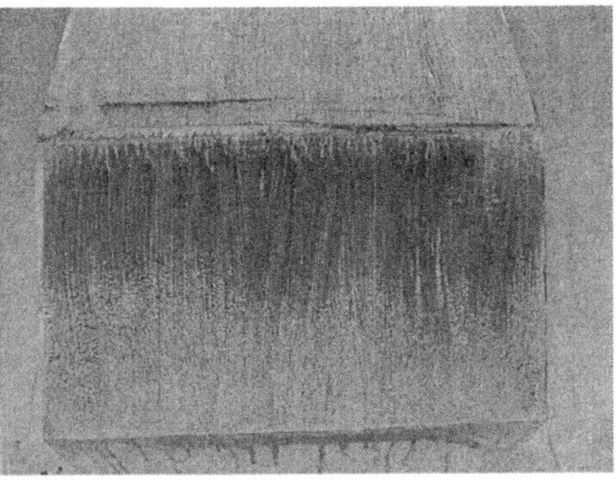

Figure 4. Oil / soot streak flow visualisation of the Ahmed model rear for a 25° slant angle.

Figure 5. Flow visualisation of the Ahmed model rear for a 35° slant angle.

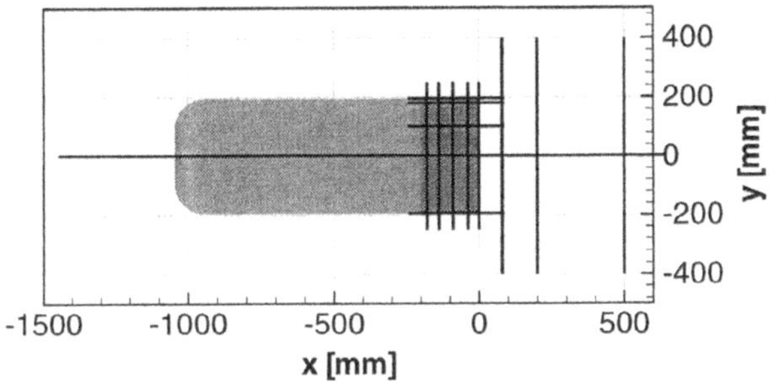

Figure 6. Measurement planes utilised to characterise the velocity distribution.

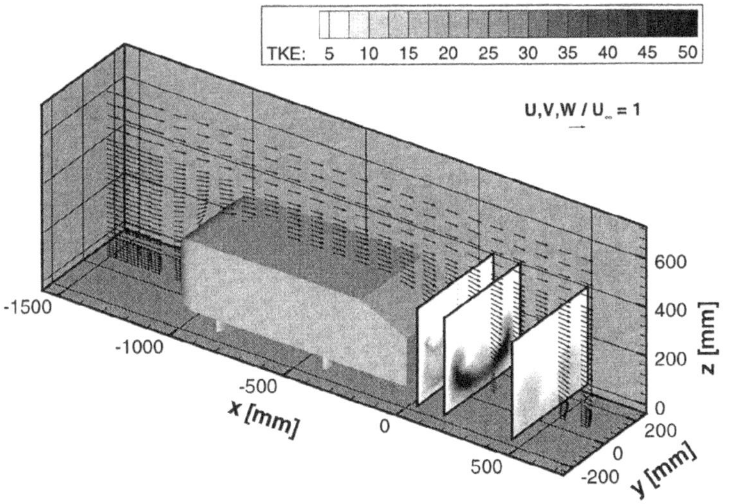

Figure 7. Overview of velocity distributions and TKE around the 25° slant Ahmed model.

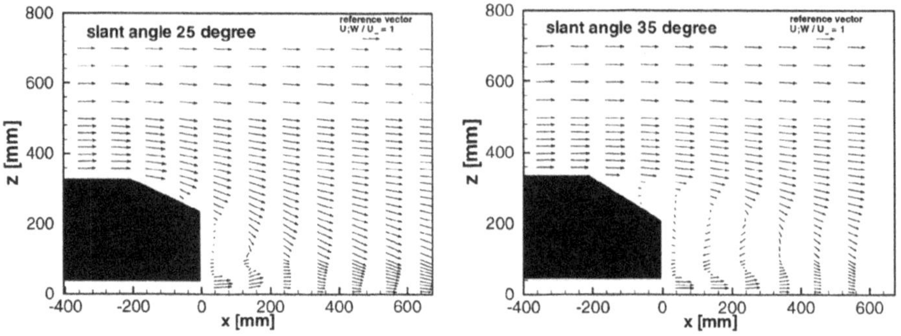

Figure 8. Comparison of the recirculating wake region and attached / detached flow for the two different rear slant angles.

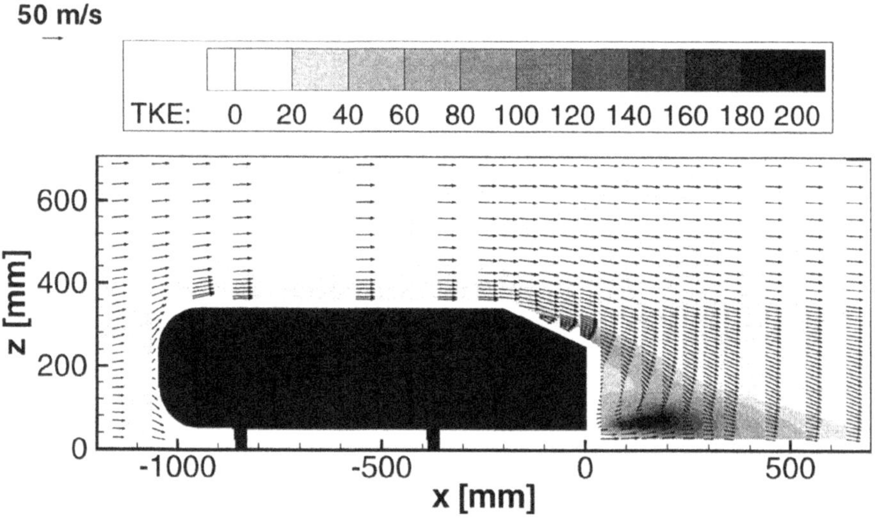

Figure 9. Contours of TKE along the symmetry plane of the 25° slant Ahmed model

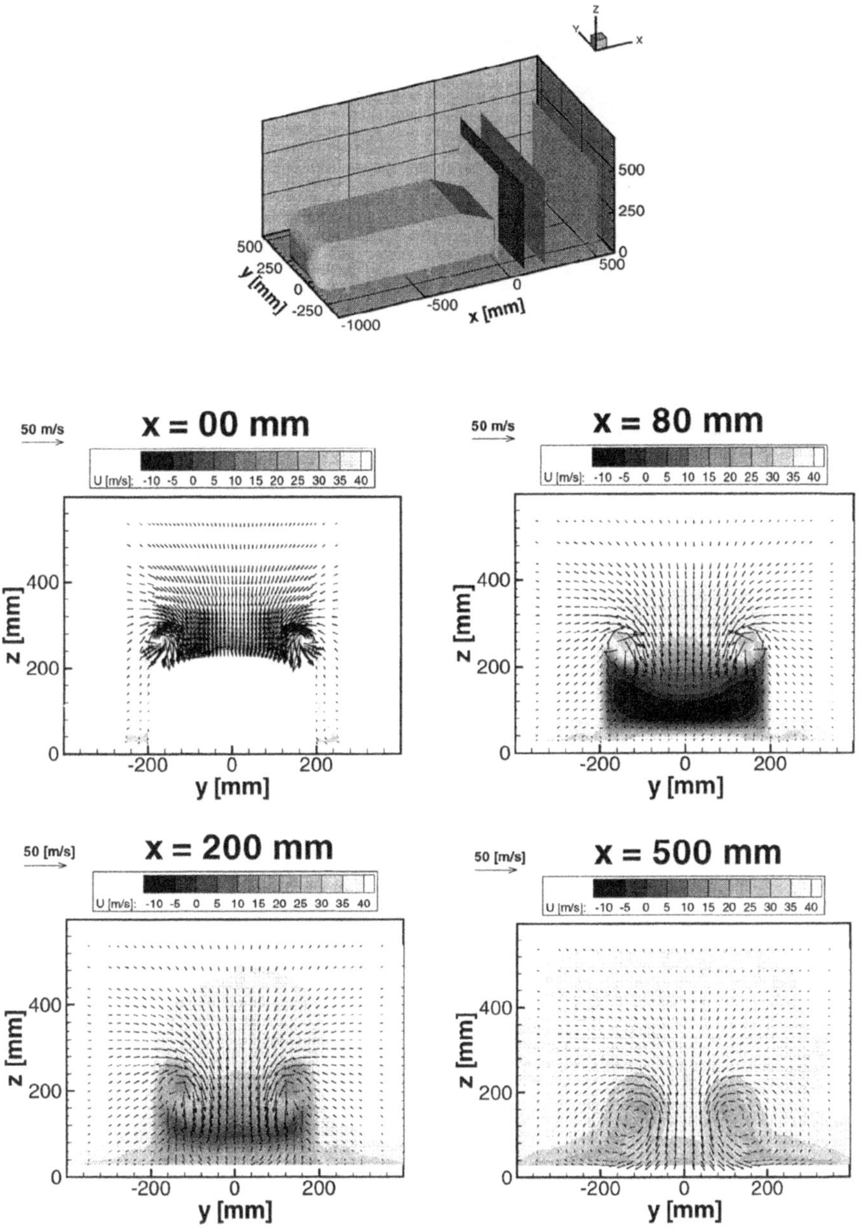

Figure 10. Development of the counter-rotating trailing vortex system within the wake of the 25° slant Ahmed model.

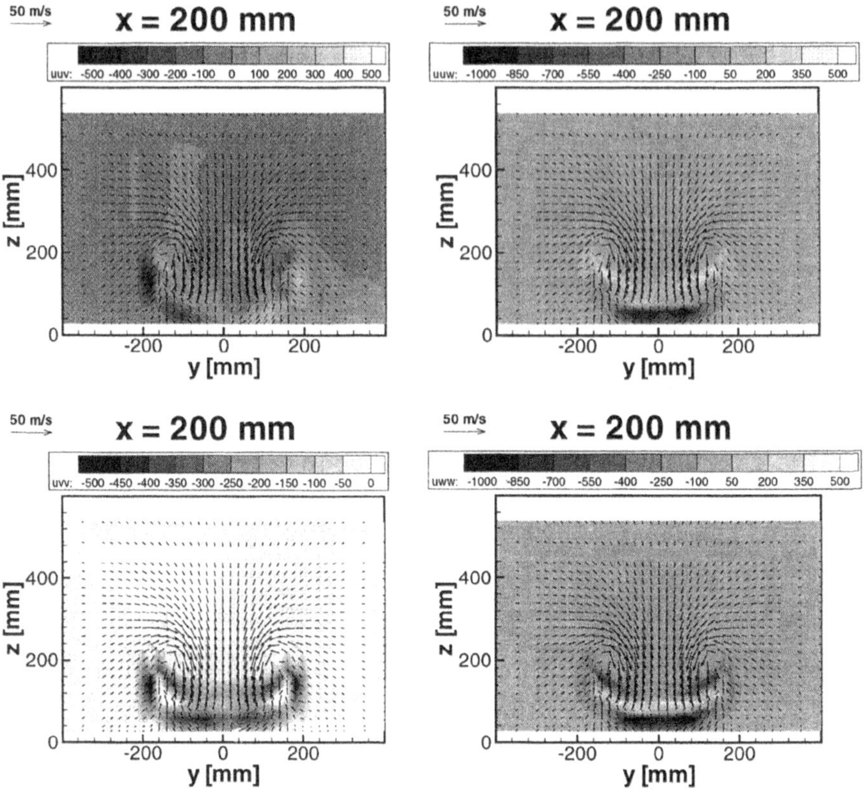

Figure 11. Profiles of higher-order statistical moments within the wake of the 25° Ahmed model.

Flow Visualisation, Pressure Measurements and Numerical Calculations on Spike-Tipped Bodies

J. SRULIJES[1], P. GNEMMI[1], K. RUNNE[2], F. SEILER[1]
[1] French-German Research Institute of Saint-Louis (ISL)
5, Rue du Général Cassagnou F-68301 Saint-Louis, France
[2] EADS Deutschland GmbH, LFK-Lenkflugkörpersysteme GmbH
D-88039 Friedrichshafen, Germany

Summary

The requirements imposed by EADS Deutschland GmbH for the design of a new high velocity missile V/SHORADS (Very/Short Range Air Defence System) are the reduction of drag and to obtain reliable information from the opto-electronic embedded sensors at even high angles of attack. The aim of our investigation is to find new spike geometries and to optimise these to reach high angles of attack without losing the drag reduction effect well-known for small angles of attack. Flow visualisation pictures taken with differential interferometry are shown. A classical disk spike geometry is compared with a new spherical spike and the advantages of the latter are discussed. For equal diameters and lengths the disk spike proves to be more efficient below 8° whereas the sphere spike is more efficient beyond this angle of attack. Pressure measurements are presented and compared to the numerical calculations.

1 Introduction

A supersonic flow around a blunt body generates a bow shock which causes high surface pressure and consequently the development of high drag. The pressure on the body surface and consequently the drag can be substantially reduced if, instead of a normal shock, an oblique shock is generated by an aero-spike. The oblique shock compresses and heats the flow less than a normal shock. Additionally, the spike produces a re-circulatory flow separation region that shields the nose from the oncoming flow. This reduces drag as well as temperature and protects the missile's surface with the embedded opto-electronic sensors. This is especially needed at high speeds to reduce the heating of the sensors. A disadvantage of using classical spikes is that the reduction on nose drag decreases substantially as the angle of attack increases. High angles of attack must be flown to fulfil the manoeuvrability requirements of state-of-the art missiles.

Preliminary experiments were published in [1-3]. The experiments discussed herein have been carried out at a Mach number 4.5 (about 1500 m/s) using one of the ISL high pressure shock tunnels. These shock tunnels enable us to produce high Mach number air flows ranging in pressure and temperature from ground level up to an altitude of about 10 km. The measuring time for the flow conditions shown in this paper is about 1.0 ms.

2 Shock Tube Facility

Two high-pressure shock tubes, A and B, have been in operation at the ISL shock tube laboratory since 1966 (see Patz [4,5] and Oertel [6]). Fig. 1 shows these two shock tubes looking towards their end, i.e., the measuring sections; on the left shock tube A and on the right shock tube B, with inner diameters of 100 mm. These shock tubes have been used to date as shock tunnels. They consist of a high pressure driver tube, a driven tube and a nozzle attached at its end. A principle sketch of the shock tunnel facility, operated in a reflected mode, is given in Fig. 2 with the high-pressure driver tube and the lower pressure driven tube containing the test gas.

A diaphragm separates the driver tube filled with hydrogen or helium from the driven tube containing the test gas which is either air or nitrogen. The driver gas fill pressure used varies from 20 MPa to 60 MPa and the test gas fill pressure from 100 kPa to 600 kPa. After diaphragm bursting, the driver gas expands into the driven tube, producing a shock wave therein and accelerating the gas behind this shock wave to either subsonic or supersonic constant flow speeds, depending on shock tube operation. A circular parallel flow nozzle having a throat diameter of 50 mm, an exit diameter of 220 mm and a length of 713.5 mm is mounted at the end of tube A, which is used for the experiments discussed in this paper. The nozzle is fixed inside a test section attached to a dump tank which contains the driver and driven gases after firing. Optical windows are placed at both sides of the test section for flow diagnostics.

3 Experiments

A generic V/SHORADS missile spherical nose model with a diameter of 70 mm was used for the tests. Two different shapes of spikes were compared, i.e., the classical disk spike shown in Fig. 3 and a new spherical spike shown in Fig. 4. The experiments were carried out using N_2 as test gas and a mixture of H_2 (83.3%) and N_2 (16.7%) as driver gas. The addition of N_2 to the driver gas delays the arrival of the reflected expansion wave at the measuring section, thus increasing the measuring time available. For the above V/SHORADS missile nose models, different spike lengths, diameters of disks/spheres and angles of attack (not shown in this paper) were experimentally tested.

To explain the advantages of the sphere spike we chose one geometry and two angles of attack.

Spike length: 70 mm

Diameter of disk/sphere: 17.5 mm

Angles of attack: 0° and 24°

Mach number: 4.5

Static pressure: 40 kPa

Temperature: 240 K

3.1 Flow Visualisation

Flow visualisation was done by means of differential interferometry as described by Smeets and George [7] and Smeets [8]. This optical measuring technique allows the visualisation of density gradient fields. We visualized the flow by a series of framing pictures taken with a rotating drum camera. For these framing pictures the flow is focused on the rotating film and eight parallax-free successive air sparks are used as light source. The time interval between the sparks is adjustable and the start is triggered by the incident shock wave. In this paper we show only one of the eight pictures taken always at about the same time after triggering.

Missile nose models with a disc spike and a sphere spike are shown in Fig. 5 and Fig. 6 for an angle of attack of 0°. The bow wave and detachment zones in front of the missile nose model can be clearly recognized by light intensity differences in the interferograms. Comparing Fig. 5 with Fig. 6 it can be seen that at an angle of attack of 0° the flow patterns in front of the bodies are quite similar, only the distance between bow wave and shoulder of the body is shorter for the sphere spike. For the same flow conditions, Fig. 7 and Fig. 8 show differential interferograms for an angle of attack of 24°. The flow pattern is different in the two cases. Differences can be clearly seen in the bow shock wave in front of the spike and near the centre of the spherical forebody. The detachment of the flow at a spike with sphere is less abrupt than at the classical disk. Consequently higher angles of attack can be flown with sphere spikes. The pressure on the shoulder of the body (for equal disk and sphere diameters) is higher for the sphere, but as the drag of a sphere is lower than that of a disk, a larger sphere will compensate this handicap with the advantage of a smoother flow detachment.

3.2 Pressure Measurements

Some problems encountered while measuring pressures at the surface of the blunt body prompted us to modify the measuring technique used. The diameter of the flush mounted PCB 113A21 pressure gauges used for the preliminary experiments [1-3] was 5.5 mm. The pressure data obtained with these gauges showed that the measuring surface was too large to provide quantitatively accurate information to validate computer codes. This is especially the case in the middle of the blunt spherical body where the pressure gradients are large. Therefore, the pressure measurements presented here were carried out with a new V/SHORADS missile nose model with the same dimensions as for the preliminary experiments, but using more and smaller pressure gauges. We now use ten pressure transducers with a diameter of 2 mm (measuring diameter 1 mm). Instead of being flush mounted, these transducers are located behind a small hole of 0.5 mm diameter and 1.5 mm length. Fig. 9 shows schematically the position of the ten pressure gauges KULITE XCQ-080-35 used. On the rear upper side of the cylindrical part of the model an additional measuring gauge is shown (PCB 113A21).

4 Numerical Calculations

Numerical calculations of the steady state flow field around the two spike-tipped missile nose models are conducted by means of the 3-D compressible turbulent CFX-TASCflow code [10]. This code is based on Reynolds averaged Navier-Stokes equations [11] and the k-ε turbulence model [12] is used to close the equations system. The computations presented herein have been carried out for the Mach number and spike geometries tested and for angles of attack of 0° to 30°.

5 Results and Discussion

Fig. 10 depicts the comparison between interferogram pictures (in grey pattern) and computations (over-printed colour constant-pressure lines) of the flow field around the missiles without incidence; the CFD code predicts the experimental flow structure accurately. Note the above mentioned difference between the bow shock geometry of the spikes, especially the location of the interaction between the spike shock and the shock in front of the hemispherical part of the missile. Fig. 11 shows the same comparison for an angle of attack of 20°; again a very good agreement between the numerical simulation and the experiment is found.

Figs. 12-14 show comparisons between calculated missile nose pressure coefficients and corresponding measured values for three angles of attack: 16°, 20° and 24°. Curves for disk and sphere spikes are shown for the windward and leeward sides as well as for the middle of the missile model. A very good agreement between the calculated curves and the experimental data, given by the solid symbols, can be seen. The main differences between the spikes are found from the nose tip of the hemispherical part of the missile (x = 70) up to around 8 mm along the missile axis, corresponding to a cap area of about 30° of the missile nose.

It is not possible to determine the missile drag coefficient by pressure measurement extrapolation due to the small number of transducers used. However this is possible with the numerical simulation. The successful experiment-computation comparison enables us to maintain that the computation predicts the missile drag coefficient qualitatively very well.

Fig. 15 depicts the computed forebody drag coefficient for all studied geometries. A drag reduction of about 50% is obtained for the disk spike-tipped missile and of about 43% for the sphere spike-tipped missile at zero angle of attack. While the disk spike reduces the missile drag up to angles of attack of 15°, the sphere spike does so up to 19°. Beyond 15°, respectively 19° the studied spikes are not efficient any more. The disk spike is (for the same diameter and length) more efficient than the sphere spike for angles of attack lower than 8°. For higher angles of attack the sphere spike becomes more efficient.

6 Concluding Remarks

Experimental investigations and numerical simulations on spike-tipped missiles have been successfully carried out at ISL. Interferogram pictures were taken to study the flow field around disk and sphere spikes of equal diameter and length. Numerical simulations were performed with a 3-D Navier-Stokes code to quantify the drag reduction obtained by the use of spikes. Pressure flow fields were compared with interferogram pictures: the results show very good agreement between experiment and numerical simulation. Numerical simulations showed for similar geometrical conditions that the disk spike contributes to a drag reduction of up to 15° and the sphere spike up to 19°. For equal diameters and lengths the disk spike proves to be more efficient below 8° whereas the sphere spike is more efficient beyond this angle of attack. The superiority of the sphere spike for high angles of attack is thus clearly evident*. Both the delayed flow detachment for higher angles of attack and the noticeably reduced aerodynamic drag coefficient [9] (C_D=0.93 for a sphere and C_D=1.7 for a disk for M=4.5) support this conclusion.

---

*EADS Deutschland GmbH and the French-Research Institute of Saint-Louis (ISL) have patented this new device.

References

[1] J. Srulijes, K. Runne and F. Seiler. "Messungen an einer Flugkörpernase mit Spike im großen Stossrohr-Windkanal des ISL". AG STAB Jahresbericht 1999, 1999.

[2] J. Srulijes, K. Runne and F. Seiler. "Flow Visualization and Pressure Measurements on Spike-Tipped Bodies". 21st AIAA Aerodynamic Measurement Technology and Ground Testing Conference, AIAA 2000-2520, 2000.

[3] J. Srulijes, K. Runne and F. Seiler. "Flow Visualization on Spike-Tipped Bodies". 9th International Symposium on Flow Visualization, Edinburgh, U.K., 2000.

[4] G. Patz. "Das Hyperschallstoßrohrlabor des ISL, 3. Teil: Stoßrohr B". ISL-Report N 30/70, 1970.

[5] G. Patz. "Das Hyperschallstoßrohrlabor des ISL, 2. Teil: Stoßrohr A". ISL-Report N 27/71, 1971.

[6] H. Oertel. "Stoßrohre". *Springer Verlag, Wien- New York*, 1966.

[7] G. Smeets and A. George. "Anwendungen des Laserdifferentialinterferometers in der Gasdynamik". ISL- Report R 28/73, 1973.

[8] G. Smeets. "Interferometry". ISL-Report CO 214/90, 1990.

[9] S.F. Hoerner. "Fluid Dynamic Drag". Published by the Author, 1965.

[10] CFX-TASCflow Computational Fluid Dynamic Software AEA Technology Engineering Software, Ltd., Waterloo, Ontario, Canada, 2000.

[11] T. Cebeci and A.M.O. Smith. "Analysis of turbulent Boundary Layers". Academic Press, Inc., New-York, USA, 1974.

[12] B.E. Launder and D.B. Spalding. "The Numerical Computation of Turbulent Flows". Comp. Meth. Appl. Mech. Eng., Vol. 3, 1974, pp. 269-289.

Figure 1 ISL shock tube laboratory

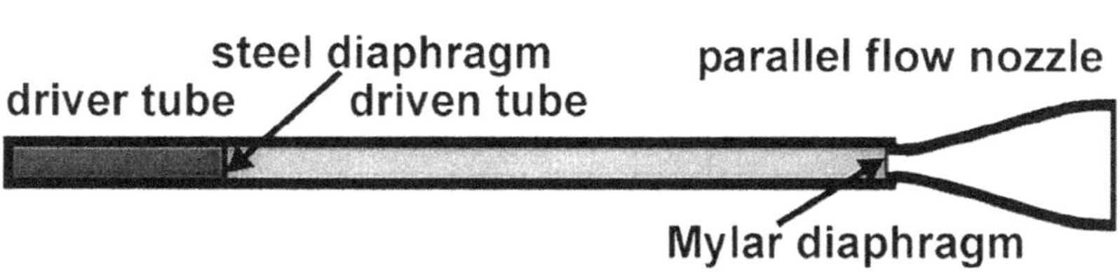

Figure 2 Principle sketch of shock tunnel facility

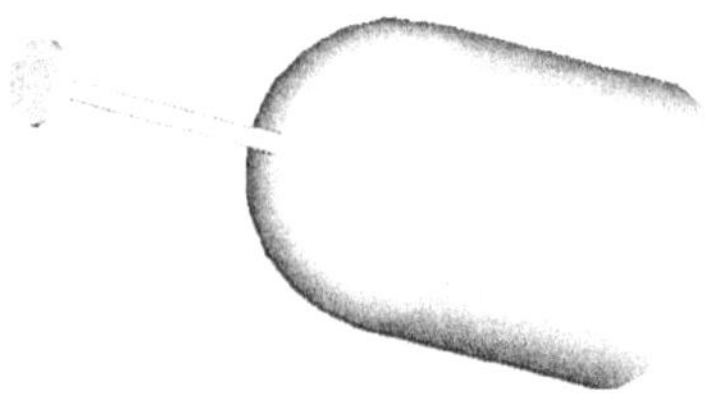

Figure 3 V/SHORADS model with disk spike

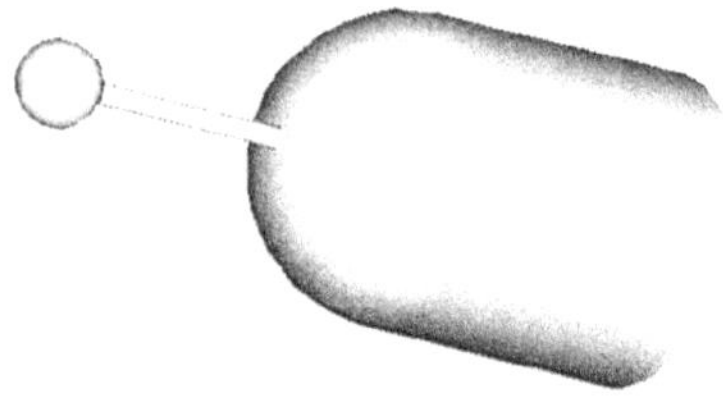

Figure 4 V/SHORADS model with sphere spike

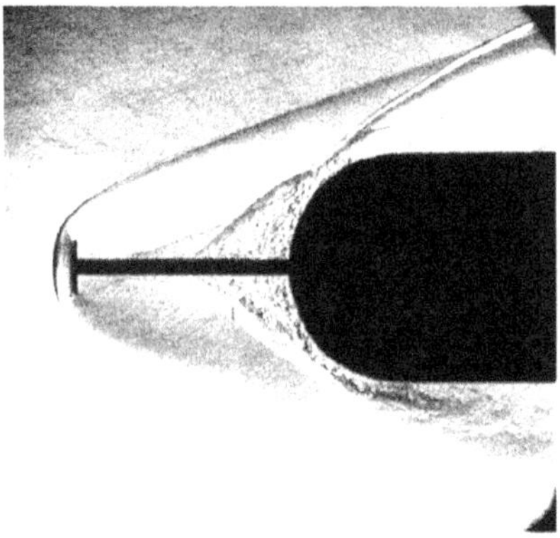

Figure 5 Disk spike - angle of attack 0°

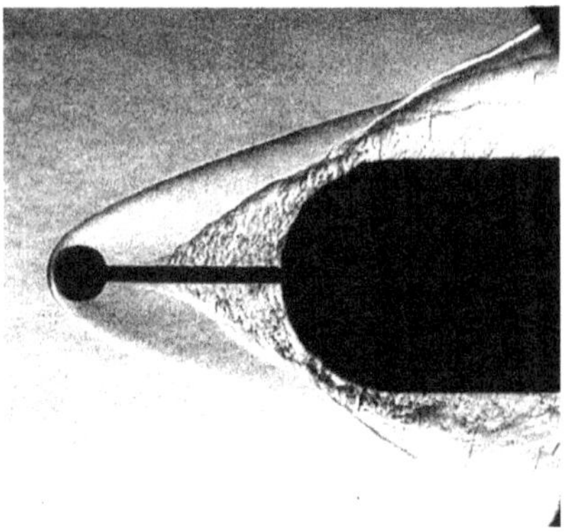

Figure 6 Sphere spike - angle of attack 0°

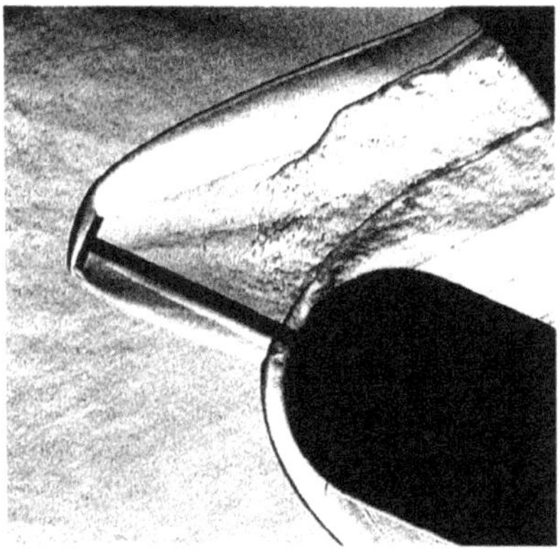

Figure 7 Disk spike - angle of attack 24°

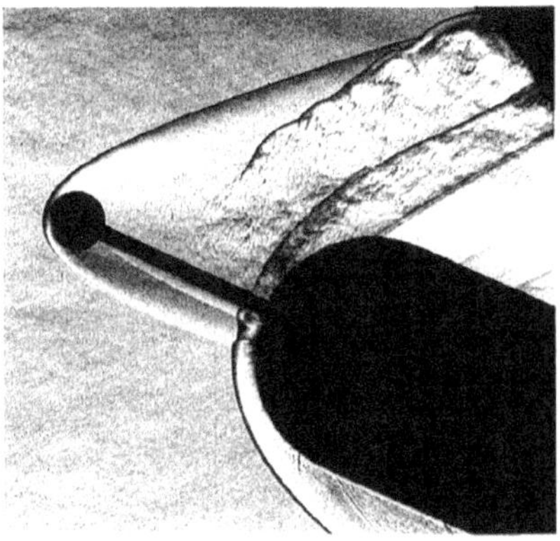

Figure 8 Sphere spike - angle of attack 24°

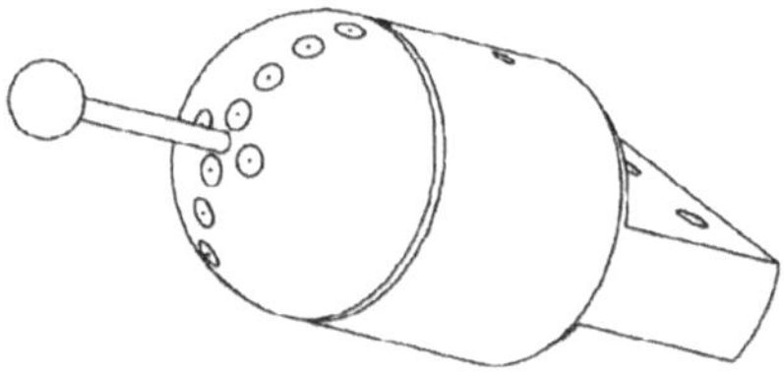

Figure 9 V/SHORADS missile model with 11 pressure gauges

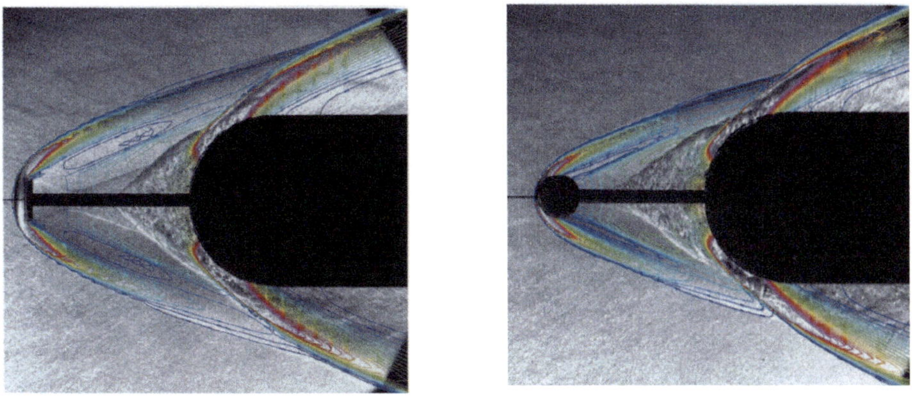

Figure 10 Comparison experiment-computation: $\alpha = 0°$

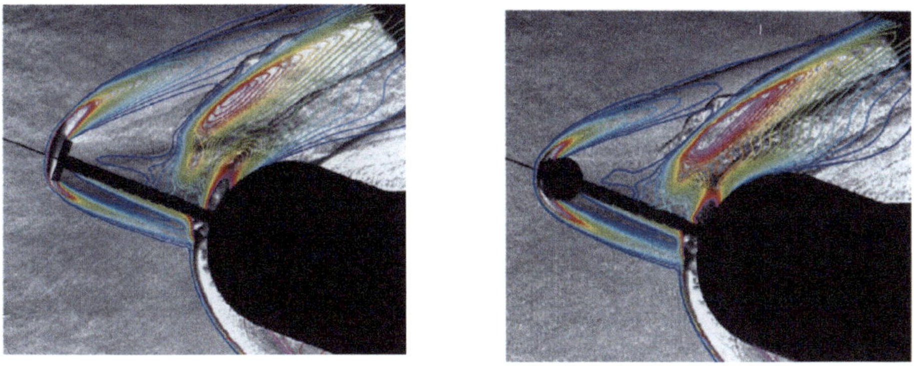

Figure 11 Comparison experiment-computation: $\alpha = 20°$

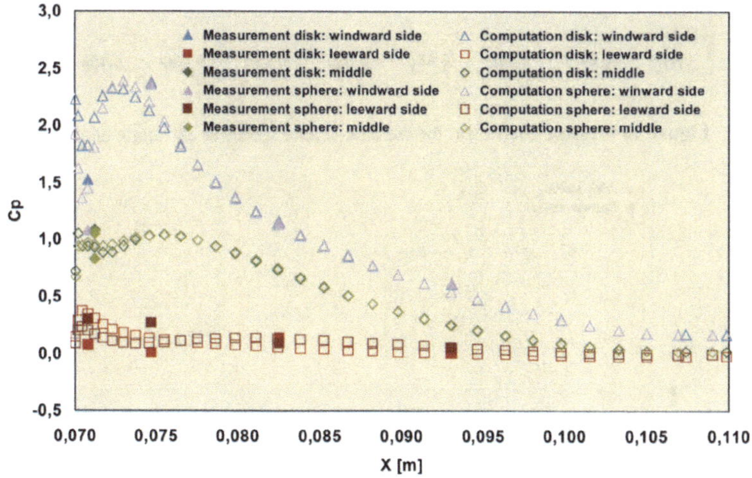

Figure 12 Pressure coefficient for the disk-sphere spikes at 16° angle of attack

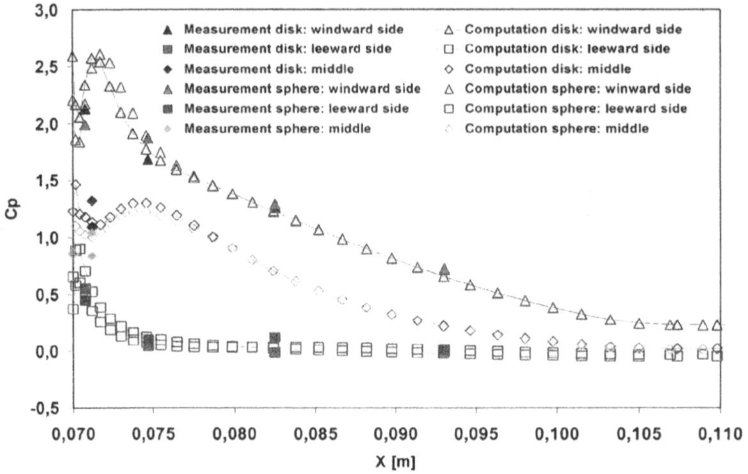

Figure 13 Pressure coefficient for the disk-sphere spikes at 20° angle of attack

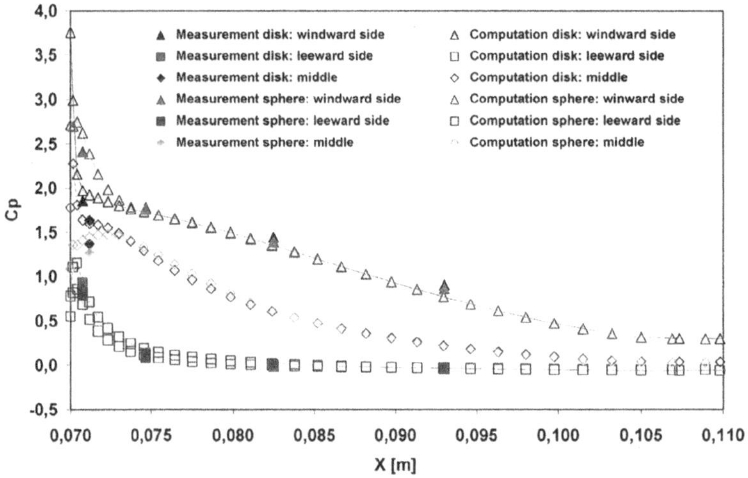

Figure 14 Pressure coefficient for the disk-sphere spikes at 24° angle of attack

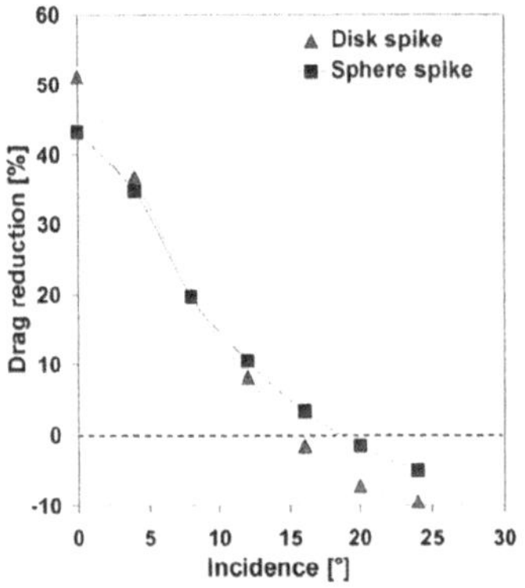

Figure 15 Drag reduction in % = (drag without spike – drag with spike)/drag without spike

Turbomachineri

Tip-Vortices in Transonic Axial-Compressors

Willy Hofmann, Josef Ballmann
University of Technology, RWTH Aachen
Mechanics Department
Templergraben 64
Aachen, 52062 FRG, Germany

Summary

A great potential regarding powerfull loss reduction in transonic axial flow compressors is located at the blades' tip. In this region losses are intensified by tip-vortices standing above the suction sides of the blades. Each vortex propagates through the blade duct in question and must cross the passage shock which is caused by the flow deflection on the pressure side of the neighbouring blade. The shock-vortex interaction leads to a sudden and strong deceleration of the flow and as consequence to strong change of the vorticity distribution in the vortex. Depending on the intensity of the shock-vortex interaction vortex breakdown can occur inducing compressor surge. In order to clarify these phenomena it is necessary to analyze the flow field in the tip region. For that purpose, numerical flow simulations for a transonic axial compressor are executed using a existing flow-solver. Detailed analyses are performed concerning tip-vortex development and topology as well as shock-vortex interaction.

1 Introduction

Advanced design for fans and axial turbo-compressors in aircraft engines aim to increase tip speed and blade loading. This leads to transonic flow fields within the first stages of compressors. Concerning subsonic axial compressors flow field instabilities such as rotating stall or rotating instability [5] cohere with operating conditions near surge. This phenomenon has been widely observed and great research efforts have been undertaken to understand the responsible mechanism. Furthermore endwall flow fields continue to be the focus of many research efforts because of their impact on compressor stability and performance [3]. For example increasing tip clearance result in efficiency penalty and reduction of mass flow range [2].

Large efficiency reduction occurs in the tip region of transonic rotors. This is due a complex combination of different loss sources present in the tip region. Tip-vortices stand above the suction side of the blades. They arise due to pressure difference between suction and pressure side of the blade inducing a circumferential high kinetic energy flow through the tip gap with low axial momentum. Typical relative Mach-numbers in the tip region of transonic axial compressor rotors are up to 1.6, whereby the axial component of the flow velocity is subsonic. Within the flow field of the rotor passage a compression shock exists. This so-called passage shock is caused by the flow deflection on the pressure side of the neighbouring blade. The aforementioned tip-vortex has to cross the passage shock. Thereby the shock-vortex interaction leads to a sudden and strong deceleration of the flow, a rapid expansion of the vortex and as consequence a strong change of the vorticity distribution of the vortex. Depending on the intensity of the shock-vortex interaction vortex breakdown may occur. Important flow parameters influencing the breakdown are the adverse pressure gradient in the streamwise direction and the swirl intensity of the vortex.

Schlechtriem and Lötzerich [8] detected tip-clearance vortex breakdown in a transonic axial compressor rotor downstream of the shock-vortex interaction. A significant indication for the occurrence of the breakdown was an abrupt deceleration of the flow within the leakage vortex across the passage shock. They suspected that a blockage resulting from the breakdown had an important impact on stall even inducing compressor surge.

2 Test Compressor Rotor

The present investigation was performed on the well known NASA rotor 37. The compressor rotor was designed as a low aspect ratio transonic inlet stage for a core compressor to study the effects of aspect ratio and loading level on the performance of speed stages. The rotor design pressure ratio is 2.106 at a mass flow of 20.19 kg/s with a measured choking mass flow of 20.93 kg/s. The inlet relative Mach-number is 1.13 at the hub and 1.48 at the tip at the design wheel speed of 17,188.65 rpm, giving a nominal tip speed of 454 m/s. The rotor has 36 blades with an aspect ratio of 1.19, hub to tip ratio of 0.70 and a tip solidity of 1.288. The running tip clearance was estimated to be 0.356 mm (0.5% span). Details of the rotor aerodynamic design were reported by Reid and Moore [7], overall aerodynamic performance of the stage by Moore and Reid [6].

The experiment was performed in the single stage transonic compressor facility at the NASA Lewis Research Center. The rotor was tested solely to avoid the interaction effects generated by the presence of an upstream inlet guide vane or downstream stator blade row. Radial distributions of static and total pressure, total temperature, and flow angle were measured at two axial stations, designated stations 1 and 4. Using the blade hub leading edge as the origin z=0 station 1 is located at z=-4.19 cm and station 4 at z=10.19 cm.

3 Flow Model and Computational Method

The computations are carried out for three dimensional flow using the commercial code EURANUS. EURANUS is an integral part of the software package NUMECA FINE. The Favre-Reynolds-averaged Navier-Stokes equations are discretized using a cell-centered explicit finite volume scheme after Jameson in relative coordinate system the rotating reference frame. Time integration is performed using four step Runge-Kutta scheme. To stabilize the time-marching scheme, scalar eigenvalue-based second or fourth order difference smoothing operators are used. To speed up convergence to steady state, local time stepping, residual smoothing, and multigrid techniques are applied. In the present work the algebraic turbulence model of Baldwin and Lomax has been used to estimate the turbulent viscosity. It is known that the Baldwin-Lomax turbulence model cannot predict the vortex flow quantitatively correct but in this paper we emphasize first on the qualitative discussion of the flow structure.

At solid boundaries, hub, blade and casing surfaces no slip and no heat transfer conditions are imposed. At the inflow boundary, flow angle, total pressure and total temperature are prescribed according to the experimental data. At subsonic outlet static pressure is prescribed at the hub and its radial variation is computed according to the radial equilibrium equation. At periodic boundaries from blade passage to adjacent blade passage periodic flow conditions are imposed.

4 Grid Generation and Computational Domain

Two different grid generators with elliptic smoothing implementations were used, IGG/ Auto Grid which is included in the NUMECA FINE software package and the code MegaCads. MegaCads is part of the MEGAFLOW project and it is used to generate structured multi-block Navier-Stokes grids for high Reynolds-number flows. MegaCads is an interactive grid generation code (no automation) with a few simple CAD functions. The resulting algebraic grids can be smoothed using elliptic techniques. IGG/Auto Grid is an automatic meshing system for turbo-machinery configurations. It takes advantage of the characteristics of turbo-machinery configurations by creating blade to blade grids by interpolating grids on hub and shroud onto surfaces of revolution in the channel.

In the present work the inlet and exit of the computational domain were chosen coinciding with stations 1 and 4. Computations were run on two different structured multi-block grids (Table 1). One grid was generated with IGG/Auto Grid and the other one with MegaCads, designated ag and mc. In both cases for the tip clearance the gap is assumed rectangular and fully gridded (Fig. 3). The minimum grid spacing for both grids was chosen such that $y^+ \leq 1$ at the walls.

Concerning grid ag the computational domain was divided into two zones. One zone was a main flow region outside the blade tip clearance, and the other was the tip clearance region. The grid belonging to the main flow region is a body-fitted I-type grid, non-matching at periodic boundaries, as shown in Figure 1.

The grid named mc consists of a composite grid assembly (Fig. 2). Near the blade an O-type body fitting grid is constructed around the blade. So orthogonal or nearly orthogonal cells along the surface are ensured. Outside this domain in the main flow region, i.e. upstream and downstream of the blade row and from blade to blade in pitchwise direction, an I-type grid is provided. The tip gap is gridded with an I-grid embedded into an O-grid (Fig. 3). Contrary to grid ag, clustering in spanwise direction is more refined in the endwall region near the casing.

5 Results and Discussion

The current study concentrates on operating conditions near choke and near peak efficiency at design speed. A understanding of the radial penetration of the tip clearance flow into the main flow part is given by considering relative Mach-number contours at 95% span and 99% span, i.e. one tip clearance height below the blades' tip. First flow fields belonging to operating conditions near choke and the computational grid mc are analyzed. Thereupon results are compared to those evaluated by using the grid ag and a back pressure which leads to operating conditions near peak efficiency.

Figure 4 shows on a S1 plane close to 95% span the relative Mach-number distribution. Flow is from the left to the right and the rotor is turning top down, as shown in the picture in the upper left. Upstream of the the leading edge a detached compression shock exists. Caused by the flow deflection on the pressure side the shock surface meets the suction side of the neighbouring blade in the middle of the passage and is therefore called passage shock. Typical relative Mach-number is about 1.5 upstream of the shock. Also visible is a second compression shock which appears at mid-chord an the pressure side to the trailing edge of the neighbouring blade. But, no evidence for the existence of the tip vortex can be detected.

The distribution an a S1 plane at 99% span (Fig. 5) is different from that at 95%. The second shock degenerates having a small extension into the passage. The Distribution of the Mach-number shows two troughs. The first begins at the leading edge and ends up in the first half of the passage shock of the neighbouring blade. The second extends from mid-chord to the second half. Further on downstream the passage shock appears with a dent followed by a roundish area of very low Mach-numbers. Caused by rolling-up of the tip leakage vortices also troughs in pressure and density can be seen. Vortices including the cores are best seen at this S1 plane.

Following Jeong and Hussain [4] vortex cores can be identified illustrating distributions of the second eigenvalue λ_2 of the symmetric Tensor $(\mathcal{V})^2 + (\mathcal{V}^T)^2$, $\mathcal{V} = grad\, \mathbf{v}$ being the gradient of the velocity field (Fig. 6). The value of λ_2 must be negative. It identifies a local pressure minimum which is interpreted as a vortex core. This contour plot is in accordance with the distribution of the vorticity (Fig. 7). Obviously the second vortex exhibits a higher vorticity values concerning its core and the vorticity of both vortices disappears downstream of the passage shock.

Figure 8 shows the distribution of the circumferential velocity component in the inertial reference frame concerning a cylindrical base system. The vortex cores consist of lower values than outside of them. The vortices proceed nearly in azimuthal direction. Hence, it can be inferred the axial velocity components referring to vortex axes are almost identical to those and the profile is wake-like.

Figure 9 shows the density distribution in a crossflow plane at about 10% chord and streamlines of the three-dimensional flow. Streamlines belonging to the suction side begin at the blade's tip and leading edge. This location is identical to that where the sonic line concerning the passage shock attaches the blade and a great pressure gradient exists. At this point the first tip leakage vortex as mentioned above comes into play (Fig. 10,a). The streamlines propagate downstream in a tube of low density, that describes the vortex core of the first tip leakage vortex. On the pressure side, oncoming streamlines from below the blade's tip turn versus the tip, enter the tip clearance very closely to the blade's tip and roll up in the vortex. The crossflow plane shows a small dent at the leading edge on the pressure side corresponding to the vena contracta phenomenon. It can be inferred this leads to a supersonic jet with low axial and high circumferential momentum through the tip clearance (Fig. 10,b). The vortex is characterized by rolling-up of a shear layer between the leakage-jet and main stream. The rolling up forms a streamwise slender vortex. As long as the vortices are rolled-up from the leakage flow the crossflow section of their vortex cores enlarge.

Knowing the relative Mach-number is supersonic, the absolute Mach-number is subsonic and the formation of the slender, tip leakage vortex is pressure driven the manner is similar to the vortices arising at subsonic leading edges of a delta wing in supersonic flow. Thereby, a compression shock exists an the pressure side and a vortex is rolled up on the suction side. Compared to a transonic axial compressor rotor the shock is similar to the passage shock. The existence of the second vortex can be explained with the flow field around kinked leading edges of delta wings. Caused by the finite pressure rise at the kink a second vortex can evolve, which does not roll up about the first one and can be stronger than the first one.

The vortices upstream of the shock region are described as regions of low Mach-numbers, yet still supersonic. The second vortex having higher vorticities than the first dominates the shock-vortex interaction and pushes away the first. When the second vortex approaches the shock front the core flow decelerates and as consequence the cross-sectional area of the vortex increases. Also the vortex induces local deformation of the shock surface. The interaction leads to a sudden and strong deceleration of the flow and as consequence to a strong change of the vorticity distribution inside the vortex. Downstream behind the shock surface the interaction causes a roundish region of very low subsonic Mach-numbers. This can be compared with vortex breakdown of slender vortices. Knowing in the present study no breakdown occurs, because there is no bubble-like recirculation zone. A decisive parameter concerning shock-vortex interaction is vorticity or circulation of the corresponding vortex.

Figures 11 a) and b) show density distribution and streamlines respectively regarding the computational grid ag and operating conditions near choke on the corresponding crossflow plane shown in Figure 9. Streamlines which take part of the vortex core start at the same point compared to grid mc. The core crossflow section is slightly smaller. Density values inside the vortex are not as low as aforementioned. Distribution of the relative Mach-number is quite similar. Due to this grid clustering in spanwise direction not only inside the tip clearance but also below the blade's tip has decisive influence on vortex simulation. In contrast, grid ac facilitates more accurate flow prediction concerning the overall performance.

Figure 12 refers to operating conditions near peak efficiency and the computational grid ag. It shows the Mach-number distribution on a S1-plane at 99% span. As a result of the higher back pressure at station 4 which induces a lower axial velocity component the angle between suction side, pressure side and shock-surface, respectively, becomes greater compared with Figure 5 and also the distance between leading edge and shock surface increases. The stagnation line on the blade's leading edge moves toward the pressure side. Hence the pressure difference between the suction and the pressure side at the blade's tip and leading edge is greater. The second vortex exist no longer, only the tip leading edge vortex. It propagates more into the passage, i.e. the angle between it's trajectory and the suction side is greater. The radial location of it's core is not noticeable changed, also the streamlines corresponding to the core (Fig. 12). Cause of the greater core cross-section and lower densities inside the vorticity is greater. Conforming, shock-vortex interaction is more intensive and the low Mach-number region downstream is more distinctive.

Conclusion and Outlook

In the present study a existing commercial 3D, RANS flow-solver has been used to simulate the stationary flow field within a transonic axial compressor rotor. The test case is well known as NASA rotor 37. Numerical results are shown for operating condition near choke and peak efficiency using two different structured multi-block computational grids concerning grid topology and clustering. Detailed analyses are performed concerning tip-vortex development and topology as well as shock vortex interaction. The turbulent viscosity has been estimated by the use of the Baldwin-Lomax turbulence model. From this point of view, vortex flow can not sufficient predict quantitatively and only qualitative discussion has been fulfilled.

Point distribution in spanwise direction below blades' tip as well as within tip clearances have a great bearing an vortex flow simulation and vorticity prediction. Tip vortex formation and development is similar to vortex evolution at subsonic leading edges of a delta wing in supersonic flow. A decisive Parameter to vortex strength is the pressure difference between the suction and pressure side. With higher back pressure at rotors' exit the tip vortex vorticity increases and the trajectory moves upstream, i.e. the angle between the trajectory and suction side becomes greater. By this means shock-vortex interaction is intensified. In the present work no vortex breakdown was observed but the same significant features. Tip vortices are slender and have a wake-like profile. When the vortex approaches the shock core flow decelerate and the core cross-section enlarges. Hence the vorticity distribution changes. The shock surface obtains a dent and downstream exists a roundish region of very low Mach-numbers.

Future work will include more quantitative conclusion using another turbulence model. For the purpose to decrease the numeric dissipation the computational grid has to be refined in spanwise as well as pitchwise direction. Simulations will be done at operating conditions near stall in order to analyze vortex breakdown.

Acknowledgment

This work has been supported by the Deutsche Forschungsgemeinschaft (DFG) in the Transfer Project "Tip Leakage Vortex - Passage Shock - Interaction concerning Transonic Compressor Rotors and It's Impact onto Surge Conditions" with industrial participation of Rolls-Royce Germany, Dahlewitz near Berlin.

References

[1] Davis, R. L., Delaney, R. A., Denton, J. D ., Giles, M, B., Strazisar, A. J., and Wisler, D. C., 1993, *CFD Code Assessment in Turbomachinery - Author's Information Package*, The American Society of Mechanical Engineers Turbomachinery Committee.

[2] Fritsch, G., Hoeger, M., Blaha, C., Bauer, D., 2000, *Viscous Three-Dimensional Simulation of Transonic Compressor Stage on Parallel Hardware*, AIAA Journal of Propulsion and Power, Vol. 16, No. 3, pp. 388-396.

[3] Furukawa, M., Saika, K., Yamada, K., Inoue, M., 2000, *Unsteady Flow Behaviour Due to Breakdown of Tip Leakage Vortex in an Axial Compressor Rotor at Near-Stall Condition*, Proceedings of ASME TurboExpo 2000, May 8-11, Munich, Germany.

[4] Jeong, J., Hussain, F., 1995, *On the identification of a vortex*, Journal of Fluid Mechanics, No. 285, pp. 69-94.

[5] Kameier, F., Neise, W., 1997, *Rotating Blade Flow Instability as a Source of Noise in Axial Turbomachines*, Journal of Sound and Vibration, 203 (5), pp. 833-853.

[6] Moore, R.D., Reid, L., 1980, *Performance of Single- stage Axial-Flow Transonic Compressor Rotor and Stator Aspect Ratios of 1.19 and 1.26, Respectively, and with Design Pressure Ratio of 2.05*, NASA TP 1659.

[7] Reid, L., Moore, R.D., 1978, *Design and Overall Performance of Four Highly Loaded, High Speed Inlet Stages for an Advanced High-Pressure Ratio Core Compressor*, NASA TP 1337.

[8] Schlechtriem, S., Lötzerich, M., 1997, *Breakdown of Tip Leakage Vortices in Compressors at Flow Conditions Close to Stall*, ASME Paper No. 97-GT-41.

[9] Strazisar, A. J., 1994, *Data Report and Data Diskette for NASA Transonic Compressor Rotor 37*, NASA Lewis Research Center, Cleveland, OH.

Table 1 Computational grid summary.

Name	ag	mc
Points, total	617.234	809.164
Spanwise, blade	65	49
Spanwise, tip clearance	17	25
Pitchwise, blade-to-blade	41	61
Pitchwise, tip clearance	13	13
Streamwise	181	213
Streamwise, blade	73	73

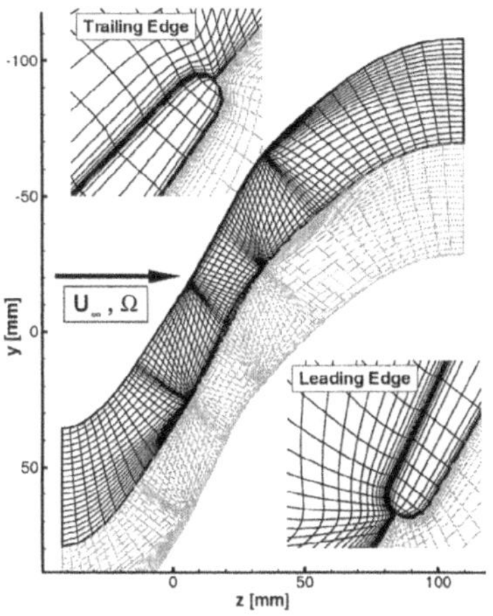

Figure 1 Computational grid ag, S1 plane, tip region.

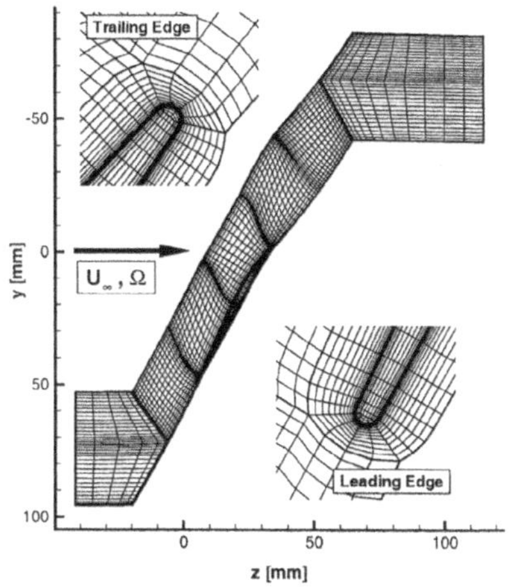

Figure 2 Computational grid mc, S1 plane, tip region.

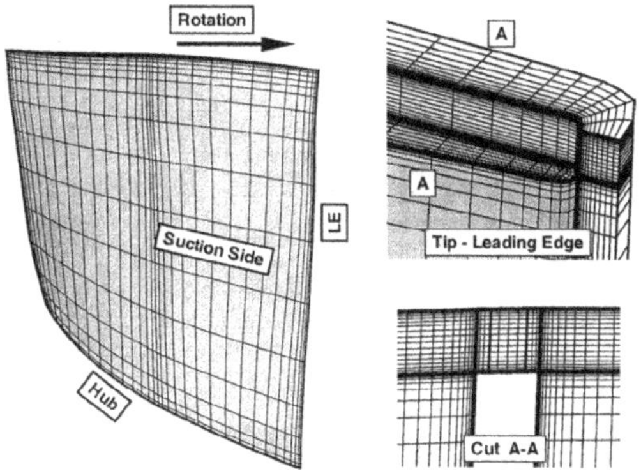

Figure 3 Computational grid mc, blade and tip clearance grid.

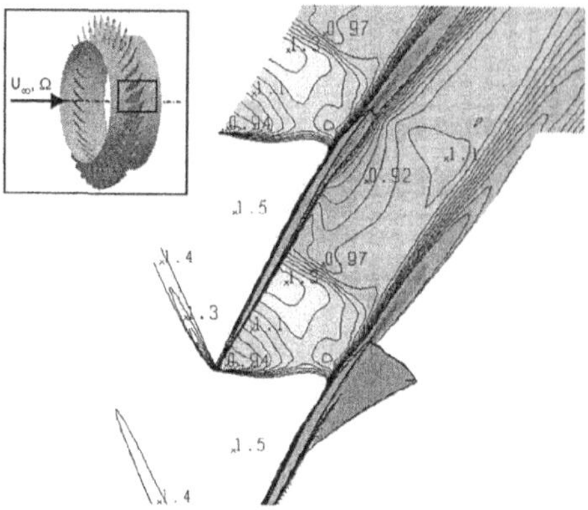

Figure 4 Grid mc, relative Mach-number, S1-plane 95% span.

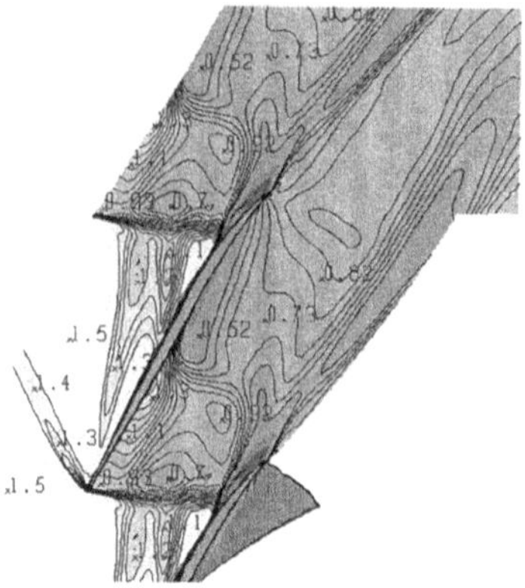

Figure 5 Grid mc, relative Mach-number, S1-plane 99% span.

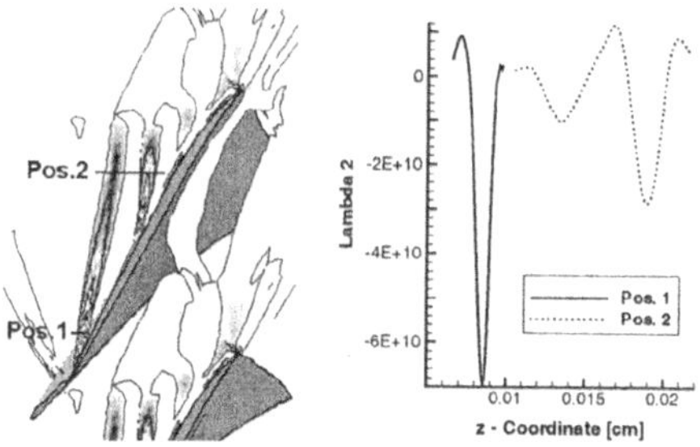

Figure 6 Grid mc, λ_2, S1-plane 99% span.

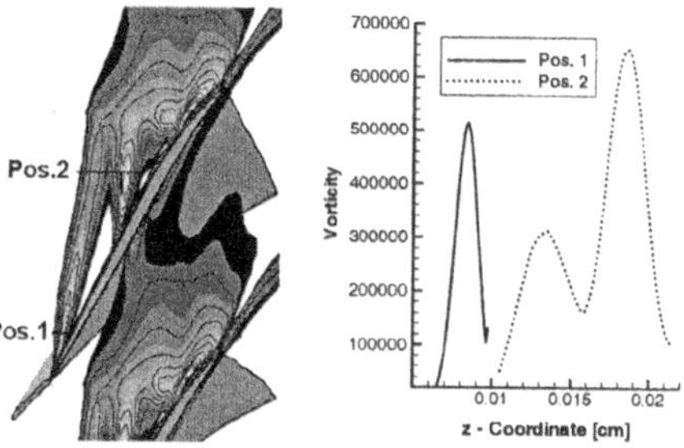

Figure 7 Grid mc, vorticity, S1-plane 99% span.

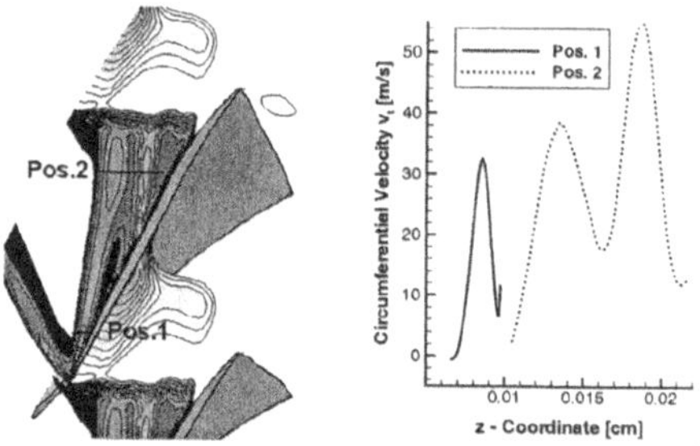

Figure 8 Grid mc, circumferential velocity component v_t, S1-plane 99% span.

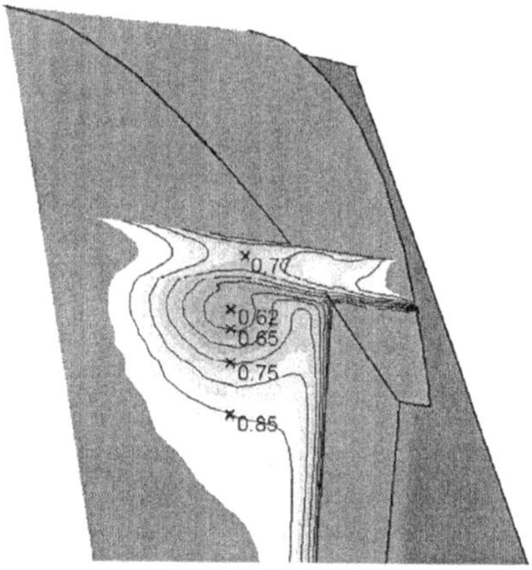

Figure 9 Grid mc, density $[kg/m^3]$, crossflow plane, tip region.

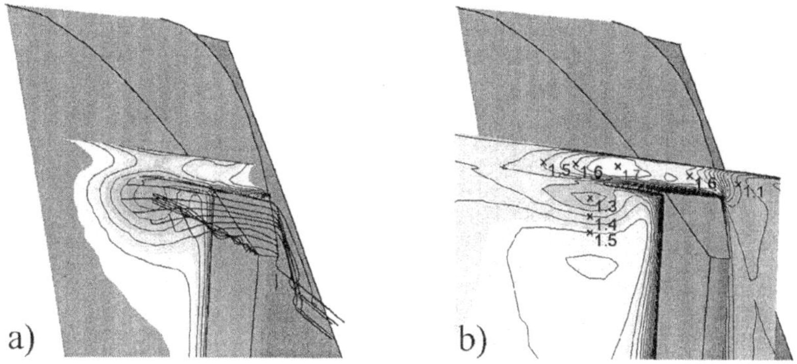

Figure 10 Grid mc, crossflow plane, tip region, a) Distribution density and streamlines, b) Distribution relative Mach-number.

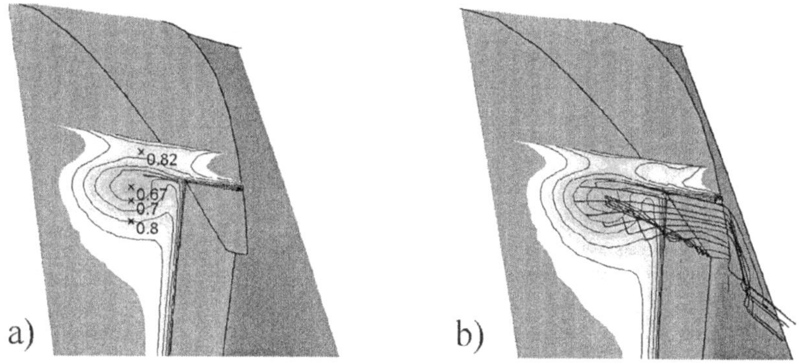

Figure 11 Grid ag, crossflow plane, tip region, density $[kg/m^3]$, a) isolines b) streamlines.

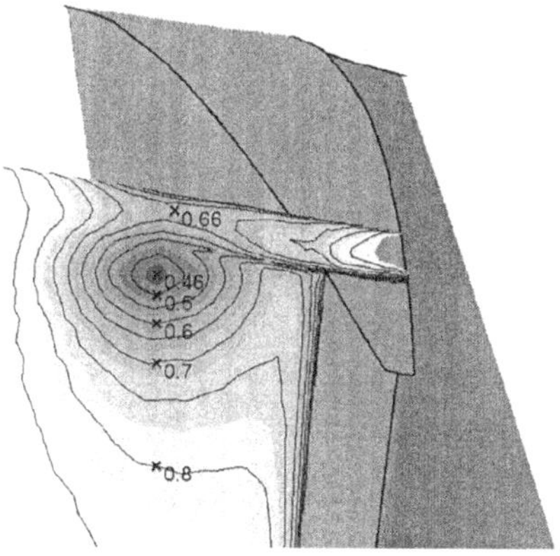

Figure 12 Grid ag, density $[kg/m^3]$, crossflow plane, tip region.

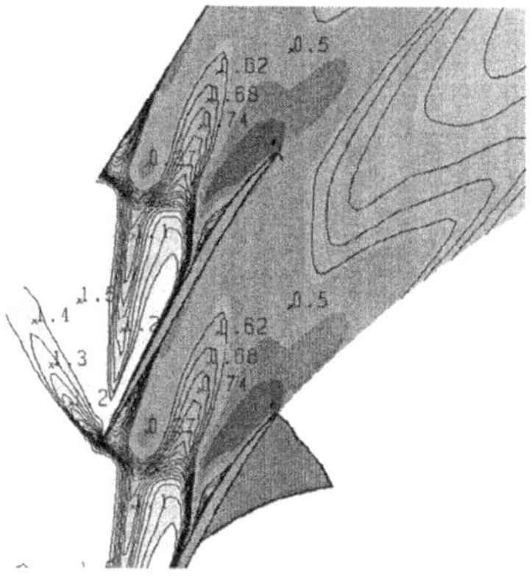

Figure 13 Grid ag, relative Mach-number, S1-plane 99% span.

Airframe Noise

Airframe Noise - Generation and Reduction
A Report on Status and Perspectives of DLR-Research

Werner DOBRZYNSKI, Hanno HELLER and Jan DELFS

DLR Institut für Entwurfsaerodynamik, Abteilung Technische Akustik
Lilienthalplatz 7, D-38108 Braunschweig, Germany
Tel.: ++49 531 295 2170, Fax: ++49 531 295 2320, e-mail: Werner.Dobrzynski@dlr.de

Summary

The proposed increase in stringency of the ICAO Chapter-3 noise certification limits and other operational restrictions due to noise necessitates the development of source noise reduction technologies for subsonic commercial aeroplanes. The noise from such aeroplanes during approach is largely dominated by airframe noise, i.e. the noise generated by the interaction of flow and those aircraft components deployed for landing, such as the landing gears and the wing high-lift devices. This paper will present highlights from recent DLR research projects, indicating the source noise reduction potential of design changes on landing gears and on wing high lift devices which could be employed without hampering the basic functionality of such components.

1 Introduction

The expected introduction worldwide of some extra ten thousand new subsonic commercial jet-aircraft over the next 20 years represents a major challenge for aviation industry. One of the "critical technologies" in this new development will be noise. Future aircraft must imperatively be quieter than today's in view of (1) ICAO's intent to make the ANNEX 16 / Chapter-3 noise limits for subsonic jet aeroplanes much more stringent, (2) the reluctance of the airports to allow any increase in the noise immission after the phase-out of the Chapter-2-aircraft in 2002, but rather their insistence to compensate any aviation traffic growth through quieter aircraft, as well as (3) the noise-based operational restrictions effective at many of the major airports.

Facing this challenge, a number of dedicated research programs on aviation noise problems were initiated worldwide. In the United States, NASA, FAA and the US-aviation industry had launched their "Aeronautics and Space Transportation Technology Enterprise", where under the designation "Roadmaps to the Future" a number of rather "sportive" objectives had been defined. Referring to global civil aviation, a 10 dB reduction in the Effective Perceived Noise Level (EPNL) within 10 years, and one of 20 dB within 25 years, referenced to today's technology, is envisioned. Unfortunately this highly successful US program has recently been curtailed, although it would be right to assume that, at the very least, aviation industry will continue to do R&D under their own auspices - simply for reasons of staying ahead in the world competitiveness.

In Europe, within the Fifth EU Framework Program under the general heading of "Competitive and Sustainable Growth" new perspectives in aeronautics have been defined with the stated objective to reduce noise impact by 10 EPNdB within a time frame of 10 years. This ambitious objective is to be attained along two avenues: within efforts on "*Critical Technologies*" noise is to be reduced at the very source (e.g. on *engines* that would encompass fan-, compressor-, core-, turbine- and jet-noise reduction drawing also on progress in nacelle technologies, and on the *airframe* that would encompass airframe noise as well as excess noise due to installation effects). Within *"Targeted Platforms"* source noise reduction elements and low-noise operational procedures are combined to lead to an overall reduction of aviation noise impact.

To appreciate the challenge of these aviation noise impact reduction objectives, a brief recapitulation of the Chapter-3 noise certification procedures is in order. These are (1) a take off at maximum power with a subsequent power cutback, where the noise is measured under the flight path at 6.5 km past break release; (2) a maximum power / no-power cutback climb flight, where the maximum sideline noise is measured at a lateral distance of 450 m, and finally (3) a 3°-slope landing approach, where the noise under the flight path is measured 2000 m before touch-down. However, an aircraft which "simply" complies with the corresponding current ICAO Chapter-3 noise limits cannot possibly be sold on a future market. Future aircraft must feature cumulative noise certification levels of between 20 to 30 dB below the limits.

It has been amply established that for take off and climb it is engine noise which dominates, while in approach - when engine-power is reduced - airframe noise sources become more important. Indeed, airframe noise, caused by the usually highly turbulent flow processes around the deployed landing gears and the deployed high-lift devices (wing leading edge slats and wing trailing edge flaps), often effectively dominates approach noise.

Flyover noise experiments conducted by Aerospatiale have conclusively shown that the deployment of landing gears on the one hand, and of slats and flaps (depending on the actual deployment angle) on the other hand can substantially raise the noise observed on the ground by several dBs. To clarify the underlying aeroacoustic airframe noise source mechanisms and to develop noise reduction means, DLR, the German Aerospace Research Center, has been and still is involved in a number of major industrial, national and European research projects. This paper will present highlights from such endeavors, to illustrate the significantly improved understanding of the underlying aeroacoustic "airframe noise"-generating mechanisms and the successful efforts to provide industry with technically feasible noise reduction technologies.

2 Airframe Noise Research

In the mid-nineties, DLR investigated the noise from commercial aircraft landing gears. More recently, DLR has entered into two formal major - still ongoing - research programs: *first*, a national program (sponsored by the German Ministry of Research and Education, BMBF) entitled "Nationales Verbundprogramm zur Eigengeräuschminderung an Verkehrsflugzeugen" (= Joint National Program on Commercial Aeroplane Airframe Noise Reduction") to specifically investigate the noise from high-lift devices, and, *second*, a European program under the auspices of the EU entitled Reduction of Airframe and Installation Noise (acronym RAIN), dealing with airframe noise from both landing gears and high lift devices.

2.1 Airframe Noise Source Identification on Overflying Aeroplanes

An approach, applied and refined by DLR-Berlin, to identify and rank-order airframe noise sources employing a highly sophisticated microphone array technique, was employed on a large number of overflying aeroplanes of the most varied type[1]. **Fig. 1** illustrates the basic principle of the array technique. Here, the combined focus of the array (consisting of between 50 to 100 individual microphones spatially extended over a surface of, say, 8 by 8 m²) electronically follows certain regions of interest on the overflying aeroplane. Thus, for example, the focus can be "attached" to an outer wing section, or to an area where one of the main gears or the nose gear is located. The spatial as well as the spectral resolution depends on the number and arrangement of the microphones and on the flyover height. In general, the more microphones and the lower the flight height, the better the resolution. Since these tests were conducted under the approach path of an airport, thus inherently catching aeroplanes in their landing configuration where landing gears and flaps are deployed and where the engine power is reduced, "classical" airframe noise sources should be readily discernible.

Such flyover noise tests reveal, on any given aeroplane, the respectively dominating source under the particular approach operational condition and frequency regime. These then could be - depending - either the engines, or the main landing gears, or the nose gear, or the flap- or the slat-region, or whatever. Suppose then, a landing gear region as a whole was identified as the dominating airframe noise source, such information will inherently not provide any clue on which particular part of the subject landing gear is responsible for the "excess noise". This is where dedicated experiments are called for on realistic landing gears or other airframe noise sources such as the wings in high-lift configuration under controlled conditions, available only in acoustic wind tunnels. In the future a valuable support of such dedicated source studies will be provided by modern computational tools.

2.2 Wind Tunnel Experiments on Landing Gears

It seems rather futile, to try and attempt to predict landing gears noise on a purely theoretical basis. A modern landing gear of a large commercial aeroplane is of such geometrical complexity that it is virtually impossible to point out the part, or parts, which under conditions of high-speed inflow would represent the dominating source.

2.2.1 Narrow-body Aircraft Main and Nose Landing Gear Investigations

Recognizing the need to then conduct experiments on realistic landing gears, DLR undertook a series of experiments, employing full-scale main landing gears (of the 2-wheel and the 4-wheel type) in the open, anechoic test section of the German Dutch Wind Tunnel, DNW-LLF[2], one of the foremost aeroacoustic wind tunnels in existence. The DNW-LLF features a nozzle of 6 (high) by 8 (across) m cross-section, allowing to test full-scale landing gears of substantial size still within the "healthy" (i.e. low turbulent) tunnel core-flow. This way, the undisturbed inflow conditions simulating approach could be attained. Moreover, the maximum free-jet tunnel flow speed of close to 80 m/s approaches actual landing speed, such that conditions are rather realistic and acoustic data should exhibit a high degree of realism.

The setup is shown schematically in **Fig. 2** to display the installation of the main landing gear bogies in the supporting side-wall construction, which is to simulate the nearby wing. Since the complete wheel-well was too large to be simulated in full-scale only the wing cavity (which would accommodate the gear leg in the retracted gear position) was reproduced and coupled to an anechoically terminated large bay. Employing the STSF (spatial transformation of sound fields) technique to visualize the acoustic source strength distribution on the landing gear bogie, dominating sources between the forward and the rearward wheel pair for the particular condition were identified. Findings such as these as well as dedicated parametric variations revealed critical areas (in terms of source noise) on the landing gear. In the course of the experiments, certain components (such as the main leg, the side-stay, then the region between the wheels and around the brakes or the wheel cavities) were successively covered to attempt to isolate the respectively dominating source region. In the end, the entire landing gear was "streamlined", not as a technically feasible means but rather to indicate the overall noise amelioration potential. Indeed, such a "complete" covering resulted in an overall A-weighted farfield noise level reduction of about 10 dB as illustrated in **Fig. 3** for the twin-wheel bogie. In the end measures were identified which still brought impressive noise reductions of up to 4 dB without hampering the functionality of the gear.

2.2.2 Wide-body Aircraft Main Landing Gear Investigations

More recently still, within the European EU-sponsored RAIN-project, another - larger - full scale landing gear was introduced into the DNW-LLF test section, to identify in yet greater detail the noise generating parts of such a landing gear of extreme complexity.

In **Fig. 4** an actual photo of the installation appears showing both the microphone array and the DNW acoustic mirror of 3 m diameter for source localization purposes. Such techniques are very helpful to further the understanding of the aeroacoustic source mechanisms since it is possible to probe into various regions of interest within the landing gear to reveal dominating sources. In this Figure, the kind of results are illustrated as obtained with the microphone array (positioned so to speak "under" the gear) and with the acoustic mirror (positioned such that it looks towards the gear from the side). Both techniques - which were employed in addition to the farfield microphones (needed to quantify the noise radiation and to allow extrapolation towards the ground-impact noise levels, or, rather the flyover noise levels) - provided valuable information on how to treat certain parts of the landing gear in a dedicated manner to achieve noise reductions (again to the tune of 3 to 4 dB overall) without adversely affecting the gear's functionality.

2.3 Wind Tunnel Experiments on Wing High Lift Devices

Within the subject German National Research Project, a number of sequential sub-projects were undertaken:

- Model tests on 1:10 scaled wing with high lift devices in DLR's small Aeroacoustic Wind Tunnel Braunschweig (AWB)
- Model tests on a 1:7.5 scaled complete aircraft in the DNW, and
- Tests employing a full scale aircraft wing with high-lift devices in the DNW

2.3.1 Tenth-scale Model Tests (in the AWB)

Fig. 5 shows a tenth-scale wing model with high lift devices in DLR's small Aeroacoustic Wind Tunnel Braunschweig (AWB)[3]. The objectives of this test series were to (1) rank order noise sources, (2) determine noise affecting parameters, (3) optimize test techniques (for later application in the DNW-tests), and, (4) check the effectiveness of noise reduction measures.

The primary measurement means was an acoustic mirror of 1.4 m diameter for detailed source studies. Here the mirror focus - looking up towards the "overflying" high-lift wing - was guided along certain supposedly critical regimes such as the slat tracks (oblique to the flow), the leading edge slat, the trailing edge flap as well as directly along the flap-side-edge, to reveal significant level increases at these various critical areas.

Specific studies of the area around the leading edge slat revealed discrete-frequency, tonal, components which could be readily "killed" by applying an appropriate tripping device. Moreover, data from miniature unsteady pressure sensors (of the Kulite-type) in the slat-cove area indicated a periodically oscillating flow pattern, right in the cove. Here, patches of turbulence released into the slot-flow from the cove-vortex in the area of the cove (which after all extends over large portions of the wing leading edge, thus representing an extended intense noise source) seemed to be responsible for the primary noise generating mechanism. Providing an appropriate cover such as to avoid this transmission of turbulence indicated a fairly substantial (about 3 dB) reduction in the farfield noise signature.

2.3.2 1:7.5 Scaled Aircraft Model in the DNW

The tests in the AWB, after all, were made on a single wing section. In order to check on the important aeroacoustic mechanisms and to test possible noise reducing devices on a complete, hence more realistic, aircraft configuration, a 1:7.5 scaled aircraft model was installed in the DNW open test section, as illustrated in **Fig. 6.** In these tests, DNW's acoustic mirror as well as farfield microphones were employed. Data reduction indicates similar effects as observed in the tenth-scale model tests in the AWB to occur. The "acoustic mirror"-traces in particular revealed - for the case considered - a strong and extended source distribution along the leading edge, with a local maximum near the slat-horn area, whereas the flap regime seems relatively less noisy, apart from a small regime near the flap side edge. Applying a "slat cove cover" results in an overall farfield A-level reduction of about 3 dB (see right part of Figure 6).

2.3.3 Full-scale Aircraft Wing in the DNW

In December of 1999, DLR within the above mentioned national airframe noise research program, inserted a full-scale aircraft wing into the open test section of the DNW for aeroacoustic studies[4]. To accommodate the outermost wing section into the clean tunnel-core-flow required an oblique installation as shown in **Fig. 7.** Again DNW's 3-meter acoustic mirror and the extended microphone array were used to great advantage for localizing noise-critical areas. Farfield microphones on the far side of the tunnel hall permitted to quantify the overall or bandwise noise radiation. Thus, noise-reduction measures towards their effect on flyover noise could be judged directly.

Data reduction revealed the slat-track openings to constitute a significant local noise source. Covering that opening for one, and affixing the slat-cove cover (as earlier demonstrated on the scaled tests) provided a local noise reduction of up to 8 dB **(Fig. 8).** Similarly, smoothing ("cleaning") the flap side edge (as compared to its baseline, fairly irregular, condition) provided a local source noise reduction of 7 dB, and additionally employing a flap side edge fence provided another 4 dB, for a combined total source noise reduction of 11 dB **(Fig. 9).** Such findings are considered paramount to the (European) manufacturer in his quest to treat recognized airframe noise sources for less noise generation and radiation.

3 Theoretical/Numerical Concepts for Airframe Noise Computation

While today's approaches of reducing airframe noise are based on developing dedicated reduction devices on existing airframe components the future research perspective encompasses the "source avoiding design". Such endeavor still requires good engineering intuition, but reliable computation tools are needed to support and guide a systematic and efficient design process.

Not only the unsteady aerodynamic processes near an airframe component need to be known, but rather their very acoustic significance in combination with the actual geometry, because this information is the key to systematic design improvement. Such information with sufficient degree in spatial/temporal detail though is very difficult, or even impossible to obtain experimentally without a) disturbing the source itself and b) involving excessive measurement efforts in terms of instrumentation and time – thus cost. In view of the above it is reasonable to develop computational tools for the numerical simulation of aeroacoustic sources.

Jointly with the universities of Aachen, Dresden and Stuttgart DLR has initiated the national joint project SWING (Simulation of WIng-flow Noise Generation) focussed on simulation concepts for noise emanating from wing high lift devices. The funding of SWING was partitioned among DFG (German Research Council), DLR and BMBF (German Ministry of Education and Research). Within the first phase of the project DLR's main task has been to develop a simulation scheme based on the spectrally highly accurate numerical solution of the perturbed Euler equations, governing the unsteady dynamics of vortical as well as acoustical perturbations in a general compressible mean flow field. Such methods are now commonly referred to as CAA (Computational AeroAcoustics) methods. Contrary to more classical approaches of solving the acoustic wave equation the CAA approach allows for a simulation of a) the mean flow effects on sound radiation and – more importantly – b) the very source mechanism of airframe noise, namely the conversion of vortical (i.e. basically hydrodynamic) flow perturbations into acoustic perturbations (sound) at the airframe component in question. In a joint effort by the partners of SWING, examples of aerodynamic trailing- and leading-edge noise generation have been successfully simulated.

As a proof of concepts, a first application of the CAA perturbation simulation was executed for a design study on the influence of airfoil thickness upon vortex-airfoil interaction noise[5]. The simulation concept here is referred to as the "vorticity test", in which initially localized vorticity (typically in the form of a small vortex or vorticity patch) is introduced into the freestream, and the corresponding acoustic response of the airfoil is simulated. A systematic variation of the airfoil thickness under otherwise constant freestream parameters (and initial vorticity disturbance) then reveals, how much less noise is generated by a thick airfoil compared to a thin airfoil. For three different airfoil thicknesses **Fig. 10** shows the respective snapshots of the pressure perturbation field $p'(t_s,x,y)$, originally generated when the vortex hits the airfoil's leading edge. In the lower part of Figure 10 the respective vorticity distribution is depicted at the same point in time t_s. Finally, in the lower right of the figure are shown directivity plots for two frequencies f extracted from the long-time histories of the pressure perturbation as would be measured on a circle of radius $r = 0.9\ c$ centered at the leading edge of the airfoil, see[5]. Clearly, the results indicate that when subject to vortical perturbations (typically originating from turbulence) the 18% thick airfoil is much quieter than the 0% (flat) profile. It is observed that the mentioned effect is predominant in the downstream direction. The example shows that CAA-schemes do have the potential for an acoustic design tool in the following sense. Airframe components (here the leading edge) are assessed with respect to their ability to convert vorticity into sound. The less efficient this conversion takes place, the quieter (and better) the design of the airframe component is. It is noted that all mean flow effects on the noise generation and radiation are *by ansatz* fully taken into account.

In future studies the above approach will be extended to more complex geometries/mean flows employing a new high resolution overlapped grids technique developed recently at DLR.

4 Achievements and Plans

Dedicated airframe noise research over the past 5 or so years by DLR, initially emphasizing the experiment have brought about significant progress in understanding the important sources of landing-gear and wing-high-lift-device airframe noise. European aviation industry was provided with first, technically feasible and quantified suggestions on reducing the dominant airframe noise sources. First computational concepts for airframe noise on the basis of CAA have been developed and successfully tested. Within its own R&D, as well as within the German National Aviation Noise Program, and in the 5th and subsequent EU Framework Programs, DLR in close cooperation with industry will intensify (a) short term experimental research and technology noise reduction demonstration efforts and (b) medium-to-long-term Computational AeroAcoustics (CAA) efforts on acoustic assessment of high-lift device configurations towards an eventual "design to noise" capability. Within the SWING-successor project SWING+, funded by DFG, the commonly developed computation tools are currently being refined and - more importantly - applied to a dedicated series of realistic airframe noise sources of ever increasing complexity.

References

[1] Michel, U., Helbig, J., Barsikow, B., Hellmig, M., "Flyover Noise Measurements on Landing Aircraft with a Microphone Array", 4th AIAA/CEAS Aeroacoustics Conference, Toulouse, France, June 1998, Paper No. AIAA-98-2336

[2] Dobrzynski, W., Buchholz, H., "Full-scale Noise Testing on Landing Gears in the German Dutch Wind Tunnel", 3rd AIAA/CEAS Aeroacoustics Conference, Atlanta/GA, USA, May 1997, Paper No. AIAA-97-1597

[3] Dobrzynski, W., Nagakura, K., Gehlhar, B., Buschbaum, A., "Airframe Noise Studies on Wings with Deployed High Lift Devices", 4th AIAA/CEAS Aeroacoustics Conference, Toulouse, France, June 1998, Paper No. AIAA-98-2337

[4] Dobrzynski, W., Gehlhar, B., Buchholz, H. "Model- and Full-Scale High-Lift Wing Wind Tunnel Experiments dedicated to Airframe Noise Reduction", 7th International Congress on Sound and Vibration, ICSV7, Garmisch-Partenkirchen, Germany, July 2000

[5] Grogger, H., Delfs, J., Lauke, Th., Lummer, M., Yin, J. "Simulation of leading edge noise of airfoils using CAA based on body-fitted grids", 7th International Congress on Sound and Vibration, ICSV7, Garmisch-Partenkirchen, Germany, July 2000 and submitted to the International Journal of Acoustics and Vibration, IJAV.

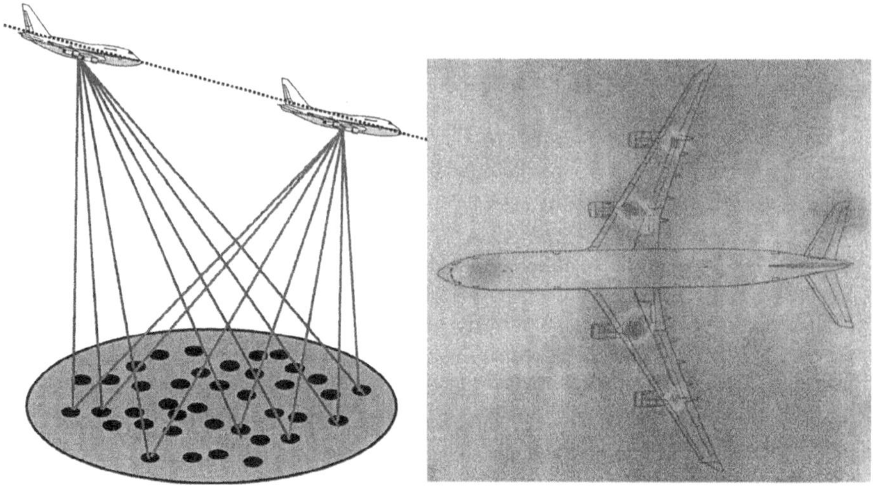

Fig. 1 Source localization employing a spatially extended microphone array where the focus „follows" the overflying aircraft (Data courtesy Ulf Michel, DLR-Berlin)

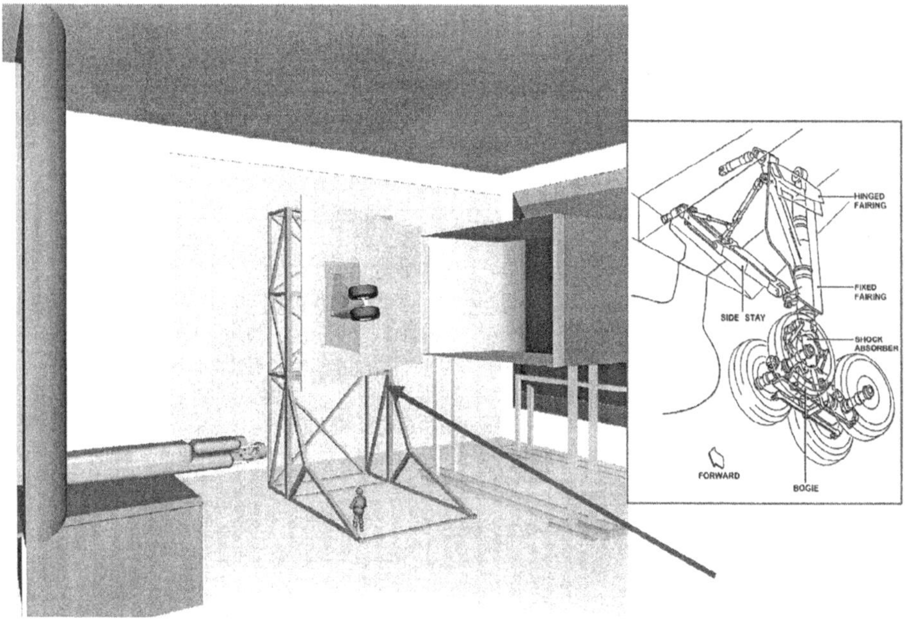

Fig. 2 Aeroacoustic Testing of full-scale Landing Gear Configurations in the DNW-LLF

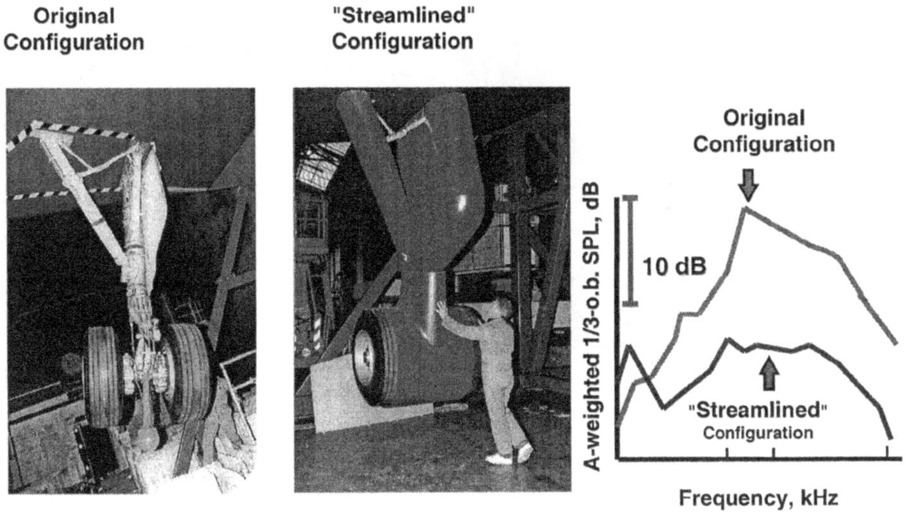

Fig. 3 Reduction Potential of Nose-Gear Airframe Noise by an „All-out“ Streamlining (Test Set-Up in the DNW-LLF)

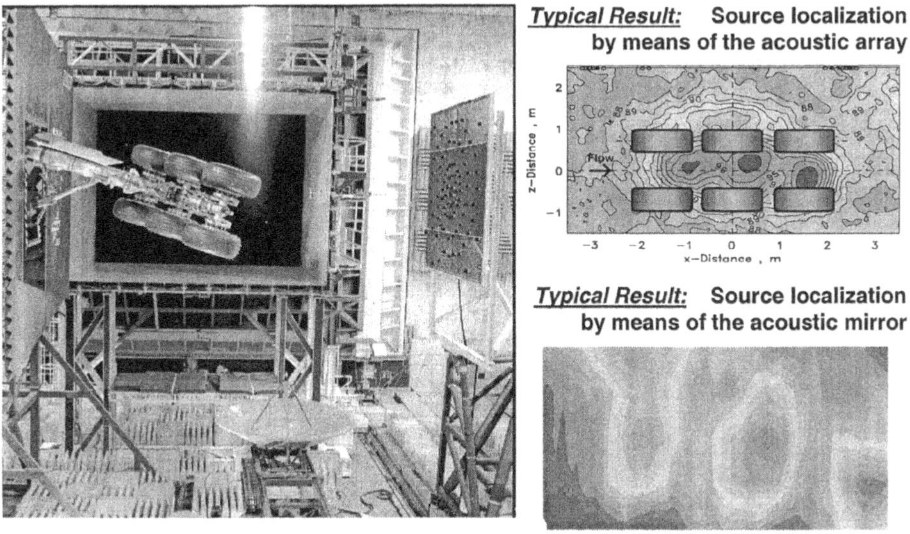

Fig. 4 Source Localization by Means of the Acoustic Mirror and the Acoustic Array on a 6-Wheel Main Landing Gear Mockup; Test Set-up in the DNW-LLF

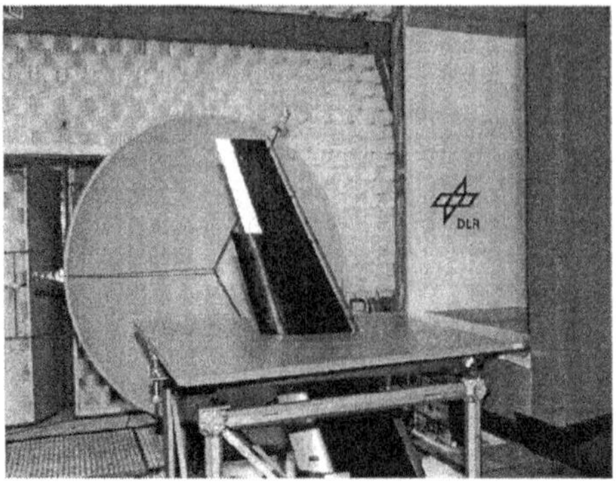

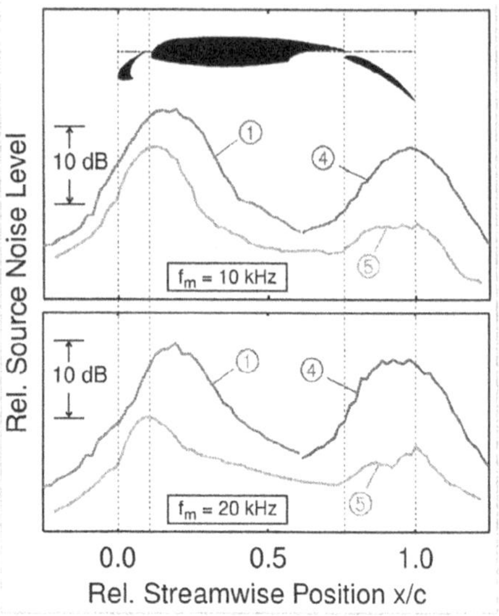

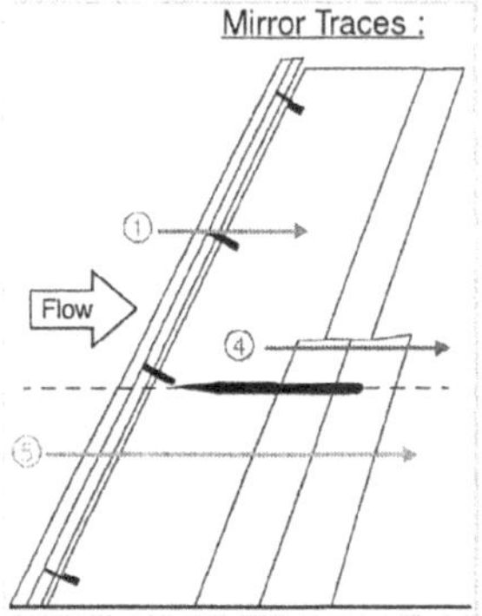

Fig. 5 Source Localization and Rank-Ordering of Wing High-Lift Airframe Noise Sources (Leading Edge Slat, Trailing-Edge Flap, Flap Side-Edge) on a Tenth-Scale Main Wing; Test Set-up in the DLR-AWB (Braunschweig)

Fig. 6 1:7.5-Scale Complete Model Aircraft in the DNW-LLF for Source-Strength Visualization employing the Acoustic Mirror

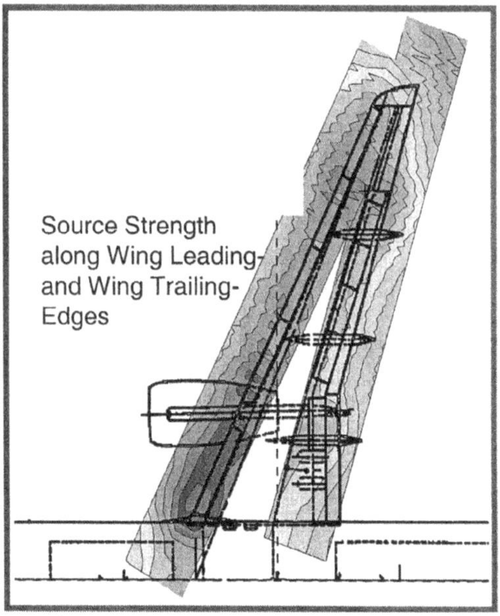

Fig. 7 Full-scale Aircraft Wing in High-Lift Configuration in the DNW-LLF for Airframe Noise Testing; Insert-Picture: Acoustic „Source Distribution" as determined by means of the Acoustic Array

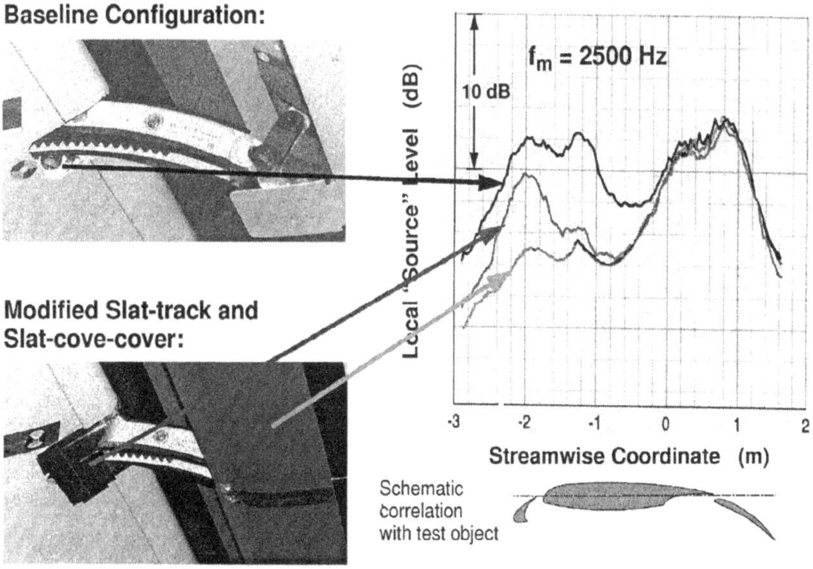

Fig. 8 Effect on Noise of Different Slat-Track and Slat Configurations

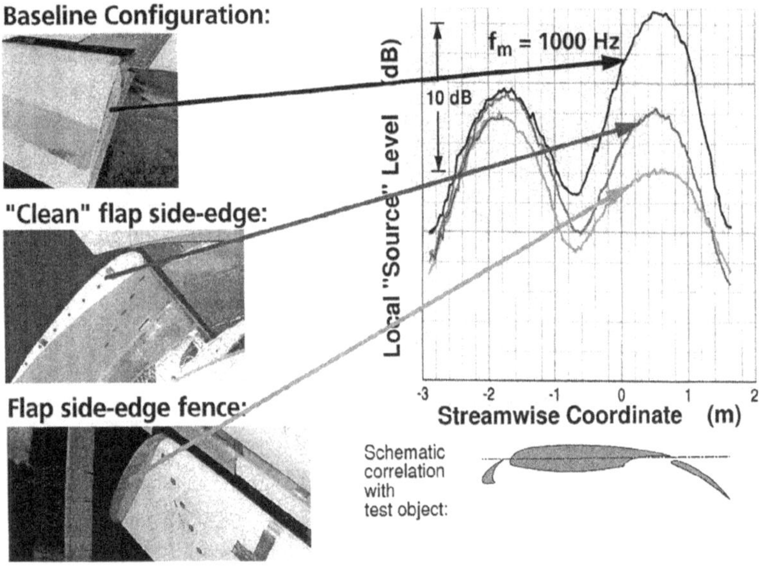

Fig 9 Effect on Noise of Different Flap Side-Edge Configurations

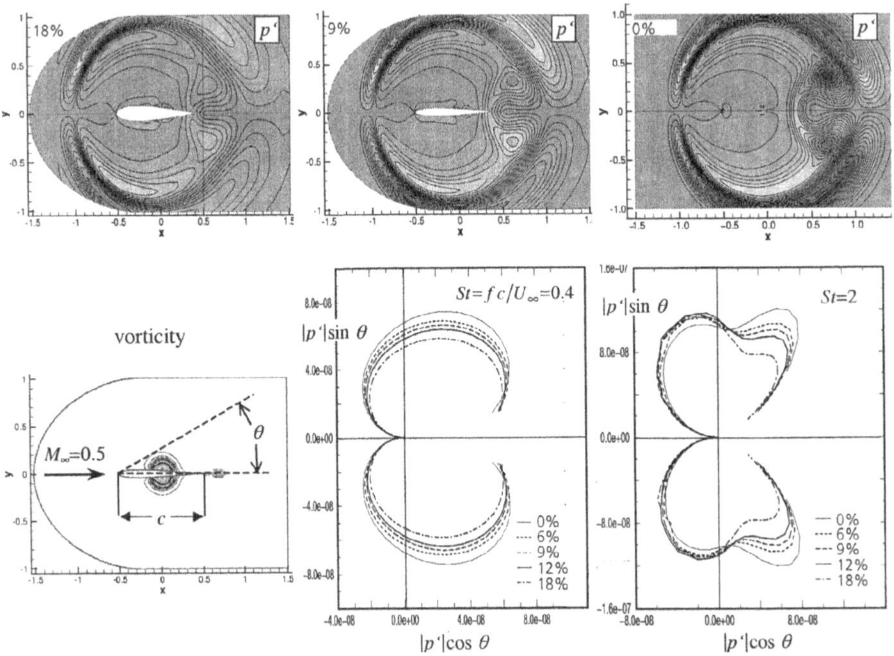

Fig 10 *Upper Part:* Pressure Perturbation Field generated by a Vortex interacting with a Wing Leading Edge; *Lower Part:* Resulting Vorticity Distribution (left) and corresponding Directivity Patterns at 2 normalized Frequencies (right)

Laminar Flow Control

Receptivity of Three-Dimensional Boundary Layers to Roughness and Suction – Linear and Nonlinear Effects

Erik Janke

German Aerospace Center (DLR), Institute of Fluid Mechanics,
Bunsenstr. 10, D-37073 Göttingen, Germany,
email: *erik.janke@dlr.de*

Summary

The paper addresses the accurate prediction of disturbance amplitudes and profiles due to localized surface roughness and wall-suction in incompressible swept-wing boundary layers subject to crossflow (CF) instability. An accurate and effective numerical approach for solving the *receptivity* problem of disturbance generation in three-dimensional boundary layers is presented. The developed solution method models correctly the far-field effects of surface roughness and suction through holes on the generated disturbances. This is shown by chosing a Falkner-Skan-Cooke (FSC) boundary layer flow and a swept-wing experiment and by comparing the numerical results for these flows with data from Direct Numerical Simulations (DNS) and measurements, respectively. The consideration of nonlinear effects and surface curvature as well as coupling the results with NOLOT/PSE-methods make this approach a useful numerical tool for designing active and passive transition control systems for swept wings.

1 Introduction

For the aerodynamic design of wings, stabilizers or high-lift systems of transport aircraft for laminar flow, the instability characteristics of the undisturbed basic flow at a given wing section are of major importance. Not only, but in particular for CF-instability dominated boundary layer flows, it has been shown that the disturbance evolution across a wing, say, can be accurately predicted by NOnLOcal Transition analysis (NOLOT) using the Parabolized Stability equations (PSE), if correct disturbance amplitudes at the inflow boundary are known (e.g. [6, 7]). These initial amplitudes can be provided by *receptivity* analysis that investigates the generation of disturbances within the boundary layer due to forcing from environmental disturbances (cf. [5, 3, 13, 2]). Moreover, the turbulent breakdown in CF-instability dominated boundary layer flows can be estimated by investigating the secondary instability characteristics of the fully developed CF-vortices using Floquet theory [11, 10, 8, 15]. Hence, information on initial disturbance amplitudes in combination with correct modelling of the instability characteristics and, finally, the detection of the relevant transition scenario can be incorporated into design-accompanying numerical tools for the layout of transition control system.

The need for effective numerical design tools becomes apparent, for example, in the case of emerging active transition control systems for swept-wing boundary layers based on variable surface roughness [14, 15]. Exploiting the physical models and numerical tools developed, expensive wind-tunnel tests for design purposes can be avoided. A step into that direction was taken by simulating the underlying nonlinear mechanism of this roughness-based transition control approach with PSE-methods based on estimated initial disturbance amplitudes for the ONERA-D wing [9, 11].

After briefly introducing the formulation for the receptivity analysis in Section 2, the paper considers two problems. First, the receptivity to suction is investigated in Section 3 using the FSC-boundary layer flow currently studied with highly accurate DNS at the University of Stuttgart [12]. Second, transition experiments in CF-instability dominated swept-wing boundary layers conducted over the past years at Arizona State University [14, 15] are used to study the receptivity to roughness in Section 4.

2 Formulation

2.1 Governing equations

When splitting the total three-dimensional flow quantity $Q(x_1,x_2,x_3,t)=\{\vec{V},P\}^T$ into a given basic flow $Q_0(x_1,x_3)=\{\vec{V}_0,P_0\}^T=\{U_0,W_0,V_0,P_0\}^T$ and disturbance quantities $q_1(x_1,x_2,x_3,t)=\{\vec{v}_1,p_1\}^T=\{u_1,w_1,v_1,p_1\}^T$, the evolution of the disturbances is governed by the Navier-Stokes equations in disturbance form (1, 2) subject to the boundary conditions as described in the following subsection. The ansatz for steady disturbances in

$$\nabla \cdot \vec{v}_1 = 0\,, \tag{1}$$

$$\frac{\partial \vec{v}_1}{\partial t} + (\vec{V}_0 \cdot \nabla)\vec{v}_1 + (\vec{v}_1 \cdot \nabla)\vec{V}_0 + (\vec{v}_1 \cdot \nabla)\vec{v}_1 = -\nabla p_1 + \frac{1}{Re}\nabla^2 \vec{v}_1 \tag{2}$$

equation 3 is identical to the PSE-ansatz, thus allowing for modeling nonlocal effects and coupling of receptivity results with NOLOT/PSE-methods. In (3), the total disturbance

$$q_1(x_1,x_2,x_3) = \sum_{m=-\infty}^{\infty} \hat{q}_{1,m}(x_1,x_3) e^{i\int \alpha_{1,m} dx_1 + im\beta_1 x_2} \tag{3}$$

is split into a phase part rapidly varying in x_1 and an amplitude part slowly varying in x_1. This prescribed wave-character can be extracted from the governing equations and results in a smooth variation of the amplitude part $\hat{q}_{1,m}$ in the chordwise x_1-direction, thus reducing the numerical resolution required to represent the disturbance quantities.

2.2 Method of Equivalent Forcing

The concept of an *equivalent forcing* was introduced for small (linear) wall-disturbances by Bertolotti [1, 2]. In here, it is extended to a nonlinear formulation that allows for modeling the effects of spanwise varying disturbance amplitudes and the influence of a meanflow distortion due to wall-forcing.

A given spanwise periodic wall-disturbance (roughness geometry or suction distribution) can be decomposed into a Fourier-series in spanwise direction and its original shape in chordwise direction. In order to use the advantages of the PSE-ansatz also for the wall boundary conditions, the *original forcing* can be tranformed into an *equivalent forcing* $\hat{g}_m(x_1)$ for each Fourier-component. For the correct modeling of nonlocal effects due to the finite extent of the wall-forcing in x_1-direction, it is important that $\hat{g}_m(x_1)$ contains the entire wave-number information from the region where $\hat{g}_m(x_1)$ is non-zero (for details see [2]). The sequence of necessary steps is depicted for a single Fourier-component in Figure 1. The sequence consists of a Fourier-transform, filtering the wave-number range of interest and an inverse Fourier-transform, applied to the *original* shape of the wall-forcing in x_1-direction. This yields the *equivalent forcing* $\hat{g}_m(x_1)$ for each Fourier-component. The total equivalent wall-disturbance is then constructed by a summation of all Fourier-components (4). A superposition of the individually transformed Fourier-components is shown in Figure 2b. In Figures 1 and 2, it can be seen that the *equivalent forcing* is smoother and can be represented with fewer points than the *original forcing*.

$$S_{equivalent}(x_1,x_2) = \sum_{m=-\infty}^{\infty} \hat{g}_m(x_1) \cdot e^{im\beta_1 x_2} \tag{4}$$

Substituting both the ansatz for the disturbances (3) and for the equivalent forcing (4) into the governing equations 1 and 2, the wall-boundary conditions for the individual Fourier-components for suction (exact) and roughness (up to $o\{\hat{g}_m^3\}$), as well as the outflow boundary conditions can be expressed as in equations 5, 6 and 7, respectively.

$$\begin{aligned}
\hat{u}_{1,m}(x_1,0) &= 0 \\
\hat{v}_{1,m}(x_1,0) &= \hat{g}_m(x_1) \cdot V_{hole,max}(x_1) \\
\hat{w}_{1,m}(x_1,0) &= 0
\end{aligned} \tag{5}$$

$$\begin{aligned}
\hat{u}_{1,m}(x_1,0) &= -\hat{g}_m(x_1) \cdot \frac{\partial U_0(x_1,0)}{\partial x_3} - \hat{g}_m(x_1) \cdot \frac{\partial \hat{u}_{1,m}(x_1,0)}{\partial x_3} - \frac{\hat{g}_m^2(x_1)}{2} \cdot \frac{\partial^2 U_0(x_1,0)}{\partial x_3^2} \\
\hat{v}_{1,m}(x_1,0) &= 0 \\
\hat{w}_{1,m}(x_1,0) &= -\hat{g}_m(x_1) \cdot \frac{\partial W_0(x_1,0)}{\partial x_3} - \hat{g}_m(x_1) \cdot \frac{\partial \hat{w}_{1,m}(x_1,0)}{\partial x_3} - \frac{\hat{g}_m^2(x_1)}{2} \cdot \frac{\partial^2 W_0(x_1,0)}{\partial x_3^2}
\end{aligned} \tag{6}$$

$$\frac{\partial \hat{v}_{1,m}}{\partial x_1}(x_{1,outflow},x_3) = 0 \ , \ \frac{\partial \hat{p}_{1,m}}{\partial x_1}(x_{1,outflow},x_3) = 0 \tag{7}$$

The solution method rests on a formulation that is fourth-order accurate in wall-normal direction and second-order accurate in chordwise direction. This results in a line-solver that can march in chordwise direction from inflow to outflow boundary until the maximum deviation in the flow-field between two iterations is less than an ϵ, say 10^{-8}.

3 The Falkner-Skan-Cooke Boundary Layer Flow

Results are presented for a basic flow that is currently investigated for its response to distributed forcing by suction through hole-arrays at the University of Stuttgart [12]. The flow parameters and hole configuration were chosen such that they closely resemble the flow conditions over the perforated leading edge region of an aicraft's wing. The scope here is the validation of the developed nonlinear receptivity formulation for suction.

3.1 Basic Flow and Instability Analysis

The most important parameters of the basic flow are U_∞^*=159m/s, L^*=9.434·10^{-3}m, $Re_\infty = U_\infty^* \cdot L^*/\nu^* = 10^5$ and Hartree Parameter β_h=0.4 (cf. [12]). Further, the spanwise hole spacing for the case considered is $\lambda_{x_2}^*$=500μm, corresponding to a spanwise wave number of $\beta=2\pi/\lambda_{x_2}$=115. Figure 3a shows the chordwise variation of the inviscid parameters and Figure 3b depicts growth rate contours based on local theory. It is seen that the chosen hole spacing corresponds to strongly amplified wave numbers.

3.2 Receptivity Analysis to Suction

This subsection considers the case of suction through one spanwise row of holes with 500μm spacing and diameters of 50μm. The suction velocity distribution in a single hole is given by $V_{hole}^*(x_1,x_2)$= -$V_{max}^* \cdot cos^3\{\pi r(x_1,x_2)/d_{hole}\}$ with V_{max}^*=0.0089·U_∞^* (see [12] for details). Figure 4a shows a comparison of the results from the DNS and the present approach (NS). Plotted are the amplitudes of the chordwise disturbance component versus a nondimensional chordwise distance $n=(x-x_0)/\Delta x$, where x_0 denotes the location of the inflow boundary and Δx=4.189·10^{-4} is the stepsize used in the DNS. Whereas the mean flow distortion (Mode 0) is overpredicted by a factor of $\sqrt{2}$, its decay rate is captured well and the deviation in the individual Fourier-components is about 10%. However, if one keeps in mind that the motivation here is to predict initial disturbance amplitudes and a 10% error amounts to a change in the N-factor of only $\Delta N \simeq 0.1$, the agreement is good. Figure 4b depicts the wave number and growth rate from the PSE without the presence of suction holes and the results from the present computation (NS). As reported from the DNS [12], suction through the holes at n=200 introduces disturbances that result in a strong deviation from the linear PSE-results in the vicinity of the hole (*near-field*), where the disturbances decay in contrast to linear theory (Figure 3b), before the most amplified disturbance according to linear theory prevails in the *far-field* and the NS-results merge with those from the PSE. The particularly long *near-field* region of Δn=1000 $\simeq$ 80·d_{hole} observed here is attributed to the very small diameter of the suction hole that excites a broadbanded modal disturbance.

4 The ASU Swept-Wing Experiment

This section considers the transition experiments in CF-instability dominated boundary layers conducted at Arizona State University (ASU). The results presented focus on swept-wing experiments using the NLF (2)-0415 airfoil at an angle-of-attack of -4° and a sweep angle of 45° [6, 14, 15].

4.1 Basic Flow

The measurements were performed in the wind-tunnel plane (swept conditions). Based on a swept chord length of C^*=1.83m, the characteristic Reynolds number is defined as $Re_c = Q^*_\infty \cdot C^*/\nu^*$. Figure 5 depicts the computed pressure distribution, the undisturbed basic flow profile at x_1/c=0.20 and the chordwise variation of the displacement thickness in comparison with experimental data taken from [6]. With a slight correction of the angle-of-attack used as input for computing C_p-distribution and basic flow from the measured value of α=-4° to α=-3.5^o, the results agree well with the experiment.

4.2 Instability Analysis

Figure 6 shows results from linear PSE-computations and the experiment for disturbances with a spanwise wavelength of $\lambda^*_{x_2}$=12mm. The strong stabilizing effect of surface curvature over the entire chordwise extent is seen both in the significant decrease of the growth rate in Figure 6a and in the resulting lower N-factors in Figure 6b. The measured disturbance evolution is well captured in the region of linear disturbance growth ($x_1/c \leq$0.30) when the effects of surface curvature are taken into account (see also [6, 14]). Further, it is noted from Figure 6a that despite the strong decrease of the growth rate, the wave number does not change significantly when considering the curvature terms.

4.3 Receptivity Analysis to Surface Roughness

The focuses of this subsection are the linear effects of surface curvature and Reynolds number on the receptivity characteristics. Receptivity factors are a geometry-independent measure for the *sensitivity* of given flow to external forcing. In general, they express the ratio of the response amplitude to the forcing amplitude. The definition of the receptivity

$$K(x_0) = \frac{A(x_0)}{h(\alpha_e(x_0))} = \frac{\hat{u}_1(x_0)/U_\infty}{h(\alpha_e(x_0))} \tag{8}$$

factor K is given in equation 8, where $h(\alpha_e(x_0))$ is the value of the Fourier-transform h of the original forcing evaluated at the wave-number of interest $\alpha_e(x_0)$ (see Figure 1).

Figure 7a shows receptivity factors versus the location of a generic roughness element for two strongly amplified disturbances with and without curvature effects. Clearly, the receptivity to forcing is highest near the leading edge and the neutral points of the disturbances (marked by arrows). This corresponds to the strong meanflow gradients in this region. Further, it turns out that the influence of surface curvature is negligible, except in a narrow region near the neutral points. There, however, the results presented are not conclusive as to whether the impact of curvature enhances or decreases the receptivity. This is in some contrast to recent results for the flow around a swept parabolic cylinder where an enhancing influence of curvature is stated [4]. In agreement with this work, however, it is concluded that curvature effects are of lower order in comparison with nonlocal effects which are incorporated in the present approach. Lastly, from Figure 7b, it is seen that the receptivity to roughness decreases with increasing Reynolds number.

The combined effects of receptivity and instability characteristics (Figures 7b and 8a, respectively) on the disturbance generation and evolution is depicted for various Reynolds numbers in Figure 8b. Plotted are rms-amplitudes $A(x_1)=|\hat{u}_1(x_1)|/U_0$ due to the presence of cylindrical roughness elements (d^*=3.7mm, h^*=6μm) at x_1/c=0.023. It is seen that the decrease in the receptivity factor with higher Re_c is outweighed by an increase of the N-factors with higher Re_c, thus resulting in an increase of the generated disturbance amplitudes with increasing characteristic Reynolds number.

5 Conclusions

The paper presents an effective numerical tool for modeling the impact of wall-forcing on the generation of disturbances within a boundary layer that is subject to CF-instability. By investigating two boundary layer flows of industrial interest, the method presented is shown to be capable of accurately predicting the disturbance amplitudes within the boundary layer due to steady wall-forcing. Besides the prediction of disturbance amplitudes for a given wall geometry or suction distribution, the receptivity factors obtained can be used effectively for evaluating the impact of any generic wall geometry on the generated disturbances. Another finding regards the influence of surface curvature on the receptivity characteristics. Despite the well-known fact that curvature effects have a strong stabilizing influence on the instability characteristics over the entire chordwise range, it was shown that their impact on the receptivity characteristics is negligible, except in a narrow region near the neutral point where it has a lower-order effect on the initial disturbance amplitudes. In summary, pre-computed receptivity factors and nonlocal instability characteristics for a given basic flow can be combined straightforwardly, thus allowing for improved transition prediction via incorporation of initial amplitude effects into variable N-factors for various levels of forcing from external disturbances. Applications are seen in design-accompanying processes for the optimization of hybrid and natural laminar flow control systems, as well as for the layout of high-lift systems.

Acknowledgements

I am indebted to Dr. Bertolotti who introduced me to the concept of equivalent forcing. Further, the cooperation with Ralf Messing from the University of Stuttgart, who kindly provided the DNS-data and valuable discussion, is highly appreciated. Finally, I have to thank EADS Airbus in Bremen for a continuous support of this work.

References

[1] Bertolotti F.P.: *A partial simulation of receptivity and transition in 3D boundary layers*, Proceedings: IUTAM Symposium on Nonlinear Instability and Transition in 3D Boundary Layers, Springer, pp. 491–498, 1994.

[2] Bertolotti, F.P. (2000). *Receptivity of three-dimensional boundary layers to localized wall roughness and suction*, Physics of Fluids, Vol. 12, No. 7, pp. 1799–1809.

[3] Bippes H.: *Basic experiments on transition in three-dimensional boundary layers dominated by crossflow instability*, Progress in Aerospace Sciences, vol. 35, pp. 363-412, 1999.

[4] Collis S.S., Lele S.K.: *Receptivity to surface roughness near a swept leading edge*, JFM vol. 380, pp. 141–168, 1999.

[5] Crouch, J.D. (1994). *Distributed excitation of cross-flow vortices in three-dimensional boundary layers*, IUTAM Symposium on Laminar-turbulent transition (ed. R.Kobayashy), Sendai Japan, Springer Verlag, pp. 500–506.

[6] Haynes T.S.: *Nonlinear stability and saturation of crossflow vortices in swept-wing boundary layers*, Dissertation, Arizona State University, 1996.

[7] Hein S., Hanifi A., Casalis G.: *Nonlinear transition prediction*, European Congress on Comp. Meth. in Appl. Sciences and Eng., ECCOMAS 2000, Barcelona, Sept. 2000.

[8] Janke E., Balakumar P.: *On the secondary instability of three-dimensional boundary layers*, Theoretical and Computational Fluid Dynamics, vol. 14 (3), pp. 167–194, 2000.

[9] Janke E., Balakumar P.: *Transition control using Leading Edge Roughness*, AIAA-Paper 99-3656, 1999.

[10] Koch W., Bertolotti F.P., Stolte A., Hein S.: *Nonlinear equilibrium solutions in a 3D boundary layer and their secondary instability*, JFM vol. 406, pp. 131–174, 2000.

[11] Malik M.R., Li F., Choudhari M.M., Chang C.-L.: *Secondary instability of crossflow vortices and swept wing boundary layer transition*, JFM, vol. 399, pp. 85–115, 1999.

[12] Messing R., Kloker M., Wagner S.: *DNS of suction through discrete holes in a 3D boundary layer*, Notes on Numerical Fluid Mechanics, vol. 72, Vieweg Press, pp. 323–330, 1999.

[13] Ng L.L., Crouch J.D.: *Roughness-induced receptivity to crossflow vortices on a swept wing*, Physics of Fluids, Vol. 11, No.2, pp. 432–438, 1999.

[14] Saric W.S., Carillo R.B., Reibert M.S.: *Leading Edge roughness as a transition control mechanism*, AIAA-Paper 98-02781, 1998.

[15] White E.B., Saric W.S.: *Application of variable leading-edge roughness for transition control on swept wings*, AIAA-Paper 2000-0283, 2000.

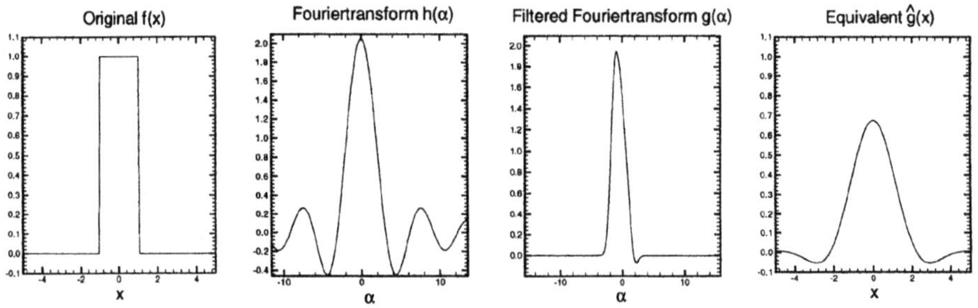

Figure 1 From left to right: Transformation of the original forcing by a spanwise row of square prisms into an equivalent forcing (linear).

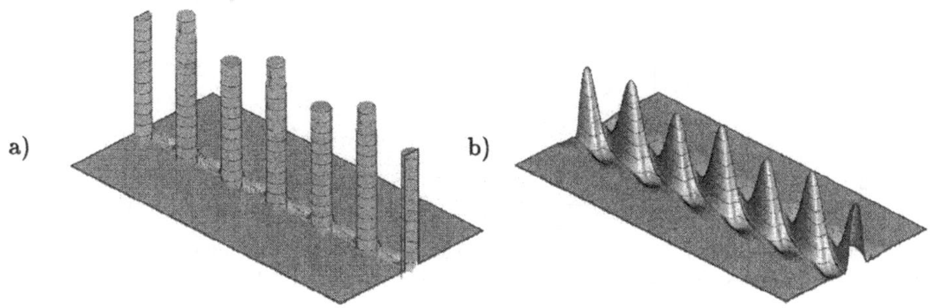

Figure 2 **(a)** Original roughness (one spanwise row of cylindrical elements) with variation in height, and **(b)** equivalent roughness (flow from lower left to upper right).

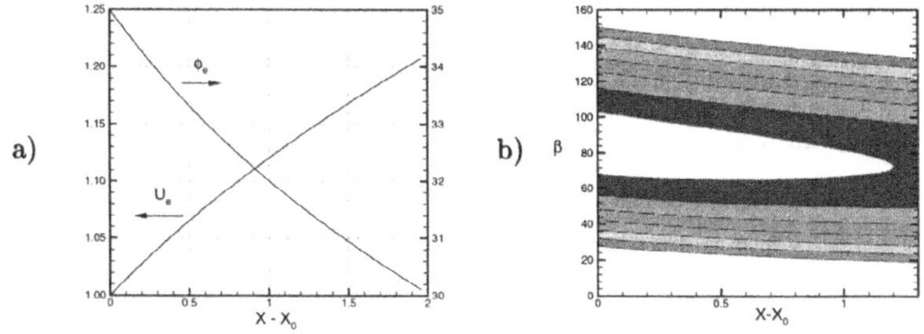

Figure 3 **(a)** Inviscid chordwise velocity and angle of inviscid streamline, **(b)** contours of growth rates σ from spatial local theory versus chordwise direction and spanwise wave number (lines are equally spaced at $\Delta\sigma$=0.2 between σ=-1.2 $\cdots$ 0).

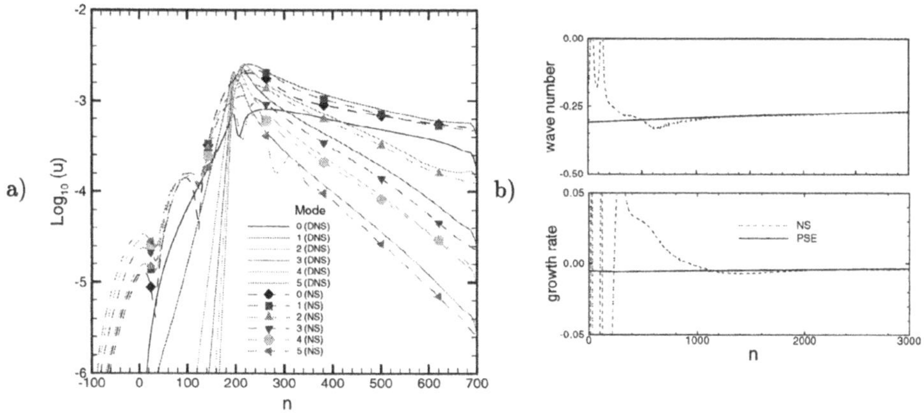

Figure 4 **(a)** Amplitude variation near a single row of suction holes at n=200, **(b)** wave number and growth rate of the fundamental disturbance with $\lambda^*_{z_2}$=500μm from PSE and NS.

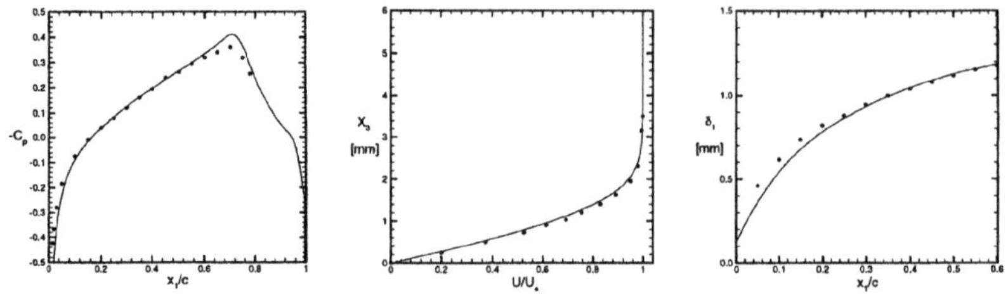

Figure 5 C_p-distribution (swept conditions), velocity profile along the inviscid streamline and displacement thickness from computation and experiment (Re_c=1.6·10^6, α=-3.5^o).

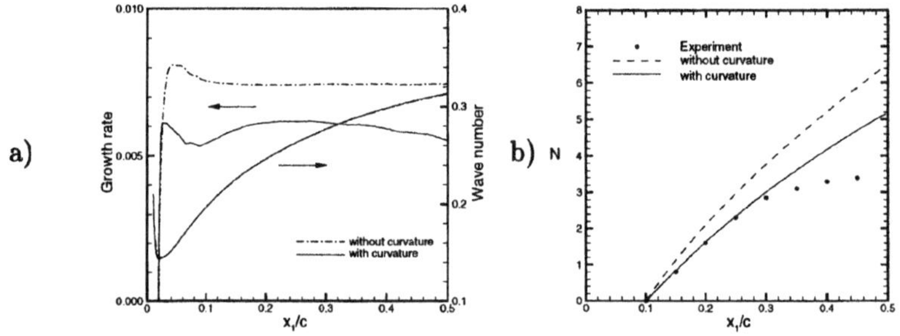

Figure 6 (**a**) Wave numbers and growth rates from linear PSE, (**b**) N-factors from linear PSE-computations and experiment (Re_c=2.4·10^6, $\lambda^*_{x_2}$=12mm, experimental data taken from [6]).

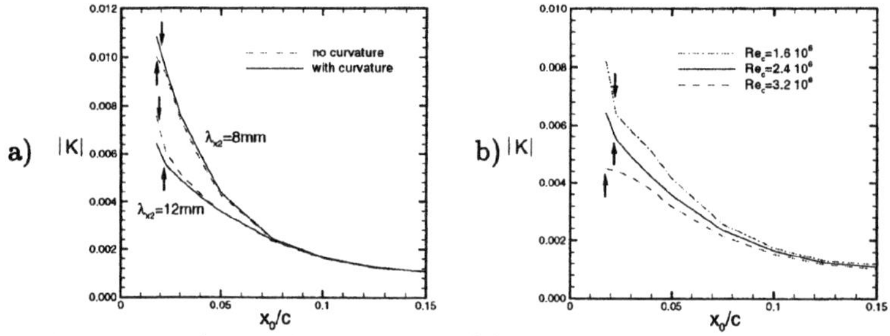

Figure 7 Receptivity factors for roughness: (**a**) With and without surface curvature for Re_c=2.4·10^6 and (**b**) at various Reynolds numbers for $\lambda^*_{x_2}$=12mm (roughness elements at x_0).

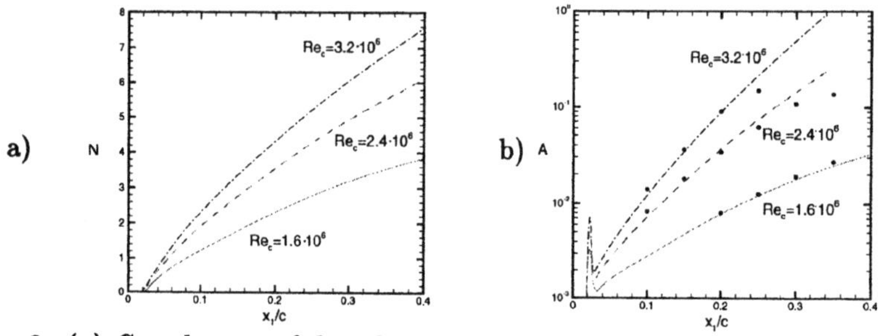

Figure 8 (**a**) Growth rates of disturbances with $\lambda^*_{x_2}$=12mm for various Re_c, (**b**) $u_{1,rms}$-amplitudes in the near and far-field of cylindrical roughness elements at x_1/c=0.23 from computation and experiment ($\lambda^*_{x_2}$=12mm, exp. data taken from [13], all results with curvature).

Active Control of Tollmien-Schlichting Instabilities by Multi-Channel Sensor Actuator Systems

D. Sturzebecher, W. Nitsche
Technische Universität Berlin, Institut für Luft- und Raumfahrt, Sekr.F2
Marchstr. 12–14, 10587 Berlin, Germany

Summary

This work describes experimental investigations intended to actively attenuate Tollmien–Schlichting(TS) instabilities on an airfoil. It is possible to superimpose harmonic TS–waves in the early linear stage with artificial damping waves in order to minimize these disturbances. Sensors and actuators integrated in the wing surface controlled by a fast signal processor are required to enable an optimum active wave control (AWC). High attenuation rates have already been achieved with a single sensor actuator system for a 2D–dominated TS–perturbation. Linear 3D–instabilities were successfully cancelled in recent experiments by a spanwise arrangement of sensors and actuators. The sensors feed a controller which drives the actuators by performing the simultaneous adaptation of the corresponding transfer functions. Further experiments have demonstrated the additional delay of transition due to the repeated use of sensor actuator systems in streamwise direction. Actuator tests were carried out on a 2D sensor actuator system in parallel with experiments on multiple systems. Slot actuators are preferably used, mainly due to their easy application in basic experiments. First tests with membrane actuators have shown a better local receptivity, especially at high amplitudes of the TS–disturbances.

1 Introduction

Recent aerodynamic research is focused on the improvement of the efficiency of future aircraft. The adaptive wing promises a large potential in this context. In addition to a continuous adaptation of the wing to varying flight conditions, drag minimizing techniques can reduce fuel consumption considerably. The skin friction presents a large part of the total drag of an aircraft, especially of a wing, but by using different active and passive methods to delay laminar–turbulent transition (e.g. boundary–layer suction at crossflow), this drag can be reduced.

One active method of sustaining a laminar boundary layer for a longer chord length is the reduction of Tollmien–Schlichting (TS) instabilities by superposition using artificially generated counter–disturbances. The feasibility of this method has already been experimentally proved in several works (LIEPMANN and NOSENCHUCK [1], PUPATOR and SARIC [2], THOMAS [3], GROSCHE et al. [4]). An efficient attenuation of natural disturbances is only possible by employing a powerful active control because of the random appearance of instable natural TS–wave packets. A fast signal processor running adaptive control algorithms has been used successfully for cancelling natural TS–instabilities on an unswept wing at TU–Berlin for some years (BAUMANN et al. [5, 6, 7]).

A considerable delay of the transition on an aircraft wing can only be achieved by repeated attenuation of the reamplified TS–instabilities in the streamwise direction. Furthermore, it is necessary to employ a spanwise multi–channel active control because of the spanwise variation of TS–waves. For these reasons, several control systems are installed based on the experience of previous experiments with 2D sensor actuator systems.

2 Sensor Actuator System

For a successful TS–wave damping, sensors to measure the TS–disturbances and actuators to generate the corresponding artificial counter–waves are required. The sensor actuator matrix must be integrated in the wing surface to avoid any roughness. The traveling TS–waves are detected by a reference sensor upstream of the actuator. The actuator signal is directly derived from the reference signal by a convolution with an adaptive FIR–Filter (FIR, finite impulse response). The FIR–filter, which models the internal transfer function of the amplified TS–wave, is continuously adapted by performing the filtered–x–LMS–algorithm (LMS, least mean square) to minimize the amplitude of the error sensor signal measured downstream of the actuator. A digital signal processor (SHARC ADSP–21060), adapted by an interface, has direct access to a 16–channel A/D converter and a 16–channel D/A converter. The advanced processor system can control up to 8 sensor actuator systems simultaneously by adapting autonomous transfer functions.

Figure 1 depicts the investigated configurations of multi–channel sensor actuator systems. The spanwise arrangement a) of autonomous systems enables the reduction of disturbances with spanwise variations of phase and amplitude by generating suitable counter–waves. The configuration b) primarily repeats the cancellation of the reamplified TS–instabilities to achieve an additional delay of the laminar–turbulent transition.

The experiments on active wave cancellation have shown that a surface hot wire is the most promising sensor to detect smallest TS–waves (STURZEBECHER et al. [8]). This sensor is a flush–mounted, welded hot wire with a narrow slot underneath (> 0.1 mm) to minimize the heat flux into the wall. This arrangement enables a much better signal–to–noise ratio than a conventional hot film. Low speed experiments have shown that the surface roughness effect of the sensor is negligible.

The actuator must be able to generate artificial TS–waves with a minimum time delay. The slot actuator was preferred in most of the experiments because of its simple construction and integration. It consists of a small, flush–mounted slot (0.2 mm) with an arrangement of loudspeakers underneath, generating an oscillating stream perpendicular to the wall. Slot actuators can be easily separated into single pressure chambers in the spanwise direction, which enables attenuation of 3D–waves such as oblique or point–induced disturbances.

3 Results

The present results on the use of single and multiple control systems were obtained in a low speed wind tunnel on an unswept test wing (modified NACA 0008, $c = 1.3$ m) at a free stream velocity of $U_\infty = 17$ m/s. It was possible to match individual sensor actuator arrangements to each experiment with an interchangeable wing surface section located between $x/c = 0.39$ and $x/c = 0.78$. An additional harmonic slot source arranged upstream of this section enabled the introduction of artificial 2D–waves, oblique waves and point source–induced instabilities.

3.1 Membrane Actuator

The slot actuator is unsuitable for use at higher Reynolds numbers, since the slot acts as a surface roughness element causing additional disturbances. A smaller slot could minimize the roughness effect, but this contradicts directly the requirement for higher counter–disturbances at higher Reynolds numbers. It is evident that flush–mounted actuators, such as membrane actuators, show a much better behaviour than slot actuators. There is no surface roughness and, in addition, the counter–wave can be introduced into the boundary layer more smoothly, even for high local TS–disturbances. The investigated membrane actuators were flush–mounted flexible surfaces driven by a solenoid. The highly elastic membrane materials allowed for appropriate wall displacement with relative short membrane lengths in the streamwise direction. The minimum achievable membrane length with conventional manufacturing methods is approximately one quarter of a TS–wavelength. Figure 2a) shows the attainable attenuations of a slot compared to a membrane ($t = {}^1/_4\lambda_{TS}$). The maximum velocity fluctuation (u'_{max}), normalized by the free stream velocity of $U_\infty = 17$ m/s, was derived from the distribution of velocity fluctuation (u') measured by a hot wire in the boundary layer at the actuator position. At lower local disturbances, both curves of the attainable damping are similar, but at higher perturbation levels, the membrane achieves a better attenuation. This behaviour is probably caused by undesirable disturbances introduced by the slot and the better local receptivity of a membrane. A membrane actuator with a length of a half TS–wavelength was also investigated and showed a behaviour similarly to the slot. This is primarily caused by the spatial integration and the characteristic eigenform of the membrane vibration. Additionally, the membrane velocity were measured by a laser vibrometer. Figure 2b) shows the relation between perturbation in the boundary layer (u'_{max}) and the required maximum membrane velocity (v') also normalized by $U_\infty = 17$ m/s. The correlation of both is obviously linear and characterized by a ratio of 0.5 (v'/u'_{max}).

Experiments with multi–channel membrane actuators, like 'traveling' waves or 3D–membrane actuators, are planned.

3.2 Investigation on a Spanwise Arrangement

The reduction of three dimensional TS–instabilities requires a spanwise arrangement of sensors and actuators with a spatial distribution corresponding to the 3D–disturbances (Figure 1a)). Surface hot wires serve as reference and error sensors in the experiments. They are arranged in spanwise arrays, each consisting of 8 sensors with equidistant spacing of $\Delta z = 14$ mm ($\approx 3/4\lambda_{TS}$), at $x/c = 0.43$ and at $x/c = 0.48$ respectively. A slot actuator, located at $x/c = 0.46$ between both arrays, has the same spacing as the sensor arrays achieved by separated spanwise pressure chambers. All chambers work autonomously in one wall–bounded slot. The distance between the actuator and the downstream error sensor array corresponds to approximately 1.5 TS–wavelengths ($\Delta x = 30$ mm), while the distance between the actuator and the reference sensor array upstream covers twice the TS–wavelength ($\Delta x = 40$ mm). In this configuration, 8 separate transfer functions were simultaneously and continuously adapted to cancel the TS–perturbations. The attenuation of naturally occurring TS–waves by such an arrangement was improved by factor 1.5 compared to a pure 2D–system. This is typical for a 2D–dominated transition occurring on an unswept wing. Figure 3 represents the time traces of the error sensor array. The uncontrolled spanwise perturbation distribution, as shown in Figure 3a), is obviously strongly two dimensional, therefore it yields a very high attenuation by introducing pure 2D–counter–waves (Figure 3b)). The remaining disturbances are small and random. These disturbances are minimized in the case of active 3D–control with a slightly increasing attenuation (Figure 3c)).

High cancellation ratios were achieved by a spanwise arrangement of sensors and actuators in the case of three dimensional instabilities. Figure 4 shows the efficiency of a multi–channel spanwise control system in the case of an artificially generated oblique wave ($\lambda_z \approx 5\lambda_{TS}$). A 2D–control system failed because of the integrating effect in spanwise direction of the averaged reference and error signal (Figure 4b)). A spanwise adapted oblique counter–wave can be generated with multiple active control to perform an optimum damping of this type of perturbation (Figure 4c)). Frequency–wavenumber spectra are also shown for each case. They are derived from the error sensor array by a Fourier transformation in the frequency and the wavenumber domain. The wavenumber is the inverse spanwise wavelength normalized by twice the sensor spacing. A wavenumber value of 1 indicates a spanwise wavelength of 28 mm and a value of 0 characterizes the pure 2D–wave. Figures 4a) and b) depict the spanwise wavelength clearly as a peak at $f = 300$ Hz and $k_z = 0.25$ ($\lambda_z = 112$ mm). In the case of multiple control (Figure 4c)), the peak as well as the coupled higher harmonic wave ($f = 600$ Hz, $k_z = 0.5$) disappear.

Point–induced instabilities shown in Figure 5 were also cancelled by the multi–channel control system. The time traces are presented as a contour plot in spanwise direction (z) and time (t). The point–induced harmonic disturbance can be recognized in the typical curved shape of the waves with an inactive control (a). Figure 5b) shows the time traces with activated multiple control. The contour plot (Figure 5a)) proves that the spanwise distribution of actuators and sensors is sufficient for the attenuation of this instability mode.

First tests were carried out with a spanwise coupling of the control systems. The coupling did not yield any higher local cancellation compared to the autonomously working control systems, which is due to the wave development on an unswept wing in the primary streamwise direction. The coupling will probably show a better performance under 'real' 3D–conditions like a spanwise pressure gradient or a swept wing.

Future investigations in the field of 3D–control will involve the integration of a 3D–membrane, other experiments will affect a spanwise coupling of the systems under 3D–conditions.

3.3 Investigation on a Streamwise Arrangement

Further experiments were carried out with a multiple arrangement of sensors and actuators in the streamwise direction. Three slot actuators were positioned one after the other with a streamwise distance of approximately four TS–wavelengths. Four spanwise surface hot wire arrays served as reference and/or control sensors located upstream of the first, downstream of the last slot and between the three actuators. First tests showed that after the second attenuation a third 2D–actuator does not yield any additional delay of transition, since repeating 2D–cancellation only leads to an attenuation of the 2D–parts of the disturbances. The remaining 3D–parts are amplified, even though any 2D–waves are already damped. This can be seen by looking at the signals of the last sensor row (Figure 6b). The signals of the outer sensors showed anticyclic behaviour. Therefore, the third actuator was divided into 5 single actuators to attenuate the dominating 3D–part of the remaining disturbances.

The adaptation of the single control paths was successively initiated. This was necessary because nonlinear disturbances were detected on the last sensor row at the beginning of the adaptation. These nonlinear disturbances can have a negative effect on the behaviour of the linear control algorithm. The local attenuation is indicated by time traces of the last sensor array (Figure 6c)). An attenuation of more than 20 is achievable at this position by employing the multi–channel control system. The comparison of the amplification trends with and without stepwise activated cancellation is shown in Figure 7. The u'–amplitudes, normalized by the free stream velocity, were measured downstream of the sensor actuator system by a traversable hot wire at a constant height of $y = 1$ mm above the wing surface. The increasing additional delay of laminar–turbulent transition is obvious. The downstream shifting of transition with only one cancellation yields approximately 100 mm ($\Delta x/c \approx 7.5\%$), while with all three control systems operating, the transition is delayed by 180 mm ($\Delta x/c \approx 14$ %). The additional benefit with the second control system is small compared to the first single control alone, because of the good attenuation of 2D–waves due to the first control system and also the more dominant 3D–disturbances. The amplification rates of the u'–amplitudes become lower in multiple control mode compared to a single cancellation. This leads to the conclusion that the whole control system is only able to cancel 2D– and 3D–disturbances with low oblique angles, while the 3D–disturbances with higher oblique angles remain in the boundary layer and grow, but with a reduced rate.

4 Conclusion

The investigations have shown the feasibility of applying multi–channel control systems for the attenuation of TS–instabilities. The laminar–turbulent transition was considerably deferred downstream with a repeating wave control ($\Delta x/c \approx 14$ %). This achieved approximately 90 % additional delay compared to the single sensor actuator system. Even though the 2D–instabilities were already optimally damped, the remaining 3D–disturbances must be also cancelled, because these disturbances can lead also to transition. Therefore, a sufficient spanwise arrangement of sensors and actuators is necessary for the streamwise repeated attenuation of TS–waves. The multiple control of 3D–disturbances with autonomously running control systems was investigated under several conditions like naturally occurring TS–instabilities, artificial oblique waves as well as point–induced disturbances. A spanwise coupling was tested in first experiments, but this did not lead to a better local attenuation. The coupling will probably show its superiority only under 'real' 3D–conditions like a spanwise pressure gradient. The use of a flush–mounted vibration membrane is justified by higher achievable attenuations and by a lower surface roughness.

Acknowledgements

The work on Active Wave Control was financially supported by the German Research Foundation DFG (Deutsche Forschungsgemeinschaft).

References

[1] Liepmann, H. and Nosenchuck, D., Active control of laminar-turbulent transition. *J. Fluid Mech.*, 118, 201–204 (1981).

[2] Pupator, P. and Saric, W., Control of random disturbances in a laminar boundary layer. AIAA Paper 89-1007 (1989).

[3] Thomas, F., Active Wave Control of Boundary–Layer Transition. AIAA (1990), Viscous Drag Reduction in Boundary Layers, Progress in Astronautics and Aeronautics,USA.

[4] Grosche, F. and Yong-guang, T., Experimente zur Dämpfung von Tollmien–Schlichting–Wellen durch aktive Anregung von Wandschwingungen. IB 222-90 A 46 (1990), Internal report of the DLR-Göttingen, Germany.

[5] Baumann, M. and Nitsche, W., Investigation of Active Control of Tollmien–Schlichting Waves on a Wing. In: R. Henkes and J. van Ingen (eds.), *KNAW*, vol. 46, pp. 89–90 (1996), Transitional Boundary Layers in Aeronautics, Amsterdam, Netherlands.

[6] Baumann, M., Sturzebecher, D. and Nitsche, W., On Active Control of Boundary Layer Instabilities on a Wing. In: H. Körner and R. Hilbig (eds.), *Notes on Numerical Fluid Mechanics*, vol. 60, pp. 22–29. Vieweg Verlag, Braunschweig (1999).

[7] Baumann, M., Aktive Dämpfung von Tollmien–Schlichting Wellen in einer Flügelgrenzschicht. In: *Fortschritt–Berichte VDI, Reihe 7*. VDI Verlag, Düsseldorf (1999), Dissertation an der Technischen Universität Berlin.

[8] Sturzebecher, D., Anders, S. and Nitsche, W., The surface hot wire as a means of measuring mean and fluctuating wall–shear stresses. *to be published* (2001).

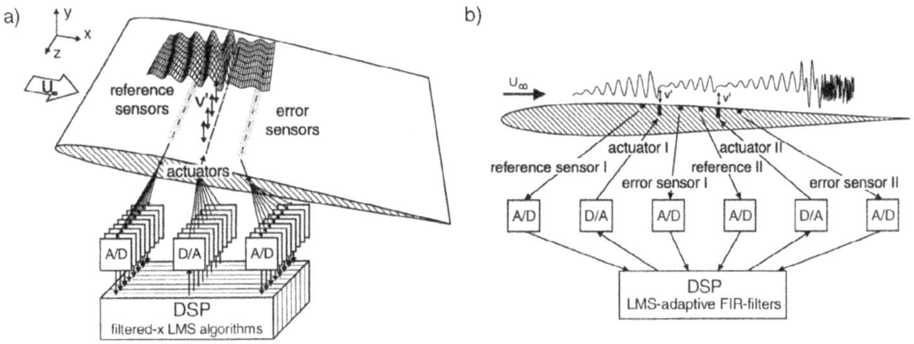

Figure 1 Basic principle of multiple control systems **a)** Spanwise arrangement **b)** Streamwise arrangement

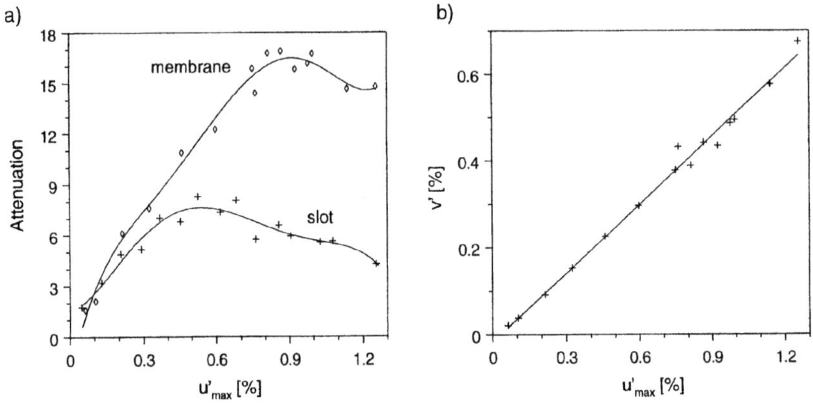

Figure 2 **a)** Achievable attenuations of slot and membrane actuators **b)** Required membrane velocity versus maximum u'–disturbance (u'_{max} and v' normalized by $U_\infty = 17$ m/s)

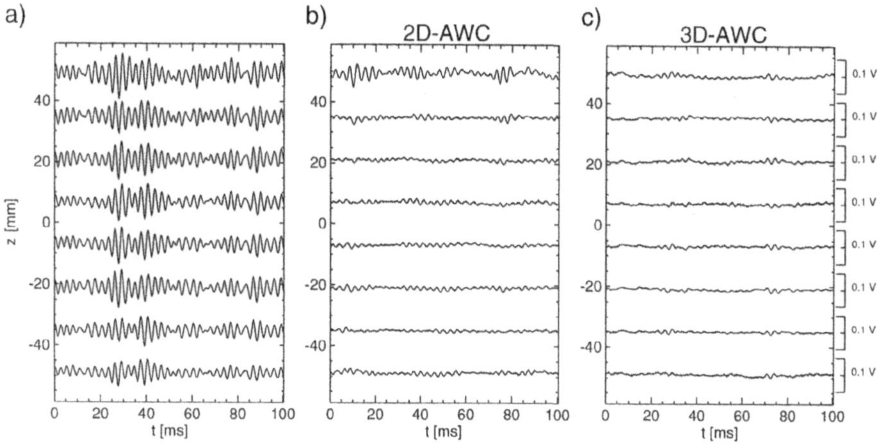

Figure 3 Time traces of the error sensor array for the case of natural disturbances (Spanwise center of the airfoil at $z = 0$ mm) **a)** Uncontrolled **b)** 2D–control **c)** Spanwise multiple control

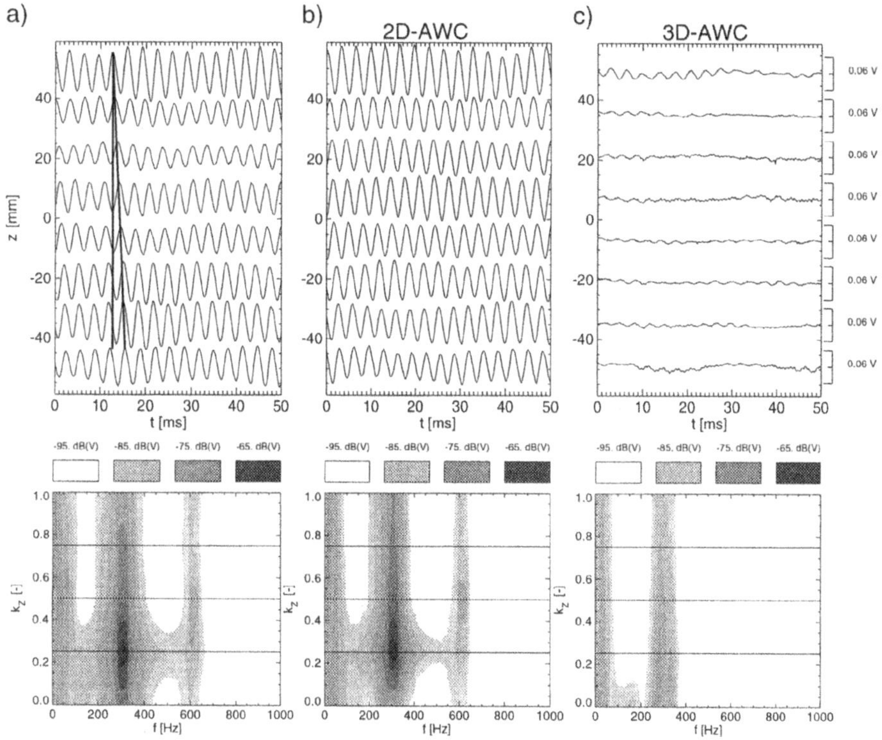

Figure 4 Time traces of the error sensor array for the case of oblique waves
a) Without control **b)** 2D–control **c)** Spanwise multiple control

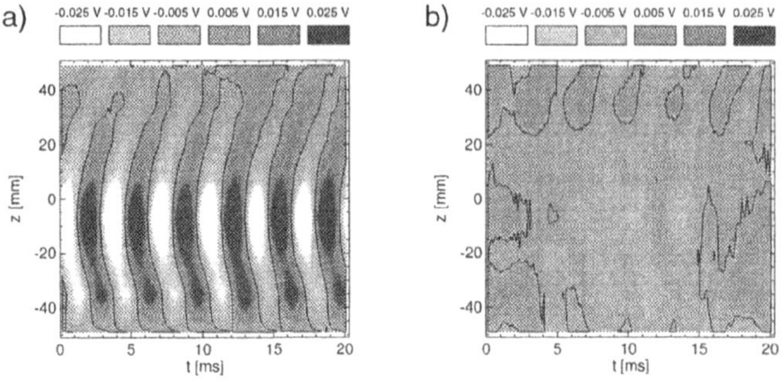

Figure 5 Time traces of the error sensor array with point–induced disturbances
a) Without control **b)** 3D–control

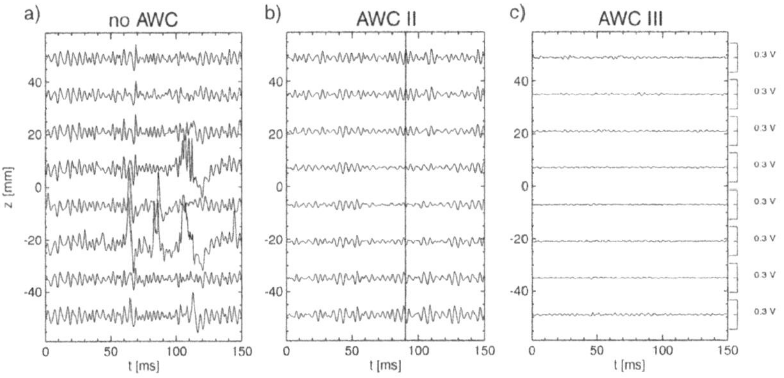

Figure 6 Time traces of the error sensor array without control and with multi–stage control

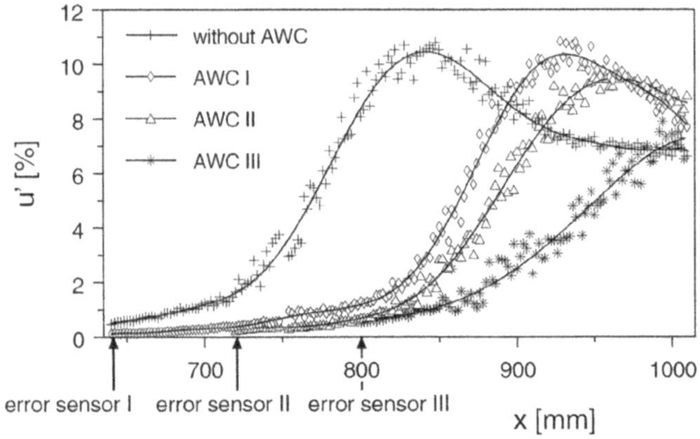

Figure 7 Streamwise u'–development with multi–stage control (Leading edge of the airfoil at $x = 0$ mm)

Measuring Techniques

Time Resolved 2D Measurement Techniques for Quantitative Determination of Flow Fields and Surface Pressure Distributions

K. A. BÜTEFISCH
DLR, Institut für Strömungsmechanik, Bunsenstr. 10, 37073 Göttingen, Germany.

Summary

For further developments of numerical predictions experimental high quality data are needed. One possibility is achieved by the application of advanced optical measurement techniques. Of course in the case of expensive facilities easy installation of the complex measurement equiment into the facility and high productivity have to be achieved. As an example Particle image velocimetry (PIV) is one of this new methods being able to provide instantaneous, planar flow velocity measurements and to produce a large amount of high quality data in a very short time. Typical features and future developments of PIV and other techniques like global Doppler velocimetry (DGV), planar intensity distributions of Rayleigh scattered laser light (PIRS), surface deformation determination by moiré technique, and surface pressure measurement by pressure sensitive paint technique (PSP) will be discussed.

1. Instantaneous Flow Field Measurements

1.1 Laser Doppler anemometry (LDA) for turbulence measurements

During the last years laser Doppler anemometry had been well developed. The DLR system in its final stage of development provides local time dependent 3 D velocity data with data rates up to 20 kHz and a spatial resolution in the order of 1/10 mm even in the vicinity of a wall. Of course to fulfil industrial requirements the system has to be robust, mobile, and easily to be installed in large wind tunnels. Fig.1 shows the application of such a fully operational 3 D LDA system in the 6m x 9m test section of the DNW-LLF taking data from a distance of 3.80 m of the flow field in the gap between the main wing and the flap on an Airbus A320 model. In order to take into account model vibration an additional sub-system simultaneously measured the position of the local measuring volume with respect to the model surface. Not only 3D velocity profiles but also profiles of the 3D turbulence were determined with high accuracy, [1, 2].

1.2 Doppler global velocimetry (DGV) for unsteady flow fields

A quite different technique to apply the Doppler effect is the Doppler global velocimetry (DGV), able to acquire the instantaneous velocity field as a whole rather than the local velocity vectors. As in the case of LDA the flow has to be seeded with particles which were illuminated by laser light with well known laser frequency. The scattered light, Doppler shifted due to the movement of the particles, is recorded by a video camera through a frequency dependent filter, i. e. the Doppler shift changes the brightness of the image in accordance to the local flow velocity. In order to correct for the brightness of the illuminated particles a second camera captures a reference image via a beam splitter. Together with a very fast computer both images were recorded and subsequently divided with a rate of 25 Hz. Hence flow velocity maps were obtained on the display at on-line conditions. Fig. 2 depicts a sequence of successive images obtained with an exposure time of 1/60 s on the unsteadiness of the core of a vortex above a delta wing. From images taken from 3 different angles of sight distributions of the 3 components of velocity and fluctuations were obtained [3]. The accuracy depends on the resolution of the Doppler shift, hence the resolution of a single measurement is in the range of 2.5 m/s.

1.3 Velocity measurements with particle image velocimetry (PIV)

From particles introduced into the flow and illuminated by a double-oscillator pulse laser as first proposed by Kompenhans [4] two images are recorded within a known time interval. From the displacement of the recorded particle pattern the local velocity of the particles and hence the velocity of the surrounding fluid element is obtained. Today up to 4000 velocity vectors within a plane of 30 cm x 40 cm are acquired representing the instantaneous velocity field during the very short time interval of typical 10 μs. Final results of high accuracy are available after a processing time of less than 10 seconds. Due to the reliable technique achieved today [5] PIV can be applied in industrial wind tunnels and offers the potential to save wind tunnel occupation time if velocity field data have to be collected.

It is obvious that PIV captures the instantaneous spatial structure of the velocity field even if the flow is fluctuating. It could be shown that PIV results agree with results obtained from conventional measuring methods like 5-holes-probe [6] if averaged PIV data are considered. Of course the capability of PIV to represent an instantaneous result has to be kept in mind.

The very short measuring time enables the acquisition of unsteady flow fields. The vortex wake behind an aircraft is a good example. Fig. 3 depicts a result obtained behind an Airbus model in co-operation with ONERA in the catapult facility of ONERA Lille, [7]. The development and decay of the vorticity distribution behind the aircraft becomes visible. In order to obtain the information of a larger flow field two cameras at the same time were applied.

At the beginning of PIV development two components of the velocity vector in a plane were recorded. The complete velocity field information was gained only if an actual 2D velocity field

$$\frac{d\vec{U}(x,y)}{d\vec{X}} = \begin{bmatrix} \frac{\partial U}{\partial X} & \frac{\partial V}{\partial X} \\ \frac{\partial U}{\partial Y} & \frac{\partial V}{\partial Y} \end{bmatrix}$$

was present:

The next step in PIV development is to apply stereoscopic PIV technique. As a result, the complete 3 components of the velocity vectors can be obtained. Fig. 4 shows a set up in the 6 m x 9 m test section of the DNW-LLF, [8]. Whereas the laser with a power of 600 mJ is sited at the wind tunnel floor the two cameras were fixed to the nozzle. All distances are around 10 m.

Fig. 5 exhibits a result on a wing of the DLR ALVAST model [9] obtained in the same wind tunnel within the frame of the EU project EUROWAKE. Again 3 components were recorded by the use of stereoscopic PIV technique. The arrows represent the third component in axial direction, the colours the total velocity of the remaining other velocity components respectively. Stereoscopic PIV captures the complete vector field but not all gradients and hence not all components of vorticity and divergence can be calculated:

$$\frac{d\vec{U}(x,y)}{d\vec{X}} = \begin{bmatrix} \frac{\partial U}{\partial X} & \frac{\partial V}{\partial X} & \frac{\partial W}{\partial X} \\ \frac{\partial U}{\partial Y} & \frac{\partial V}{\partial Y} & \frac{\partial W}{\partial Y} \end{bmatrix}$$

In order to obtain the complete 3D field data multiplane PIV was developed. Two or even more laser light sheets illuminate the particles and corresponding sets of cameras record the particle images. Fig. 6a depicts the set up together with a result, Fig. 6b, obtained within the boundary layer of a flat plate. Clearly the vortical structure of the flow field is visible. Now all gradients are available [10]. A sequence of such PIV recordings would yield the time dependent velocity gradient tensor:

$$\frac{d\vec{U}(x,y,z,t)}{d\vec{X}} = \begin{bmatrix} \frac{\partial U(t)}{\partial X} & \frac{\partial V(t)}{\partial X} & \frac{\partial W(t)}{\partial X} \\ \frac{\partial U(t)}{\partial Y} & \frac{\partial V(t)}{\partial Y} & \frac{\partial W(t)}{\partial Y} \\ \frac{\partial U(t)}{\partial Z} & \frac{\partial V(t)}{\partial Z} & \frac{\partial W(t)}{\partial Z} \end{bmatrix}$$

The present stage of development is as follows:

Today PIV has matured from first laboratory set ups to reliable whole field flow measurement technique. The desired application in large, industrial wind tunnels poses a number of special requirements:

large observation areas	One or more cameras (up to 4 at Catapult ONERA)
large observation distances	One or more lasers for illumination (2 systems at DNW-LLF)
time constraints in the set up of the system	Only several hours in the case of 6mx9m test section DNW-LLF
strict safety requirements for laser, people, model, wind tunnel components	
need to reach a high productivity despite high operational costs of the wind tunnel.	Today 3 recordings/s independently from number of cameras

Accuracy is always an important question but answers are extremely dependent on the experimental requirements. PIV provides a short time measurement technique together with the detection of large and small scale spatial structures and small and large velocities in the flow at the same time. Locally velocities may be resolved with an accuracy of better than 1%. Of course the result is extremely depending not only on the optical equipment but also from the whole area and the desired dynamic velocity range covered by one individual PIV recording.

1.3 Density measurements with Rayleigh scattered light (PIRS)

Powerful lasers with wavelength in the UV range enable the collection of Raleigh scattered light with reasonable signal to noise ratio and hence the determination of density distributions in a plane. An optical set up equivalent to that of laser light sheet technique provides the wanted density information from the light intensity distribution scattered from the molecules of the fluid – nitrogen and oxygen- rather than from particles artificially brought into the flow. Problems connected with seeding the flow by the use of small particles are avoided. Of coarse sophisticated evaluation procedures have to be developed in order to correct for erroneous intensity contributions originating from solid surfaces. Fig. 7 depicts a final result when several individual instantaneous images are recorded, arithmetically averaged, and converted into density information by the use of calibration recordings [11]. The use of the powerful interpolation schemes of the DLR flow field analysis and visualisation tool COMADI results in the density presentation in a volume of the flow field above the delta wing, [12].

2. Planar model surface measurements

If data about the deformation of the wing due the aerodynamic load is needed the model deformation measurement system (MDMS), based on Moiré technique [13], can be applied, Fig. 8.

3. Surface pressure distribution measurements by pressure sensitive paint (PSP)

Today surface pressure data are available by a new optical method [14]. It is based on a thin layer of special paint molecules on the surface of the model. The paint is illuminated by UV light and emits fluorescence light in the visible range. As the fluorescence power of the molecules is lowered by the amount of oxygen molecules at the surface of the paint the local fluorescence intensity of the paint measures the density of the air or at known temperatures the local surface pressure. Pressure data are therefore available all over the surface rather than at the positions of individual pressure taps brought into the surface of the model.
It is obvious that the ability to obtain the complete surface pressure distribution in a period of time of about 10s is a big step forward to improve wind tunnel testing. Even if the accuracy of a local pressure information is not as good as in the case of pressure transducers the new kind of pressure information is an impact in the field of experimental simulation technique. Results are much more comparable with results known from numerical treatments. Actually the amount of experimental data exceeds the number of numerical data as each "pressure image" consists of about 500 000 local pressure data in comparison to about 50 000 in the case of a numerical result.

Of course aerodynamic research needs very accurate data. From the conventional pressure measurement technique at transonic Mach numbers an accuracy of a local pressure value in the range of Δc_p=0.01 is usual. If a model contains a total number of pressure taps of about 100 the statistical error drops to about Δc_p=0.001. Although the local accuracy of PSP technique may also achieve Δc_p=0.01 at parts of the model surface the local error may increase to about Δc_p =0.1 due to erroneous light intensity contributions. But if the overall statistical error is considered, the huge amount of data results in an error in the range of 1/700 of Δc_p =0.1, hence it will be below the statistical error of conventional technique.

Fig. 9 shows a PSP result [15,16] on a delta wing in comparison to a RANS calculation of Fritz [17]. As the PSP result combines the recordings of two cameras the pressure seems to locally jump at the centre line. New evaluation procedures will remove these local distortions. Never the less the overall statistical error of this pressure distribution is comparable with that of conventional pressure measurement techniques.

4. Conclusions

The 2D measurement technique provides a new quality of experimental result. This is because not only a certain amount of quantitative data are available. The new quality is based on the fact that 2D experimental results have a data density equivalent to that of numerical treatments or even more. Fig.9 yields the pressure distributions on a delta wing at M=0.8 and an angle of attack α=17°. Both figures show results close together. It is impossible to distinguish them from the point of view of data density. Actually the data density in the case of the experiment is much higher and had to be reduced in order to distribute them on the surface grid of the numerical treatment. The comparable availability of data allows for much better comparisons as both experimental and numerical data are present everywhere and may be visualised by the same software tool.

The other new quality of experimental research is to use mobile optical measuring systems together with advanced software and experienced experts in facilities wherever they are available. By this strategy economic conduction of tests with high degree of complexity are feasible. Following the strategic concept of "Measuring Teams", experts together with their equipment are requested by the organisation of a facility without the appropriate instrumentation. They prepare, conduct and evaluate tests in accordance with specifications of the customer.

DLR works already along such concept for several years. New optical techniques, first developed under laboratory conditions have then been applied in large facilities all over the world:

Flow visualisation by laser light sheet technique (LLS):

Germany:	DNW-NWB, DNW-TWG, DNW-KKK, ETW
The Netherlands:	DNW-LLF
France:	S1 ONERA Modane

Laser Doppler Anemometry (LDA):

Germany:	EGG, RGG, DNW-NWB, DNW-TWG, DNW-KKK, NGAV Uni Hannnover, LSWT DASA Bremen, NW TU Berlin, FK DaimlerChrysler, EWT Ford
The Netherlands:	DNW-LLF
France:	A 320 CF56 engine AI Toulouse,

Particle Image Velocimetry (PIV):

Germany:	DNW-NWB, DNW-TWG, LSWT DASA Bremen, NW TU Berlin, LW TU Stuttgart, VS Fa. Voith, VS Fa. Bayer
The Netherlands:	DNW-LST, DNW-LLF
France:	Catapult ONERA Lille

Pressure Sensitive Paint (PSP):

Germany:	DNW-NWB; DNW-TWG, LSWT DASA Bremen, FK DaimlerChrysler, MTU
The Netherlands:	DNW-HST
India:	NAL Bangalore
Japan:	NAL

Model deformation measurement system (MDMS) by Moiré Interferometry:

Germany:	DNW-TWG, ETW
The Netherlands:	DNW-LLF

References

[1] Seelhorst, U.; Bütefisch, K.-A.; Weiland, M.; 3 Component LDV for Near Wall Measurements on a Multi-Element High Lift Airfoil Configuration. 16th ICIASF Congress, Dayton (USA), pp. 167-173 (1995).

[2] Möller, T.J., Bergmann, A.; Experimental investigation of the flow close to the surface of a high lift airfoil configuration in a large wind tunnel, STAB Symposium, Stuttgart 2000.

[3] Förstermann, M.; Messung der Momentanwerte aller drei Geschwindigkeitskomponenten eines Wirbelsystems mittels der Doppler Global Velozimetrie. DLR-IB 223-98 A 32 (1998).

[4] Kompenhans J., Reichmuth J.; Particle imaging velocimetry in a low turbulent wind tunnel and other flow facilities. Proc. AGARD Conference on Advanced Instrumentation for Aero Engine Components, 19-23 May, Philadelphia (AGARD-Cp 399-35) (1986).

[5] Raffel, M.; Willert, Ch.; Kompenhans, J.; Particle Image Velocimentry - A Practical-Guide. Springer Verlag Berlin, Heidelberg, New York (1998).

[6] Hegen, G. H.; Puffert-Meisner, W.; Dieterle, L.; Vollmers, H.; Flow field investigation on a wing installed Counter Rotating Ultra-high-bypass Fan engine simulator in the low speed wind tunnel DNW-LST. 7th European Propulsion Forum on Aspects of Engine/Airframe Integration, Pau (F), 10.-12. März 1999, National Aerospace Laboratory-NLR (1999).

[7] Dieterle, L.; Stuff, R.; Vollmers, H.; Coton, P.; Wake Vortex Studies in the ONERA Catapult Facility and in the DNW Wind Tunnels. 1st ONERA-DLR Aerospace Symposium, Paris (F), ONERA (1999).

[8] Kompenhans, J., Raffel, M., Dieterle, L., Dewhirst, T., Vollmers, H., Ehrenfried, K., Wilert,C., Pengel, K., Schröder, A. and Ronnberger, O.; Particle Image Velocimetry in Aerodynamics: Technology and Applications in Wind Tunnels. The Visualization Society of Japan and Ohmsha, Ltd. Journal of Visualization, Vol. 2, Nos. 3-4, pp. 229-244 (2000).

[9] Kompenhans, J., Dieterle,L., Vollmers, H., Stuff, R., Schneider, G., Dewhirst, T., Raffel, M., Kähler,C., Monnier, J.-C., Pengel, K.; Aircraft wake vortex investigations by means of particle image velocimetry: Measurement technique and analysis methods. Proceedings 3rd International Workshop Particle Image Velocimetry. Santa Barbara, California (1999).

[10] Kähler, C. J.; Adrian, R.J.; Willert, C.E.; "Turbulent boundary layer investigations with conventional- and stereoscopic particle image velocimetry", Proc. 9th Intl. Symposium on Appl. of Laser Techniques to Fluid Mechanics, Lisbon, paper 11.1. (1998).

[11] Bütefisch, K.-A.; Schmidt, V.; Vollmers, H.; Wolfrum, G.; Krämer, E.; Density Measurements on a Delta Wing by Means of Planar Intensity Distribution of Rayleigh Scattered Laser Light (PIRS) in the 1 m x 1 m Transonic Windtunnel Göttingen (TWG). Notes on Numerical Fluid Mechanics, Vol. 60, pp. 87-94 (1997).

[12] Vollmers, H.; COMADI – Handbuch 1998. DLR-IB 223-98 A 30 (1998).

[13] Baumann, P. H.; Bütefisch, K.-A.; Measurement of hinge moments and model deformations in wind tunnels by means of Moiré interferometry. SPIE International Symposium on Optical Science, Engineering and Instrumentation, San Diego (USA), pp. 16-32 (1995).

[14] Engler, R. H., Hartmann, K., Troyanovski, I. & Vollan, A.; Description and assessment of a new optical pressure measurement system (OPMS) demonstrated in the high speed wind tunnel of DLR in Göttingen, DLR-FB 92-24 (1992).

[15] Engler, R. H.,Fonov, S., Klein, Ch., Bütefisch, K.-A., Bock, K.W., Weiskat, D., Fritz, W.; Study on unsteady vortex behaviour of a rotating 65° delta wing at M=0.8 using Pressure Sensitive Paint (PSP), SYMPOSIUM on Advanced Flow Management: Vortex Flows and High Angle of Attack - Military Vehicles -, Loen, Norway 7-10 May (2001).

[16] Bruse, M.; Bock, K-W.; Tusche, S.; Jacobs, M; Unsteady Measurements with the Continuously Rotating DLR-PSP-Model at the Transonic Windtunnel Göttingen (DNW-TWG), STAB Symposium, Stuttgart (2000).

[17] Fritz, W., Arthur, M. T., Brandsma, F. J., Bütefisch, K.-A., Ceresola, N.; Time Accurate Euler Calculations of Vortical Flow over a Delta Wing in Rolling Motion Wing in Rolling Motion, SYMPOSIUM on Advanced Flow Management: Vortex Flows and High Angle of Attack - Military Vehicles -, Loen, Norway 7-10 May (2001).

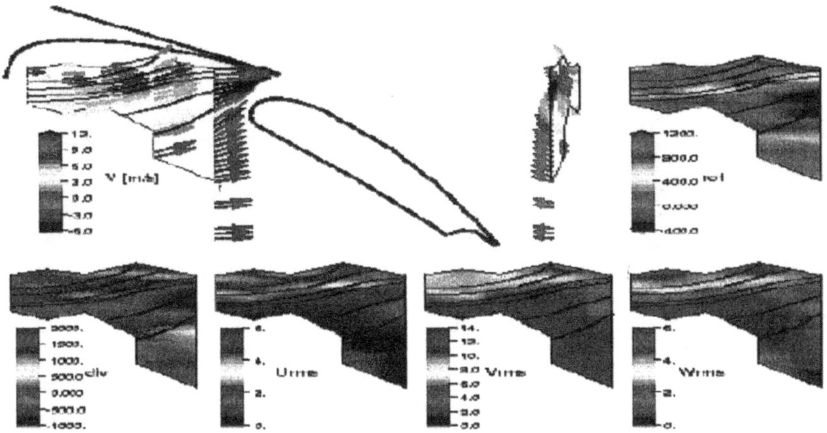

Fig. 1: 3 D LDA in the 6m x 9m test section of the DNW-LLF taking data on a Airbus A320 model.

Measuring distance: 3.80 m
Spatial resolution: 0.25 mm in the vicinity of the wing and flap.

Results: 3 D velocity profiles, distributions of 3D turbulence; curl, divergence, and stream lines between main wing and flap

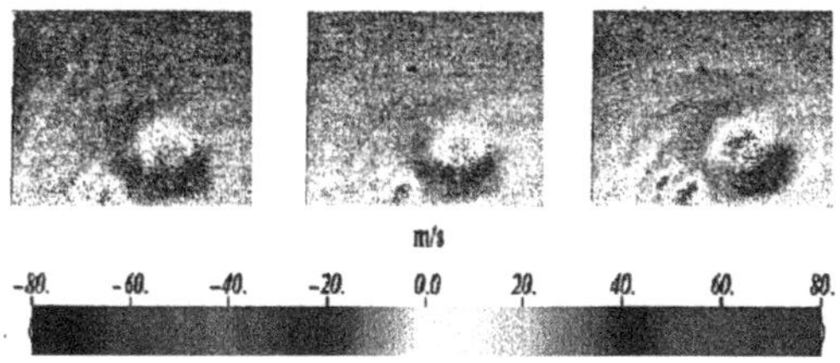

Fig. 2: A sequence of successive DGV images of a vortex core above a delta wing obtained with an exposure time of 1/60.

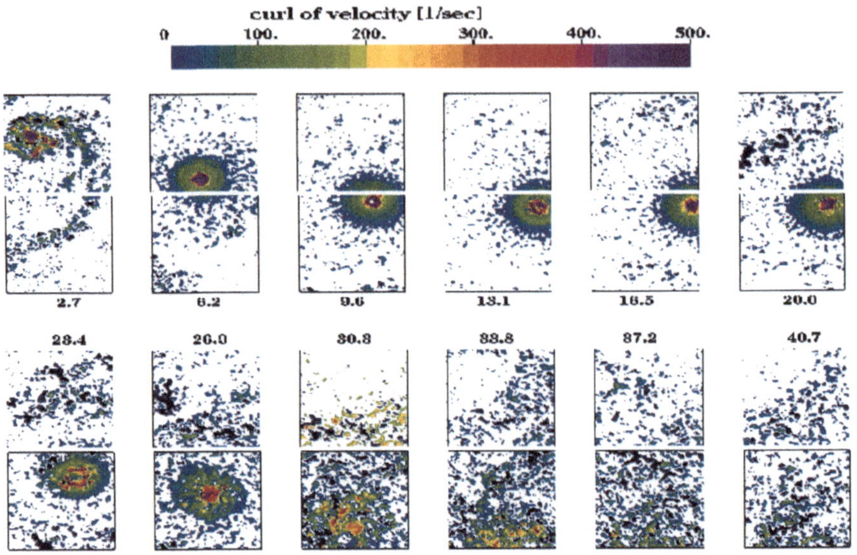

Fig. 3: Vorticity distribution of a vortex at several spanwise stations behind an Airbus model obtained in co-operation with ONERA in the catapult facility of ONERA Lille.

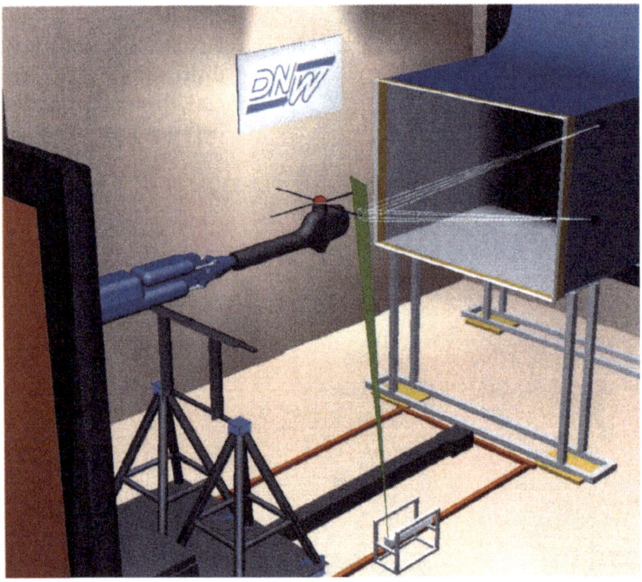

Fig. 4: PIV set up in the 6 m x 9 m test section of the DNW-LLF. The laser is sited at the wind tunnel floor, two cameras were fixed to the nozzle. Velocity measurements in the plane of rotation of a helicopter model in 10 m height above the ground.

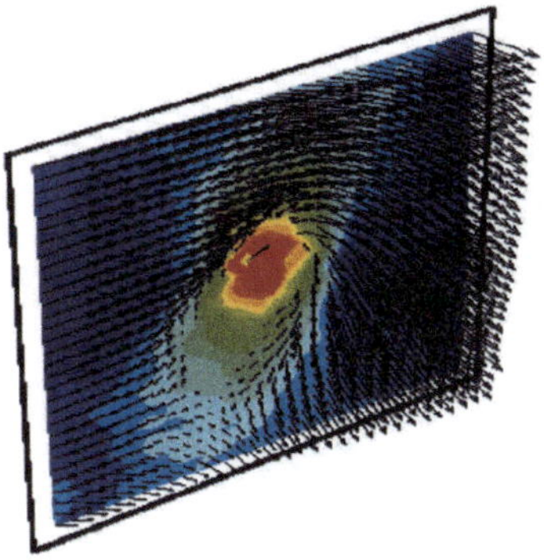

Fig. 5: Result of 3 components (stereoscopic) PIV technique showing the tip vortex of a wing. The arrows represent the third component in axial direction, the colours the total velocity of the remaining other velocity components respectively. The result has been obtained in the DNW-LLF tunnel within the frame of the EU project EUROWAKE on a wing of the DLR ALVAST model.

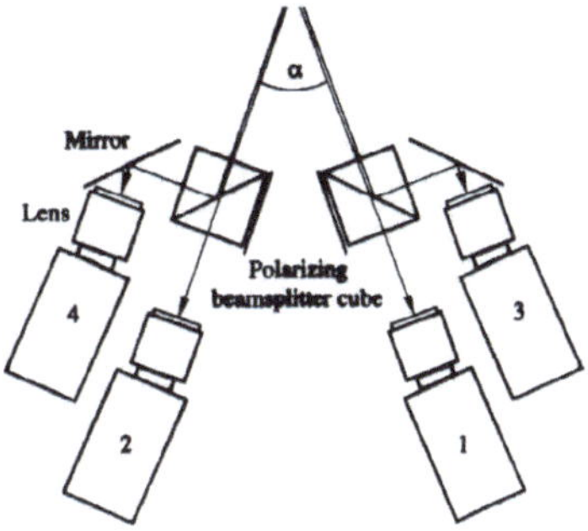

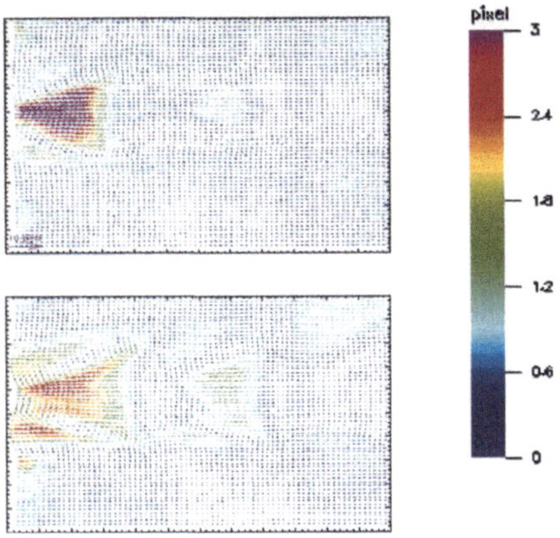

Fig. 6a: Set up for multiplane PIV measurements in the boundary layer of flat plate flow.
Fig. 6b: Velocity field with Λ-vortices recorded simultaneously. (upper field closer to wall). Flow from right to left: U_∞=10 m/s. Mean Velocity subtracted. 3 pixels correspond to a velocity magnitude interval of 3m/s.

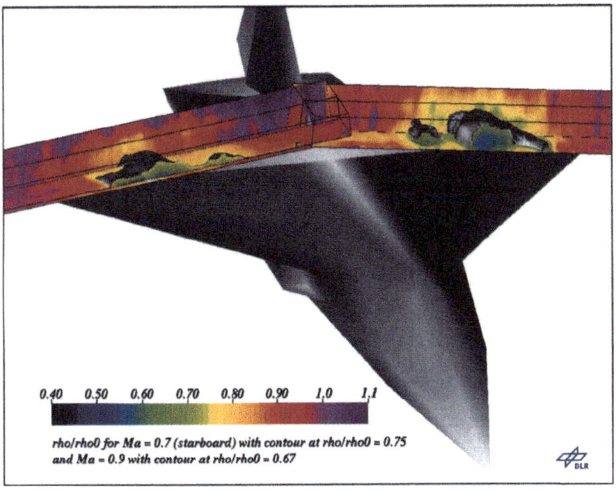

Fig. 7: Several individual instantaneous PIRS images, arithmetically averaged, converted into density information, processed by powerful interpolation software COMADI yield the 3D density distribution of the flow field above the delta wing.

M=0.7 left

M=0.9 right

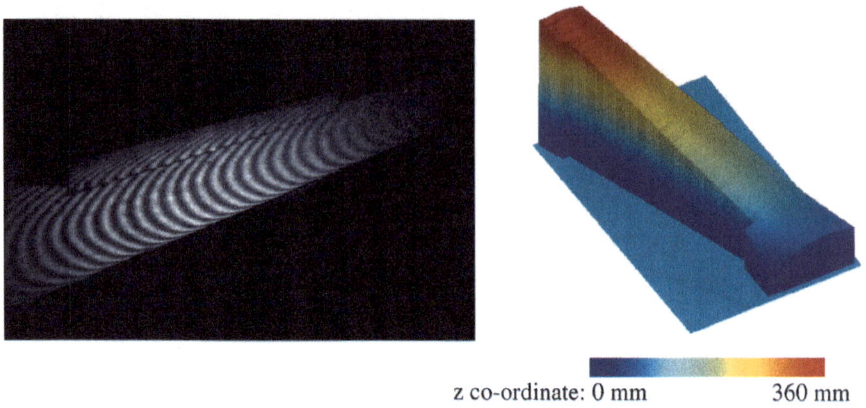

Fig. 8: Interferogram (left) and reconstructed surfaces (right) of the wing of DLR ALVAST model in the DNW-LLF at angle of attack of 16°. Displayed area in xy-plane: 1.3 m x 0.9 m.

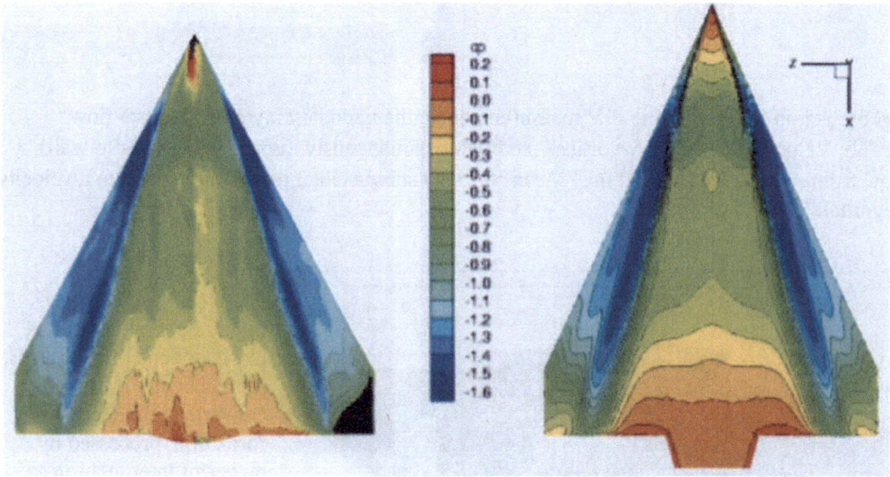

Fig. 9: PSP result
EADS-M [17] at the same conditions with Wilcox k-ε turbulence model.

Investigation of Transitional Structures in Artificially and Disturbed Excited Flat Plate Boundary Layer Flows Using Stereo and Multi-Plane PIV

Andreas SCHRÖDER and Jürgen KOMPENHANS
Institut für Aerodynamik und Strömungstechnik, DLR,
Bunsenstr. 10, 37073 Göttingen, Germany
Email: Andreas.Schroeder@dlr.de

1 Summary

Stereoscopic (SPIV) and Multi-plane PIV (MPIV) measurements on controlled excited transitional boundary layer flows along a flat plate will be presented. The experimental results are focused on the structures of the late stages of K- and oblique type transition and on the substructures and the development of a turbulent spot induced by a short blow injection. Instantaneous velocity fields are examined in order to quantify the local and global dynamics of the structures leading into breakdown. For the K-type transition the multi-spike characteristics of the wall normal motion on top of the Λ-vortices is quantified as one of the first sources for shear stress decomposing the transitional structures towards turbulence. For the oblique type of transition one source for shear stress is found in the large velocity gradients at the crossings of high- and low-speed streaks. The analysis of the substructures of the turbulent spots leads to a model of a cascade of vortices with a significant ω_y-component leading to turbulence with characteristic streaky structures inside the spot.

2 Introduction

The fluid-mechanical research on transition of boundary layers has become an important field for many different time and space resolving measurement techniques and especially for (spatial) Direct Numerical Simulations in the last decades. In addition to all sensor based time recording and averaging methods the non-intrusive Stereo and Multi-plane PIV technique is able to determine all three velocity components in a plane of the flow instantaneously with a high spatial resolution. Aligning a second light sheet plane with perpendicular light polarization it is possible to get additional PIV information about the flow field in this plane which can be seperated to the first plane in time or space. Particularly for the sensitive non-linear stage of the boundary layer transition the non-intrusive PIV technique is a suitable and reliable instrument for the quantitative visualization of transitional flow structures and their behaviour close to the turbulent breakdown. For many different boundary conditions the transitional process itself seems to generate similar Λ-shaped structures with a characteristic behaviour. The SPIV technique is able to give information about instantaneous velocity distributions of typical transitional flow structures as there are Λ-vortices, streaks, shear layers, local spikes or the global subdivision in valley- and peak-planes. However, the development of these structures forward to breakdown, which is connected to the turbulence initiating high frequency processes, can only be detected as a result of intrinsic flow processes taking place before. In the literature four different artificially excited boundary layer transition scenarios have been distinguished. The fundamental, harmonic or K-type, the subharmonic, H- or N-type, recently the oblique-type transition and as an example for Bypass-transition the *turbulent spot*. Each scenario is characterised by a typical arrangement of flow structures as presented in the sketch of Fig. 1. Beside investigations of paths to transition with small amplitude disturbances, where the linear growth of Tollmien-Schlichting-waves leads to secondary instabilities with strong amplification rates due to non-linear interaction of different modes and finally to breakdown, the effect of high amplitude and localized disturbances are examined widely under the title of Bypass-transition. Mostly *turbulent spots* are generated by pulsed disturbances introduced into the boundary layer flow by acoustical, mechanical or electrical means. Artificially generated turbulent spots grow downstream of the leading edge even in cases of subcritical Reynolds numbers. They are also appearing in all "natural" transition-scenarios in a stochastic manner caused by small local disturbances which grow rapidly in the unstable boundary layer flow, which explains the continuous fluid mechanical interest in their characteristics. DNS of turbulent spots have up to now only been made for the early stages of development [1] and for plane Poiseuille or Couette flow [2] [3], but with remarkable similarities to boundary layer spots. In the literature the turbulent spot in a laminar boundary layer has often been investigated by means of visualization techniques und hot-wire anemometry [4] [5] [6]. The models for an understanding of the development of spots are so far restricted to phenomenological viewpoints and the results of time or ensemble averaged measurements. The Multi-plane Stereo PIV technique [7] is able to quantify the velocity distribution instantaneously and give e.g information about the Reynolds stresses and the acceleration of the substructures with a high spatial resolution. The aim of this short paper is to show prelimenary results of the MPIV measurements on turbulent spots and give an overview of the possibilities of our quantitative visualization technique for late transitional structures. A fuller paper, including a detailed description of the MSPIV technique and the experimental setup, with all results and derived fluid mechanical statements will be presented soon.

3 Experimental set-up and evaluation

The experimental set-up for K- and oblique transition has been realized in the low turbulence wind tunnel (TUG) at DLR Göttingen. The vertically mounted flat plate had an elliptic leading edge and spanwise embedded slots, a 420 mm long slot at x = 194 mm for 2D excitation and 40 10 mm wide slots at x = 206 mm for oblique waves excitation. The slots are connected to pressure chamber loudspeakers in order to introduce controlled acoustic sinusoidal wave disturbances into the boundary layer flow (see Fig. 2). The turbulence level for the used free stream velocity of 12 m/s was at 0.065 %. The K-type transition was initially disturbed by a dominating 150 Hz sinusoidal 2D wave with two weak 60° phase shifted oblique waves, which fixed the spanwise positions of the Λ-vortices at a distance of $z \approx 32$ mm. The Oblique-type transition was performed through two 90 Hz 60° phase shifted oblique waves, equal to $(1, \pm 1)$-modes in the frequency-spanwise wave number spectrum. The measurement volume was situated between 570 and 640 mm downstream of the real leading edge and its field of view varies between 42×36 mm and 80×62 mm. The Laser light sheet position was adjusted at about 0.9 and 1.2 mm wall distance, at a boundary layer region with the highest u_{rms}- values during the transitional process. In order to measure the development of a *turbulent spot* a Multiplane PIV setup was arranged with a 1 msec pulsed strong acoustic disturbance injected into the boundary layer flow. The small 2×0.3 mm slit source was situated 206 mm downstream of the leading edge. The free stream velocity in this experiment was 7 m/s and a slightly adverse pressure gradient along the plate has been adjusted, to accelerate the generation of the turbulent spot in the measurement volume between 205 and 425 mm downstream the disturbance source, which corresponds to a small range of Re_x from $1,91 \times 10^5$ to $2,99 \times 10^5$ where x is the distance from the virtual leading edge of the plate. The laser light sheet has been adjusted at 4.5 mm wall distance for 2D-MPIV and 6 mm for the Stereo MPIV measurements. The stored stereoscopic and multiplane PIV data-sets have been de-warped and evaluated with a FFT based cross-correlation method with a 32×32 window size and a Levenberg-Marquardt peak fitting algorithm, which determines a vector field with over 12000 vectors in sub-pixel accuracy and minimizes the peak-lockingëffect [8]. In order to determine all three velocity components in a plane of the flow, a reconstruction of the back-projection of the particle image displacement into the light sheet plane has to be done after de-warping and evaluation of the single PIV images of each camera. Physically unreasonable vectors (less than 1% for each data set) were deleted and re-interpolated by a fit to their next neighbors. As the last evaluation step the conversion into physical units was calculated.

4 Results

In general the described PIV results of K- and oblique-transition, which are quantitative visualizations of structures of the late non-linear stage in one wall-parallel plane, confirm the problem of receptivity even in this case of controlled acoustical excitation and fixed experimental boundary conditions: The periodicity and especially the stadium of amplification of the late transitional structures are, within given limits, affected sensitively and stochastically by unknown initial boundary conditions and the growing (0,0)-mode. Therefore, reliable statistical statements even for the phase locked SPIV measurements need a higher number of images than could be captured in the present work. Nevertheless, the results comprise information about different stages of coherent structure development, effects of asymmetry and all three velocity components in a plane parallel to the plate. The results enable a quantitative comparison with point measurements of time recording methods or the necessarily symmetrical data of DNS, but this requires careful considerations. The turbulent spot measurements on the other hand are not very sensitive to slightly changing conditions, because the growing turbulent spot already "forgot" the feature of the initial disturbance and seems to have a stable and reproducible path of development, which varies only in shape and velocity for different pressure gradients and changes in free stream velocity. The destabilization of the unstable laminar boundary layer in the vicinity of the turbulent spot, especially the downstream transitional wake region we detected, may be an effect of the special boundary conditions.

4.1 K- and Oblique-type transition

Since fundamental work on the flat plate boundary layer transition in the mid of the last century the K-type transition [9] [10] is known as an artificially induced breakdown scenario with streamwise arranged Λ-vortices. The simple wave structure model of the disturbances and their downstream effects gave researchers the possibility to analyze the characteristics of growth-rates, non-linear secondary instabilities [11] and the resonant nature [12] of the transitional structures by abstracting from irrelevant and not known boundary conditions in the transitional regime. The presented SPIV results directly extend the experimental work of Wiegel [13] under the same conditions but for the later non-linear stages of K- and oblique-type transition generated through higher disturbance amplitudes and now employing Stereo PIV. Vector fields with subtracted u-components (noted as U_{ref}) in order to visualize the flow structures are presented in the following figures. A beginning two-spike development on top of the downstream growing Λ-vortex is shown in the vector field of Fig. 3. In such a spike low speed fluid from near-wall regions is transported in wall normal direction producing strong shear layers (velocity changes of 5 m/s within 0.5 mm interval in this wall parallel plane). These spikes are the first step of an energy transfer from the mean flow in the direction of the wall and coincide with turbulence initiating high-frequency processes. For the fundamental or K-type transition the location, multi-spike characteristics and wall normal motion on top of the Λ-vortices as one of the first sources for shear stress decomposing the transitional structures towards turbulence agree quite well with former investigations and similar spatial DNS results [14]. Also, a good correspondence is found for the oblique type transition [15], where non-linear generated high- and low-speed streaks, equal to (0, 2) modes in the frequency-spanwise wave numbering, cause shear stress due to large velocity gradients at their crossings. The oblique type of transition has the special feature that only 3D disturbances, in the streamwise-spanwise wave numbering (1, ±1), cause the non-linear development of growing high and low-speed streaks (0, 2) as shown in Fig.3. The structure of the S-shaped alternating streaks changes from a wavy to a streaky appearance by an increase of the disturbance amplitude. Λ-shaped structures inside the low-speed regions can be detected in a global arrangement, which is comparable to the H-type transition. This spanwise displaced arrangement of Λ-shaped structures is divided by high-speed streaks, but streamwise vortices at the crossings of the streaks as assumed in [15] cannot be found (no significant wall-normal velocity).

4.2 Turbulent spots

Turbulent spots are known since the mid of the last century [4] as arrowhead-shaped growing turbulent regions in a laminar boundary layer with a stochastical appearance in "natural" transitional flows. Complementary to former investigations with qualitative visualization techniques and time-resolving point measurement methods we want to present first data of the MPIV measurements of a turbulent spot in a slight adverse pressure gradient flat plate boundary layer flow. The substructures of the spot are wavy and streaky as seen in Fig. 4. The results of the data-analysis lead to the assumption that an absolute instability at the trailing edge of the spot is the reason of its growing: Hairpinlike vortices are generated at wallnear streaks in the upstream region of the spot and have a favorable upstream-effect on the following vortex development, it seems to be a *self-inducing* process. For this experimental setup a spreading half angle of 7.7° and a trailing edge velocity of 0.62 U_∞ have been evaluated for the spot. Stereo MPIV results with a view into the substructures of the turbulent spots show cascades of vortices transporting low speed fluid in wall normal direction interrupted by planes with a flow direction from the mean flow to the wall (see Fig. 5). At present data evaluation is still in progress. Different time-separated measurements in the same plane enable to obtain acceleration data of the substructures and time-space correlations of structure-ensembles. The assumption of Sankaran et al. [16] that the hairpin vortices originating at the trailing edge are the fundamental structure-forming process were confirmed with this data set. More details on turbulent spot results and their interpretation will be presented in future.

5 Discussion

SPIV and MPIV measurements on excited and disturbed boundary layer flows for different transitional scenarios have been performed. The K- and oblique-type transition were generated at a flat plate with zero pressure gradient and turbulent spots were produced on the same plate with a slightly adverse pressure gradient. The SPIV data-set contains information about the instantaneous velocity distributions of all three velocity components in a plane of the transitional flow, complementing time-resolving methods and spatial DNS. Although the breakdown processes for K- and oblique transition starts with the non-linear amplification of different waves or wave-packages a non-vanishing similarity has been found in our data set: A differently weighted appearance of streaks and Λ-vortices can be detected, whereby the oblique-type is dominated by streaky structures and the K-type by Λ-structures. The turbulent spot measurements are the first quantitative spot visualizations using the MPIV technique. The preliminary results confirm former investigations and enable a connection to basic equations of fluid mechanics, which will be presented in a fuller paper soon.

References

[1] Singer, B., Characteristics of a young turbulent spot. *Phys. Fluids.*, 8 (2), 509–521 (1996).

[2] Henningson, D. and Kim, J., On turbulent spots in plane Poiseuille flow. *J. Fluid Mech.*, 228, 183–205 (1990).

[3] Lundbladh, A. and Johansson, A., Direct simulation of turbulent spots in plane Couette flow. *J. Fluid Mech.*, 229, 499–516 (1991).

[4] Emmons, H., The Laminar-Turbulent Transition in a Boundary Layer. *J. Aero. Sci.*, 18, 490–498 (1951).

[5] Gad-el-Hak, M., Blackwelder, R. and Riley, J., On the growth of turbulent regions in laminar boundary layers. *J. Fluid Mech.*, 110, 73–95 (1981).

[6] Wygnanski, I., Zilberman, M. and Haritonidis, J., On the spreading of a turbulent spot in the absence of a pressure gradient. *J. Fluid Mech.*, 123, 69–90 (1981).

[7] Kähler, C. and Kompenhans, J., Multiple plane stereo PIV-Technical realization and fluid-mechanical significance (1999), Paper PIV96, 3rd Int. Workshop on PIV, 16-18 Sep. 1999, Santa Barbara, USA.

[8] Ronneberger, O., Raffel, M. and Kompenhans, J., Advanced evaluation algorithms for standard and dual plane particle image velocimetry. p. 10.1 (1998), 9th Int. Symp. on Appl. of Laser Tech. to Fluid Mech., Lisbon.

[9] Schubauer, G. and Skramstad, H., Laminar boundary-layer oscillations and transition on a flat plate. 38 (1947), National Bureau of Standards, Research Paper RP1772.

[10] Klebanoff, P., Tidstrom, K. and Sargent, L., The three-dimensional nature of boundary-layer instability. *J. Fluid Mech.*, 12, 1–39 (1962).

[11] Herbert, T., Secondary instability of boundary layers. *Ann. Rev. Fluid Mech.*, 20, 487–526 (1988).

[12] Kachanov, Y., On the resonant nature of the breakdown of a laminar boundary layer. *J. Fluid Mech.*, 184, 43–74 (1987).

[13] Wiegel, M., Experimentelle Untersuchung von kontrolliert angeregten dreidimensionalen Wellen in einer Blasiusgrenzschicht (1997), VDI, Reihe 7, Nr.312.

[14] Meyer, D., Rist, U. and Wagner, S., DNS of generation of secondary lambda vortices in a transitional boundary layer. pp. 97–100 (1998), Advances in Turbulence 7, Uriel Frisch (Ed.), Vieweg Verlag.

[15] Berlin, S., Wiegel, M. and Henningson, D., Numerical and experimental investigations of oblique boundary layer transition. *J. Fluid Mech.*, 393, 23–57 (1999).

[16] Sankaran, R., Sokolov, M. and Antonia, R., Substructures in a turbulent spot. *J. Fluid Mech.*, 197, 389–414 (1987).

Figures

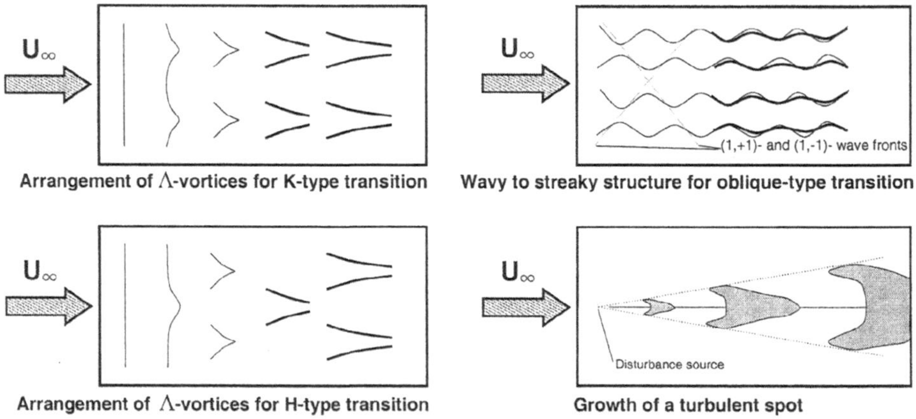

Figure 1: Flow structure arrangement of four different types of transition

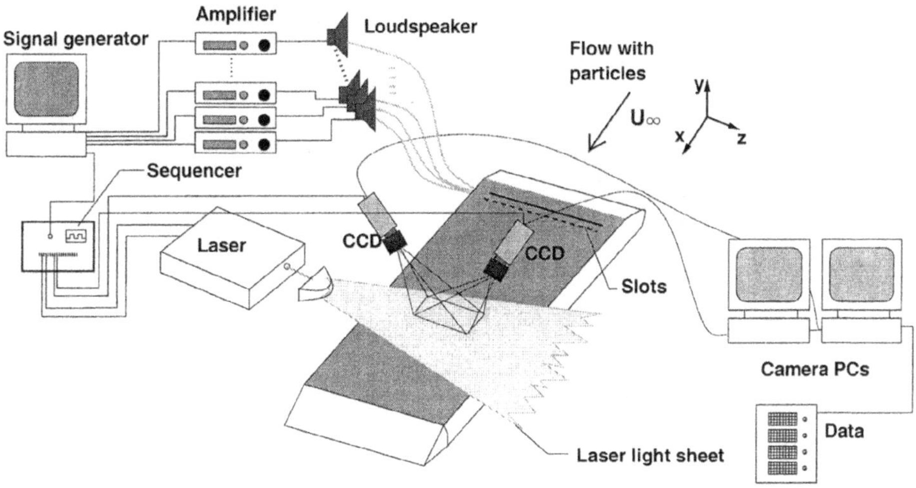

Figure 2: Experimental setup at the Low Turbulence Wind Tunnel (TUG)

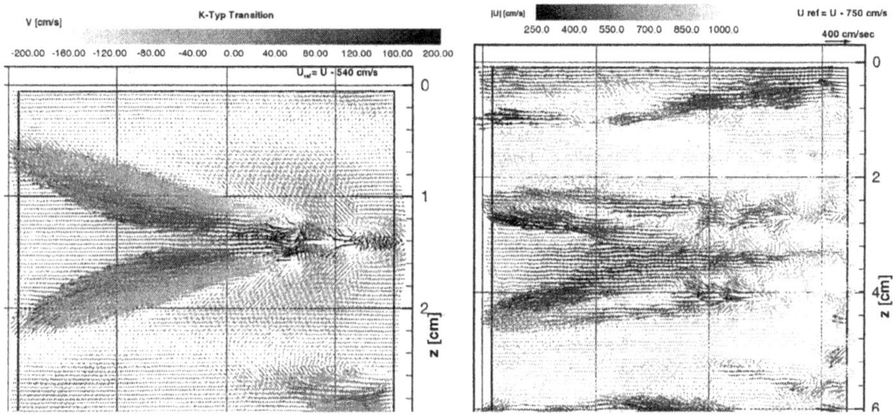

Figure 3: Velocityfields of K-type (left) and oblique-type transition (right)

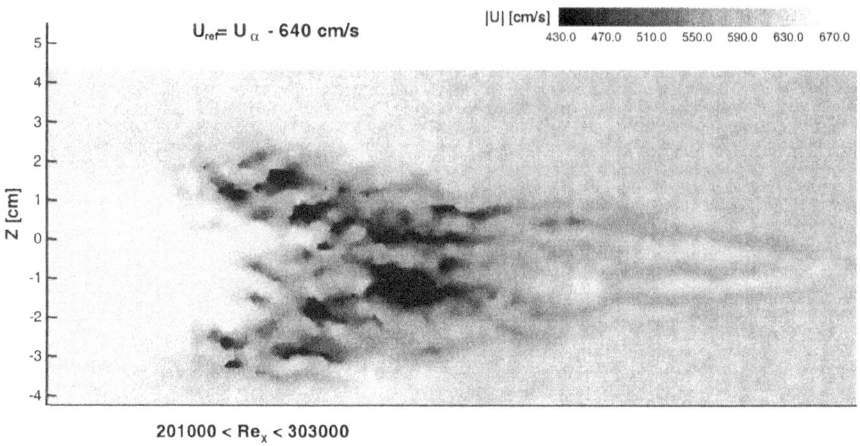

Figure 4: Velocity contour field of a turbulent spot, $U_\infty = 700$ cm/s

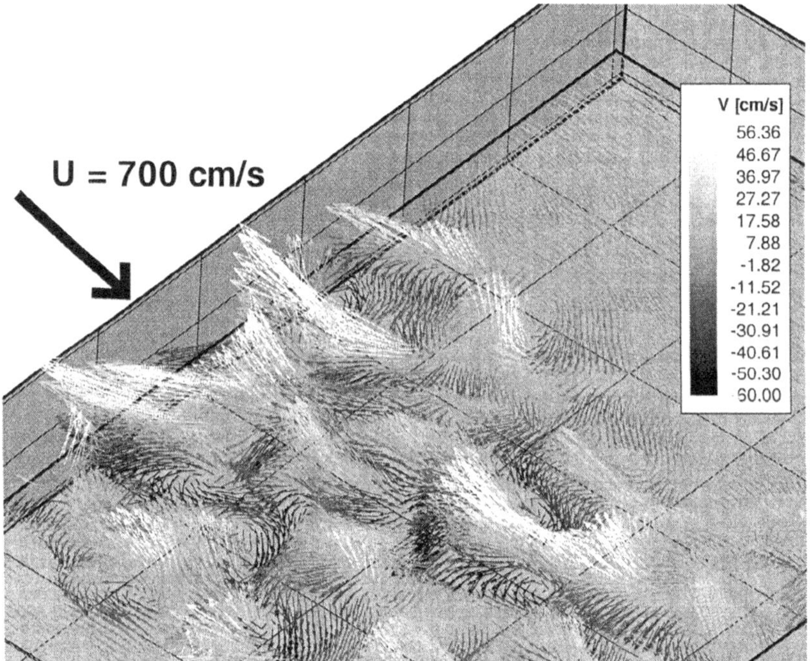

Figure 5: SPIV view into the substructures of a turbulent spot (v-com. b/w)

The Meaning of Disturbance Fields in Transition Experiments and Their Detection in the Test Section Flow of a Short Duration Wind Tunnel

U. Gaisbauer, H. Knauss, S.Wagner, J. Weiss
Institut für Aero- und Gasdynamik, Pfaffenwaldring 21, 70550 Stuttgart, Germany

Summary

A short clarification about the meaning of wind tunnel noise on transition measurements is given. Possible sources of additional disturbance modes in a short duration facility are discussed. The validation of a specially designed CTA fast scanning device for the detection of all possible flow disturbance modes in the facility was performed. First measurements in the wind tunnel revealed in a confined modal analysis a dominating acoustic disturbance field only. A temperature fluctuation could not be quantified up to now but seems to be of a lower order of magnitude like possible vorticity. Differences in static pressure fluctuations and deduced by the means of the different detection techniques are going with the uncertainty of the methods given in literature.

1 Introduction

In the design of a hypersonic vehicle equipped by an airbreathing propulsion system, boundary layer (BL) development and transition location play a dominant role in the shape finding of the forebody. The latter is influencing skin friction as well as heat load onto the structure. An unambiguous wind tunnel simulation, transferable to free flight conditions is extremely difficult, if not even impossible [1]. The reason for the transmission problem is due to the receptivity behavior of the model BL to disturbance fields superimposed in the test section flow, which will not exist in such a kind under atmospheric flight conditions or at least with a much lower intensity [2], [3]. Besides smaller transition Reynolds numbers (Re_t) by a factor of $3 \div 4$ in comparison to free flight condition also the transition extent is reduced from about $85 \div 50\%$ in a conventional tunnel [4]. On the other hand also a geometry dependent anomaly was found under so called "noisy" and "quiet" test section flow conditions, respectively in conventional and low disturbance wind tunnels. Thus, while Re_t-values are lower in the conventional tunnel for a planar BL in comparison with the conical BL, in the quiet tunnel under low noise condition just the opposite trend exists. Therefore a designer relying on the conventional tunnel data would create a conical forebody (i.e. Sänger, first stage) in order to delay transition, while the designer taking the quiet data will prefer a more 2-dim., wedge shaped forebody (Hyper-X 43). Consequently the significance of the disturbance fields in the test section flow becomes obvious especially in transition measurements concerning forebody design [5].

Disturbance modes in wind tunnel test section flow are sound waves, entropy spottiness and vorticity. The first mentioned mostly have their origin in the turbulent boundary layer of the test section walls and their intensity is proportional to square of the free stream Mach number [6]. All three kinds of disturbance modes can also have their origin in the "prehistory" of the flow preparation, i.e. in the control valve for mass flow rate or

in flow stratifying construction elements in the settling chamber for a homogenization of the influx to the test section. In such a situation the disturbance fields are carried by the core flow into the test section [7].
The receptivity behavior of the model BL to the different disturbance modes is depending on a lot of parameters, up to now unclear and therefore subject to numerous experimental and theoretical research. For a meaningful comparison of results achieved from both procedures, wind tunnel experiments must include "complete boundary conditions" involving the "escorting disturbance fields" their kind of character as well as their intensity and spectral distribution. Only by means of such additional information of the test section flow disturbances, Re_t-values found are useful and can be put into relation to Direct Numerical Simulations (DNS). At the Institute of Aero- and Gasdynamic (IAG) at Stuttgart University, experimental investigations of flows about compression surfaces with integrated compression ramps in forebodies for hypersonic vehicles are subject of research. Because BL thickness plays a dominant role in the shock BL interaction of the ramp flow and is dependant on transition location too, such kind of experiments involve also the difficulty of a transferable Re_t-value simulation. For a better interpretation and validation of Re_t-values, therefore escorting disturbance field detection must be performed simultaneously. Such a requirement presents under existing circumstances a challenging measuring task of its own and is subject of this paper.

2 The short duration wind tunnel

2.1 Main features of Shock Wind Tunnel

The aforementioned forebody studies are performed in the so called shock wind tunnel (SWK) at IAG of Stuttgart University [8], [9]. This short duration facility is in principle a shock tube with a Laval nozzle and produces a sequence of two steady flow states of $2 \times 120ms$. Because of the large test section (nozzle exit: $1.2 \times 0.8m^2$) investigation of models with large dimensions can be carried out in a Mach number range of $M = 1.75 \div 4.5$ and corresponding unit Reynolds numbers of $R/m = 70 \div 30 \times 10^6$. Reynolds numbers, based on the characteristic length (forebody tip to compression ramp), equivalent to original flight conditions can be simulated. The pure size of the tunnel is promoting for a noise level reduction in the test section [10], and the use of a rapid expansion nozzle design additionally minimizes at least sound emission radiated from a turbulent BL along the nozzle contour [11].

2.2 Analysis of possible disturbance sources in the SWK

In consequence of the working principle for flow generation by means of the unsteady propagating expansion and shock wave like in a Ludwieg tube, the sequence of the flow is controlled only by a gas-dynamic process. Starting and preparation of the flow in the driver tube upstream of the test section, determining the stagnation condition of the flow, is induced in a pure isentropic change of state coming out from the initial condition after the charging period of the tube. No entropy spottiness, vorticity or sound wave formation is associated in this process and the induced flow reaches without special treatment the test section where a stationary expansion to supersonic flow succeeds in the nozzle. Because of the large tube diameter ($2.1m$) and by comparison of a small thickness of the unsteady growing BL in the wake of the expansion wave, reaching only 2% of the diameter, there is reason to believe that a laminar state exists in the core flow of the driver tube.
But such a presumption implied that there are no disturbances at all in the state of

"quiescence" immediately after the charging period of the driver tube, which must be put in question out of different reasons:
1.) Typical loading cycles of the driver tube are always combined by uncontrolled temperature rise ($\approx 50°C$) in reference to the ambient temperature which is reduced by convective heat exchange with the pipe wall. Consequently convection current is induced and convective cells and vortex structures of large scale are possible. Turbulence production and irregularities in the temperature distribution result in the "state of quiescence".
2.) Additionally in the influx of the compressed air, turbulence is produced mechanically, which will be compared with that of a convective origin of a much smaller scale. The decay time till down to energy dissipation (Kolmogorov scale) is unknown and difficult for an estimation. Therefore in the relevant periods of time from the loading process till to the release of the run, a remaining finite vorticity and entropy spottiness can be expected in the core of the test section flow.
These two uncontrollable phenomena possibly leading to temperature fluctuation and turbulence must be determined at least to their order of magnitude. Only with a possible neglecting of these almost uncontrollable disturbance modes, a reproducibility of the "complete boundary conditions" for transition experiments can be guaranteed. Because the disturbance fields produced in a typical loading cycle are unique and of a stochastic nature, their detection must be performed within one run ($120ms$) to enable a reliable statement about random series and their influence to the flow quality.

3 Detection techniques for flow disturbance fields

For the detection of disturbance modes in supersonic flow up to now no standardization exists concerning the performance of such kind of measurements which allows a reliable comparison of disturbance levels measured in different wind tunnels [12]. In principle two techniques are used for fluctuation measurements in the free stream.

3.1 Pitot pressure fluctuation measurements

In measuring this quantity no unique relationship exists between measured Pitot pressure and the static pressure fluctuation which is the decisive quantity to describe acoustic waves [13]. In fact one cannot extract anything useful out of the fluctuating voltage output of a Pitot probe unless certain assumptions about the character of the flow itself are made which only can be obtained by an alternative detection technique described further below. A classic interpretation of the fluctuating Pitot pressure signal comes from a quasi static analysis applied to the Rayleigh supersonic Pitot formula. Concerning this transformation between <*effective*> static and Pitot pressure fluctuation and refering to [9], [14], following "quasi steady" and "unsteady" relations can be defined respectively

$$< p_\infty >_{steady} = \left(\frac{p'_{\infty,rms}}{\overline{p_\infty}}\right)_{steady} = K\left(\frac{p'_{t2,rms}}{\overline{p_{t2}}}\right)_{steady} = K < p_{t2} >_{steady} \quad (1)$$

$$< p_\infty >_{unsteady} = \left(\frac{p'_{t2,rms}}{\overline{p_{t2}}}\right)_{unsteady} = 2\left(\frac{p'_{t,rms}}{\overline{p_{t2}}}\right)_{steady} = 2K < p_{t2} >_{steady} \quad (2)$$

A K-value of unity [15] and the quasi steady relation (1) can be applied, when peak energy in the pressure fluctuation is below $30 \div 40kHz$ [16]. To sum up, this procedure of measuring Pitot pressure fluctuation alone is not sufficient, because it is restricted on the detection of pure sound waves and no information is given about additional simultaneously existing disturbance fields possibly even more predominant. Moreover an uncertainty factor of 2 always is inherent in this method [2], [5].

3.2 Hot wire technique and modal analysis

A more meaningful procedure is given by means of hot wire (HW) anemometry in connection with the *modal analysis* by [17]. The method allows a detection of all possible, simultaneously existing disturbance modes (vorticity, entropy spottiness and sound waves) in the flow. The principle of HW signal decomposition into various "modes" is based on dissimilar ratios $r = -F/G$ of the sensitivities in the heat exchange of the probe element in reference to velocity, density and temperature of the flow when "overheating parameter" $a = (R_w - R_0)/R_0$ defined by the ohmic resistance or assigned "temperature loading" $\tau = (T_w - T_e)/T_0$ is varied during the flow measurement. Subscripts "w" and "0" are pertaining to states of heated and unheated wire probe. T_0 and T_e are assigned to stagnation and recovery conditions. In the present situation with the "specific" type of problem one gets in some difficulties and compromises had to be made to realize the method. Most serious problem is the limited measuring time in the wind tunnel and to define the disturbance field of a specific run with strictly speaking "unique" initial conditions, sufficient information must be caught within one flow event. A *C*onstant *T*emperature *A*nemometer (*CTA*) device was developed to operate the HW in a sequence of different overheat parameters within the available short time as shown in fig. 1. The first compromise was made in the determining of a CTA instead of a *C*onstant *C*urrent *A*nemometer (*CCA*). Although CCA is more suitable for a modal analysis in the relevant low overheat ratio $a = 0.01 \div 0.5$ [19], its use in a short duration measurement is tedious and difficult when overheat is changed. On the other hand because the heat flux into the flow by a HW, when operating at higher overheat ratio in CT-mode, is more influenced by mass flow density (ρu) as by total temperature T_0, CT-mode is predestined for the measurement of corresponding mass flow density fluctuations $(\rho u)'$, more likely as T_0'. Moreover the dimensionless mass flux sensitivity coefficient $F(\tau)$, is almost independent of τ (fig. 2) in contrast to CCA [19]. Because of a finite temperature sensitivity coefficient $G(\tau)$ of the HW, also existing with increasing τ (fig. 2), a full calibration of the used probes allows the elimination of temperature influence and the performance of a "confined" modal analysis. This seemed to be important for a very first judgement and classification of the main feature of the SWK's test section flow disturbances.
A second problem are the very harsh conditions in the breakdown phase of the test section flow which have restricted the use of a fragile 5μ DANTEC HW probe only to lower unit Reynolds numbers ($Re \leq 13 \times 10^6/m$, $M = 2.5$). The use of a DANTEC 55R31 wedge shaped hot film (HF) probe as a second compromise could be a subject of criticism, as outlined in [9], but also substantiated by a full calibration of the probe (fig. 2) and a corresponding interpretation of the signal.
The basic dynamic equation, relating instantenous HW fluctuation signal e' to corresponding fluctuation quantities $(\rho u)'$ and T_0' is according [9], [19] and [20] for CTA mode

$$\frac{e'}{E} = F\left(\frac{\rho u'}{\overline{\rho u}}\right) + G\left(\frac{T_0'}{\overline{T_0}}\right), \tag{3}$$

where E, $\overline{\rho u}$ and $\overline{T_0}$ are assigned temporal mean values. Dividing (3) by G, squaring and taking mean results, a relationship for determining corresponding *effective* fluctuation values $< rms >$ results in

$$(\Theta_{rms})^2 = \left(\frac{e'_{rms}}{EG}\right)^2 = r^2\left(\frac{m'_{rms}}{\overline{m}}\right)^2 + 2rR_{m'T_0'}\left(\frac{m'_{rms}}{\overline{m}}\right)\left(\frac{T'_{0,rms}}{\overline{T_0}}\right) + \left(\frac{T'_{0,rms}}{\overline{T_0}}\right)^2 \tag{4}$$

where $r(\tau) = -F(\tau)/G(\tau)$ is the so called modal parameter. The determination of the 3 unknowns can be solved by the formation of a linear equation system, defined by measurements with three different overheat ratios. Because of scattering by random errors, redundant measurements are necessary and in general the graphical method by [17] [18] is used to determine *effective* values of the disturbance modes. In case of a high degree of correlation between mass flow and temperature fluctuation, which is typical for a pure acoustic field with a corresponding correlation coefficient $R_{m'T'_0} \approx -1$, equation (4) is reduced to a linear relation

$$\Theta_{rms}(\tau) = r(\tau)\left(\frac{m'_{rms}}{\overline{m}}\right) + \frac{T'_{0,rms}}{\overline{T_0}} = r(\tau) < m > + < T_0 > \qquad (5)$$

Consequently, when there is a linear arrangement of measured effective fluctuation signals $\Theta(\tau)$ versus assigned modal parameter $r(\tau)$, a mean *effective* mass flow fluctuation value $< m >= m'_{rms}/\overline{m}$ can be defined from the slope in a linear regression. The intercept presents the correlating and escorting also normalized total temperature fluctuation $< T_0 >= T'_{0,rms}/\overline{T_0}$ but which must be seen reservedly when determined in a CT-mode.

3.3 Scanning device, comparative modal analysis for validation

For detailed description about the main technical features of the present COSYTEC CTA scanning device, reference is made to [21] and here only most relevant specifications are given: cut-off frequency of the bridge circuit (1:10) is comparable to DANTEC 55M10 CTA device when using also a DANTEC 55P11 standard HW-probe. Logic of the scanning system allows a setting of max. 10 overheat ratios within a maximum adjustable sequence time of $2.5s$. Resolution in temporal adjustment is $1ms$ and in overheat 0.005. Installed AC/DC signal conditioning allowed for present measurements in A/D conversion (sample rate $800kHz/s$) the exploitation of the complete voltage range of a 14 bit A/D converter.

To test the ability of the scanning device comparative modal analysis must performed. The disturbance field of a $M = 2.5$ test section flow in the middle sized supersonic wind tunnel HMMS of IAG ($0.18 \times 0.2m^2$), operating in a suck down mode, was used as a reference. A detailed mode diagram found by means of the conventional DANTEC 55M10 system is shown in fig.3. It reveals an almost linear dependency in the range of $0.4 \leq r \leq 0.9$ with a mean slope of $< m >= m'_{rms}/\overline{m} \approx 0.63\%$. The slightly decreasing slope with diminishing modal parameter below $r \leq 0.4$ can be interpreted as a finite vorticity and temperature fluctuation in the flow. Extensive measurements revealed a standard deviation in the reproducibility of this reference disturbance field with $< m >\approx 0.63\%$ of $\sigma = \pm 1\%$. In fig. 4, two mode diagrams as mean values of 10 measurements each, produced by the DANTEC and COSYTEC CTA in conventional operation are plotted. The difference in the mean slope between both systems is only 1.6%.

In fig.5, 4 mean mode diagrams from the COSYTEC device for a HW and HF probe, produced in a conventional and in a scanning mode are presented for a comparison and depicts following facts: on one hand there is a considerable difference between mean fluctuations determined by the HF-sensor with only 0.26%, 0.22% in comparison to the HW values 0.62%, 0.64% measured respectively in conventional and in a scanning mode each. This is equivalent to an "attenuation factor" (AF) of $AF \approx 0.4$. On the other hand there is a significant departure of +7% for the HW in the scanning mode and even −17% in corresponding comparative HF measurements when there is a standard deviation of

$\sigma \approx \pm 1\%$ within 10 measurements each. No uniform trend becomes apparent when the scanning mode is used for both type of sensors. It is noteworthy that the sequence time of scanning for both probes was about the same (HW:81ms; HF:95 ms) and adapted to the SWK measuring condition. Within this period of time for HW and HF probe a minimum sample time per overheat ratio of $\Delta t_s = 9ms$ and $19ms$ respectively could be defined. This time span was found each as the correspondent residue of the selected time interval $\Delta t_s = 13ms$ and $23ms$ for HW and HF respectively, where a constant temporal mean value in the DC output signal was approx. established. For the HF probe there is still a change of about 1% in the DC-signal E. On the other hand it is noteworthy that a different change in frequency response of the anemometer system exists in combination with the two sensors each, and a scanning procedure can be performed by a pre-adjustment of the bridge, optimized for a single overheat ratio only. Here further investigations are going to improve this situation. A post-correction by means of a known transfer function for each selected overheat ratio seems realizable.

4 Fluctuation measurements in the SWK flow

Fig.6 presents mass flow fluctuation measurements from the SWK test section flow, at a position near the nozzle exit, where preceding correspondent Pitot pressure and mass flux detection by means of conventional methods have already been performed in a sequence of individual runs [9]. Most important feature of all mode diagrams shown is a linear character which means that dominating disturbance modes are sound waves, presumably radiated from the turbulent BL of the test section walls. This certainty is of special importance because in consequence vorticity and entropy spottiness are of a lower order in occurrence and intensity and what is not trivial for the type of facility. A series of HW scannings (1) reveals a mean effective fluctuation value of $< m > \approx 0.24\%$ which can be decomposed [6], [22] to a corresponding static pressure fluctuation value $< p_\infty > \geq 0.36\%$. On the other hand measured Pitot pressure fluctuations of $< p_t > = 0.18\% \div 0.28\%$ [9], can be transformed to correlative static pressure fluctuations $< p_\infty >_{steady} = 0.18\% \div 0.28\%$ or $< p_\infty >_{unsteady} = 0.36\% \div 0.56\%$ according the uncertainty factor of 2. The higher value of HW data compared to corresponding static pressure values deduced from "steady" Pitot pressure can be clarified [23]: Due the poor high frequency response of the Pitot probes, because of their much larger size and therefore a smaller spatial resolution than the HW, acoustic disturbances of smaller amplitude cannot be resolved by the larger Pitot probes and that portion of the signal is not sensed by the pressure transducer.
HF scanning (2a) results after a scaling (1:0.4) in $< m > \approx 0.38\%$ (2b) which is considerably higher compared to (1). Otherwise by means of the DANTEC CTA system, in a sequence of 6 individual runs (3a) with a different HF sensor and a scaling factor (1:0.75) [9], this sensor signals transformed to "HW" fluctuations results in $< m > \approx 0.32\%$ (3b). This value is also close to the value of (2b) and confirms at least qualitatively a dominance of acoustic disturbances and in addition that differences in the disturbance fields of individual runs, caused by different "unique" initial conditions should not be to large to put a repeatability of the "boundary conditions" in question.

Acknowledgements

The authors would like to thank the DFG for grants within SFB 259 established at Stuttgart University in the partial project TPC 5 under which the project was supported. The authors are also grateful to Dr. A.D. Kosinov from ITAM (Novosibirsk) who has performed the full calibration of the HW and HF probes.

References

[1] Stetson, K.F.: *Hypersonic Transition Testing in Wind Tunnels. In Instability and Transition.* H.Y. Hussaini a. R.G. Voigts Eds. Band I, Springer Verlag, New York, 1990, pp 91-100.
[2] Schneider, S.P.: *Effects of High-Speed Tunnel Noise on Laminar-Turbulent Transition.* AIAA-paper 2000-2205, June 2000.
[3] Beckwith, I.E.; Miller III, C.G.: *Aerothermodynamics and Transition in High-Speed Wind Tunnels at NASA Langley.* Annual Review of Fluid Mechanics, 22; 419-439, 1990.
[4] Chen, F.-J.: *Boundary-Layer Transition Extent Measurements on a Cone and Flat Plate at Mach 3.5.* AIAA Paper 93-0342, January 1993.
[5] Reed, H.; Kimmel, R.; Arnal, D.; Schneider, S.P.: *Drag Prediction and Transition in Hypersonic Flow.* AGARD CP-600, C15-1 - C15-17, 1997.
[6] Laufer, J.: *Aerodynamic Noise in Supersonic Wind Tunnels.* Journal of Aer. Sci., Vol. 28, No. 9, Sept. 1961, pp 685-692.
[7] Morkovin, M.V.: *On Supersonic Wind Tunnels with Low Free Stream Disturbances.* Journal of Applied Mechanics, Vol. 26, No. 3, Sept. 1959, pp 319-323.
[8] Weise, A.; Schwarz, G.: *The Shock Wind Tunnel of IAG at University Stuttgart.* ZfW 21 (1973), Heft 4, pp 121-131, (in German).
[9] Knauss, H.; Riedel, R.; Wagner, S.: *The Shock Wind Tunnel of Stuttgart University - A Facility for Testing Hypersonic Vehicles.* AIAA 99-4959, Norfolk, Virginia, 1999.
[10] Pate, S.A.; Schueler, C.J.: *Radiated Aerodynamic Noise Effects on Boundary Layer Transition in Supersonic and Hypersonic Wind Tunnels.* AIAA Journal, Vol. 7, No. 3, March 1969, pp 450-457.
[11] Anders, J.B.; Stainback, P.C.; Beckwith, J.E.: *A New Technique for Reducing Test Section Noise in Supersonic Wind Tunnels.* AIAA Journal Vol. 18, No. 1, Jan. 1980.
[12] Wolf, S.W.D.; Laub, J.A.: *Characteristics of the NASA Ames Laminar Flow Supersonic Wind Tunnel for Unique Mach 1,6 Transition Studies* AGARD CP585, pp 30.1-30.17.,1997
[13] Demetriades, A.; Brogan, T.: *Influence of Sidewall Transition on Measured Free Stream Noise in a two-Dimensional Supersonic Tunnel.* 20th AIAA Conference, June 1998, Albuquerque, NM.
[14] Stainback, P.C.; Wagner, R.D.: *A Comparison of Disturbance Levels Measured in Hypersonic Tunnels Using a Hot Wire Anemometer and a Pitot Pressure Probe.* AIAA Paper No. 72-1003, Palo Alto, CA, Sept. 1972.
[15] Laufer, J.: *Some Statistical Properties of the Pressure Field Radiated by a Turbulent Boundary Layer.* Physics of Fluid, Vol. 7, No. 8, August 1964, pp. 1191-1197.
[16] Chen, F.-J.; Beckwith, I.E.; Creel jr., T.R.: *Correlations of Supersonic Boundary Layer Transition on Cones Including Effects of Large Axial Variations in Wind-Tunnel Noise.* NASA TP 2229, 1984.
[17] Kovasznay, L.S.G.: *The Hot Wire Anemometer in Supersonic Flow.* Journal of the Aero Sci, Vol. 17, No. 9, 1950, pp 565-584.
[18] Morkovin, M.V.: *Fluctuation and Hot Wire Anemometry in Compressible Flow.* AGARDograph, 24 Nov. 1956.
[19] Smits, A.J.; Dussauge, J.P.: *Hot Wire Anemometry in Supersonic Flow.* AGARDograph No. 315, 1989.
[20] Smits, A.J.; et al: *Constant Temperature Anemometer Practice in Supersonic Flow.* Part I, The normal Wire. Experiments in fluids, 1983, pp- 83-91.
[21] Gaisbauer, U.; Knauss, H.; Wagner, S.; Weiss, J.: *Measurement Techniques for Detection of Flow Disturbances and Transition Localisation in a Short-duration Wind Tunnel* Proceedings of 10th ICMAR, Novosibirsk, July 2000.
[22] Lebiga, V.A.; Zinoviev, V.N.: *Hot-Wire Measurements in Compressible Flow.* AGARD-CP-585, pp. 31.1 - 31.9, 1996.
[23] Beckwith, I.E.; Creel, T.R. jr.; Chen, F.-J.; Kendall, J.M.: *Free-Stream Noise and Transition Measurements on a Cone in a Mach 3.5 Pilot Low-Disturbance Tunnel.* NASA Techn. Paper 2180, 1983.

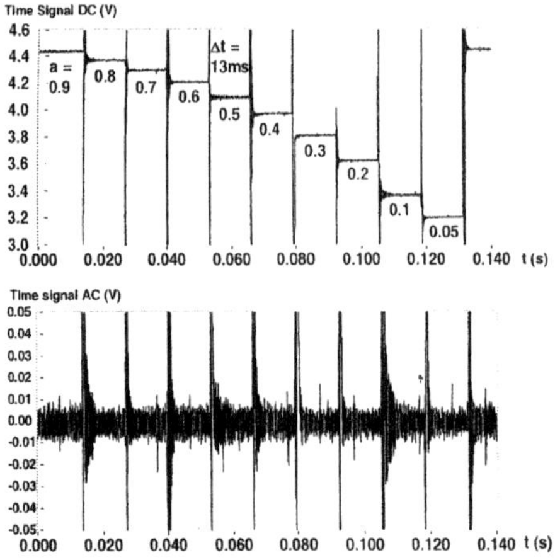

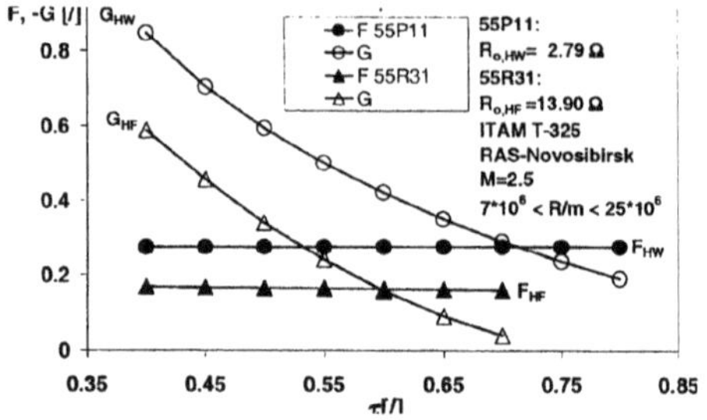

Fig. 2: *Sensitivity coefficients of HW and HF probe, by Dr. A.D. Kosinov ITAM.*

Fig. 1: *DC- and AC-output signal of COSYTEC CTA scanning device.*

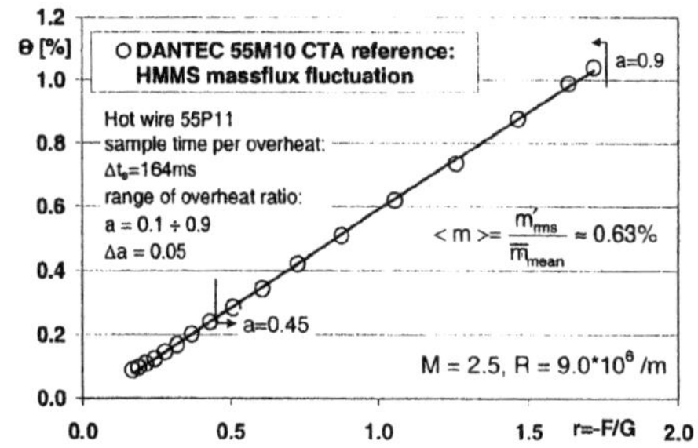

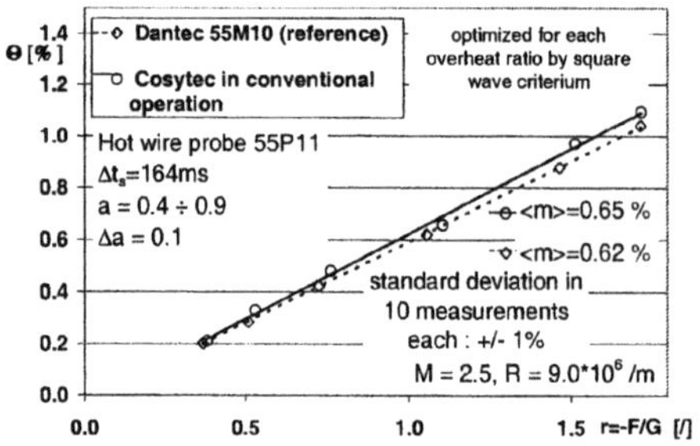

Fig. 3: *Mode diagram of reference disturbance field in HMMS test section flow determined by DANTEC 55M10 CTA-system.*

Fig. 4: *Mean mode diagram by COSYTEC CTA device (conventional operation) in comparison with DANTEC CTA system.*

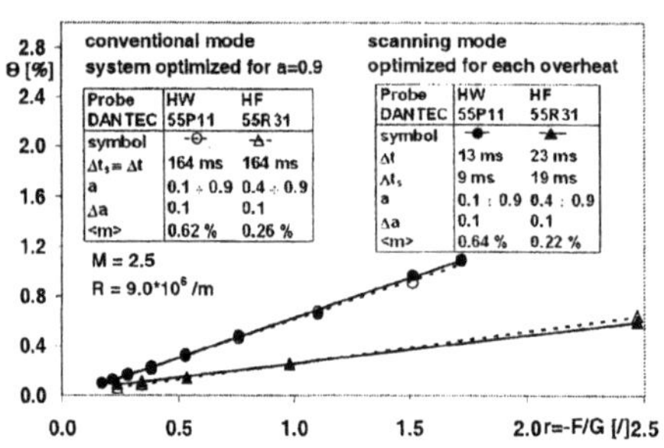

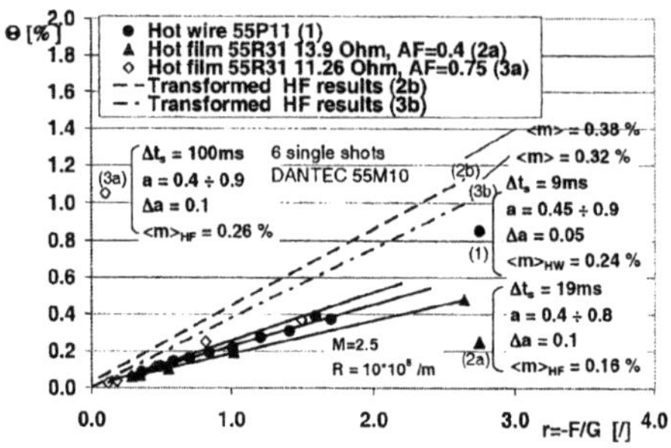

Fig. 5: *Comparison of different mode diagrams: HW and HF in conventional and scanning mode.*

Fig. 6: *Mode diagrams of SWK test section flow attained by different methods.*

Surface Fence with an Integrated, Piezoresistive Pressure Sensor for Measurements of Static and Dynamic Wall Shear Stress

O. BURKHARDT, U.G.S. DINATA, W. NITSCHE
Berlin University of Technology, Institute of Aeronautics and Astronautics
Marchstr. 12-14, D-10587 Berlin, Germany

Summary

A surface fence device used for wall shear stress measurements was improved by integrating a small, piezoresistive pressure sensor placed very close to the fence. The piezoresistive sensor was used to measure a pressure difference of the fence that can be correlated to the local wall shear stress based on the similarity law of the viscous sublayer. The piezoresistive sensor made it possible to use the fence device in measuring both static and dynamic wall shear stresses. This had been tested at fluctuating wall shear stresses in a small circular duct as well as in turbulent boundary layer flows in a wind tunnel up to Mach number of 0.9. The analysed sensor characteristics were calibrations, time traces, power spectra and frequency responses to show typical application ranges and to compare the device with a conventional obstacle wire, a surface hot film and a skin friction balance.

1 Introduction

A wall shear stress measurement technique can be realised using a small obstacle (fence or wire) fixed on a surface within the boundary layer flow. The pressure difference in front of and behind the obstacle can be correlated to the local wall shear stress based on the similarity law of the viscous sublayer. A surface fence for use in measuring wall shear stress was first applied by Konstantinov and Dragnysh in 1960 [1], later by Head and Rechenberg [2] and described in detail by Vagt and Fernholz [3]. Such a device was compared with other skin friction measurement techniques by Nitsche, Haberland and Thünker [4] and modified by using a wire as an obstacle, by Weiser, Nitsche and Renken [5].

For a conventional obstacle device, two taps of the device are connected with long tubes (see Figure 1) to a pressure transducer. However, such tubes may produce an additional damping effect causing a low sensitivity in wall pressure and shear stress fluctuation signals. This disadvantage can be minimised by placing the transducer very close to the taps and fence as shown in Figure 2. In the framework of the AEROMEMS research project, this type of design was realised by integrating a small, piezoresistive pressure sensor based on MEMS (Micro-Electro-Mechanical System) [6,7] in order to improve the frequency response of the surface fence.

2 Surface Fence and Piezoresistive Pressure Sensor

The surface fence principle is based on the similarity law of a viscous sublayer for a low obstacle height, h_f. The pressure difference of the fence is proportional to the characteristic flow velocity, u_f, at half the obstacle height ($y_f=h_f/2$), while this velocity can be correlated with a velocity of wall shear stress, u_τ, as well as the wall shear stress, τ_w, using the boundary layer law. A calibration function of the local wall shear stress against the pressure difference, Δp, can be formulated by an empirical relationship with dimensionless parameters written as

$$\tau^+=A(\Delta p^+)^n \tag{1}$$

where A and n are empirical constants. The dimensionless parameters are calculated using

$$\tau^+ = \tau_w h_f^2/(4\rho v^2) \tag{2}$$

$$\Delta p^+ = \Delta p\, h_f^2/(4\rho v^2) \tag{3}$$

where ρ and v are fluid specific density and kinematic viscosity, respectively [5].

Most flows occurring in practical applications are turbulent which means that an irregular motion is superimposed on the main flow. A typical feature of the turbulent motion is that wall pressure and shear stress do not remain constant with time, but undergo irregular fluctuations, so an instantaneous wall shear stress can be given by a mean, $\overline{\tau}_w$, and fluctuation, τ'_w, part written as

$$\tau_w = \overline{\tau}_w + \tau'_w \tag{4}$$

Therefore it is important to know the dynamic response of a sensor in measurements of the fluctuating parameter, particularly in high frequency fluctuations.

The transducer used to improve the surface fence is a commercially available piezoresistive pressure difference sensor from Honeywell™. A piezoresistance property of a semiconductor can be described as the change in resistance caused by an applied strain of the diaphragm. The sensing element used consists of four nearly identical piezoresistors buried in the surface of the thin silicon diaphragm, which is formed by chemically etching a square cavity into the surface opposite the piezoresistors. The unattached portion of the silicon slice provides a rigid boundary constraint for the diaphragm and a surface mounting to the seals. A pressure applied to the sensor causes the diaphragm to flex resulting in a stress or strain in the diaphragm as well as in the buried piezoresistors [8].

Figure 3 shows a sketch of the sensor diaphragm of the piezoresistors in a sensor die with a diaphragm deflection caused by the pressure difference. The four piezoresistors are built into a full Wheatstone bridge connected to an amplifier and a signal conditioner (Figure 4). $R \pm \Delta R$ represents the actual resistor values at the applied pressure. R is the resistor value for the undeflected diaphragm, where all four resistors are nearly equal in value and ΔR represents the change in resistance due to the applied pressure. All four resistors will change by approximately the same value under pressure force. Two resistors will increase and another two decrease depending on their orientation with respect to the crystalline direction of the silicon material. Signal voltages produced by the Wheatstone bridge are proportional to the amount of applied pressures, which generate the resistance changes [8].

The sensor used (26pc type) has a typical specification including an unamplified sensitivity (output voltage per pressure difference) of $\Delta E/\Delta p$=16.7mV/psi under a bridge voltage of E_o=10V, a pressure range of ±6.9kPa and a response time of t=1ms [8].

3 Test Set-up

The surface fence with the piezoresistive sensor was calibrated using a skin friction balance and compared with a conventional obstacle wire, which has nearly the same specification but with a non-integrated pressure sensor. The piezoresistive surface fence was also tested and compared to a surface hot film to investigate sensor sensitivities. These tests were carried out in the transonic wind tunnel of Berlin University of Technology in turbulent boundary layer flows up to $M_\infty=0.9$ ($Re=14.5\times10^6$). The four surface devices were flush-mounted in a flat wall (see Figure 5) of a test section to minimise sensor-induced disturbances and placed spanwise at a position of 0.8m from the wall leading edge to be able to carry out simultaneous measurements. The measured steady state wall shear stress reached up to 80N/m² approximately with a calculated friction velocity, u_τ, of below 10m/s.

The obstacle wire was equipped with 0.1mm-diameter wire and 0.1mm-diameter taps connected by 1.5-m plastics tubes (inner dia.=1.6mm) to a pressure transducer (from PSI™, Series 3200), which delivers mean output voltages. The skin friction balance (from Electrotechnologies Selem™, SM-251 type) had a floating head with a diameter of 10mm, a ratio of output voltage to wall shear stress of 1mV/mgf, a resolution of less than 1mgf and an accuracy of less than 0.5% (±1gf). The surface hot film used (TAO Senflex™) was made of Nickel with a cold resistance of $R_{o(20°C)}=6\Omega$ and a sensor length and width of 0.057 and 0.004 inches, respectively. This thermal sensor was operated using a TSI IFA-100™ anemometer with an overheat ratio of 1.5.

Furthermore, investigations on the time trace and frequency response of the piezoresistive surface fence were conducted using excitations of acoustic plane-waves in a small circular duct (Kundt's tube). The power spectra of the piezoresistive sensor at high speed turbulent flows were also investigated at various flow velocities up to $M_\infty=0.9$. All dynamic responses were recorded using a dynamic signal analyser (Hewlett-Packard™, HP 3562A model).

4 Test Results and Discussions

4.1 Calibration

Figure 6 compares the typical calibration curve of the new fence device with that of surface wire device that is shown in the logarithmically scaled dimensionless parameters (τ^+ vs. Δp^+). For the tested range ($M_\infty<0.9$), the calibration of the piezoresistive surface fence is indicated to have the same trend as that of the conventional obstacle wire designed with the same specifications. Moreover, the diagram shows almost the same constants (A, n) of the typical calibration functions for both obstacle devices.

4.2 Static Response

The modified surface fence and the surface hot film were tested to obtain static response characteristics influenced by various wall shear stresses. Test results of the surface devices are presented in Figure 7 with output voltage, E, and pressure difference of the surface fence, Δp, against wall shear stress, τ_w. This diagram shows increasing sensor output voltages as wall shear stresses increase but with different trends. The hot film follows the King's law relationship with a curve-fitting equation of $E=2.198+0.253\tau_w^{0.33}$, while the characteristics of the piezoresistive surface fence has a third order polynomial trend with a fitted equation of $E=-0.032602+0.049211\tau_w+0.00276\tau_w^2-0.000014\tau_w^3$.

It is interesting that these trends affect sensor sensitivities obtained from a derivative of a static response function ($dE/d\tau_w$). Normalised sensitivity curves for both surface devices are shown in Figure 8, plotted against τ_w. This diagram indicates that at increasing wall shear stresses, the hot film sensitivity gets lower due to a decrease in the sensor heat loss rate, while that of the piezoresistive sensor increases until a maximum of about $67N/m^2$ is reached.

The different sensitivity trends representing ranges of sensor applications on wall shear stress measurements. This comparison suggests that the hot film sensor is useful, especially for measurements of the lower wall shear stresses, but the surface fence using the piezoresistive sensor is suitable for measuring the higher wall shear stresses.

4.3 Dynamic Responses Due to Excitations

In-situ determination of a frequency response is necessary because an accurate modelling of the inherent mass, compliance, and fluidic damping of a whole sensor system is difficult to achieve in practice. An acoustic plane wave generator, which is capable of providing a sinusoidal shear input into the sensor, is employed to describe these effects. The sensor is tested to obtain a time trace and frequency response due to a mechanical excitation by providing the system with a known input and observing the output. In such a test, purely travelling acoustic plane-waves generate oscillating wall shear stresses [9].

The piezoresistive surface fence was excited with a fluctuating flow driven by a 16Ω-impedance loudspeaker placed at 137mm from the sensor in a 25.4mm-diameter brass tube (Figure 9). First, the excitations were conducted using single frequency sine-waves (5, 10, 20kHz) to determine time trace responses following the fluctuating wall shear stresses, and then, using sweeping sine-waves of 0.1-100kHz, to obtain a broadband frequency response to compare with that of the hot film sensor. The employed input voltage for the loudspeaker was peak-to-peak 10V.

The time trace results at several fixed frequency excitations are shown in Figure 10, which indicates that the sensor signals are clearly higher than its noise and precisely follow the sinusoidal excitation signals with delays mainly caused by the exciter-sensor distance and electronic devices used. Additionally, it is evident that the amplitude at 20kHz is lower than at 5 and 10kHz. Subsequently, the sensor output signal at 5kHz-excitation is also verified using a power spectrum diagram (Figure 11) showing two signals, where the sensor signal at the fixed excitation is precisely overlapped by the noise signal. This diagram shows the 5kHz-peak of the sensor signal and some harmonic amplitudes as well. According to the time trace and power spectrum diagrams, the piezoresistive surface fence is well capable of measuring wall shear stress fluctuations.

Figure 12 represents the frequency response of the piezoresistive surface fence that was obtained by using sweeping sinusoidal wall shear stresses in comparison to that of the surface hot film. Based on this result, two typical broadband widths of approximately 11 and 30kHz are obtained for the piezoresistive sensor and hot films, respectively. In addition, this diagram verifies that at above 10 kHz, the surface fence signal amplitudes are attenuated.

4.4 Dynamic Wall Shear Stress at Turbulent Flows

The piezoresistive surface fence was tested in the transonic wind tunnel on the test section flat wall, which provides a relatively simple and clean case for a turbulent boundary layer.

Figure 13 shows time traces of output voltage fluctuations of the piezoresistive sensor, which indicate that the device is able to be used in measuring dynamic wall shear stresses in higher speed flows. It can be observed that, as the flow velocity was increasing, the device could detect increasing amplitude levels of the wall shear stress fluctuations in the boundary layer flow. Moreover, Figure 14 shows typical power spectra at different flow velocities for frequencies up to 40kHz that show a clear influence of flow velocities on sensor signal levels. An increase in the flow velocity, indicating an increasing τ_w fluctuation, results in an increase in the power spectrum level of sensor output signals.

Additionally, it can be seen that the signal levels for free stream velocities are well above the electronic noise at no-flow condition, which suggest a high signal-to-noise ratio (SNR) of the whole sensor system (including surface fence, sensing elements and amplifier). The turbulent flow spectra show that the sensor output signals are dominated by wall shear stress fluctuations at frequencies of below 11kHz. This is also already verified by the test result of the broadband width of the piezoresistive sensor signal through the oscillating wall shear stresses.

5 Conclusions

A surface fence device developed with an integrated, piezoresistive pressure sensor and its result of static and dynamic wall shear stress measurements up to Mach numbers of 0.9 ($\tau_w \leq 80 N/m^2$) has been presented. The typical calibration characteristics have shown that the piezoresistive surface fence has a third order polynomial trend while that of the hot film trend follows the King's law relationship. Based on the calibration results, the sensitivity of the piezoresistive surface fence rises up to a maximum value but that of the hot film gets lower with increasing wall shear stresses for the range tested.

The frequency response of the piezoresistive surface fence due to oscillating wall shear stresses with purely travelling acoustic plane-waves was also obtained. The result shows that the frequency response of the excited piezoresistive surface fence has a broadband width of about 11kHz. Furthermore, the achieved time traces and power spectra indicate the ability of the piezoresistive surface fence to measure dynamic wall shear stresses. As the flow velocity increases, the device detects increasing amplitudes of the wall shear stress fluctuation and the output signal levels are well above the electronic noise indicating a high signal-to-noise ratio.

References

[1] N.I. Konstantinov and G.L. Dragnysh. "The measurement of friction stress on a surface", English Translation, DSIR RTS 1499, 1960.

[2] M.R. Head and I. Rechenberg. "The Preston tube as a means of measuring skin friction", J. of Fluids Mechanics 14, 1962, pp. 1-17.

[3] J.D. Vagt and H.H. Fernholz. "Use of surface fences to measure wall shear stress in three-dimensional boundary layer", Aeronautical Quarterly XXIV, 1973, pp. 87-89.

[4] W. Nitsche, C. Haberland and R. Thünker. "Comparative investigations on friction drag measuring techniques in experimental aerodynamics", Proceeding of 14th Congress of the international council of the aeronautical sciences (ICAS), Toulouse, France, 1984, pp. 391-402.

[5] N. Weiser, W. Nitsche and F. Renken. "Wall shear stress determination by means of obstacle wires", Proc. of 8th Symp. on Turbulent Shear Flows, Technical University of Munich, Germany, 1991, pp. 4-5-1– 4-5-6.

[6] O. Burkhardt, U.G.S. Dinata, C. Warsop and W. Nitsche. "New developments in surface flow sensor technology within the framework of AEROMEMS", to be published in Notes of Numerical Fluid Mechanics, Vieweg, Germany, 2000.

[7] AEROMEMS Homepage. "Investigation of the viability of MEMS technology for boundary layer control on aircraft aerospace technology, industrial manufacture, materials technology", Update Date: 4 Apr. 1998. http://www.tra3.com/dragnet/aeromems.html

[8] Honewell™. "Micro switch, sensing control, environmental condition sensors", Honeywell Catalogue 15, Honeywell Inc., Freeport, Illinois 61032, USA, 1998.

[9] A. Padmanabhan, M. Sheplak, K.S. Breuer and M. A. Schmidt. "Micromachined sensors for static and dynamic shear-stress measurements in aerodynamic flows", Department of Aeronautics and Astronautics, Massachusetts Institute of Technology, Cambridge, USA, 1998.

Figures

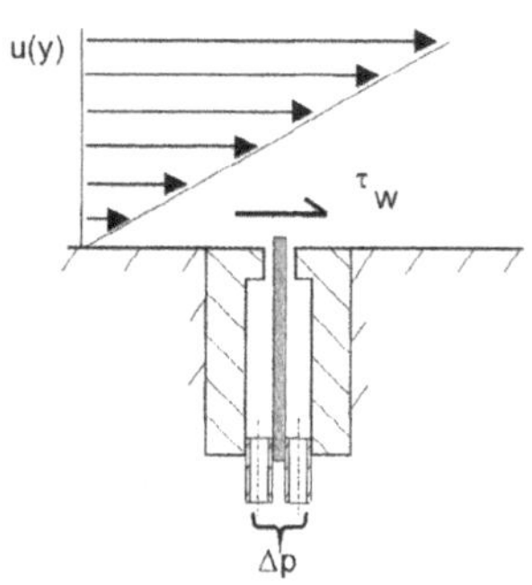

Figure 1 Conventional surface fence [5]

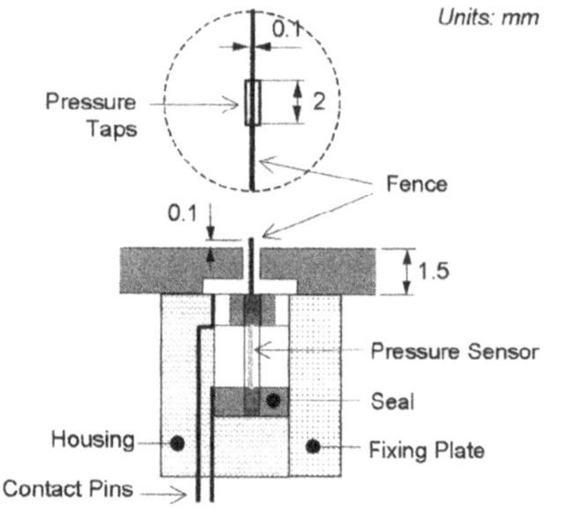

Figure 2 Fence with integrated pressure sensor [6]

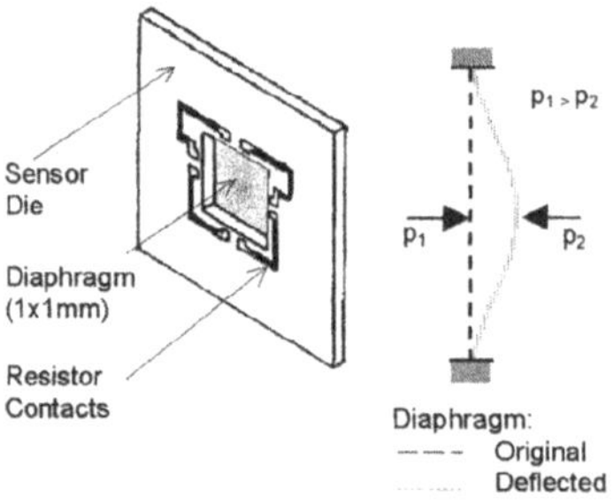

Figure 3 Diaphragm in a sensor die [8] and its deflection due to a pressure difference [6]

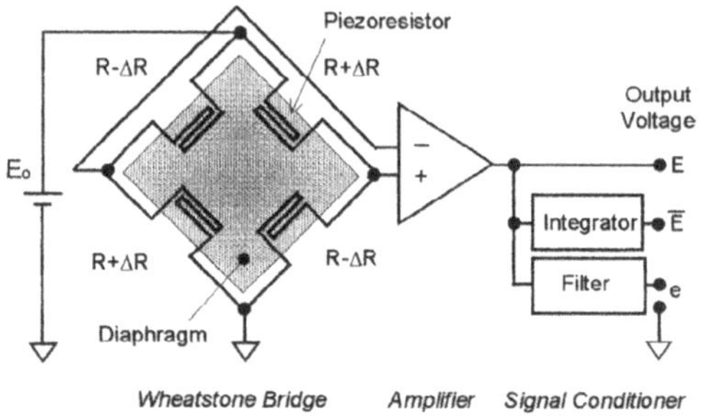

Figure 4 Piezoresistor bridge connected with an amplifier and a signal conditioner [6]

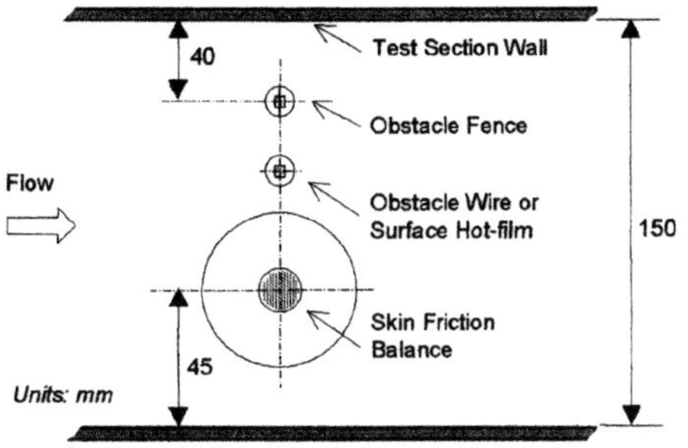

Figure 5 Tested sensors on a flat wall

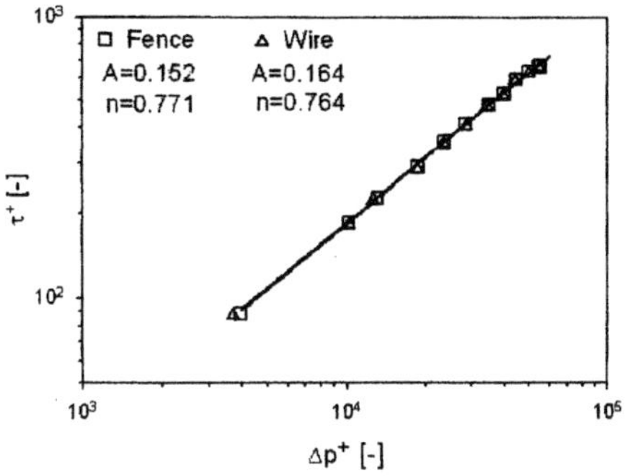

Figure 6 Typical calibration results of the obstacles

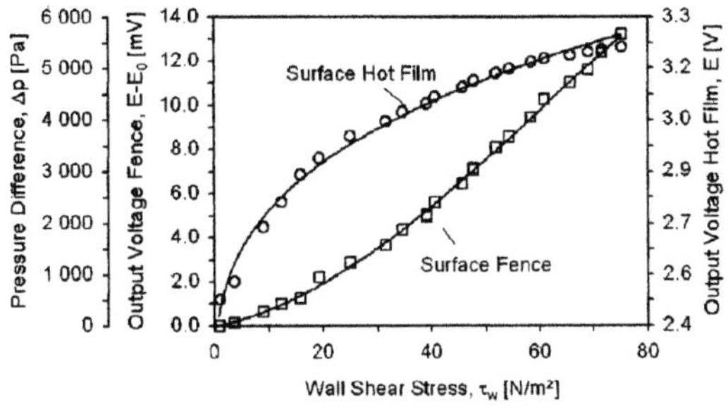

Figure 7 Static responses of the surface sensors

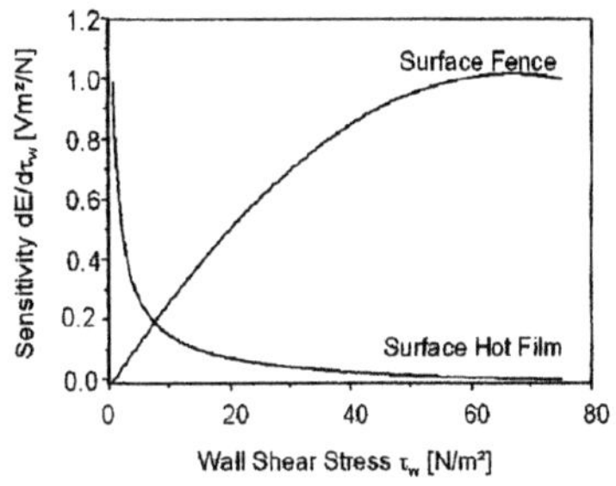

Figure 8 Normalised sensor sensitivities

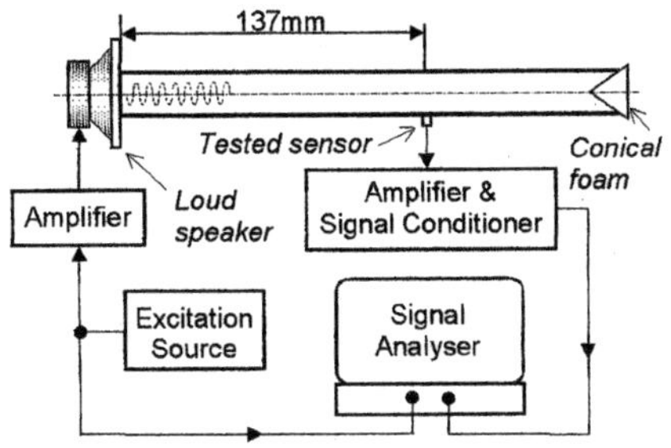

Figure 9 Test set-up for dynamic responses [9]

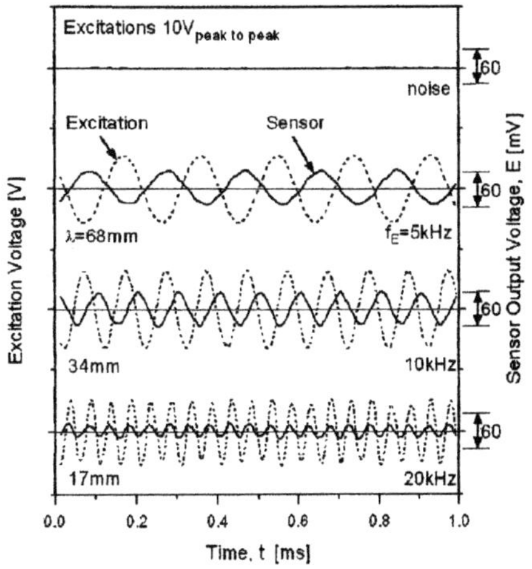

Figure 10 Sensor signals at sine-wave excitations

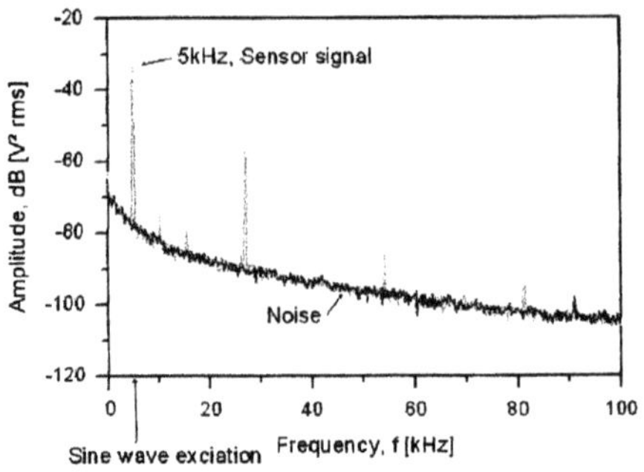

Figure 11 Sensor noise and output signals at single frequency sine-wave excitation of 5kHz

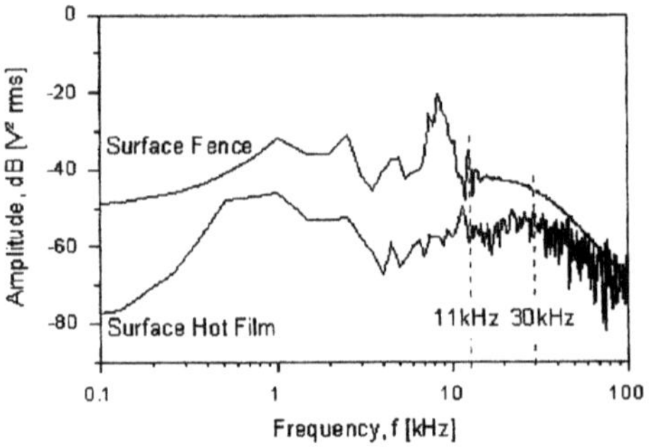

Figure 12 Typical frequency response at sweeping sine-wave excitation up to 100kHz

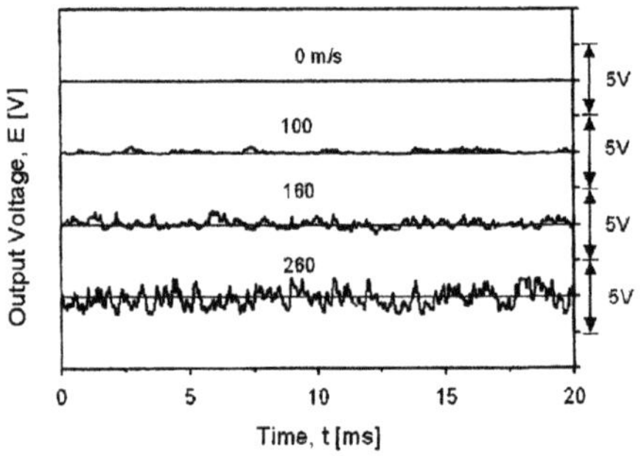

Figure 13 Time traces of sensor signal at turbulent flows in the transonic wind tunnel

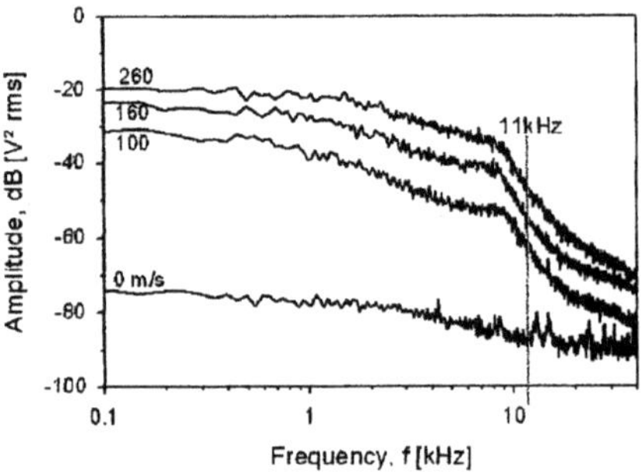

Figure 14 Typical power spectra of the sensor signal at the turbulent flows

Experimental Investigation of the Flow Close to the Surface of a High Lift Airfoil Configuration in a Large Wind Tunnel

T.J. MÖLLER[1], A. BERGMANN[2]

[1]Deutsches Zentrum für Luft- und Raumfahrt e.V. (DLR), Institut für Strömungsmechanik, Bunsenstr.10, 37073 Göttingen, Germany

[2]Deutsch-Niederländische-Windkanäle DNW-NWB, Lilienthalplatz 7, 38108 Braunschweig

Summary

In the present investigation the flow close to the surface of a high lift configuration has been studied experimentally to obtain data suitable to validate advanced numerical models for complex flow around a high lift airfoil configuration. The investigations have been performed in the closed test section of the DNW-NWB wind tunnel at a constant flow velocity of 50 m/s. The flow velocity has been determined by means of 3D Laser Doppler Anemometry (LDA). The investigated model is a Half - Model and a corresponding Reynolds Number of 1.5×10^6. IR imaging has been employed to study transition which was also investigated by the oil painting method.

1. Introduction

For the design of modern transport aircraft wings it is necessary to have reliable numerical tools. In the development for such numerical codes for the prediction of complex aerodynamic flows there is a strong need to obtain reliable flow field data which are suited to validate turbulence models used in numerical calculations. For the design of high lift configurations numerical tools utilising time averaged Navier Stokes calculations are commonly used. To evaluate this kind of results, three dimensional flow data are necessary.

Tinapp and Nitsche [1] presented investigations at Reynolds numbers of 150.000, but their main interest was the extinction of flow separation. For the validation of turbulence models higher Reynolds numbers are required. Furthermore, three dimensional flow should be investigated, since two dimensional data are available from literature [2] as well as data for low Reynolds numbers [3]. In the present investigation the flow close to the surface of a high lift configuration has been studied experimentally to obtain data suitable to validate advanced numerical models for complex flow around a high lift airfoil configuration [4]. In literature such experiments investigation at high Reynolds numbers are rarely found.

The objective of this paper is to demonstrate the successful application of LDA to obtain velocity data close to the model surface. This was achieved by an optimised optical set up which reduces the stray light and the application signal processing hardware which is able to determine LDA signals with low signal-to-noise-ratio (SNR). Furthermore, a system to determine the vibration of the wind tunnel model is applied, to determine the possible influence of motions of the model on the boundary layer. IR imaging with a system developed by DLR Braunschweig [5] has been employed to study transition which was also investigated by the oil painting method. The present paper shows first results of the investigation of the flow close to the surface of a high lift airfoil configuration.

2. Experimental Set-up

The experiments were carried out in the low speed wind tunnel in Braunschweig (DNW-NWB) [6]. The DNW-NWB is a continuously working wind tunnel with a closed, slotted or an open test section. During the experiments the closed test section has been applied. The nozzle exit area is 3.25 m x 2.8 m. The wind tunnel has been operated by a constant flow velocity of 50 m/s. The flow can be observed through a glass window on top of the test section.

The investigated model is a Swept Constant Chord Half (SCCH) model provided by DaimlerChrysler Aerospace Bremen with a span of 1.14 m and cord length of 0.45 m. The experiments were conducted by constant flow velocity of 50 m/s. This corresponds to a Reynolds number based on the model chord length of 1.5×10^6. The angle of attack was 10° and 22° for the presented experiments. The flap angle has been fixed at 20°. Figure 2 shows a sketch of the wind tunnel model and the measurement stations. In the experiments a co-ordinate system normal to the model surface has been used.

The DLR LDA system is a three component system [7]. Figure 3 shows a sketch of the applied set-up. The set-up has been specially adapted to the environment of the DNW-NWB. Furthermore CAD simulations of the optical path have been performed before the LDA system has been set up in the DNW-NWB to define an optimal optical set-up which allows measurements close to the surface of the model. The LDA operates in a back scatter set-up. The present set-up has a focal length of 3.8m for the receiving and the sending optics and an angle between the individual laser beams of 3°. The probe volume is typically 220 μm in diameter, with 10 μm fringe spacing. This leads to typical frequencies of the Doppler signal of 45 MHz, including the Bragg frequency. The scattered light is collected by an optical system with 350mm aperture. The system includes a signal processor DSA model 3000 manufactured by Aerometrics [8]. A seeding system developed by DLR Göttingen has been employed during the experiments. The mean diameter of the particles produced by the atomiser is 1μm. In [9] the principle of this atomiser is explained in detail. The particles are added to the flow in the settling chamber of the DNW-NWB, where the seeding could be traversed to optimise the seeding conditions.

To determine the vertical position of the model the DLR position monitoring system [10] has been employed in the DNW-NWB. The application in an wind tunnel is presented in [11]. The system is capable to acquire data with frequencies up to 500 Hz. The spatial resolution of the system is better than 0.02 mm. For LDA measurements position data have to be recorded simultaneously with LDA data storage which are controlled by data rate and not by fixed time intervals. The digital position data is therefore continuously converted into analogous voltage which can be recorded by most LDA processors. This permits storage of the latest position data synchronously with LDA data. The DLR LDA system can record three analogous signals, hence it is possible to record the displacement of the model in all six degrees of freedom. In the present experiments the vertical position of the model was monitored.

3. Results

The analysis of the LDA data has been performed by use of DLR software, which could also process the data obtained by the position monitoring system. The first step to evaluate the LDA data is to transform the velocity data from the non-orthogonal LDA co-ordinate system into the wind tunnel co-ordinate system. This was done by using the well-known geometrical data of the LDA set up, which was obtained by a calibration procedure.

The data from the position monitoring are recorded simultaneously with the LDA measurements. In the experiments no time depending vibration of the wind tunnel model could be determined. Hence the influence of the motion of the model on the flow close to the surface of the model can been neglected in the present investigation. Nevertheless a constant static displacement due to aerodynamic forces is found. This has to be considered in the evaluation of the LDA data, i.e. the actual co-ordinate of the probe volume has to be corrected by this shift.

The measurement should give information about the boundary layer condition and the development of the wake of the slat and the main airfoil. In figure 4 the velocity profile at measurement station 5 and for α=10° is shown. The wake of the slat could not be determined in the velocity profile. The LDA, IR imaging and oil painting data yielded that the boundary layer is still laminar for α=10° at this station. This changes in the case of α=22° '(figure 5). The wake of the slat can be detected clearly at z=11mm. This result has not been expected. Also the beginning of the boundary layer has been detected at this station. The data of LDA, IR imaging and oil painting result in a turbulent boundary layer condition for this case.

Figure 6 shows the wake of the main airfoil in the middle of the flap (station 7) for α=10°. Here the wake of the main airfoil is clearly found, which corresponds with increasing RMS values in this region (figure 7). More interesting are the results for α=22°. (figure 8). Additionally to the wake of the main airfoil at z=12mm, the wake of the slat is still found at z=28mm. The RMS values, shown in figure 9, match this result. An increase of the RMS values for both regions, z=12mm and z=28mm can clearly be seen.

The results from IR-imaging, oil painting method and the data from the LDA measurements allow to determine the condition of the boundary layer for the entire configuration at both α=10° and α=22°, respectively (figure 8). In the case of α=10 the boundary layer laminar on the slat, the main airfoil and in the beginning the of the flap. The boundary layer becomes turbulent for α=22° on the main airfoil.

4. Conclusion

The flow close to the surface of a high lift airfoil configuration has been measured by means of 3 component Laser Doppler Anemometry. Furthermore a position monitoring system has been applied to investigate possible vibration of the wind tunnel model. It is shown the 3D LDA system and the position monitoring system has been successfully been adapted to this problem. Velocity data as well as turbulent quantities have been successfully acquired even close to the surface of the model. Additionally IR-Imaging and oil painting have been applied in this experiment.

The presented first experimental data show that the velocity fluctuations are significantly different for all three velocity components. This result has to be considered when making numerical predictions for this kind of flow, i.e. it is necessary to apply advanced turbulence models which can also predict an isotropic turbulence. It has been found the influence of the wake of the slat is much more important than expected. The obtained results demonstrate that the 3D LDA set up has been successfully adapted to investigate the flow close the surface of a high lift airfoil configuration in the closed test section of the DNW-NWB.

The data obtained by the position monitoring system show that no relevant vibration amplitude occurs during the experiments. Nevertheless, a static displacement, depending on the angle of incidence of the model has been determined which has to be considered in the post processing.

Acknowledgement

The authors would like to thank Daimler Chrysler Aerospace Bremen for providing the SCCH Model.

References

[1] Tinapp, F., Nitsche, W., LDV-Measurements on a high-lift configuration with separation control, Proceedings 9. Int. Symp. of Applications of Laser Technology to Fluid Mechanics, Lisbon, S. 19.1.1-19.1.8, 1998.

[2] R.C. Hastings, B.R. Williams, Studies of the flow near a NACA 4412 airfoil at nearly maximum lift, Aeronautical Journal, January, 1997

[3] Seelhorst, U., Bütefisch, K.A., Weiland, M.: 3 component LDV for near wall measurements on a multi element high lift airfoil configuration, ICIASF record 95, 55.1-55.7, 1995

[4] Erickson, G.E.: High angle of attack aerodynamics, Annual. Rev. Fluid Mech., 27, pp. 45-88, 1995

[5] A. Quast: Detection of Transition by Infrared Image Technique. International Congress on Instrumentation in Aerospace Simulation Facilities, Williamsburg Va, June 22-25, Record 125, 1987.

[6] Kausche et. al.: Der Niedergeschwindigkeits-Windkanal der DFVLR in Braunschweig, DFVLR-Mitt. 88-25, 1988.

[7] Seelhorst, U., Bütefisch, K.A., Sauerland, K.H.: Three component Laser-Doppler Anemometer development for large wind tunnel. ICIASF'93 record, 33.1-33.7, 1993

[8] Aerometrics Doppler Signal Analyzer Installation Manual. Sunnyvale. CA, USA, Aerometrics Inc., 1991

[9] Raffel, M., Willert, Ch., Kompenhans, J.: Particle Image velocimetry – A practical guide. Springer 1998

[10] Grauer-Carstensen, H.: A Versatile Model Monitoring System, Rev. Sci. Instrum., 65 (2), 481-484, 1994

[11] Möller, T.J., Wiedemann, A., Bütefisch, K.A., Micheli, F., Scholz, D., Sjörs,K.: Application of 3 component Laser-Doppler Velocimetry in transonic flow over a Delta wing, ICIASF'99 record, 1999.

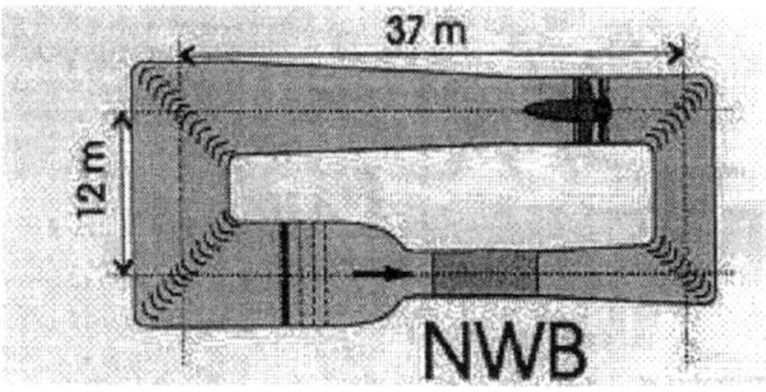

Figure 1 sketch of the DNW-NWB

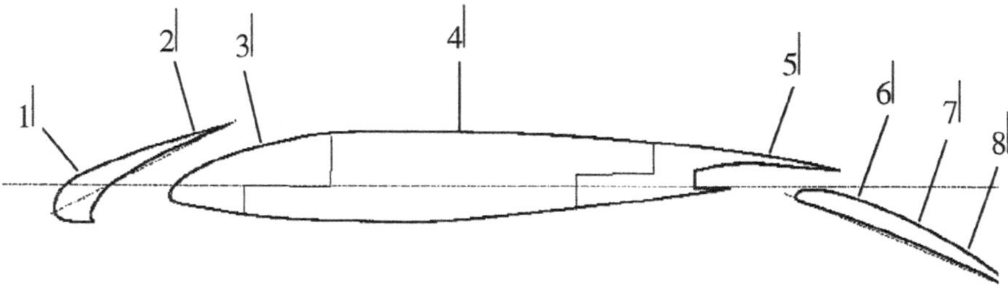

Figure 2 Schematic contour of the SCCH model and the measurement stations

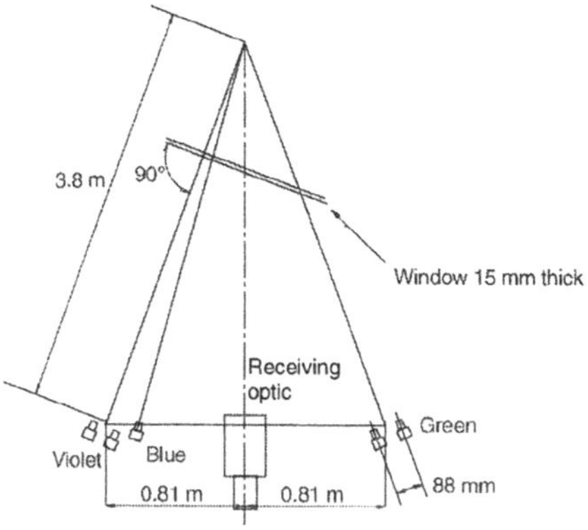

Figure 3 Sketch of the DLR three component LDA system

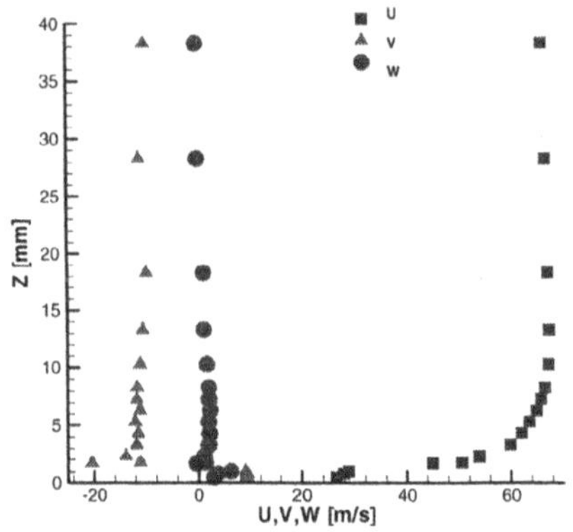

Figure 4 Velocity profile α=10°, station 5

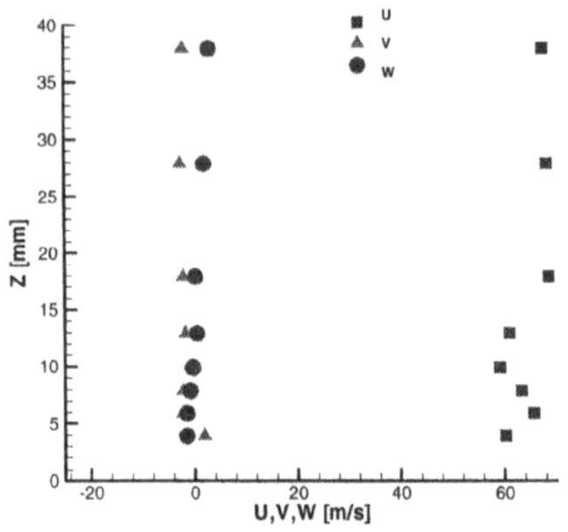

Figure 5 Velocity profile α=22°, station 5

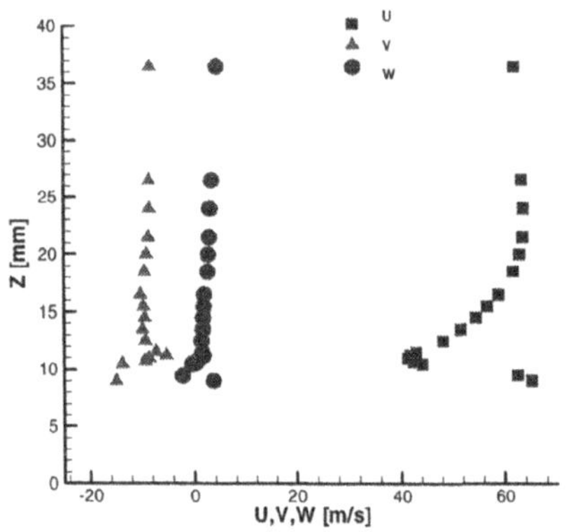

Figure 6 Velocity profile α=10°, station 7

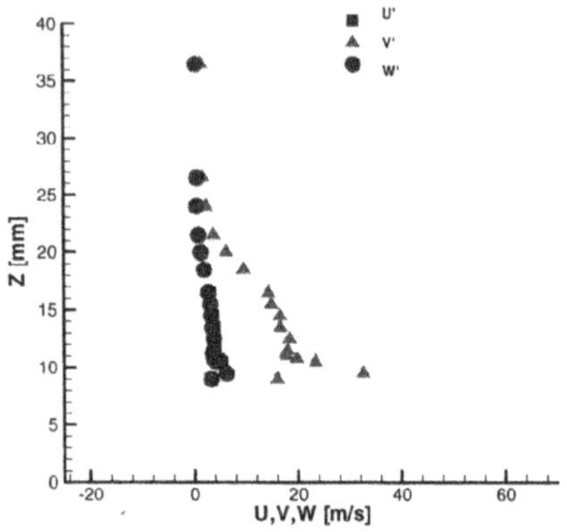

Figure 7 RMS values for α=10°, station 7

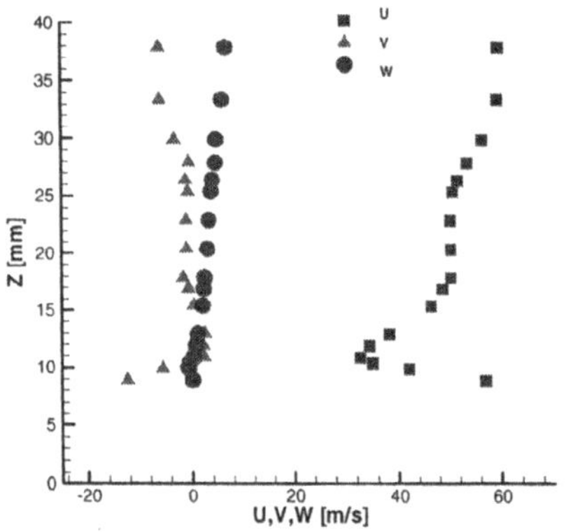

Figure 8 Velocity profile α=22°, station

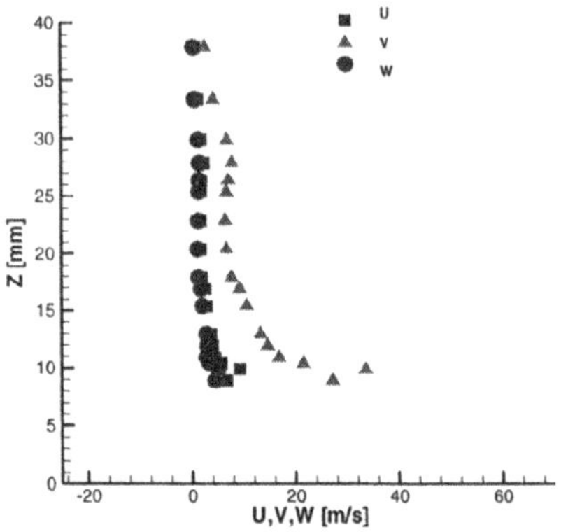

Figure 9 RMS values for α=22°, station 7

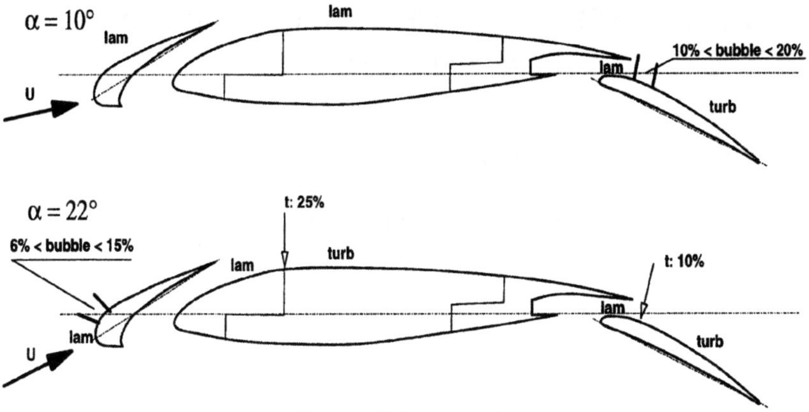

Figure 10 boundary layer types

Detection of Density Gradients at Helicopter Rotor Blades in Flight

M. Raffel, H. Richard

Institut für Strömungsmechanik, DLR
Bunsenstraße 10, 37073 Göttingen, Germany

Summary

Application of a novel schlieren technique, referred as Background Oriented Schlieren (BOS), with two different helicopter tests are presented. Additionally, another extended method, called Reference-free Background Oriented Stereoscopic Schlieren (BOSS), and the differences between these two techniques will be discussed. Experimental studies have been carried out to investigate details of the vortex formation from blade tips of two different helicopters, an Eurocopter BK117 and a large US utility helicopter. The background oriented techniques appears to complement well other optical techniques like shadowgraphy or focussing schlieren methods and yield additional quantitative information. Furthermore, these techniques allow - in contrast to laser based techniques - the effect of Reynolds numbers on vortex development from blade tips in full-scale flight tests to be studied more easily.

1 Introduction

The objective of this paper is to show the quality and kind of information that can be obtained from full-scale helicopter flight tests using density gradient measurement techniques. The details of these optical techniques can be found in Raffel et al 2000, Richard et al. 2000 and Meier 2000. However, first, some general background to the investigations will be outlined in the following.

With increasing use of civil helicopters, the problem of noise emission of helicopters has become increasingly important. Blade vortex interactions (BVI) have been identified as major responsable for the acoustic nuisance which occurs especially during low speed descending flight conditions, like on a landing approach, or in maneuvers. Moreover, the interactions affect blade loads and rotor performance. Therefore, many aeronautical research organizations develop and run numerical codes in order to predict the rotor wake (see e.g. Beaumier et al. 1994). These aerodynamic results can then be used for further numerical investigations of the acoustic near and far field of BVI. (e.g. Burley et al. 1991, Ehrenfried et al. 1991).

Theoretical and experimental studies were conducted in order to validate models for predicting BVI and to determine the model-to-full-scale acoustic scaling conditions (Schlinker and Amiet 1983, Splettstößer et al. 1984), which only confirmed the necessity and importance of wind tunnel tests. It has been found that high speed noise at the advancing blades can be well scaled, whereas the scalability of BVI noise is subject to certain restrictions. It is assumed that Reynolds number effects do not allow direct scaling of the vortex dimensions. Therefore, aerodynamic rotor model investigations have been undertaken at large scales e.g. in the large low speed facility of the Dutch-German wind tunnel DNW-LLF (e.g. Splettstößer et al. 1997).

By reviewing the literature on experimental studies over the last decade, it can be seen, that particle-based flow velocity measurements have most frequently been applied (e.g. Sullivan 1973, Biggers and Orloff 1974, Raffel et al. 1998), most likely, because they deliver a lot of quantitative information needed for BVI prediction.
Density measurements of the BVI phenomena, on the other hand, are also desirable, because the vortices under investigation are compressible and density gradient techniques can therefore be used to visualize flow structures. Density measurements have also been performed quite frequently, but mostly for two-dimensional basic investigations (e.g. Meier et al. 1998, Chandrasekhara 1994). Density gradient visualization by the classical schlieren method has been conducted by Tangler (1977) and reported on more recently by Bagai and Leishman (1993). For large or full–scale investigations on rotating systems, the traditional quantitative density techniques like Mach-Zehnder or point diffraction interferometry seem to be less feasible, since they require delicate optical systems. Due to the aforementioned reason of limited scalability and the known difficulties of particle-based optical techniques in resolving the vortex core in detail, further large or full scale tests of an easy-to-use optical technique are desirable.

2 Background oriented schlieren (BOS)

The principle of the background oriented schlieren (BOS) technique, which we used for our experiments can best be compared with the density speckle photography as described by Debrus et al. (1972), Köpf (1972), and in an improved version by Wernekinck and Merzkirch (1987). A description of the density speckle photography and the novel BOS technique is given by Meier (2000). In the following, a short overview of the underlying principal will be given.
The main difference of the technique proposed in this article is, that in contrast to the conventional schlieren techniques, the BOS technique does not require any optical devices on the transmitting side. The only optical part needed is an objective lens mounted, for instance, on a video camera on the receiving side. The camera is focused on a random dot pattern in the background, which generates an image quite similar to a particle image or speckle pattern. For this reason we refer to this approach as background oriented schlieren (BOS). This procedure results in significantly reduced efforts in implementing the technique. However, the optical paths over which the density effects are averaged are divergent with respect to each other. This can result in a clear disadvantage when large viewing angles have to be used, but is of little influence for the recording distances of 30 meters or less, used for the tests described in this article. For a later extension towards "Background Oriented Optical Tomography" (BOOT), which is proposed by G.E.A. Meier (Meier 2000), the divergence of the optical paths should be no problem, since convolution methods for divergent beams are already known and described in the literature (see e.g. Herman 1980).

3 Background oriented stereoscopic schlieren (BOSS)

The previous technique can be extended by using two cameras simultaneously in a stereoscopic configuration. Both cameras look onto the same background through the flow under investigation (see figure 2). Due to the different camera positions, the density gradient is not at the same position in the image plane of each camera, so the two resulting images can be cross-correlated with each other: the image of one camera can be defined as the reference for the other one and vice-versa.

The advantage of this extension is that no reference image has to be taken at a different time which makes the technique well suitable for moving objects. In the case of a helicopter in flight, two cameras can be very easily installed aboard, looking through the blade and focused on a far distance background which offers sufficient contrast. This concept is referred to as the reference-free BOSS technique in the following.

4 First application to a helicopter in flight

First tests have been performed in order to verify the feasibility of BOS and BOSS for large scale aerodynamic investigations.

The subject, a helicopter - BK117 from Eurocopter - departed from the ground of DLR center Göttingen. Two progressive scan CCD cameras have been mounted in a window of a building at a horizontal distance of 32 meters from the helicopter and 11.2 meters above the ground (see figure 3). The two cameras belong to DLR's PIV equipment and have a 1024 x 1280 pixel resolution. They were closely spaced, mounted 23 cm apart, in order to try different magnifications of the same dot pattern simultaneously, and they were synchronized with a trigger signal of 3 Hz. A 100 mm lens and a 180 mm lens were mounted and directed to the same area on the ground. A random dot pattern was generated by splashing tiny droplets of white wall paint with a brush onto the concrete ground. The pilot gave us time to record more than 50 digital images within 20 seconds of hover flight. The reference recordings were made directly after the departure of the helicopter. A sketch of the observation area on the ground, the rotor blades, and the vortices along the optical axis are shown in fig. 3.

Even if acceptable results could be obtained by using standard cross-correlation displacement measurement software, as developed for PIV, more sophisticated programs helped to adapt the peak-fitting routine to the size of the dot images. Finally, we obtained the best result by using an iterative Levenberg-Marquardt fit to a 10x10 pixel area, where the correlation values are weighted according to the Fisher transform (for details see Ronneberger et al. 1998). The size of the interrogation window was 20 by 20 pixels searched in a window of 64 x 64 pixels. The evaluation led to the vector plots shown in figure 4, they have been obtained by a massive oversampling using a 5 pixel step-width, resulting in an improved visibility of the flow structures under investigation.

In figure 5 at least the young vortex shedding from the blade that just passed the observation area – trailing edge visible on the right hand side – can easily be detected. It is located nearly directly on the blade tip path, which we were able to reconstruct from different recordings. The elliptical shape of the path is due to the underlying projection as shown in figure 3. Further structures were produced by the previous blades: after their generation, the blade tip vortices do not dissipate or diffuse for many rotor revolutions, but are not as well organized as the theoretical vortex trajectories shown in figure 3.

A grey level coded density distribution can be derived from the displacement field. It has been computed by integration along vertical and horizontal lines, which is the simplest method but produces line noise. Such a distribution can also be derived using the Poisson equation, but this method had the disadvantage of strongly smoothing the result.

After scaling the displacement field, the size of the compressible vortex part can be estimated as shown in figure 5. The size decreases with the age of the vortex: the vortex generated by the blade that just passed has a compressible part of around 9.6 cm in diameter while the older vortices have diameters of around 6 cm. The use of two cameras closely spaced allows us to use reference-free BOSS technique after re-sizing the image from the 100mm lens camera. The result of the cross-correlation between the two instantaneous pictures is shown on the right hand side in figure 4.

The elliptically shaped structures appear clearly, but they are not as well defined as for the previous results and particularly for the youngest vortex shedding. This is due to the fact that the position of these structures are nearly overlapping in the camera field of view. For future investigations, the optical path should be chosen in such a way, that the vortical structures under investigation do not overlap. Then, the stereoscopic, as well as the reference-free technique, would increase in resolution and accuracy.

5 Larger scales, more complex blade tip geometry

Further tests have been performed very recently in close cooperation between the Aeromechanic Branch of the US Army and DLR at NASA Ames research center. The tests here were focussed on the structure of the tip vortices immediately after generation in the vicinity of the blade. The background of those investigations is the need to identify and localize the structure and center of the blade tip vortices in order to model and predict their interaction with the following blades. Complex tip geometries are frequently used in order to displace the vortices with respect to the tip path plane and to reduce the velocity gradients of the vortices by generating vortices of larger core sizes. Those tip geometries are mostly used on large and heavy helicopters, because the BVI noise amplitude increases with increasing air loads on the blades. Since larger helicopters require larger observation distances and the spatial resolution relative to the blade tip size had to be increased, the tests at Ames were quite challanging compared to the tests at the BK177 in Göttingen.

The equipment used was again the DLR's PIV cameras. In this case they were connected to a mobile computer system and equipped with different long focal lenses.

A random dot pattern was again generated by splashing tiny droplets of white wall paint, but in a second test a grass area approximately 100 meters behind the helicopter has also been successfully used as a 'background pattern'.

The formation of two individual vortices, rolling up to one larger vortex, the distance the vortices need to merge, and the radial position of the resulting vortex can clearly be identified.

6 Conclusions

Large scale flight tests with two helicopters confirmed the feasibility of the BOS and BOSS techniques to visualize different blade tip vortices. It has been shown that geometric parameters, like the location of the vortex relative to the rotor plane, orientation of the vortex axis in space, spatial information on vortex generation and roll up, and some information about the vortex size, can be derived. In spite of the difficult experimental conditions density gradient data were obtained, which allow a visualization of the density fields with encouraging spatial resolution. Compared to previous measurements, the time needed for the setup and data acquisition can be considerably decreased. Further investigations have to be performed using multiple cameras in order to obtain more complete three-dimensional data. After having demonstrated the feasibility of the concept by application to a technological relevant, but fluid mechanical quite complex problem, more detailed studies e.g. of tip vortices from stationary blades, should be performed in order to reduce the complexity of the vortical structures under investigation.

However, the feasibility of using a 'natural background', like grass in a relative far background distance, and the feasibility of the reference-free stereoscopic technique that enables the cameras to be mounted on board of the helicopter during future tests, confirm the worthiness of the technique.

Acknowledgements

The authors would like to thank the pilot, engineers, and technicians of Eurocopter Deutschland (ECD) and DLR's Institute for Aeroelasticity for their cooperation. Many thanks also to Ken McAlister, Alex Sheikman and the people of the flight crew like Pilot Dave Arterburn at NASA Ames. Special thanks to C. Tung, Y. Yu, G.E.A. Meier and S. Sampath for supporting our work.

References

A. Bagai, J.G. Leishman: "Flow visualization of compressible vortex structures using density gradient techniques", *Exp. in Fluids*, Vol. 15, 431-442, 1993.

P. Beaumier, J. Prieur, G. Rahier, P. Spiegel, A. Demargne, R. Kube, B. van der Wall, B.G. Schultz, W. Splettstößer, C. Tung, J. Gallman, Y.H. Yu, T.F. Brooks, C.L. Burley, D.D. Boyd: "Effect of higher harmonic control on helicopter rotor blade vortex interaction noise", *75th Fluid Dynamics Symposium on Aerodynamics and Aeroacoustics of Rotorcraft,* Berlin, October 1994.

J.C. Biggers, K.L. Orloff: "Laser velocimeter measurements of the helicopter rotor induced flow field", *30th Annual National Forum of the American Helicopter Society*, reprint No. 800, May, 1974.

C.L. Burley, H.E. Jones, M.A. Marcolini, W.R. Splettstößer: "Directivity and prediction of low frequency rotor noise", *AIAA 91-0592, Aerospace Science Meeting, Reno*, USA 1991.

M.S. Chandrasekhara, D.D. Squires, M.C. Wilder, L.W. Carr: "A phase-locked high-speed real-time interferometry system for large amplitude unsteady flows", *7th International Symposium on Applications of Laser Techniques to Fluid Mechanics*, Lisbon, Portugal, 38.3.1-38.3.8, July, 1994.

S. Debrus, M. Francon, C.P. Grover, M. May, M.L. Robin: "Ground glass differential interferometer", *Applied Optics* 11, 853, 1972.

K. Ehrenfried, G.E.A. Meier, F. Obermeier: "Sound produced by vortex-airfoil interaction", *Proc. of the 17th European Rotorcraft Forum*, paper 63, Berlin, 1991.

G.T. Herman: "Image reconstruction from projections", Academic Press, New York, 1980.

U. Köpf: "Application of speckling for measuring the deflection of laser light by phase objects", *Optics Communications*, Vol. 5, No. 5, 347-350, 1972.

G.E.A. Meier: "A BOOT for optical Flow Diagnostic", *9th International Symposium on Flow Visualization, Edinburgh*, August 22-25th 2000

G.E.A. Meier: "New optical tools for fluid mechanics", *Proc. 8th International Symposium on Flow Visualization*, paper 226, Sorrento, Sept. 1998.

M. Raffel, U. Seelhorst, C. Willert: "Vortical flow structures at a helicopter rotor model measured by LDV and PIV", *Journal of the Royal Aeronautical Society*, 221-227, April, 1998.

M. Raffel, F. De Gregorio, K. Pengel, C. Willert, C. Kähler, T. Dewhirst, K. Ehrenfried, J. Kompenhans: "Instantaneous flow field measurements for propeller aircraft and rotorcraft research", *9th International Symposium on Applications of Laser Techniques to Fluid Mechanics*, Lisbon, Portugal, 19.6.1-19.6.8.10, (Bound volume in press), July, 1998.

M. Raffel, H. Richard, G.E.A. Meier: "On the applicability of background oriented optical tomography", *Experiments in Fluids 28 (2000)*, pp477-481

H. Richard, M. Raffel, M. Rein, J. Kompenhans, G.E.A. Meier: "Demonstration of the applicability of background oriented schlieren (BOS)", *Int. Symp. On Appl. Of Laser Techniques to Fluid Mechanics*, Lisbon, July 2000

O. Ronneberger, M. Raffel, J. Kompenhans: "Advanced evaluation algorithms for standard and dual plane particle image velocimetry", *9th International Symposium on Applications of Laser Techniques to Fluid Mechanics*, Lisbon, Portugal, 10.1.1-10.1.8, July, 1998.

R.H. Schlinker, R.K. Amiet: "Rotor vortex interaction noise", *NASA CR 3744*, Oct. 1983.

W.R. Splettstößer, K.J. Schultz, D.A. Boxwell, F.H. Schmitz: "Helicopter model rotor blade/vortex interaction impulsive noise: scalability and parametric variations", *NASA TM 86007*, 1984.

W.R. Splettstößer, R. Kube, W. Wagner, U. Seelhorst, A. Boutier, F. Micheli, E. Mercker, K. Pengel: "Key results from a higher harmonic control aeroacoustic rotor test (HART)", *Journal of the American Helicopter Society*, Vol. 42, No. 1, 58-78, 1997.

J.P. Sullivan: "An experimental investigation of vortex rings and helicopter rotor wakes using a laser Doppler velocimeter", MIT Technical Report 183, June. 1973.

J.L. Tangler: "Schlieren and noise studies of rotors in forward flight", *33rd Annual National Forum of the American Helicopter Society*, Reprint 77.33-05, May, 1977.

U. Wernekinck, W. Merzkirch: "Speckle photography of spatially extended refractive-index fields", *Applied Optics* Vol. 26, 31-32, 1987.

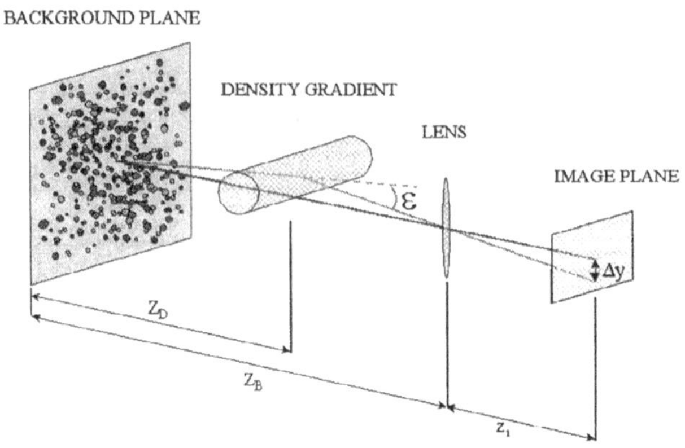

Figure 1: Optical path for density gradient measurement by light deflection

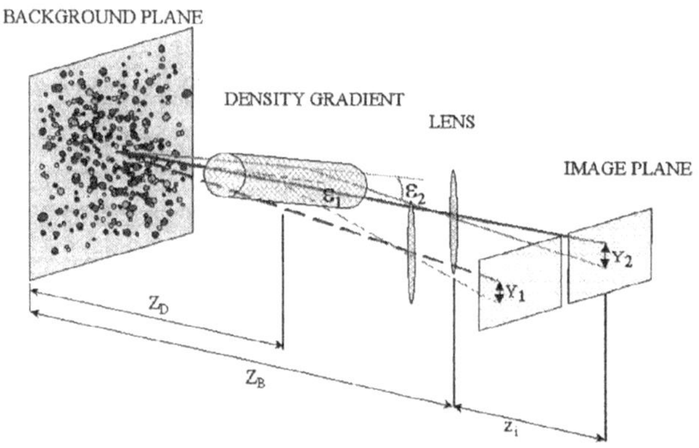

Figure 2: Optical path with 2 cameras

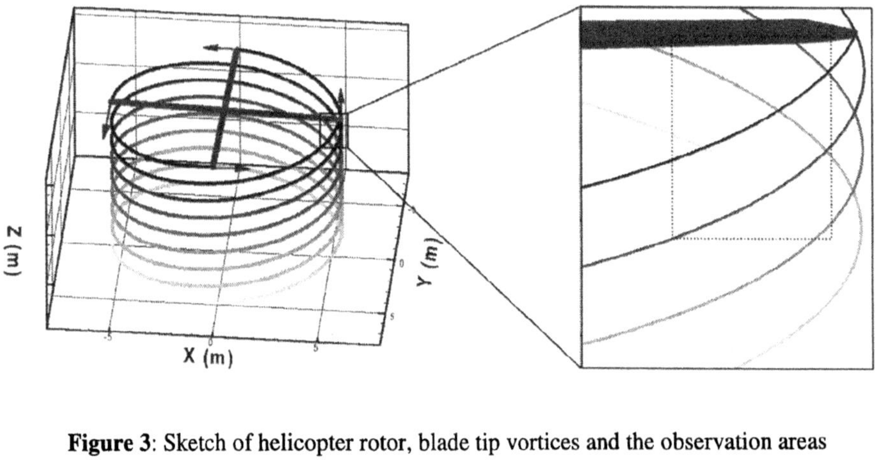

Figure 3: Sketch of helicopter rotor, blade tip vortices and the observation areas

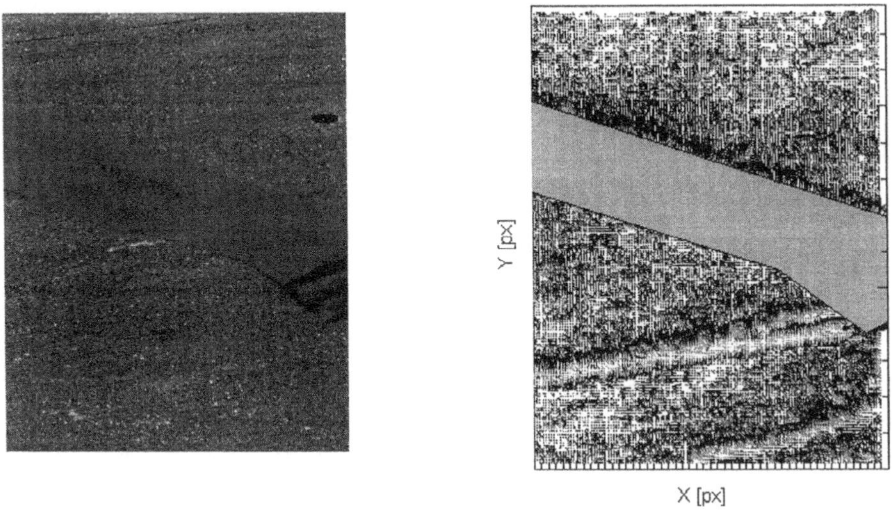

Figure 4: Raw image from transonic blade tip over the background pattern (white dots on grey concrete) and the corresponding density gradient plot,with grey level coded gradient magnitude.

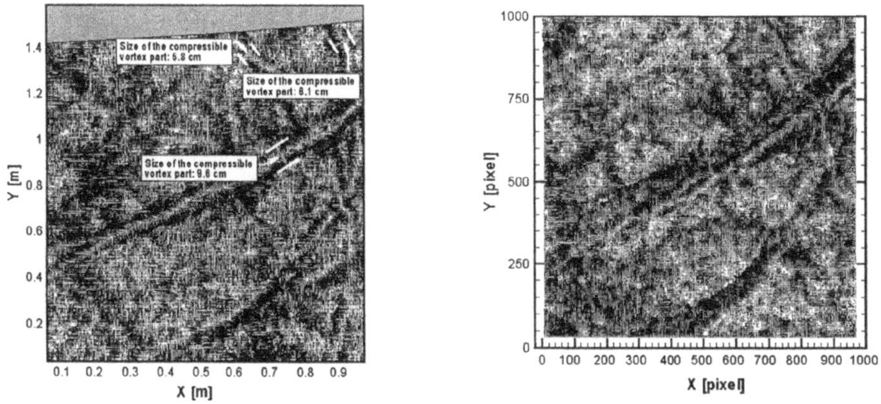

Figure 5: Scaled displacement data and estimation of size of the compressible vortex parts. (parallel lines, left hand side) and stereoscopic results (right hand side)

Notes on Numerical Fluid Mechanics

Available volumes

Volume 78: M. Hafez, K. Morinishi, J. Periaux (eds.): Computational Fluid Dynamics for the 21st Century. Proceedings of a Symposium Honoring Prof. Satofuka on the Occasion of his 60th Birthday, Kyoto, Japan, 15-17 July 2000. ISBN 3-540-42053-3

Volume 76: P. Thiede (ed.): Aerodynamic Drag Reduction Technologies. Proceedings of the CEAS/DragNet European Drag Reduction Conference, 19-21 June 2000, Potsdam, Germany. ISBN 3-540-41911-x

Volume 75: E.H. Hirschel (ed.): Numerical Flow Simulation II. CNRS-DFG Collaborative Research Programme, Results 1998–2000. ISBN 3-540-41608-0

Volume 74: D. Henry, A. Bergeon (eds.): Continuation Methods in Fluid Dynamics. Contributions to the ERCOFTAC/EUROMECH Colloquium 383, Aussois, France, 6.–9. September 1998. ISBN 3-540-41556-4

Volume 73: M. Griebel, S. Margenov, P. Yalamov (eds.): Large-Scale Scientific Computations of Engineering and Environmental Problems II. Proceedings of the Second Workshop on "Large-Scale Scientific Computations" Sozopol, Bulgaria, June 2–6, 1999. ISBN 3-540-41551-3

Volume 72: W. Nitsche, H.-J. Heinemann, R. Hilbig (eds.): New Results in Numerical and Experimental Fluid Mechanics II. Contributions to the 11th AG STAB/DGLR Symposium Berlin, Germany 1998. ISBN 3-540-41552-1

Volume 70: W. Hackbusch, G. Wittum (eds.): Numerical Treatment of Multi-Scale Problems. Proceedings of the 13th GAMM-Seminar, Kiel, January 24–26, 1997. ISBN 3-540-41562-9

Volume 69: D. P. Hills, M. R. Morris, M. J. Marchant, P. Guillen (eds.): Computational Mesh Adaptation. ECARP – European Computational Aerodynamics Research Project. ISBN 3-540-41548-3

Volume 67: C. H. Sieverding, G. Cicatelli, J. M. Desse, M. Meinke, P. Zunino: Experimental and Numerical Investigation of Time Varying Wakes behind Turbine Blades. Results of the Industrial and Material Technology Aeronautics Research Project AER2-CT92-0048, 1992–1996. ISBN 3-540-41546-7

Volume 66: E. H. Hirschel (ed.): Numerical Flow Simulation I. CNRS-DFG Collaborative Research Programme. Results 1996–1998. ISBN 3-540-41540-8

Volume 64: R. Friedrich, P. Bontoux (eds.): Computation and Visualization of Three-Dimensional Vortical and Turbulent Flows. Proceedings of the Fifth CNRS-DFG Workshop on Numerical Flow Simulation, München, Germany, December 6 and 7, 1996. ISBN 3-540-41557-2

Volume 63: M. Fiebig, N. K. Mitra (eds.): Vortices and Heat Transfer. Results of a DFG-Supported Research Group. ISBN 3-540-41558-0

Volume 62: M. Griebel, O. Iliev, S. D. Margenov, P. S. Vassilevski (eds.): Large-Scale Scientific Computations of Engineering and Environmental Problems. Proceedings of the First Workshop on "Large-Scale Scientific Computations" Varna, Bulgaria, June 7–11, 1997. ISBN 3-540-41560-2

Volume 57: H. Deconinck, B. Koren (eds.): Euler and Navier-Stokes Solvers Using Multi-Dimensional Upwind Schemes and Multigrid Acceleration. Results of the BRITE/EURAM Projects AERO-CT89-0003 and AER2-CT92-00040, 1989–1995. ISBN 3-540-41535-1

GPSR Compliance

The European Union's (EU) General Product Safety Regulation (GPSR) is a set of rules that requires consumer products to be safe and our obligations to ensure this.

If you have any concerns about our products, you can contact us on ProductSafety@springernature.com

In case Publisher is established outside the EU, the EU authorized representative is:

Springer Nature Customer Service Center GmbH
Europaplatz 3
69115 Heidelberg, Germany

Batch number: 10755329

Printed by Printforce, the Netherlands